Two Manuals Combined into One Book

Static Line Parachuting

And

FM 7-85 Ranger Unit Operations

Department of Defense

Pentagon
Publishing

***FM 3-21.220**(FM 57-220)/
MCWP 3-15.7/AFMAN11-420/
NAVSEA SS400-AF-MMO-010

FIELD MANUAL
No. 3-21.220(57-220)/
MCWP 3-1.5.7/AFMAN11-420/
NAVSEA SS400-AF-MMO-010

HEADQUARTERS
DEPARTMENT OF THE ARMY
Washington, DC, 23 September 2003

STATIC LINE PARACHUTING TECHNIQUES AND TACTICS

CONTENTS

DISTRIBUTION RESTRICTION: Approved for public release; distribution is unlimited.

*This publication supersedes FM 57-220/MCWP 3-1.5.7/TO 14D1-2-1-121, 19 August 1996.

Part Five. DROP ZONES

PREFACE

This manual contains basic and advanced training and techniques for static line parachuting. It is designed to standardize procedures for initial qualification and training of personnel in their duties and responsibilities in airborne operations. The jumpmaster, assistant jumpmaster, safeties, DACO, DZSTL, and DZSO occupy key positions in airborne operations. This manual contains the initial training and qualifications of the personnel designated to occupy these critical positions.

SOF unit personnel must meet the requirements for static line parachuting contained in this manual as well as provide special training and instruction for nonstandard equipment, aircraft, and personnel procedures. These procedures are documented in FM 31-19, TC 31-24, TC 31-25, and USASOC Reg 350-2.

Individual service components that deviate from this manual will use approved procedures, techniques, equipment, and equipment-attaching methods specified by their respective service. All deviations must be approved in writing by the using unit commanders.

The proponent of this publication is HQ U.S. Army Infantry School. Submit changes for improving this publication to doctrine@benning.army.mil or on DA Form 2028 (Recommended Changes to Publications and Blank Forms) directly to Commandant, U.S. Army Infantry School, ATTN: ATSH-TPP-A, Fort Benning, Georgia 31905-5593.

Unless this publication states otherwise, masculine nouns and pronouns do not refer exclusively to men.

NOTE: The terms *jumper* and *parachutist* are used interchangeably in this manual.

PART ONE

Basic Airborne Techniques and Training

CHAPTER 1

AIRBORNE TRAINING

The purpose of airborne training is to qualify personnel in the use of the parachute as a means of combat deployment. This training also develops leadership, self-confidence, and aggressive spirit through tough mental and physical conditioning.

1-1. STANDARDS

Airborne training initiates and sustains a high standard of proficiency through repetition and time-proven techniques. Valid results are obtained when the following training standards are employed:

- Strict discipline.
- High standards of proficiency on each training apparatus and during each phase of training.
- A vigorous physical conditioning program to ensure parachutists are capable of jumping with a minimum risk of injury.
- A strong sense of esprit de corps and camaraderie among parachutists.
- Emphasis on developing mental alertness, instantaneous execution of commands, self-confidence, and confidence in the equipment.

1-2. PHASES

The three-week airborne course is divided into two training phases. Weeks 1 and 2 form the ground and tower training phase, and Week 3 is the jump training phase.

a. **Ground and Tower Phase**. Each of the five basic jump techniques pertains to a particular area of military parachuting and provides a sequence for dividing the ground phase into six instructional segments.

(1) ***Actions Inside the Aircraft.*** To ensure that the maximum number of parachutists can safely exit an aircraft, a means of controlling their actions inside the aircraft just before exiting is necessary. The jumpmaster maintains control by issuing jump commands. Each command calls for specific action on the part of each parachutist.

(2) ***Body Control Until Opening Shock.*** Due to aircraft speed and air turbulence around the rear of the aircraft, the parachutist must exit properly and maintain the correct body position after exiting. This action reduces spinning and tumbling in the air and allows for proper parachute deployment.

(3) ***Parachute Control During Descent.*** Parachute control is essential to avoid other parachutists in the air and to avoid hitting obstacles on the ground.

(4) ***Parachute Landing Fall Execution.*** The PLF is a landing technique that enables the parachutist to distribute the landing shock over his entire body to reduce the impact and the possibility of injury.

(5) ***Parachute Control on Landing.*** The parachutist releases one canopy release assembly after landing. Winds on the drop zone may cause a parachutist to be injured from being dragged along the ground.

(6) ***Physical Training.*** Prior to reporting for airborne training, volunteers must achieve APFT standards for the 17- to 21-year-old level (Table 1-1 and DA Pam 351-4). Physical training is included in each day of ground training. Students who cannot progress in daily physical training are referred to a board that decides either to recycle them or to return them to their unit. Daily exercises are designed to condition the muscle groups that play a significant part in jumping (Table 1-2).

EVENT	REPETITIONS		TIME LIMIT
	MALE	FEMALE	
PUSH-UPS	42	19	2 MINUTES
SIT-UPS	53	53	2 MINUTES
TWO-MILE RUN		MALE	15.54 MINUTES
		FEMALE	18.54 MINUTES

Table 1-1. APFT standards for the 17- to 21-year-old level.

WARM-UP EXERCISES	SETS	REPETITIONS
CHIN-UPS (MALE)	1	10
CHIN-UPS (FEMALE)	1	10
ROTATION EXERCISES	**SETS**	**INTERVAL**
NECK ROTATION	1	10 seconds
ARM/SHOULDER ROTATION	1	10 seconds
HIP ROTATION	1	10 seconds
KNEE/ANKLE ROTATION	1	10 seconds
STRETCHING EXERCISES	**SETS**	**INTERVAL**
ABDOMINAL STRETCH	1	15 seconds
OVERHEAD ARM PULL	1	15 seconds
UPPER BACK STRETCH	1	15 seconds
CALF STRETCH	1	15 seconds
HAMSTRING STRETCH	1	15 seconds
HIP/BACK STRETCH	1	15 seconds
GROIN STRETCH (SEATED)	1	15 seconds

Table 1-2. Daily physical training exercises.

NOTE: The following calisthenic exercises are conducted with a 15-second break between each exercise.

CALISTHENIC EXERCISES CATEGORY A	SETS	REPETITIONS
SKI JUMPER	1	15
FOUR-COUNT PUSH-UP	2	15
SIT-UP	2	15
MOUNTAIN CLIMBER	2	15
KNEE BENDER	2	15
CALISTHENIC EXERCISES CATEGORY B	**SETS**	**REPETITIONS**
SKI JUMPER	1	15
EIGHT-COUNT PUSH-UP	2	15
SUPINE BICYCLE	2	15
SQUAT BENDER	2	15
FLUTTER KICK	2	15

RUN DISTANCES AND TIMES PER MILE			
WEEK	**DISTANCE**	**TIME PER MILE**	**TOTAL TIME**
GROUND	3.2 MILES	Min. 8:45 - Max. 9:15	Min. 28:00 - Max. 29:36
TOWER	4.0 MILES	Min. 8:45 - Max. 9:15	Min. 35:00 - Max. 37:00

8:45-MIN MILE		9:00-MIN MILE		9:15-MIN MILE	
.2 MILE	1 MIN 45 SEC	.2 MILE	1 MIN 50 SEC	.2 MILE	1 MIN 51 SEC
.4 MILE	3 MIN 30 SEC	.4 MILE	3 MIN 35 SEC	.4 MILE	3 MIN 42 SEC
.6 MILE	5 MIN 15 SEC	.6 MILE	5 MIN 15 SEC	.6 MILE	5 MIN 33 SEC
.8 MILE	7 MIN 00 SEC	.8 MILE	7 MIN 15 SEC	.8 MILE	7 MIN 24 SEC
1 MILE	8 MIN 45 SEC	1 MILE	9 MIN 00 SEC	1 MILE	9 MIN 15 SEC

Table 1-2. Daily physical training exercises (continued).

During ground week, students must complete three 3.2-mile runs at a 8:45- to 9:15-minute pace. The runs are completed in formation and the student must not fall more than three steps behind the original formation. During tower week, students must complete two 4-mile runs at the same standard (Table 1-2) and one 5-mile off-track run.

b. **Jump Phase**. Students who meet training proficiency in the basic jump techniques and physical fitness requirements during ground and tower week training are advanced to the jump training phase. During jump phase training, the student makes five qualifying jumps from aircraft at an altitude of 1,250 feet AGL (Table 1-3, page 1-4).

JUMP NUMBER	EQUIPMENT	TYPE EXIT
1	Ballistic helmet	ADEPT Option 1 Individual Exit
2	Ballistic helmet, combat equipment (HSPR, ALICE pack, and M1950 weapons case)	ADEPT Option 2 Both Doors
3	Ballistic helmet	Mass Exit Both Doors
4	Ballistic helmet	Mass Exit Both Doors
5 (Night)	Ballistic helmet, combat equipment (HSPR, ALICE pack, and M1950 weapons case)	Mass Exit Both Doors

Table 1-3. Typical jump week schedule.

1-3. PREJUMP ORIENTATIONS

Students are thoroughly briefed before performing their qualification jumps. The topics include—

- A review of the five points of performance, collisions and entanglements, towed parachutist, malfunctions, activation of the reserve, and emergency landings.
- Maintenance of the T-10-series or MC1-series parachute to include shakeout and storage after landing.
- Donning the parachute "by the numbers" on the first jump. Additional instructors are available for close supervision and JMPI.
- Aircraft orientation to include enplaning and jump procedures.
- Drop zone and approximate point of impact.

CHAPTER 2

PERSONAL EQUIPMENT

This chapter discusses the types of parachutes, donning the parachutes, and the headgear used in airborne training.

Section I. PARACHUTES

The T-10-series and MC1-series parachutes are used during static line airborne operations. The T-10-series is a nonsteerable canopy and the MC1-series is a steerable canopy. The main parachute consists of five major components—the harness assembly, the riser assembly, the deployment bag, the pack tray, and the canopy assembly. A reserve—either the modified improved reserve parachute system (MIRPS) or the T-10 troop chest reserve parachute—is used in conjunction with the main parachute.

2-1. HARNESS ASSEMBLY

The T-10-series harness assembly (Figure 2-1, page 2-2) is used with the T-10-series and MC1-series main canopies.

a. **Components**. The harness assembly is made of a flexible framework of Type XIII nylon webbing. The components attached to it are as follows:

(1) Female fitting, canopy release assembly with a safety clip, cable loop, and latch. Rated capacity of the assembly is 5,000 pounds.

(2) Canopy release assembly pads permanently attached behind the canopy release assembly.

(3) Main lift web constructed of two plies of Type XIII nylon with a rated capacity of 6,500 pounds.

(4) Chest strap with an ejector snap with activating lever, ball detent, and opening gate with a rated capacity of 2,500 pounds.

- An ejector snap pad is attached behind the ejector snap.
- The quick-fit V-ring has a rated capacity of 2,500 pounds.
- A webbing retainer is used for stowing excess webbing.

(5) D-rings used for attaching the MIRPS or T-10 reserve and additional combat equipment, with a rated capacity of 5,000 pounds.

(6) Triangle links sewn into the main lift web of the harness (about 5 1/2 inches below the D-rings) for attaching equipment and lowering lines.

(7) Leg straps with an ejector snap with activating lever, ball detent, and opening gate with a rated capacity of 2,500 pounds. The parts of the leg straps (ejector snap pads, quick-fit V-rings, and webbing retainers) are identical to the chest strap.

(8) Saddle (Type XIII nylon) with two attached leg straps.

(9) Diagonal back straps with six sizing channels (S, 1, 2, 3, 4, and L).

(10) Back strap adjusters with attached free-running ends of the horizontal back strap (rolled and sewn).

b. **Adjustment Points**. The harness assembly has five points of adjustment: the chest strap, two leg straps, and two free-running ends of the horizontal back strap.

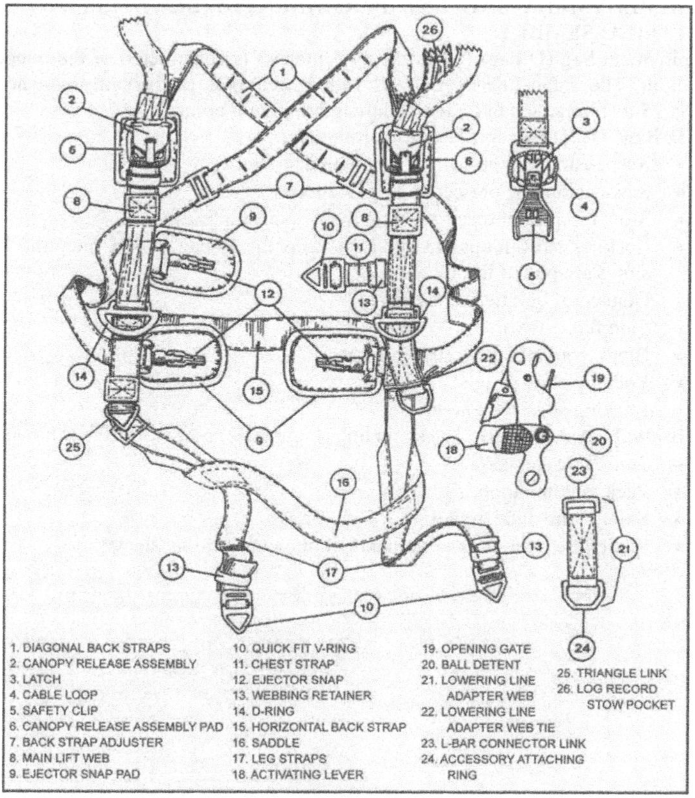

Figure 2-1. T-10-series harness assembly and nomenclature.

2-2. RISER ASSEMBLY

The riser assembly on the T-10-series consists of the following:

- Two riser assemblies, with a finished length of 30 inches (Type XIII nylon) and a tensile strength of 6,500 pounds. A male fitting canopy release assembly is permanently attached to the center of the webbing. When attached to the canopy, the riser assemblies provide four individual risers.
- Log record stow pocket. Opening the log record book by non-parachute rigger personnel is prohibited.
- Connector link loops.
- L-bar connector links.

The riser assembly on the MC1-series is identical to the T-10-series, but it also has a guide ring retainer strap, a guide ring, and upper and lower control line channels.

2-3. DEPLOYMENT BAG AND PERMANENTLY SEWN 15-FOOT STATIC LINE ASSEMBLY

The deployment bag (D-bag) (18 by 12 by 5 inches) is constructed of 8.8-ounce cotton sateen cloth. The static line (Type VIII yellow nylon) is permanently attached to the D-bag, is 15 feet long, and has a tensile strength of 3,600 pounds.

a. **D-Bag**. The D-bag consists of the following:

- Suspension line protective flap with data block.
- Suspension line protective flap tie loop.
- Stow loop panel (used to retain the suspension lines).
- Locking stow loops (two, which keep the D-bag closed until the first two stows are pulled free).
- Connector link tie loops (four).
- Side flaps (two).
- Break cord attaching strap pocket.
- Locking stow panel.
- Locking stow loop hood.

b. **Static Line Assembly**. The static line (Figure 2-2) consists of the following:

- Static line sleeve.
- Pack opening loop.
- Safety wire and lanyard.
- Static line snap hook with locking button and sliding sleeve.

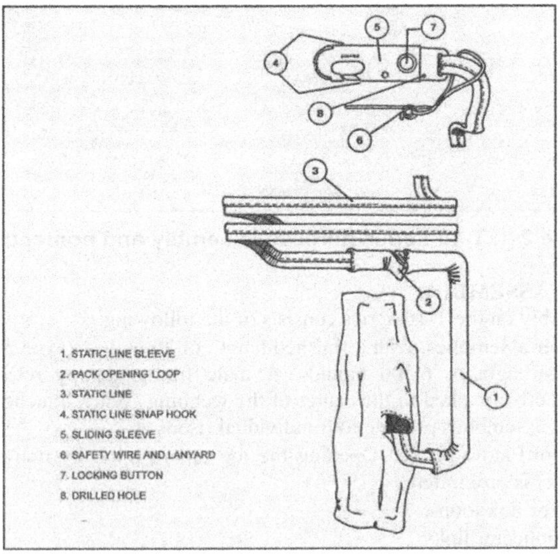

Figure 2-2. Static line and nomenclature.

2-4. DEPLOYMENT BAG WITH UNIVERSAL STATIC LINE ASSEMBLY

The deployment bag (D-bag) (Figure 2-3) consists of the following:

- Suspension line protective flap with data block.
- Suspension line protective flap tie loop.
- Stow loop panel (used to retain the suspension lines).
- Two locking stow loop, which keep the D-bag closed until the first stows are pulled free.
- Four connector link tie loops.
- Two side flaps.
- Break cord attaching strap pocket.
- Locking stow panel.
- Locking stow loop hood.

a. **Universal Static Line.** A 15-foot universal static line (USL) is secured to the D-bag by a girth hitch. The opposite end of the USL has a snap hook, which is also secured by a girth hitch. To configure the USL to the 20-foot length, the snap hook is removed and the extension is secured to the USL by a girth hitch. The snap hook is then secured to the opposite end of the extension. The length of the 15-foot static line after being secured to the deployment bag and snap hook is 14 feet 7 inches. The 20-foot configuration is 20 feet 1 inch long. There are four separate components to the USL, which must be purchased separately: deployment bag, static line, static line extension (Figure 2-4), and static line snap hook (Figure 2-5, page 2-6).

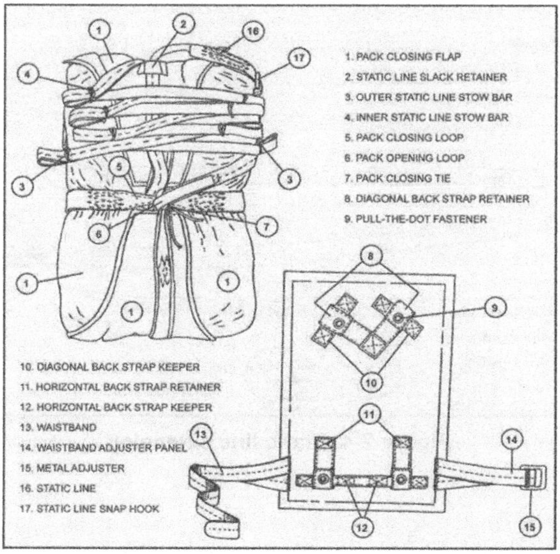

Figure 2-3. Deployment bag components with universal static line.

b. **Deployment Bag.** The deployment bag is currently used on the T10- and MC1-series personnel parachutes. The deployment bag comes without the static line and snap hook.

c. **Static Line.** The static line is 14 feet 10 inches long (+ - 1 inch). The static line webbing material is 3/4-inch wide, nylon, tube edge, class 1A (nylon 6.6), class R (resin treated), with "Dooley" finish. It has a 4,000-pound breaking strength and 30 percent minimum elongation. It is yellow in color with one green color marker yarn located at the center of the warp.

d. **Static Line Extension.** The static line extension is 70 1/2 inches long (+ - 1/2 inch) (Figure 2-4). It is made from the same webbing as the static line.

e. **Snap Hook.** The snap hook is 5 1/2 inches long, approximately 1/4-inch thick, and weights 0.67 pounds (Figure 2-5, page 2-6). It is made from alloy steel with Dacroment 320, grade A plus L finish for abrasion/humidity/salt/fog protection. The finish is environmentally compliant and a standard in the automotive industry. The snap hook gate and rivet pin are made from the same alloy steel, whereas the pullback spring is made from steel spring wire. The snap hook is rated to withstand a 1,750-pound load and has an ultimate failure rate of no less than 8,000 pounds.

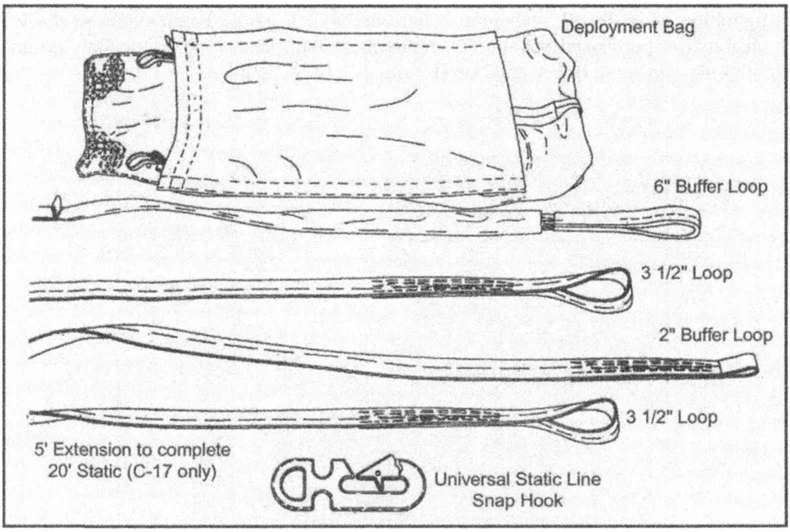

Figure 2-4. Static line extension.

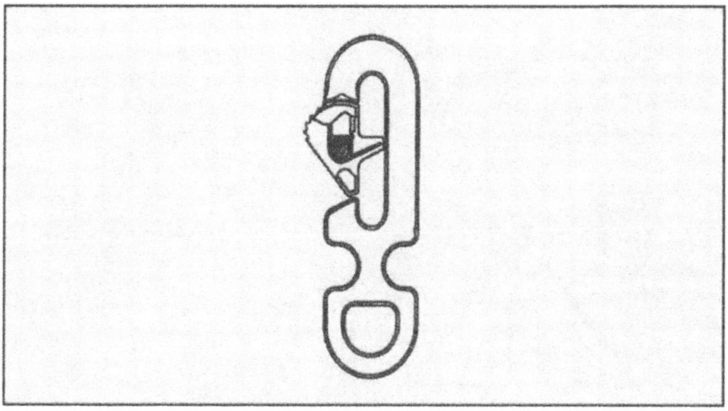

Figure 2-5. Static line snap hook.

2-5. PACK TRAY

The pack tray (20 by 14 by 5 inches) (Figure 2-6) is constructed of 7.25-ounce nylon duck material and consists of the following:

- Pack closing flaps (four): right and left side flaps, and upper and lower end flaps.
- Pack closing loops (four): right and left side pack closing loops, and upper and lower end pack closing loops.
- Static line stow bar.
- Waistband adjuster panel.
- Metal adjuster.
- Waistband (43 inches long).
- Pack closing tie (one turn 1/4-inch cotton webbing) tied in a surgeon's knot and a locking knot; the knot is between the 3 and 6 o'clock positions.
- Diagonal back-strap retainer.
- Diagonal back-strap keeper.
- Horizontal back-strap retainer.
- Horizontal back-strap keeper.

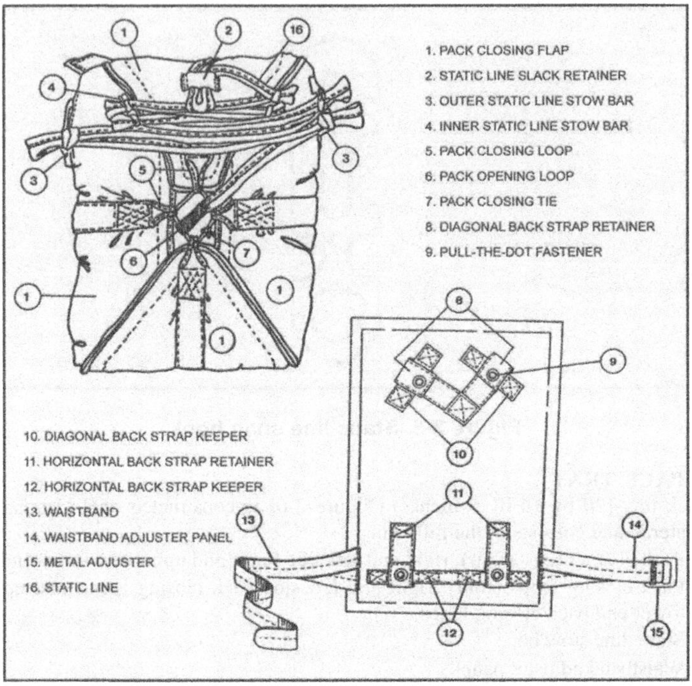

Figure 2-6. T-10-series parachute pack tray and nomenclature.

2-6. CANOPY ASSEMBLY

The T-10- and MC1-series parachutes are static line deployed. The parachute deployment sequence is the same for both types of canopies.

- The parachutist falls to the end of the static line. His body weight breaks the pack closing tie, and the deployment bag is pulled from the pack tray.
- Two connector link ties break, and the suspension lines are pulled from the deployment bag.
- Two locking stows disengage, and the canopy is pulled from the deployment bag to its full length.
- The break cord tie securing the apex of the canopy to the static line ending loop breaks, and the parachute begins to inflate, retarding the parachutist's rate of descent.

a. **General Characteristics**. Following are characteristics of both T-10-series and MC1-series canopies:

(1) ***Shape and Weight.*** Shape is parabolic; weight is between 28 and 31 pounds.

(2) ***Rates of Descent.*** Depending on the jumper's total weight and relative air density, the average rates of descent for the different canopies are as follows: MC1-series, 14 to 22 feet per second and T-10-series, 19 to 23 feet per second.

(3) ***Diameter.*** Nominal diameter is 35 feet (measured 3 feet up from the skirt) and 24.5 feet at the skirt.

(4) ***Anti-Inversion Nets.*** The anti-inversion net is sewn 18 inches down on each suspension line and is made of 3 3/4-inch square mesh, knotless, braided nylon.

(5) ***Shelf and Service Life.*** Combined shelf life and service life is 16.5 years; service life is 12 years, and shelf life is 4.5 years.

(6) ***Repacking.*** Both canopies are repacked every 120 days.

(7) ***Use.*** Both canopies are suitable for airdropping personnel from as high as 10,000 feet MSL.

b. **MC1-1B Parachute Design Characteristics**. The MC1-1B has an estimated 8.8-second turn rate.

(1) The bridle loop is 3 inches in diameter; it is made of Type VIII nylon with a tensile strength of 3,600 pounds.

(2) The 15 apex vent lines are 19 inches long and made of Type II nylon cord with a tensile strength of 375 pounds. The apex vent lines with centering lines keep the bridle loop in place and the canopy even during deployment.

(3) The apex vent cap has a 3-inch diameter opening, which reduces oscillation and assists in a positive opening of the canopy.

(4) The upper lateral band is 1-inch tubular nylon with a tensile strength of 4,000 pounds. It is sewn in as reinforcement, since the first 64 inches of the canopy is a high-pressure area.

(5) Each canopy has 30 gores with five sections in each gore. Each gore is numbered 1 through 30, and each section is sewn diagonally to prevent rips and holes from spreading throughout the canopy.

(6) The 30 radial seams are 9/16 inch wide and 17 feet 2 7/32 inches long, measured from the upper lateral band to the lower lateral band.

(7) The T-U shaped configuration has 11 gores removed (25 to 5) from the rear (100.4 square feet of canopy), which enables the canopy to turn 360 degrees in 8.8 seconds and gives a forward thrust of 8 knots (9.5 mph), or 14 feet per second.

(8) Two 28-foot-long control lines are attached to a control bridle that in turn is attached to radial seams 5 and 6. They are 6 feet long and attached to seams 25 and 26. They run down and out to the front of the rear set of risers and through the control line channel and control line guide ring. They are attached to a toggle that is a 5/8-inch diameter hardwood dowel.

(9) The lower lateral band is a 1-inch nylon tape with a tensile strength of 525 pounds.

(10) The 15 pocket bands ensure positive opening of the canopy.

(11) The 30 V-tabs are 9/16 inch wide and are sewn over the suspension lines to the lower lateral band for reinforcement.

(12) The 30 suspension lines are Type II nylon with a tensile strength of 375 pounds. They are 25 feet 6 inches long when measured from the lower lateral band to the L-bar connector link.

c. **MC1-1C Parachute Design Characteristics**. The MC1-1C canopy has the same basic design as the MC1-1B with the following exceptions:

(1) It has an estimated 7.7-second turn rate.

(2) It is made of nonporous (NOPO) material (0-3 CFM).

(3) The vent cap is removed.

(4) The suspension lines are shortened to 22 feet.

(5) The H-TC modification is a 60-square-foot opening.

d. **T-10-Series Parachute Design Characteristics**. The T-10-series parachute is designed with the following characteristics:

(1) The bridle loop is 3 inches in diameter and made of Type VIII cotton or nylon with a tensile strength of 3,600 pounds.

(2) The 15 apex vent lines are 19 inches long and made of Type II nylon cord with a tensile strength of 375 pounds.

(3) The two apex centering loops are 9 inches long and made of Type II nylon cord with a tensile strength of 375 pounds.

(4) The apex vent is 20 inches in diameter and expands to 22 inches in diameter when the canopy is inflated.

(5) The upper lateral band is 1-inch tubular nylon with a tensile strength of 4,000 pounds.

(6) The 30 gores have five sections each.

(7) The 30 radial seams are 9/16 inch wide and 17 feet 2 7/32 inches long.

(8) The lower lateral band is made of 1-inch nylon tape with a tensile strength of 525 pounds.

(9) The 15 pocket bands have been lengthened to 7 1/2 inches to provide a more positive opening and a 4.37 foot (overall) increase in the canopy to reduce descent to about 15 feet per second.

(10) The 30 V-tabs are 9/16 inch wide.

(11) The 30 suspension lines are 25 feet 6 inches long.

2-7. MODIFIED IMPROVED RESERVE PARACHUTE SYSTEM

The MIRPS (used with the T-10- and MC1-series) is an emergency-type parachute designed to be activated by the parachutist if the main parachute malfunctions. It is chest-mounted, manually operated, and loaf-shaped when packed. It weighs between 13 and 13.5 pounds, has a combined shelf life and service life not to exceed 16.5 years, and is 24 feet in diameter. It is repacked every 365 days. The MIRPS consists of six major components: pilot parachute, bridle line with weight and apex weight, ejector spring, canopy assembly, pack assembly, and rip cord assembly.

a. **Pilot Parachute Assembly**. The pilot parachute assists in deployment of the parachute canopy by serving as an air anchor. It is a five-foot, flat, circular parachute with bridle line and is constructed from zero-porosity nylon parachute cloth and marquisette netting. The netting is reinforced with six radial tapes that form the bridle attachment loop. A centerline is attached to the parachute cloth to speed pilot chute inflation and also forms part of the bridle attachment loop. The pilot chute does not have suspension lines. It resembles a large ball.

b. **Bridle Line Assembly**. The bridle line assembly is 13 feet long and is constructed from two-inch-wide polyester webbing with a four-inch loop at each end. One end of the bridle line is fitted with an apex sock that aids in pressurizing the reserve main canopy during low-speed deployments. Adjacent to the apex sock, the bridle line is fitted with two curved metal pins that are used to secure the canopy staging flaps located in the pack assembly. The other end of the bridle line is fitted with a five-ounce deployment weight. The deployment weight provides the necessary mass to cause positive launch of the pilot chute once the ejector spring reaches full extension.

c. **Deployment Assistance Device**. The ejector spring is a 30-inch-long helical spring encased in marquisette netting and fitted with an end cap at each end. On one end cap, two grommet tabs are attached. These are used only during packing to keep the spring compressed. Before the final closing of the pack, the spring compression aid is removed, and the grommet tabs are no longer used.

d. **Canopy Assembly**. The canopy assembly is a 24-foot, flat, circular parachute constructed of 1.1-ounce olive drab ripstop nylon parachute cloth. Depending on the jumper's total weight, its rate of descent varies from 15 to 22 feet per second. It is described as follows:

(1) An apex vent, 20 inches in diameter.

(2) Twelve suspension lines, 57 feet 6 inches long (measured from connector snap to connector snap), made of Type III nylon cord with a tensile strength of 550 pounds. On this parachute, the suspension lines serve three purposes:

- From the connector snaps that double as connector links, the lines are 20-foot suspension lines.
- Where the suspension lines go through the radial seams of the canopy, they become canopy lines.
- Across the apex vent, the canopy lines become apex vent lines until they again go into the upper radial seam; there they again become canopy lines.

(3) An upper lateral band of 1-inch tubular nylon with a tensile strength of 4,000 pounds.

(4) A lower lateral band of 1-inch nylon tape with a tensile strength of 525 pounds.

(5) Twenty-four gores with four sections in each gore.

(6) A 9/16-inch nylon V-tab reinforcement sewn into the lower lateral band of the parachute and wrapped around each suspension line.

(7) Twenty-four pocket bands.

e. **Pack Tray**. The MIRPS and the T-10 troop chest reserve use different pack trays that are almost identical on the outside (Figure 2-7 and Figure 2-8, page 2-12). Thus, an identifying yellow binding tape on the rip cord protector flap distinguishes the MIRPS. The major parts of the pack tray used for the MIRPS are the main panel and two end panels. The main panel forms a top and bottom flap, and the end panels form a right end flap and a left end flap. A rectangular metal pack frame is enclosed in a pocket formed in the bottom. The pack has two holes with lift-the-dot strap fasteners for attaching the MIRPS to the primary parachute harness using the connector snaps. A suspension-line free-bag deployment pouch is located on the inside of the container on the pack bottom. Canopy staging flaps are attached to the inside of the top and bottom flaps; these provide canopy retention.

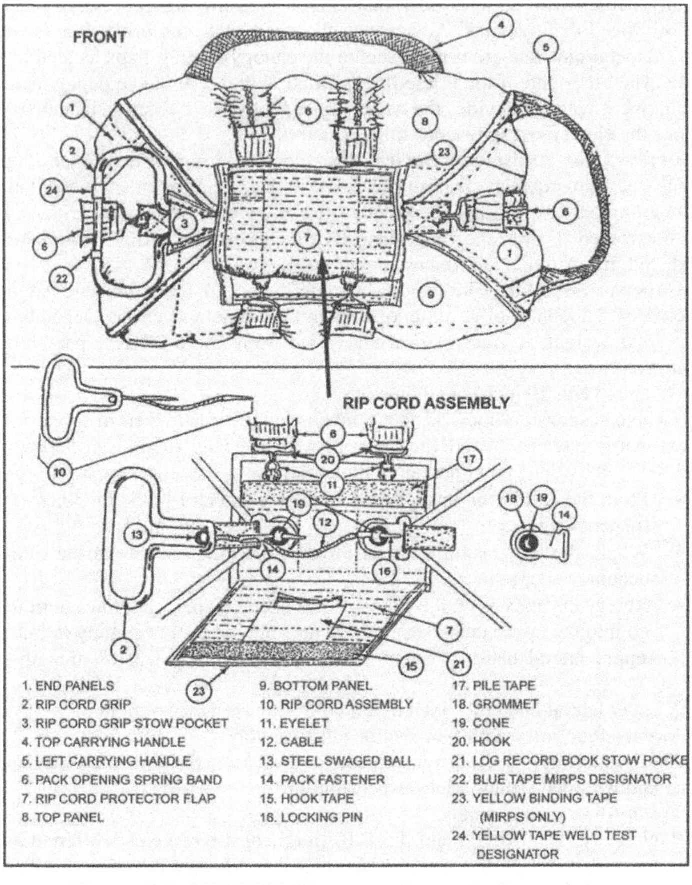

Figure 2-7. MIRPS/T-10 reserve parachute pack tray and nomenclature (front).

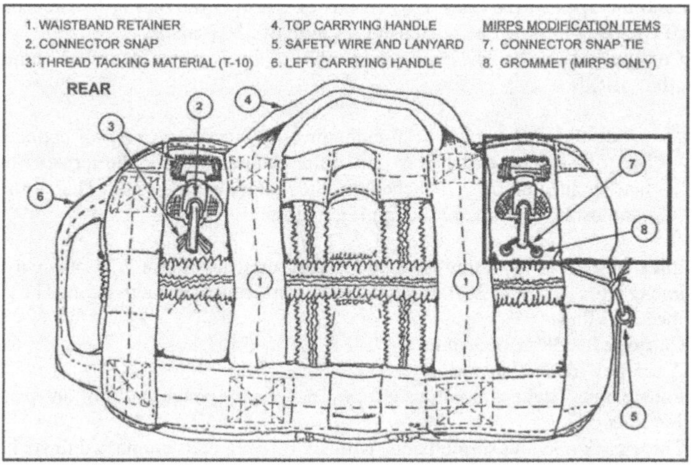

Figure 2-8. MIRPS/T-10 reserve parachute pack tray and nomenclature (back).

f. **Rip Cord Assembly**. The rip cord assembly is used to activate the reserve parachute and requires about 27 pounds of pull pressure. It is composed of a cloverleaf-shaped, stainless steel rip cord grip attached to a 7.25-inch-long flexible steel cable by means of a steel swaged ball. The cable has two permanently fastened locking pins (1 1/4 inches long) that are inserted into holes in the pack tray bottom panel cones; they keep the reserve pack tray closed until activation.

2-8. T-10 TROOP CHEST RESERVE PARACHUTE

The T-10 reserve parachute (used with T-10- and MC1-series) is chest-mounted, manually operated, and loaf-shaped when packed. It weighs about 12 pounds, has a combined shelf life and service life not to exceed 16.5 years, and is 24 feet in diameter. It is repacked every 365 days. It is an emergency-type parachute designed to be activated by the parachutist if the main parachute malfunctions. The T-10 reserve parachute consists of four major components: spring-activated pilot parachute assembly, canopy assembly, pack tray assembly, and rip cord assembly. The rip cord grip must be marked with a blue and yellow piece of tape. Blue signifies that it is for the MIRPS and yellow shows that it has been weld tested.

a. **Pilot Parachute Assembly**. The pilot parachute acts as an air anchor to assist the canopy in deploying quickly. It is spring-activated, made of 1.1-ounce ripstop nylon parachute cloth, and is 40 inches in diameter. It is described as follows:

(1) Octagon-shaped, with reinforcements sewn to the inside of the canopy.

(2) Four pockets sewn on the inside of the canopy to stow the spring framework.

(3) Eight suspension lines of Type I nylon cord. Each pair of suspension lines is formed by one continuous line that runs through the canopy and is stitched together at the lowest point to form the connector loop. The connector loop is used to attach the pilot

parachute to the apex of the reserve by means of a bridle line that is 15 inches long and made of Type III nylon cord with a tensile strength of 550 pounds.

b. **Canopy Assembly**. The canopy assembly in the T-10 reserve is the same canopy used for the MIRPS.

NOTE: Inside the pack tray is a 10-inch spreader bar. If one reserve connector snap becomes disconnected from one of the harness D-rings, the spreader bar keeps that connector snap attached to the parachute harness. This arrangement enables the reserve to function if a malfunction occurs.

c. **Pack Tray**. The pack tray is made of 12.29-ounce nylon or 7.25-ounce nylon on a rigid frame (Figures 2-7 and 2-8). It contains the parachute until activation. The pack tray is described as follows:

(1) Carrying handles (top and left).

(2) A safety wire and lanyard.

(3) Four panels (right end panel, left end panel, top panel, and bottom panel) with attached eyelets.

(4) Three pack opening spring bands with six running ends connected to six hooks.

(5) A rip cord grip stow pocket sewn to the right-end panel.

(6) A rip cord protector flap sewn to the bottom panel, with log record stow pocket on the inside of the rip cord protector flap.

(7) Pack fasteners sewn to the left- and right-end panels.

(8) Grommets and cones that keep the pack closed by inserting the locking pins of the rip cord assembly through the cones. (Grommets are on the top panel and cones are on the bottom panel.)

(9) Two connector snaps to attach the reserve to the D-rings.

(10) Two nylon waistband retainers through which the waistband is routed. The waistband retainers consist of a single piece of material sewn into the pack tray, which forms the reserve top carrying handle.

d. **Rip Cord Assembly**. The rip cord assembly for the T-10 reserve is the same as for the MIRPS with one exception. The length of the cable is .25 inch shorter (7 inches rather than 7.25).

2-9. BA-18 BACK AUTOMATIC PARACHUTE

The USAF emergency parachute BA-18, back automatic parachute, is used by non-jumping jumpmasters and safety personnel onboard USAF aircraft. This USAF parachute has a repack cycle of 180 days and must pass a routine inspection at the home unit every 30 days. Once the parachute is activated, it must be destroyed. The BA-18 is equipped with either an FLB-Model 7000 or FLB-Model 11000 automatic release system (ARS). The owning USAF unit's life support department sets the time delay release (from 1 to 13 seconds) and then attaches a small white tag on the back of the BA-18 to indicate what delay time has been set (3 seconds, 4 seconds, 5 seconds, and so on). To employ the BA-18, the jumper either pulls the red arming cable knob to activate the time delay feature or overrides the ARS by pulling the T-shaped blast handle. Personnel scheduled to wear the BA-18 must inspect it before each use. If they note any discrepancies during the inspection, they must not use the parachute.

a. **BA-18 Inspection Criteria**. The inspection procedures are simple, and their sequence is not important. However, the parachute must be inspected before each donning.

(1) Check the canopy release assemblies in the same manner as on the T-10-series or MC1-series main parachutes.

(2) Ensure that no straps (chest, leg, and horizontal back strap) are cut or frayed.

(3) Ensure all three ejector snaps are serviceable.

(4) Between the parachute harness assembly and the pack tray, a pad is held with four pull-the-dot fasteners (one at each corner). Open the pull-the-dot fasteners, expose the long zipper that is behind the pad, and unzip the zipper. Ensure that the locking pins are not bent and are routed through the white nylon cord loops, and that the cable is free of rust or other corrosion. Re-secure the zipper and pad once the inspection has been completed.

(5) Inspect the risers and ensure they are not cut or frayed and are secured to the canopy release assemblies.

(6) Make an overall inspection on the outside of the pack tray to ensure that the pack tray is free of rips, tears, oil, grease, dirt, or water and that no canopy or suspension lines protrude from the pack tray.

(7) Open the slide fastener on the back of the pack tray. Ensure that the spring has both hooks (one at each end) routed through the white nylon cord loops; then secure the slide fastener.

(8) Ensure the rip cord is properly stowed and free of rust or other corrosion.

(9) Ensure the red time delay arming cable knob is properly stowed.

(10) If the parachute is equipped with a personnel lowering device, inspect it to ensure that the associated hardware is secure inside the stow pocket. Inspect the lowering line from where it exits the stow pocket to the point it disappears into the pack tray.

b. **Planned and Unplanned Exits**.

(1) ***Above 14,000 Feet.***

(a) Once the last jumper has exited the aircraft, pull the red automatic release arming knob to arm the ARS, and exit the aircraft.

(b) When you reach an altitude of 14,000 feet, the canopy will open. However, at anytime during descent, you can activate the rip cord and override the ARS.

(2) ***Below 14,000 Feet.***

(a) If the parachute is equipped with an ARS set for a delay of 4 seconds or more, pull the automatic release arming knob and exit the aircraft.

(b) If the parachute is equipped with an ARS set for less than 4 seconds, exit the aircraft, assume the correct body position, and pull the automatic release arming knob. To assume the correct body position, place your chin on your chest, your elbows into your sides, your feet and knees together, and visually locate and place your hand(s) on the rip cord. Allow sufficient time (one second, minimum) to clear the aircraft, then pull the rip cord.

NOTE: When bailing out of an aircraft at extremely low levels, the most important factor is to pull the rip cord as soon as you clear the aircraft, regardless of your body position.

c. **Proper Method to Pull the Rip Cord.**

(1) The automatic release may be overridden by pulling the rip cord. Grasp the "T" handle with your right hand and guide or assist it with your left hand. Pull the rip cord down toward your feet and away from your body hard and fast. Pull it to arm's length and then bring your arms back close to your body immediately. Ensure the rip cord clears the housing.

(2) After exiting the aircraft, complete the normal remaining four points of performance—check canopy and gain canopy control, keep a sharp lookout during your entire descent, prepare to land, and land (see Chapter 3). To avoid obstacles and other jumpers in the air, or maneuver the canopy, pull a vigorous two-riser slip in the desired direction of travel. To execute a two-riser slip with this canopy, reach up high onto the risers to the elbow-locked position, grasp a set of risers in the direction of the desired movement, and pull them down to your chest.

d. **Personnel Lowering Device (PLD).** If you become hung in wires, trees, or rough terrain and the parachute is equipped with a PLD, use it to climb down to safety.

(1) Visually check to see that you are securely hung in the obstacle.

(2) Take the hardware out of the stow pocket. Grasp the braking device and snap it into the top portion of the V-ring of the chest strap with the hook side facing your midsection.

(3) Pass the snap hook at the end of the tape through the "V" of both risers.

(4) Attach the snap hook to the O-ring (which should be in front of your face).

(5) Grasp the tape with your left hand and disconnect the right canopy release assembly with the right hand. Exchange hands on the tape, with your left hand disconnecting the left canopy release assembly.

(6) Lower yourself to the ground by feeding the tape through the braking device.

(7) To stop the descent, pull the tape up vertically with your right hand.

2-10. CARE OF THE PARACHUTE BEFORE JUMPING

Troop parachute assemblies and reserves may be issued in kit bags to aid handling and to prevent damage or unintentional opening. Until removed for fitting by parachutists, parachutes and reserves should remain in the kit bags and protected from moisture during storage to prevent mildew. (Kit bags are not waterproof and do not provide adequate protection from wet weather or damp ground.) Parachutes must be stored in weatherproof areas such as adequate storage buildings, trucks, tents, or transport aircraft.

2-11. CARE OF THE PARACHUTE AFTER JUMPING

The parachute is recovered and properly cared for to minimize damage. Upon landing, parachutists activate the canopy release assembly (while lying on their backs to observe other landing parachutists), take off the harness, and place the parachute in the kit bag using either the tactical recovery method or by executing a series of figure-eight folds with the arms (Figure 2-9, page 2-16). The specific actions are as follows:

a. Remove all air items. Place harness in the aviator kit bag with the smooth side up, leaving the waistband out. Place released riser underneath the harness.

b. Move to the apex of the canopy, grasp the bridle loop, elongate the parachute into the wind to straighten the canopy and suspension lines, and remove all foreign objects and debris from suspension lines and canopy.

c. Fold the canopy and suspension lines into a series of figure eights, using both arms. Do **not** twist the canopy unnecessarily because friction can cause the nylon to fuse.

d. Lay the canopy on the top of the harness and, before closing the bag fasteners, ensure the bridle loop is on top of the canopy and the waistband is routed through the bridle loop. Close bag fasteners; **do not zip** the bag because the canopy may become entangled in the zipper and damaged. Attach a reserve connector snap to each kit bag handle. Carry the equipment so the reserve parachute is to the (parachutist's) front and the kit bag to the rear. (Reverse the carry when jumping with combat equipment.)

NOTE: If it is necessary to activate both canopy release assemblies upon landing, then fold the canopy in figure eights by itself and place it in the kit bag on top of the harness.

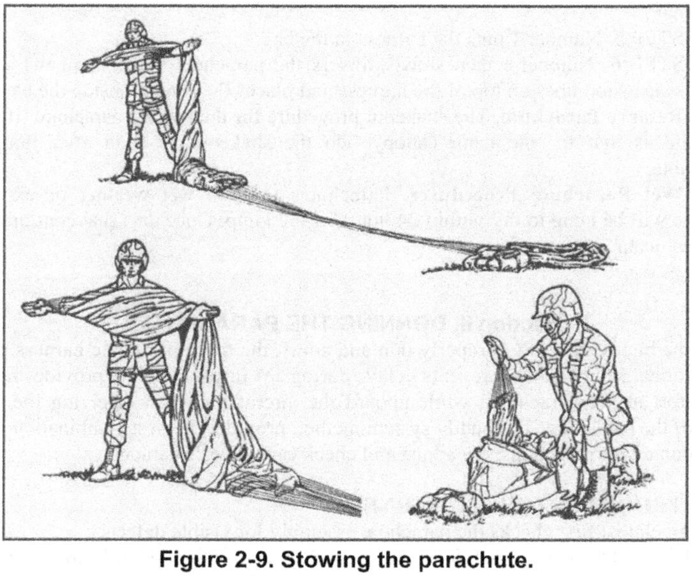

Figure 2-9. Stowing the parachute.

2-12. SHAKEOUT PROCEDURES

The parachute is suspended from a rope passed over a pulley suspended from a ceiling (or from pole's) high enough to allow the canopy to clear the surface area.

a. **Main Canopy**. A two-man shakeout team is recommended. Number 1 holds the bridle loop, while number 2 fastens the rope to the loop. Number 2 pulls the rope until the skirt is about 1 foot above number 1's head. They accomplish the rest of the shakeout by taking the following steps:

STEP 1: The team leaves the bulk of the suspension lines and the parachute harness in the kit bag. Number 2 grasps the rope attached to the suspended canopy while number 1 shakes the parachute.

STEP 2: Number 1 grasps two adjacent suspension lines at the lower lateral band, one in each hand, and vigorously shakes the gore, making certain no grass, twigs, insects, or other foreign matter are left on the fabric or tangled in the anti-inversion net.

STEP 3: Number 1 then transfers both suspension lines to his left hand, grasps the suspension lines of the next gore with his right hand and continues as in step 2, working counterclockwise, until each gore has been shaken and all suspension lines are in his left hand. He must pay particular attention to the anti-inversion net to ensure no foreign material remains. Debris left in the net can result in a total malfunction.

STEP 4: Number 2 begins to slowly pull the canopy up, elongating the suspension lines. Number 1 shakes the suspension lines and dusts them by hand and then turns the kit bag inside out and cleans it thoroughly to ensure no debris is in the bag.

STEP 5: Number 1 puts the harness in the bag.

STEP 6: Number 2 then slowly lowers the parachute while number 1 coils the suspension lines on top of the harness and places the canopy inside the bag.

b. **Reserve Parachute**. The shakeout procedure for the reserve parachute (if used) is the same as that for the main canopy. Do the shakeout as soon after jumping as practicable.

c. **Wet Parachute Procedures**. Parachutes used in wet weather or exposed to moisture will be hung to dry within 24 hours of the jump. Once dry, shakeout procedures will then occur.

Section II. DONNING THE PARACHUTES

Using the buddy system to properly don and adjust the troop parachute harness provides an additional safety check, prevents delays during JM inspection, and provides minimum discomfort to the parachutist while aboard the aircraft or when receiving the opening shock of the parachute. The buddy system method provides the best combination of speed and accuracy for parachutists to adjust and check each other's parachutes.

2-13. TROOP PARACHUTE HARNESS

Each parachutist first checks the parachute assembly for visible defects.

STEP 1: The parachutist lays the assembly out with the pack tray face down. Then he—

(1) Activates the waistband quick-release and pulls up each of the activating levers on the ejector snaps, releasing the leg straps and the chest strap.

(2) Checks for appropriate size and, if necessary—

- Lifts the pull-the-dot fasteners on the diagonal back-strap retainer and frees the diagonal back straps from the sizing channel (using fingers only; do not use tools).
- Sizes the parachute to one of the six sizing channels (S, 1, 2, 3, 4, and L).
- Rechannels the diagonal back strap retainer and fastens the pull-the-dot fastener.

- Lets out about half of the slack in the horizontal back strap, leg straps, and chest strap; straightens the leg straps and chest strap; and folds the kit bag, leaving the outermost carrying handle extended.

STEP 2: The parachutist (number 1) bends slightly forward at the waist to don the parachute. A second parachutist (number 2) holds the parachute assembly by the main lift web under the canopy release assemblies and places it on the back of number 1.

STEP 3: Number 1 remains bent forward at the waist; number 2 pushes the pack tray high on number 1's back and pulls the saddle well down over the buttocks. As the adjustment is being made, number 1 fastens the chest strap and ensures that the activating lever is closed over the ball detent.

STEP 4: Number 2 calls out "Left leg strap," grasps the leg strap by the quick-fit V-ring with one hand, and with his other hand starts from the saddle (with thumb and index finger) and feels the length of the leg strap, removing any twists and turns, and hands the left leg strap to the jumper. Number 1 inserts the left leg strap through (over the bottom and under the top) the kit bag carrying handle and snaps the quick-fit V-ring into the left ejector snap. The right leg strap is passed over the other end of the kit bag (securing it in place), and the quick-fit V-ring is snapped into the right ejector snap. The parachutist ensures that both the left and right activating levers are closed over the ball detents.

STEP 5: Number 1 stands erect and checks to ensure the canopy release assemblies are in the pockets of the shoulders.

STEP 6: Number 2 locates the free-running ends of the horizontal back strap and tightens the harness until number 1 indicates it fits snugly and comfortably. The horizontal back strap is the main point of adjustment for the harness. After final adjustment, number 1 should be able to stand fully erect without straining (Figure 2-10).

Figure 2-10. Troop parachute harness fitted.

STEP 7: Number 1 and number 2 then change positions and repeat steps 1 through 6. When both parachutists have donned their parachute harnesses, they face each other and make a visual inspection. They correct any discrepancies before securing the reserve parachute.

All excess webbing is stowed in webbing retainers.

2-14. MIRPS/T-10 RESERVE PARACHUTE

The parachutist attaches the reserve parachute by cradling the parachute in his left arm with the connector snaps up and the rip cord grip in the palm of the left hand.

STEP 1: Using the right hand, start at the pack tray and run out the waistband between the thumb and index finger to remove any twists or turns.

STEP 2: Thread the waistband through the two reserve waistband retainers and fasten the right connector snap to the right D-ring. Insert the safety wire in the right connector snap and then bend the wire down to safety it. Then connect the left connector snap to the left D-ring.

STEP 3: Parachutists help each other in securing the waistband and forming the quick-release. They ensure that all slack is pulled out of the waistband, and the slack in the quick-release loop is about the width of two to three fingers.

NOTE: During the initial periods of airborne training, students receive thorough training in the nomenclature, fitting, and wearing of the parachute assemblies. Demonstration, followed by student participation, is the key to the success of this instruction. Instructors constantly check to ensure students know the proper nomenclature as well as the proper methods of wearing and fitting the parachutes.

Section III. PROTECTIVE HEADGEAR

The ballistic helmet or the advanced combat helmet (ACH) is used during airborne operations. The helmets must be fitted properly to ensure they stay on the parachutist's head during deployment of the parachute and during the parachute landing fall.

2-15. BALLISTIC HELMET DESCRIPTION

The ballistic helmet is a laminated, one-piece helmet. It has a low profile, a close fit, and a low center of gravity. The helmet is available in four sizes: extra small, small, medium, and large. It weighs from 2 pounds 8 ounces (extra small) to 3 pounds 4 ounces (large).

a. **Parachutist Helmet Modifications**. The ballistic helmet is modified for airborne operations by adding a parachutist retention strap and a foam impact pad (Figure 2-11, page 2-20). These items provide maximum safety and stability.

NOTE: Be sure the hook-pile tape on the ends of the parachutist retention strap faces the rear of the helmet.

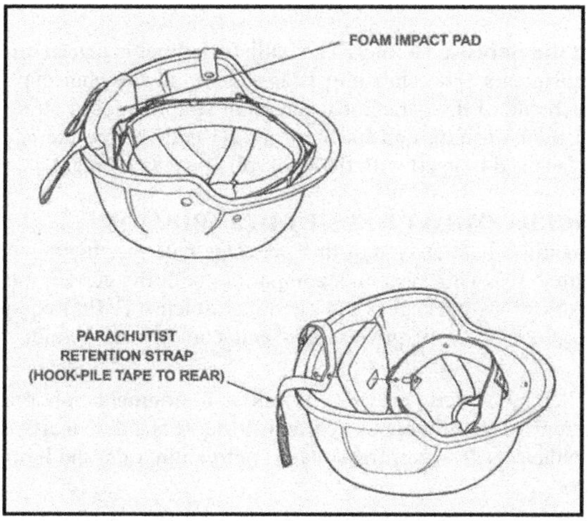

Figure 2-11. Ballistic helmet with retention strap and foam impact pad.

b. **Camouflage Cover**. A camouflage cover (Figure 2-12) is secured to the helmet by placing the cover over the helmet and threading both sides of the unfastened chin strap through the corresponding slits in the cover. The two rear attachment tabs of the cover are tucked between the cover and the helmet. The two ends of the parachutist retention strap are threaded through the corresponding slits in the rear of the cover. The four remaining attachment tabs are fastened around the nylon webbing of the suspension band with drawstring and adjustable tab.

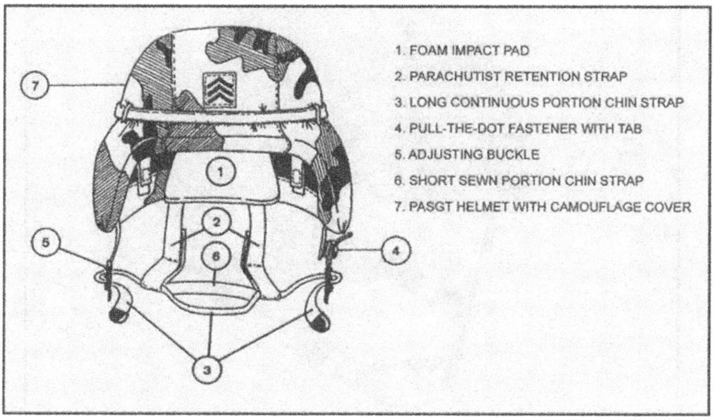

Figure 2-12. Ballistic helmet and nomenclature.

c. **Donning the Ballistic Helmet.** The ballistic helmet is placed on the head to make any comfort adjustments. The chinstrap is fastened and the chin cup is adjusted to fit snugly. Then, each end of the parachutist retention strap is placed alongside the face and wrapped around the chinstrap, and its free-running end (below the adjusting buckle) is pulled tight and fastened to itself with the hook-pile tape attachment.

2-16. ADVANCED COMBAT HELMET DESCRIPTION

The ACH is a modular helmet system that provides ballistic, fragmentation, and impact protection (Figure 2-13). This system is compatible with the current night vision devices (NVGs), communications packages, and chemical defense (NBC) equipment. The ACH is intended to replace standard government and commercial helmet systems currently in use.

- The helmet system provides 9-mm and fragmentation protection within a spectrum of environments (cold [-40 degrees Fahrenheit]; hot [140 degrees Fahrenheit]; salt water; fresh water; petroleum, oils, and lubricants [POL]; and so on).
- The ACH allows maximum sensory awareness for the operator. which includes an unobstructed field of view and increased ambient hearing capabilities.
- The helmet's retention/suspension system provides unsurpassed balance, stability, and comfort. This unique system provides for proper size, fit, and ventilation.
- The pad suspension system provides superior impact protection throughout all operational requirements.

Figure 2-13. Advanced combat helmet.

a. **Helmet Assembly.** The ACH has a unique, fully adjustable pad suspension system. All seven pads are worn with the helmet (Figure 2-14).

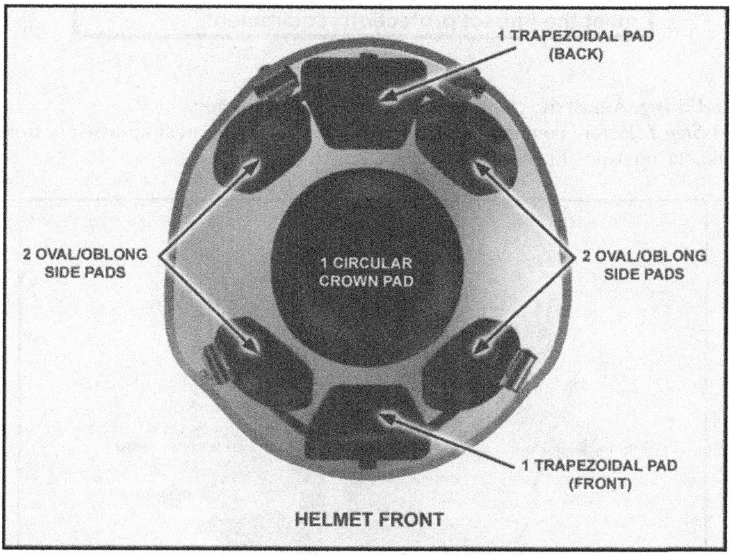

Figure 2-14. ACH pads.

(1) The placement of the oblong side pads must cover the four bolt ends in the four internal buckles in the helmet shell.

(2) The crown pad may be replaced by two oblong/oval pads. The direction of the side oblong/oval pads may be changed to maximize comfort: they may be routed vertically from bolt to crown, which maximizes airflow for better temperature regulation; or, they may be routed horizontally to make a seal around the user's head, which is better for cold weather environments.

NOTES:

1. If hot spots or discomfort are experienced, try rearranging the pad system to accommodate a more comfortable fit. If discomfort persists, try resizing the shell.
2. See the sizing and fitting troubleshooting guidelines (paragraph c, page 2-26) for problems with fit, such as tightness or looseness, or if the helmet profile is too high or too low.
3. When trying on the helmet for the first time in a cold environment, wear the helmet for a few minutes to allow the pads to warm up and conform to the shape of your head.

WARNING

Helmet must be worn with crown pad or two properly placed oblong/oval pads (size 6 or 8) to meet the impact protection requirement.

b. **Fitting.** Adjust the chinstrap to optimize fit and comfort.

(1) ***Step 1.*** Before donning the helmet, loosen all adjustment straps (two in front, two in back, one on nape of neck) (Figure 2-15).

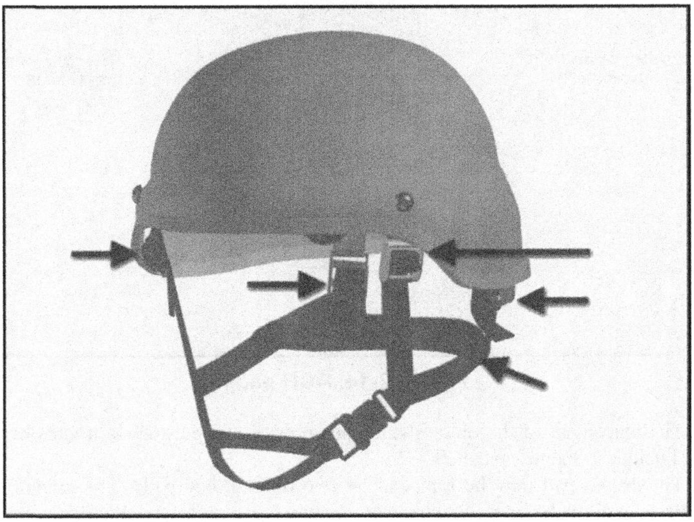

Figure 2-15. ACH adjustable straps.

(2) ***Step 2.*** Position the helmet on the head and hold in place with one hand on top of the helmet for initial adjustment (Figure 2-16, page 2-24).

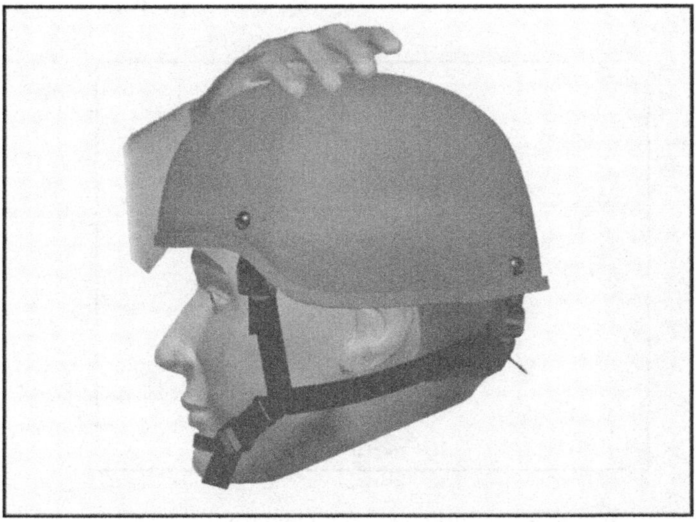

Figure 2-16. ACH held in place for initial adjustment.

(3) ***Step 3.*** Partially tighten the two back adjustment straps (one side at a time) (Figure 2-17).

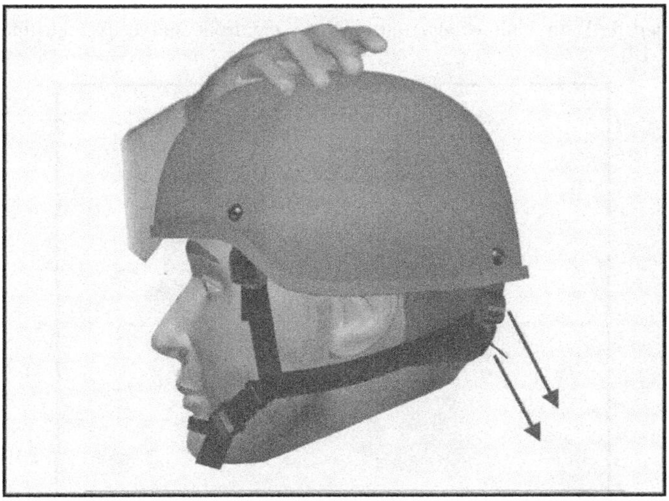

Figure 2-17. Partially tighten two back straps.

(4) ***Step 4.*** Partially tighten the two front adjustment straps (one side at a time) (Figure 2-18).

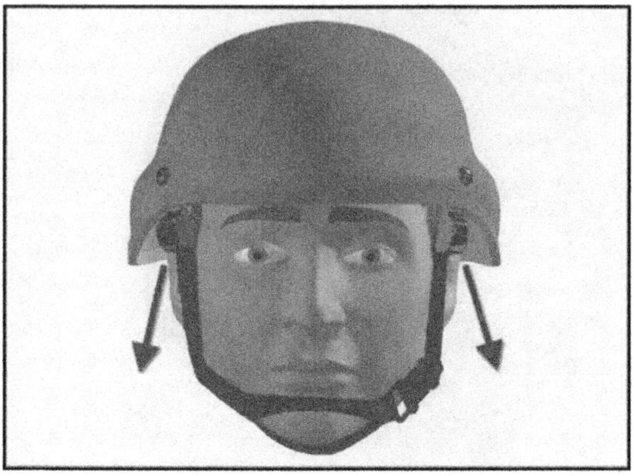

Figure 2-18. Partially tighten two front straps.

NOTE: If any one strap is pulled too tightly during steps 3 and 4, the helmet may become uncomfortable and tilted on the head.

(5) ***Step 5.*** With both hands, fully tighten the front and back adjustment straps (Figure 2-19).

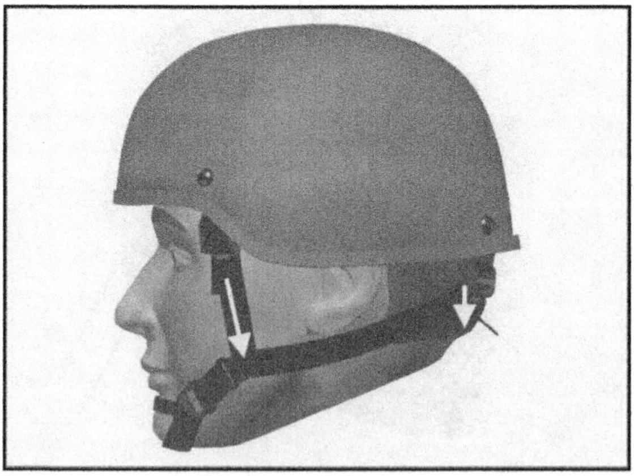

Figure 2-19. Fully tighten front and back straps.

(6) ***Step 6.*** Position the nape pad up and down according to personal comfort (Figure 2-20).

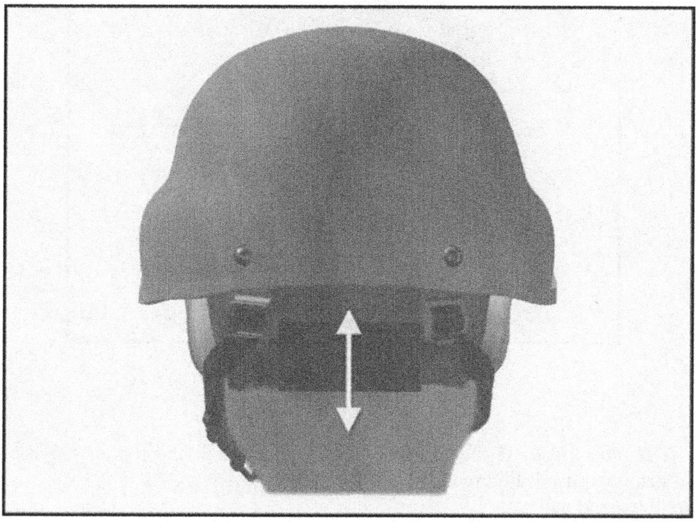

Figure 2-20. Position nape pad.

c. **Sizing and Fitting Troubleshooting Guidelines.** Use the following guidelines to solve the problems discussed herein.

(1) ***ACH Too Tight.*** If the ACH feels too tight—

- Try arranging the side pads in a vertical configuration.
- Try removing the back pad and rearranging the side pads.
- Try to create space in the area that feels tight.
- Try the next smallest pad size if rearranging the pads does not alleviate the tightness.

(2) ***ACH Too Loose.*** If the ACH feels too loose (Figure 2-21)—

- Try a larger sized pad set or increase the current number of pads in use.
- If wearing a size 8 pad set and a size large shell, downsize to a size medium shell. If the helmet is still too loose, increase the number of pads.

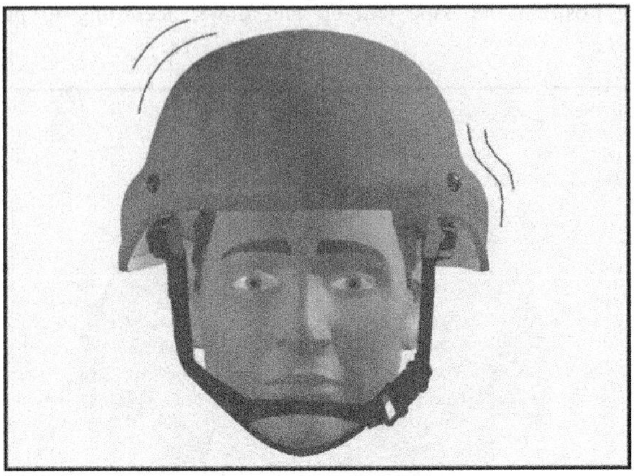

Figure 2-21. ACH too loose.

(3) ***ACH Too High.*** If the crown pad does not touch the top of the head or the forehead is too exposed (Figure 2-22)—
Try a smaller sized pad set.
Try a larger shell size if smaller pads do not work.

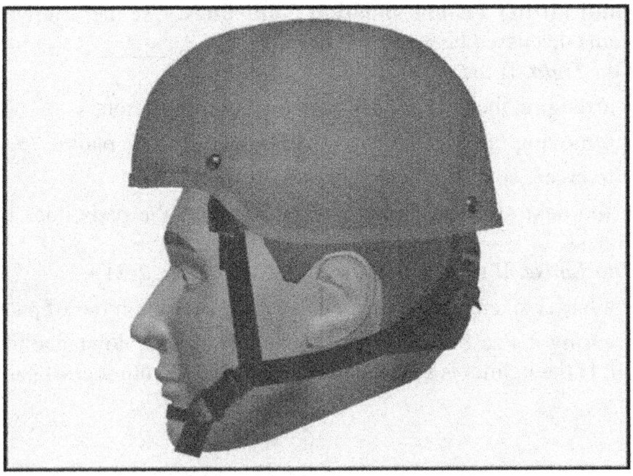

Figure 2-22. ACH sits too high.

(4) ***ACH Too Low.*** If the ACH sits too low on the brow, is not compatible with eyewear, or other compatibility issues exist (Figure 2-23)—

- Try larger suspension pads.
- Try a smaller ballistic helmet shell if the larger pads do not solve the problem.

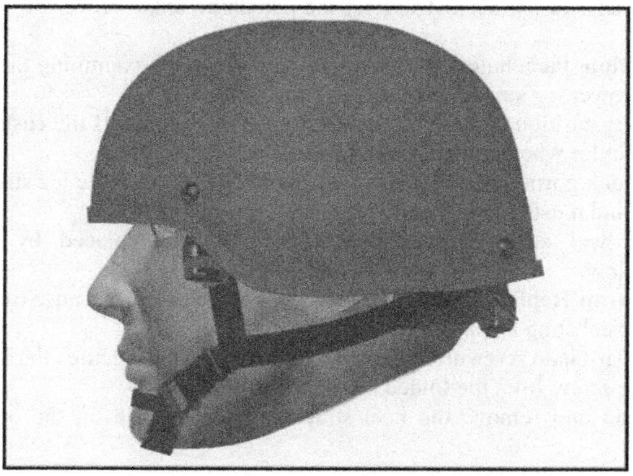

Figure 2-23. ACH sits too low.

Section IV. PARACHUTIST ANKLE BRACE

The parachutist ankle brace (PAB) stabilizes the jumper's ankle during PLFs and reduces the potential for ankle injury. The PAB, which is worn outside the combat boot, consists of sidewalls that extend vertically to encircle the ankle and the lower leg. The side and back portions are connected by a U-shaped support that fits over the boot heel. The support has a hook-pile strap system that attaches the brace to the leg and boot. The PAB is lined with a three-section aircell to cushion the lower leg.

NOTE: The PAB is not a required item of equipment.

2-17. OBTAINING THE PARACHUTIST ANKLE BRACE

The unit orders the PAB or the heel strap replacement kit directly from AIRCAST, P.O. Box 709, Summit, NJ 07902-0709, commercial 1-800-526-8785. All sizes are based on men's shoe sizes. The size codes for ordering are small, less than men's size 8, 02G (NSN 8465-01-416-6217); medium, men's sizes 8 through 11, 02H (NSN 8465-01-417-4002); or large, larger than men's size 11, 02I (NSN 8465-01-416-6218). European sizes are small, less than 42; medium, 42 through 45; and large, larger than 45. The cost is about $60 per pair.

NOTE: The PAB must be worn with standard-heeled boots. When worn with boots with flat or rippled soles, the PAB tends to slip and does not provide the proper support for the jumper's ankles.

2-18. INSPECTING THE PARACHUTIST ANKLE BRACE

Both jumpmaster and parachutist inspect the parachute ankle brace for serviceability as follows:

a. **Inspection**. Parachute ankle braces are inspected by examining the PAB cushion, the upper and lower leg straps, and the heel strap.

(1) If the leg cushion is missing, the PAB is unserviceable. If the cushion is torn, the jumpmaster decides whether the PAB is serviceable.

(2) If the hook portion of the hook-pile tape is missing from the leg straps, the PAB is unserviceable and must be replaced.

(3) If the heel strap is unserviceable, it can be replaced by following the instructions below.

b. **Heel Strap Replacement**. If the PAB is unserviceable because of a worn or torn heel strap, the heel strap is replaced as follows:

(1) Use a flat-blade screwdriver to remove the screw that secures the heel strap to the brace. Save the screw from the folded tab on the heel strap.

(2) Unthread and remove the heel strap from the buckle on the opposite side of the brace.

(3) Fasten the hook-pile tab to the brace.

NOTE: Ensure the new heel strap size corresponds to the brace size (small, medium, or large).

(4) Flip the brace over so the heel strap buckle is facing up. Thread the heel strap through the buckle top slot and pull through the slack. Loop the strap through the bottom slot and pull until snug.

(5) Flip the brace back over, remove the hook-pile tab to expose the screw hole, fold the strap end over, line up the screw holes, insert the screw through the strap, and refasten the screw to the brace.

2-19. DONNING THE PARACHUTIST ANKLE BRACE

To reduce the potential for jumper ankle injury, the parachutist ankle brace must be adjusted to fit snugly around the foot, ankle, and lower leg. The PAB is adjusted to the jumper's foot as follows:

a. Loosen the two leg straps and the heel strap, step into the PAB with the heel strap under the boot's instep (the area in front of the boot heel), and tighten the heel strap and then the two leg straps. The bottom of the shell should align with the top of the sole.

b. Fasten the leg straps using only the outer portion of the leg strap buckle. If both inner and outer portions of the buckle are used, the PAB could be difficult to remove in a tactical situation.

NOTE: The PAB is designed so the jumper can run short distances while wearing the PAB if necessary due to the tactical situation.

2-20. DOFFING THE PARACHUTIST ANKLE BRACE

The parachutist ankle brace is removed by loosening the leg straps and pulling the PAB off the leg and foot. The PAB is an air item and must be retained by the jumper for use during subsequent jumps.

CHAPTER 3

FIVE POINTS OF PERFORMANCE

The five points of performance are specific actions the parachutist performs between the time of exit from the aircraft and the recovery after landing. These points of performance are individual actions and are essential on every parachute jump. Failure to perform any one point correctly could result in a jump injury. They are stressed during jumper training, and each point is taught using one or more of the training apparatuses.

3-1. FIRST POINT OF PERFORMANCE: PROPER EXIT, CHECK BODY POSITION AND COUNT

A proper exit, body position, and count are essential to lessen the possibility of a parachute malfunction/bodily injury during the deployment and inflation of the parachute. The duration of the 4000-count corresponds to the approximate time it takes the main parachute to fully deploy when used by a jumper exiting an aircraft flying 130 knots per hour. The following must be trained reflex actions as the parachutist exits the aircraft:

a. The parachutist starts the 4000-count at ONE THOUSAND and snaps his feet and legs together, locking his knees and pointing his boot toes toward the ground. He lowers his head and places his chin firmly against his chest.

b. At the same time, he rotates his elbows firmly into his sides (with the palms of his hands on the ends of the reserve parachute, fingers spread, and right hand over the rip cord grip), and he bends his body forward at the waist to look over the reserve and to see his boot toes while he continues to count, TWO THOUSAND, THREE THOUSAND, FOUR THOUSAND, at normal cadence. (He keeps his eyes open to react to situations around him.)

WARNING

If no opening shock is felt by the parachutist at the end of the 4000-count, he must activate the reserve parachute as for a total malfunction.

3-2. SECOND POINT OF PERFORMANCE: CHECK CANOPY AND GAIN CANOPY CONTROL

When he finishes the 4000-count, the parachutist feels the parachute open, checks the canopy for malfunction/damage, and controls the parachute.

a. **T-10-Series**. He grasps the risers (thumbs up), spreads the risers apart, and throws his head back to inspect the entire canopy.

b. **MC1-Series**. He throws his head back to inspect the entire canopy and at the same time grasps the control line toggles, with his elbows well back, for immediate canopy control.

c. **Twists**. The main parachute may have twisted suspension lines, risers, or both. This condition may be caused by a single action or a combination of actions. The most common causes include the following:

- The deployment bag spinning before the canopy deploys.
- The canopy spinning when it comes out of the deployment bag and before it inflates.
- The parachutist tumbling or spinning (caused by improper exit and body position) during his descent.

If the suspension lines are twisted and the parachutist cannot raise his head enough to check the canopy properly, he compares his rate of descent with that of nearby parachutists.

(1) ***Rate of descent same as others around him.*** If his descent is the same as other jumpers around him, the parachutist untwists his suspension lines by reaching behind his neck, grasping each pair of risers (thumbs down, knuckles to the rear), and exerting an outward pull on each pair. He kicks his legs in a bicycle motion, continues to pull outward on the risers, and kicks until the twists are out of the suspension lines. When the twists are out of the lines, he checks the canopy and gains canopy control.

(2) ***Partial malfunction and rate of descent too fast.*** If the parachutist's main canopy has a partial malfunction and his descent is too fast (when compared to nearby parachutists), he activates the reserve parachute. (If using the T-10 reserve parachute, he uses the down-and-away method of activation.) See Chapter 6 for more information on activation of the reserve.

(3) ***No comparison can be made.*** When other parachutists are not close enough to compare rates of descent, he activates the reserve parachute (down-and-away method if using the T-10).

3-3. THIRD POINT OF PERFORMANCE: KEEP A SHARP LOOKOUT DURING THE ENTIRE DESCENT

The ability to hit a specific landing spot and to avoid other parachutists during descent is essential to successful airborne operations.

WARNING

During descent, the parachutist must watch to avoid collisions and entanglements with other parachutists and to avoid obstacles on the DZ. Jumpers stay 25 feet away from other jumpers.

a. **T-10-Series Parachute**. The degree of maneuverability with a T-10-series parachute is limited compared to the MC1-series. The jumper maneuvers the T-10-series parachute using slips.

(1) ***Use of Slips.*** The parachutist performs slips to avoid other parachutists, to avoid obstacles on the ground, and to prepare to land.

(2) ***Types of Slips.*** The two types of slips are the **two-riser** and the **one-riser**. For an effective slip of either type, the parachutist must ensure his hands are not placed through or behind the riser(s).

(3) ***Execution.*** When slipping, the parachutist looks in the direction that the slip is being made, makes a sharp initial pull to effectively spill air from the canopy, and releases the riser(s) slowly to prevent spinning or oscillations.

(a) Two-riser slip. A two-riser slip is made by reaching up to the elbow-locked position, grasping a pair of risers in the direction of the desired movement, and pulling them down to your chest.

(b) One-riser slip. A one-riser slip is made by pulling down three full arm lengths of the one riser nearest the desired direction of movement with a hand-over-hand motion.

b. **MC1-Series Parachute**. Depending on the wind conditions and his skill, the parachutist can steer his parachute to a selected point of impact on the DZ to avoid other parachutists in the air, to avoid obstacles on the ground, or to use a preferred PLF. To control MC1-series movement, the parachutist must know the principles by which the canopy operates and the factors that govern its control. The movement of the canopy is controlled by the action of the wind, the position of the canopy orifice (opening) relative to wind direction, and the way in which the control lines are manipulated.

WARNING

Due to the canopy's steerability, parachutists should stay at least 50 feet apart in the air to prevent collisions.

(1) ***Opening Away from Wind.*** When the orifice is located directly opposite the wind, the thrust of the orifice will be acting against the wind. This reduces the effect of wind velocity on the canopy and will retard the lateral movement of the canopy in the direction of the wind. This technique is called **holding**.

(2) ***Opening with the Wind.*** When the orifice is located directly with the wind, the thrust of the orifice combines with the thrust of the wind to speed the movement of the canopy in the direction of the wind. This technique is called **running**.

(3) ***Opening at an Angle.*** When the orifice is at an angle to wind direction, then the force of the wind from one direction and the thrust of the orifice at the angle moves the canopy in a direction near to a right angle to the direction of orifice thrust. The direction of movement varies with wind velocity and the angle at which the orifice is pointed. This technique is called **crabbing**.

WARNING

Before attempting any maneuvers, the parachutist must check around him to prevent collisions with other parachutists.

(4) ***Canopy Manipulation.*** Properly executed MC1-series maneuvers require correct canopy manipulation to combine the force of the wind and the thrust of the canopy orifice to move the parachute in a given direction. To maneuver the parachute to a certain point on the ground or to avoid ground obstacles, the parachutist may have to turn and hold into the wind, run with the wind, or crab to the left or right while running or holding.

(a) *Turning.* This is accomplished by pulling down on one control line toggle. The farther down the toggle is pulled, the faster the turn. Pulling the right toggle causes a right turn. Pulling the left toggle causes a left turn. Pulling both at the same time reduces forward speed but increases the rate of descent. This is called **braking**. To deliberately lose altitude quickly, the parachutist pulls down on both toggles. This maneuver should be stopped before he is less than 250 feet above the ground.

(b) *Holding into the Wind.* Holding into the wind is done by rotating the MC1-series canopy until the orifice is on the downwind side. Thereafter, the parachutist manipulates the control line toggles to retain this position.

(c) *Running with the Wind.* Running with the wind is accomplished when the parachutist rotates the canopy until the orifice is on the upwind side. Thereafter, control line toggles are manipulated as needed to retain the position.

WARNING

Running with the wind just prior to landing can cause injury and must be avoided below 100 feet above the ground.

(d) *Crabbing.* Maneuvering while holding into, or running with or at an angle to the wind, is performed by rotating the canopy to the left or right. As the canopy begins to move in the desired direction, the parachutist manipulates the control line toggles to maintain this direction.

(e) *Maneuvering with a Broken Control Line.* If a right or left control line is broken, the canopy can still be maneuvered, though more slowly. The parachutist reaches high on the right or left rear riser, on the same side as the broken control line, and pulls down.

(f) *Maneuvering with an Inversion.* If the canopy has inverted while opening, the parachutist reverses the maneuvering technique. To turn left, the parachutist pulls down on the right control line; to turn right, he pulls down on the left control line.

(g) *Maneuvering with a Tangled Control Line.* If the control line becomes tangled in a suspension line, the parachutist uses the same procedure as with a broken control line.

c. **Collisions and Entanglements**. A collision is the physical impact or contact, however slight, of one parachutist or parachutist's equipment with that of another parachutist. An entanglement is the entwining or attachment of a parachutist or parachutist's equipment with that of another parachutist during descent, whether or not the entanglement lasts until the parachutists contact the ground.

(1) ***Collisions.*** Parachutists must be alert in the air and warn each other of impending collisions. If a collision cannot be avoided by slipping or turning, the parachutist attempts to bounce off the other parachutist's suspension lines or canopy by spreading his arms and legs just before making contact.

(2) ***Entanglements.*** If a parachutist becomes entangled with one or more suspension lines of another parachute, the parachutist does one of the following, depending on the type of parachute being used.

(a) *T-10-Series.* Release entanglements as follows:

- The upper parachutist firmly grasps a portion of the lower parachute and moves hand under hand down the suspension lines of the lower parachute until each parachutist can grasp and hold the main lift web of the other's parachute, being careful not to grip the canopy release assemblies.
- If neither parachutist has a fully inflated canopy, both parachutists will push away from each other and activate their reserves using the pull-drop method.
- When the balls of the feet strike the ground, both parachutists release their grips and make either right, left, or rear PLFs away from each other. No front PLFs will be made.
- With the T-10-series, both jumpers can ride one good canopy to the ground. If both canopies collapse, both jumpers must activate their reserves using the pull-drop method.

(b) *MC1-Series.* Both jumpers remain where they are and activate their reserves for a partial malfunction. When using the MIRPS, jumpers must have a clear path to their front for the spring-assisted deployment device.

d. **Stealing Air**. A descending parachute causes an area of partial air compression immediately below the canopy and an area of partial vacuum and descending turbulent air above the canopy. **This turbulent air extends about 50 feet above the canopy**.

(1) A parachute falling into an area of partial vacuum (from a parachute below) does not capture enough air to stay fully inflated. The top parachute may partially collapse and drop below the other parachutist's canopy until the force of unaffected air reinflates it. Then this canopy, being lower, "steals" the air from the canopy above; this causes the canopy above to partially collapse and the jumper to drop past the lower canopy. This "leap-frogging" action continues unless corrective action is taken by the parachutist. Depending on the type of parachutes involved, the parachutist does one of the following:

(a) *T-10-Series.* He slips vigorously to maintain a lateral distance of at least 25 feet between the parachutes.

(b) *MC1-Series.* He turns in the opposite direction to provide at least a 50-foot distance between the parachutes. (When facing another parachutist, both parachutists execute a right turn.)

(2) When 250 feet or less above the ground, parachutists must exercise care to avoid stealing air from another parachute, because a deflated canopy will not be high enough above the ground to reinflate completely. If this situation occurs, the parachutist immediately prepares to land and to execute a PLF.

3-4. FOURTH POINT OF PERFORMANCE: PREPARE TO LAND

A proper landing attitude is necessary to lessen the risk of injury to the parachutist when he hits the ground (Figure 3-1, page 3-6). The preliminary movements of the parachutist vary, depending on the type of parachute used. However, lowering his individual equipment is the same with either parachute. He lowers the equipment on a lowering line when he is between 200 and 100 feet above the ground.

a. **T-10-Series**. When he is about 100 feet above the ground, the parachutist checks the direction of drift and pulls a two-riser slip into the wind. He holds the risers firmly against his chest and presses his elbows against his body. He keeps his head erect with his eyes on the horizon. He keeps his legs slightly bent and knees unlocked, and he keeps his feet and knees together with the balls of his feet pointed slightly toward the ground. He maintains moderate muscular tension in the legs, which absorb a significant portion of the landing impact, and he avoids becoming stiff or tense.

b. **MC1-Series**. When he is about 100 feet above the ground, the parachutist turns and holds into the wind. When nearing the ground, he holds the control line toggles at eye level. On impact, he holds the toggles, rotates his arms into his body, and executes a PLF.

c. **Obstacles**. The parachutist slips or turns to avoid obstacles. If obstacles (trees, water, or high tension wires) cannot be avoided, the parachutist takes the following precautions.

Figure 3-1. Landing attitude.

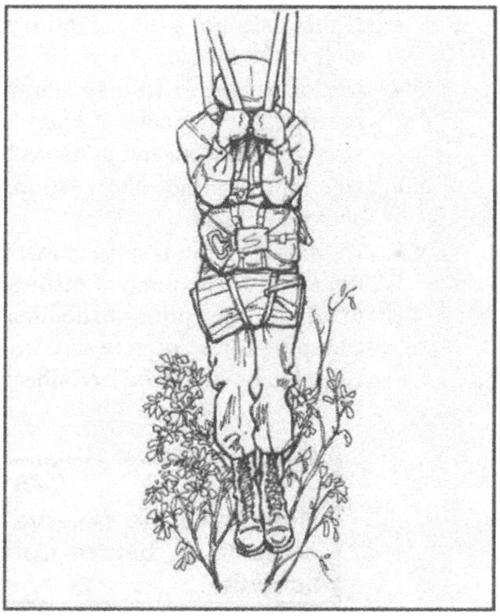

Figure 3-2. Tree landing attitude.

(1) ***Tree Landing.*** Initial precautions taken depend on the type parachute worn.

(a) *T-10-Series.* The parachutist continues to execute a diagonal slip to avoid the trees. Once he sees he cannot avoid them, he immediately assumes a normal prepare-to-land attitude. Just before he makes contact with the trees, he brings his hands up in front of his head and elbows in front of his chest while he continues to grasp the risers (Figure 3-2). He keeps his equipment attached. If he has lowered his equipment before realizing a tree landing is imminent, he looks below prior to jettisoning his equipment and then jettisons the equipment. He continues to watch the ground and trees. He keeps his feet and knees together and prepares to make a good PLF should he pass through the trees.

(b) *MC1-Series.* The parachutist keeps his feet and knees together and toes pointed downward. He continues to control the canopy to avoid the trees if possible. He keeps his equipment attached and wears his ballistic helmet. If he has lowered his equipment before realizing a tree landing is imminent, he looks below prior to jettisoning his equipment and then jettisons the equipment. He continues to watch the ground and trees. Just before impact, he assumes a normal prepare-to-land attitude, but he rotates his arms inward with his elbows high. Upon impact with a tree, the jumper places his hands in front of his face. He must be prepared to execute a parachute landing fall. If he gets hung up in a tree, he takes the following action:

- He reaches up high on both sets of risers and tugs on them three or four times to determine if he is securely hung. He prepares to do a good PLF in case he drops.
- He tries to reach the tree trunk or a large limb to allow him to climb down to the ground.
- If this does not work, he pulls the saddle down and over his buttocks and sits well in the saddle.
- He locates the D-ring attaching straps on his combat equipment and looks to see if it is clear below. Then he pulls down and out on the D-ring attaching straps and lowers and jettisons his combat equipment.
- He releases the chest strap by pulling outward on the ejector snap activating lever.
- He activates the reserve parachute by pulling the rip cord grip. (When using the MIRPS, the jumper ensures his left hand covers the ripcord protector flap to control the spring-loaded deployment assistance device.) He helps feed the canopy of the reserve out to ensure that all of the suspension lines are completely out of their retainers.

WARNING

Make sure the reserve reaches the ground, or is close to it, before continuing with the following actions.

- He activates the quick release in the waistband and frees it from the metal adjuster.
- He unfastens the left connector snap of the reserve from the left D-ring and pushes the reserve behind his right arm.
- He seats himself well into the saddle.
- He wraps his legs around the suspension lines of the reserve parachute and then carefully gets out of the harness.
- With one hand, he grasps the main lift web and holds it firmly.
- With the other hand, he grasps the activating lever of either the left or right leg strap and pulls outward, releasing the leg strap. He releases the other leg strap in the same manner.

- He climbs down the suspension lines and canopy, staying to the outside of the canopy.

(2) ***Water Landing.*** As soon as the parachutist realizes he is going to land in water, he does the following:

- He tries to slip, or steer, away from the water.
- He looks below to be sure the area is clear. Then he jettisons his helmet.
- He releases all equipment tie-downs.
- He looks below to be sure the area is clear. Then, he lowers any attached equipment.
- He activates the waistband quick release.
- He unhooks the left connector snap of the reserve parachute from the D-ring and rotates the reserve parachute to his right side.

(a) *Without a Life Preserver.* When wearing the troop parachute harness and a water landing **without** a life preserver is imminent, the parachutist does the following (Figure 3-3):

- He pulls the saddle well under his buttocks.
- He releases the chest strap by pulling on the activating lever of the ejector snap.
- He makes all possible attempts to remove the pistol belt and all equipment attached to his body that may hinder movement in the water.
- He releases the leg straps ejector snaps when his feet touch the water.
- He prepares to make a PLF in case the water is shallow (2 feet or less in depth).

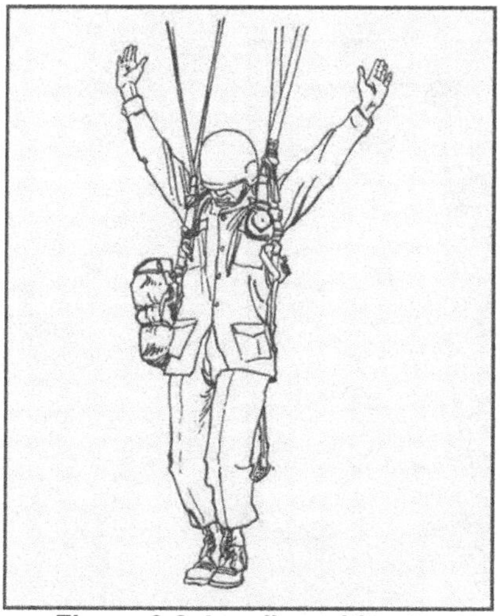

Figure 3-3. Landing without a life preserver.

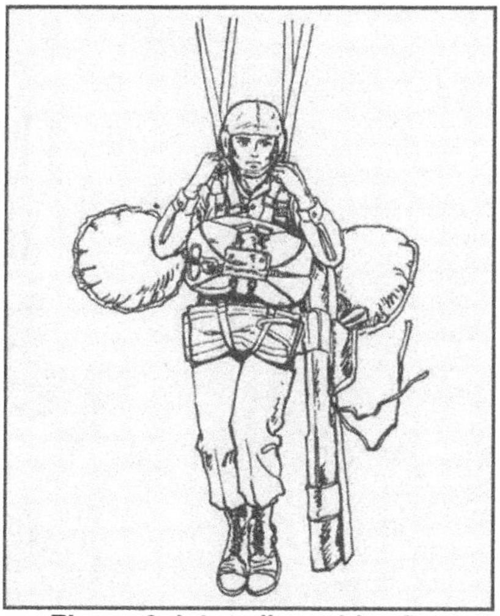

Figure 3-4. Landing with a B-7 life preserver.

(b) *With a Life Preserver.* When wearing a troop parachute harness and jumping **with** a B-7 or service approved life preserver (Figure 3-4), the parachutist does the following:

- He activates the B-7 after checking the canopy. If the B-7 fails to inflate, the parachutist inflates the B-7 manually by blowing air into the inflation valve hose. The parachutist activates one canopy release assembly after entering the water and signals "All okay" to the recovery boat.
- He pulls the safety clip out and away from his body (exposing the cable loops) and activates the canopy release assembly using one of the two methods used in the recovery from the drag as his feet touch the water.
- He does not remove the harness and equipment, since the B-7 will support up to 500 pounds.

NOTE: For more information on life preservers, see Chapter 12, Section II.

WARNING

When wearing the B-5, the parachutist does not inflate the B-5 until the parachute harness is removed. If restricted by the harness, the inflation force may crush his ribs.

(3) ***High Tension Wire Landing.*** The parachutist does the following if unable to avoid high tension lines when landing (Figure 3-5):

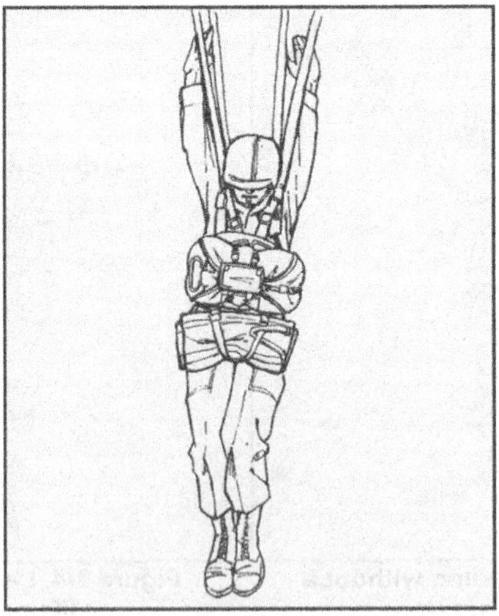

Figure 3-5. Wire landing attitude.

- He tries to slip away from the wires.
- He keeps his feet and knees together and toes pointed downward.
- He looks below and checks for fellow jumpers.
- He jettisons his combat equipment.
- He holds his hands high, inside the front set of risers with palms out and thumbs behind the risers, elbows back, with the fingers extended and joined.
- He keeps his chin on his chest, his body straight, with an exaggerated bend to his knees.
- He prepares to make a normal parachute landing fall.

If he contacts the wires, he begins a rocking motion with his body by pushing forward on the front risers and kicking back with his legs; this may keep him from getting entangled in the wires. He prepares to execute a PLF should he pass through the wires.

NOTE: If the jumper becomes entangled in the wires, he makes no attempt to climb down, but waits to be rescued by a recovery team.

3-5. FIFTH POINT OF PERFORMANCE: LAND

Most jump injuries occur because of improper PLF techniques. To lessen the possibility of injuries, the parachutist is trained to absorb the impact of landing by executing a proper PLF. To do this, the following five fleshy portions of the body must contact the ground in sequence: balls of feet, calf, thigh, buttock, and pull-up muscle(s). The three basic PLFs are **side** (right or left), **front** (right or left), and **rear** (right or left). The type of fall to be made is dictated by the direction of the wind drift. Before the landing attitude is assumed, the parachutist judges the direction of drift by looking at the ground. Then he prepares to make the appropriate PLF.

a. **Side PLF**. As the balls of his feet strike the ground, the parachutist begins several actions at the same time. As the fall continues, he does the following to complete a left-side PLF. (The right-side PLF is similar to the left-side PLF, except the points of contact on the right side of the body are used.)

(1) He lowers his chin firmly to his chest and tenses his neck. He brings his hands up in front of his head and elbows in front of his chest, continuing to grasp the risers (T-10-series) or the toggles (MC1-series). Then he bends and twists his torso sharply to the right. This movement forces the body into an arc. The twisting motion of the hips pushes both knees to the left as the fall continues, and it exposes the second through the fifth points of contact (calf, thigh, buttock, side).

(2) As the PLF is completed in the direction of drift, the parachutist maintains tension in his neck to prevent his head from striking the ground. The momentum caused by drift brings his feet around to the right and into the line of drift. After completing the PLF, he activates the canopy release assembly to keep from being dragged.

b. **Front PLF**. The two types of front falls are **right-front fall** and **left-front fall**. The right-front fall is used if the direction of (wind) drift is slightly to the right. The left-front fall is used if the direction of drift is slightly to the left. If the direction of drift is directly to the front, the parachutist selects either PLF. For a left-front PLF, he rotates from the waist down 45 degrees to his right, exposing his second and third points of contact to the line of drift. Upon contact, he continues to rotate his body to the right, exposing the second, third, fourth, and fifth points of contact. (When executing the right-front PLF, he rotates to the left.)

c. **Rear PLF**. The two variations of the rear PLF are **right-rear PLF** and **left-rear PLF**.

(1) The parachutist determines what PLF to make by checking the direction of drift. If the drift is directly to the rear, he selects the appropriate PLF.

(2) For a left-rear PLF, he rotates from the waist down 45 degrees to his left, exposing the second and third points of contact to the line of drift. Upon contact, he continues to rotate his body and bend his upper torso away in the opposite direction, exposing the second, third, fourth, and fifth points of contact. When executing the right-rear PLF, he rotates to the right.

CHAPTER 4

TRAINING APPARATUSES

The apparatuses described in this chapter are used in basic airborne training. They allow the student to demonstrate proficiency in the tasks necessary to complete the course.

Section I. PARACHUTE LANDING FALL DEVICES

The three types of PLF training devices are the **2-foot high platform**, the **lateral drift apparatus**, and the **swing landing trainer**. The 2-foot high platform may be portable or permanently fixed. A soft landing area of pea gravel or like material is used with all of the training apparatuses. These devices are used to teach front, side (right and left), and rear PLFs. They are high enough to simulate the shock the student will feel when he contacts the ground during parachute jumps. Initial instruction for each PLF is given without using the apparatuses. Once students are familiar with the techniques, they progress to the 2-foot high platform, the lateral drift apparatus, and the swing landing trainer. The swing landing trainer provides a means for gaining forward momentum and simulating the lateral movement experienced during a parachute landing.

4-1. INSTRUCTOR CRITIQUES

Table 4-1, on page 4-3, lists common PLF errors, their causes, and ways to correct them. Instructor PLF critiques should be brief and clear, and should emphasize the following points to students (Figure 4-1, page 4-2).

a. When the balls of your feet make contact with the ground, several actions occur at the same time: Place your chin on your chest, tense your neck muscles, bring your elbows high in front of your face, and expose the second (calf) and third (thigh) points of contact by shifting and bending your knees, maintaining pressure with your opposite knee.

b. Rotate your upper body (from the waist up) around toward the opposite direction of drift. Your body should be contorted in an arc, and the four remaining points of contact should be exposed at this time.

c. Lay the points of contact down on the ground in sequence and complete the fall by bringing your feet up and around your opposite shoulder, completing the fall on your back.

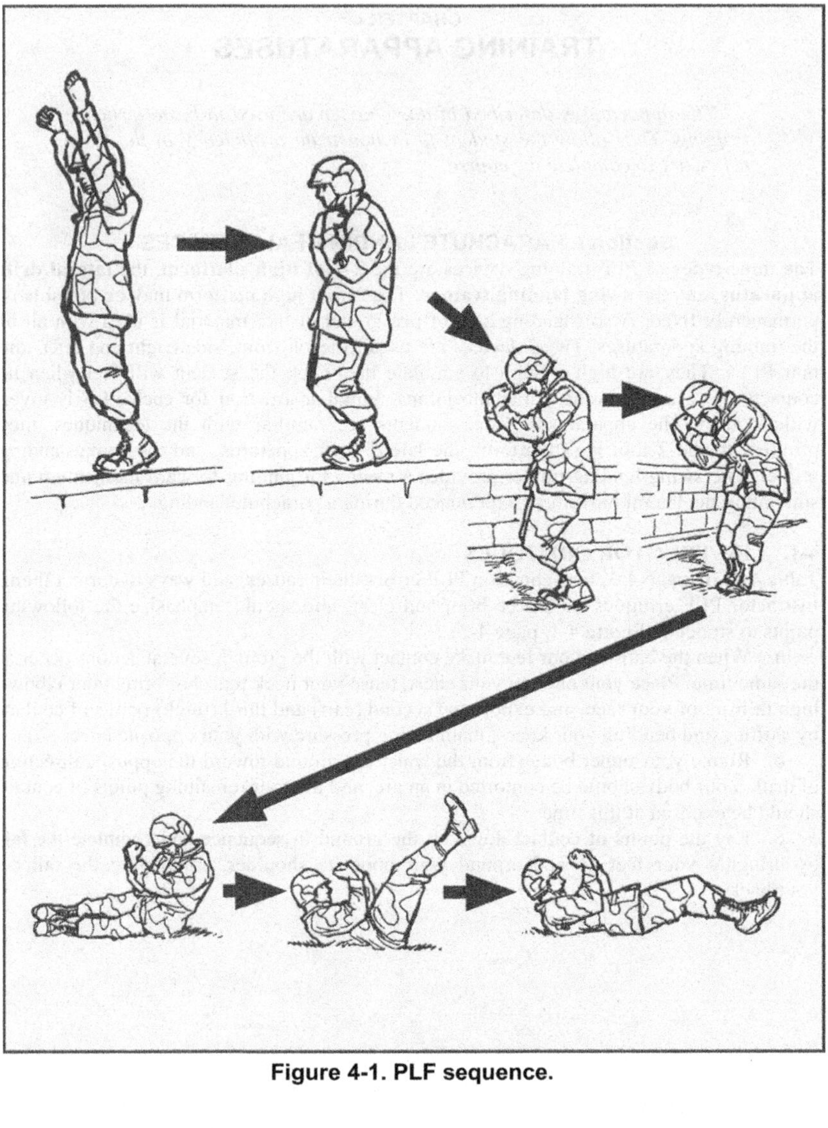

Figure 4-1. PLF sequence.

ERROR	CAUSE	CORRECTION
Feet apart	Anticipation of landing.	Keep moderate muscle tension in the legs. Keep feet and knees together.
Drawing the legs up under the buttocks	Anticipation of landing. Relaxing the knees. Exaggeration of the bend in the knees. Pulling feet up upon landing.	Keep moderate muscle tension in legs. Maintain a proper prepare-to-land attitude, naturally exposing the balls of the feet to the ground.
Missing the second (calf) and third (thigh) points of contact	Feet and knees apart. Straightening the legs. Failure to shift the knees.	Bend the knees slightly. Keep feet and knees together. Shift and bend the knees throughout the fall.
Knees into the ground	Hesitation upon landing. Pushing knees forward toward ground. Excessive bend in knees. Turning the feet toward the direction of drift.	Do not hesitate upon landing. Shift knees to the side. Keep moderate muscle tension in legs. Maintain the proper prepare-to-land attitude.
Elbows hit the ground	Leaning forward toward ground. Failure to rotate upper body away from ground. Breaking fall with elbows.	Rotate upper body away from the ground with elbows up in front of face.
Head strikes the ground	Relaxing the neck. Arching the back during the prepare-to-land attitude. Rotating the upper body into the ground. Missing the points of contact.	Keep chin on chest. Tense neck muscle throughout the PLF. Assume proper prepare-to-land attitude. Rotate upper body away from the ground.

Table 4-1. Common PLF errors.

4-2. TWO-FOOT HIGH PLATFORM

Each platform is divided into dismount points. One instructor controls each point.

a. On the command READY, the student rocks up on the balls of the feet with feet and knees together and knees slightly bent, naturally exposing the balls of the feet to the ground. Arms are skyward, simulating the grasping and spreading of the risers for the T-10-series parachute, or, for the MC1-series, palms are facing forward at eye level with elbows out to the side and back.

b. On the command SLIP (TURN for MC1-series), the student assumes a landing attitude by simulating the grasping of a set of risers opposite the direction of drift and pulling down into his chest with his head and eyes on the horizon, back straight, elbows tight into the side, and feet and knees together. If executing front or rear parachute landing falls, the student rotates his lower body (from the waist down) 45 degrees to expose the second and third points of contact.

c. On the command LAND, the student jumps straight off the platform, executes the PLF, and makes a quick recovery.

d. The instructor critiques each PLF immediately, emphasizing the significance of the five points of contact.

4-3. LATERAL DRIFT APPARATUS

A platoon is distributed among several apparatuses. Each apparatus requires a parachutist, a ropeman, and a safety officer. The safety officer is positioned on the top step of the platform to catch the trolley when it is returned by the ropeman, who is located to the left side at the base of the platform (Figure 4-2). The parachutist mounts the platform, grasps the bar with his palms facing toward his face, and assumes a good landing attitude. On the command CLEAR THE PLATFORM, the parachutist maintains a grasp on the trolley with both hands, picks up his feet, and drifts off the platform. On the command LAND, the parachutist releases the bar and executes the PLF.

Figure 4-2. Lateral drift apparatus.

4-4. SWING LANDING TRAINER

The swing landing trainer (SLT) apparatus is suspended above a 12-foot-high platform from which students, wearing a parachute harness, descend to practice PLFs (Figure 4-3). The apparatus provides a downward motion and oscillation similar to that experienced during a parachute jump. The suspension is placed so that the student swings when stepping off the platform. Using a control line, the instructor controls the rate of descent. Students receive practical work in two fundamental training objectives: assuming the correct landing attitude (T-10-series and MC1-series) and executing front, side, and rear PLFs.

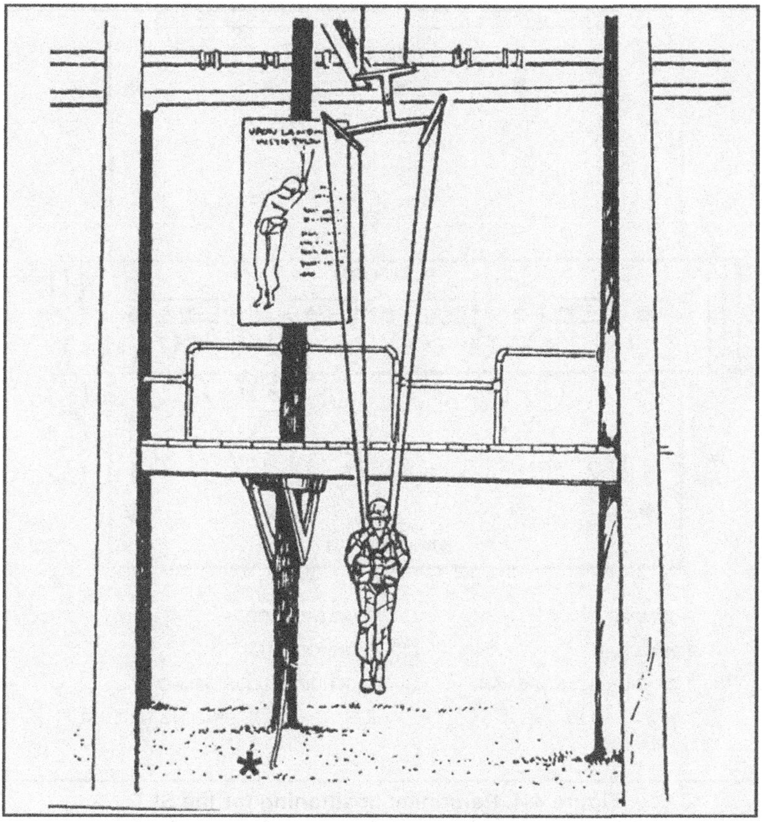

Figure 4-3. Swing landing trainer.

a. **Personnel and Equipment Requirements.** Personnel required to train on the SLT apparatus include one instructor for every two dismount points in use and four detail personnel for each unit (Figure 4-4, page 4-6). Enough harnesses to accommodate the students are also required.

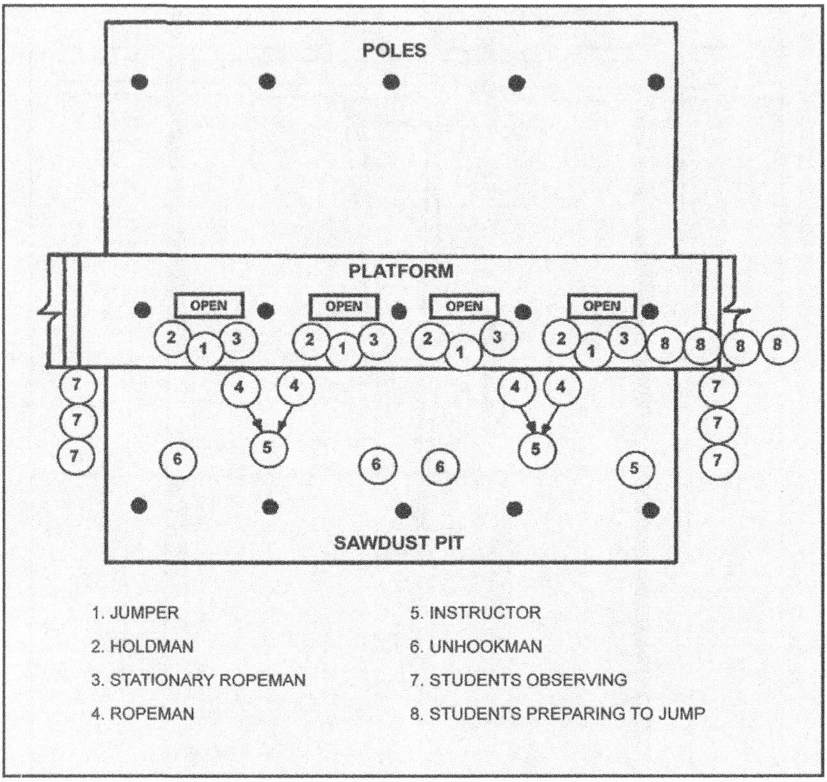

Figure 4-4. Personnel positioning for the SLT.

b. **Training.** The swing landing trainer provides practical exercise in the various PLFs and may be used to practice the last two points of performance. Students must—

- Secure a harness and reserve parachute, and put them on using the buddy system.

NOTE: The instructor inspects students before they mount the steps.

- Move to and mount the stairs at an assigned point. One student covers each open point. If a point is not open, the students wait on the stairs. Students stand on alternate steps of the stairs while waiting.
- Jump from two even or odd points in a section. They do not change sections or switch sides of the apparatus unless directed to by an instructor.

Detail personnel hook up students, who await commands from instructors.

c. **Platform Personnel Duties.** The following detail personnel are needed for training with the SLT.

(1) ***Stationary Ropeman***. The stationary ropeman stands by the stationary rope and—

- Pulls up the stationary rope with risers hooked to it.
- Unhooks the risers from the rope.
- Passes one set of risers to the holdman.
- Hooks one set of risers to one of the jumper's D-rings and butterfly snaps.
- Grasps the horizontal back strap of the parachutist to prevent him from clearing the platform before being told to do so. The stationary ropeman grasps the platform handhold with his free hand.
- Releases the parachutist on the command, CLEAR THE PLATFORM.

(2) ***Holdman.*** The holdman on the platform stands beside the rail padding, allowing space for the parachutist between himself and the stationary ropeman. He—

- Receives one set of risers from the stationary ropeman and removes any twists.
- Hooks the risers to the jumper's D-rings and butterfly snaps.
- Grasps the horizontal back strap of the parachutist to prevent the parachutist from clearing the platform before being told to do so. The holdman grasps the platform handhold with his free hand.
- Releases the parachutist on the command, CLEAR THE PLATFORM.

(3) ***Unhookman.*** The unhookman stands in the pit at parade rest near the outside pole at each point. The unhookman—

- Commands the ropeman, TAKE UP THE SLACK, ROPEMAN.
- Unhooks the parachutist's risers after each PLF.
- Hooks the risers to the stationary rope and returns to his assigned position.

(4) ***Ropeman.*** The ropeman stands beneath the platform and—

- On the command TAKE UP THE SLACK, ROPEMAN, grasps the rope in both hands and moves toward the instructor while taking up the slack.
- Gives the rope to the instructor and sounds off, "Rope, Sergeant"; he then returns to his assigned position.

d. **Instructor Duties.** When ready to drop a student, the instructor takes the rope from the ropeman and tells the parachutist the direction of drift and type of parachute (T-10-series or MC1-series) before commanding CLEAR THE PLATFORM. Then the following occurs:

- The platform detail personnel release the parachutist.
- The parachutist executes a half chin-up on the risers, clears the platform, and assumes the correct landing attitude.
- The instructor lowers the student to the ground.
- At the completion of the PLF, the parachutist activates one canopy release assembly, makes a quick recovery, and reports to the instructor for a critique and grade.
- The parachutist sounds off "Clear" or "Not clear" at the completion of the critique and moves directly out of the pit.
- The detail personnel perform their duties in sequence to prepare another student for the exercise.

4-5. SAFETY CONSIDERATIONS

The following precautions are taken to ensure the student's safety.

- The landing area must be constructed of at least 12 inches of pea gravel or like material.
- The pea gravel must be loosened by raking before each period of instruction.
- The ropes on the apparatus must be checked daily for wear.
- The spreader bars and risers must be checked for wear.
- The harnesses and canopy release assemblies must be checked for completeness and serviceability.
- The student must not be dropped from more than 3 feet.
- The student must not be dropped on the initial oscillation or when unprepared for the PLF.
- The student must be dropped at a point in oscillation that aids in executing the desired PLF.
- Any student with a prior head injury will have the letter H on his helmet to allow close instructor monitoring.

Section II. MOCK DOOR

The mock door is a replica of the cargo/troop compartment of a troop carrier aircraft. This apparatus includes openings about the size of the aircraft door and anchor line cables for each door (Figure 4-5). For training purposes, the instruction is divided into a basic phase and an advanced phase. The basic phase teaches basic jump techniques and familiarizes students with equipment, aircraft terms, and safety procedures. The advanced phase provides instruction in the sequence of jump commands and the mass exit technique.

Figure 4-5. Mock door apparatus.

4-6. BASIC PHASE

Students are taught the terms and nomenclature of military parachuting and the use of aircraft equipment. Teaching objectives include the shuffle position, static line control, the STAND BY position, the exit and body positions, and the 4000-count.

a. **Personnel and Equipment Requirements.** One instructor is required for each mock door in use. One section of static line (about 4 feet long and not attached to the parachute assembly) is required for each student. Loudspeakers will be needed if the class or facilities are large.

b. **Training.** The initial instruction in the mock door apparatus includes a lecture and demonstration followed by practical exercises.

(1) Each student is given the commands STAND BY and GO. The student is critiqued and corrected by the instructor on the movement to the door, his exit, his body position, and the 4000-count.

(2) Repetition is the key to this training; however, its value is decreased if allowed to become boring. Correct and automatic reactions by each student is the goal.

(3) The "Hit It" exercise is given as a test of mental alertness and for practical work in assuming the proper body position. On the command HIT IT, the student—

- Snaps into the proper body position and at the same time commences the 4000-count.
- Remains in the body position until commanded to RECOVER, or until he is told to CHECK CANOPY and GAIN CANOPY CONTROL. The student simulates checking the canopy. Then he is told to RECOVER or that there is a MALFUNCTION. In the latter case, the student returns to a modified body position and simulates activation of the reserve. The command RECOVER is given.
- Receives instruction on the right and left jump doors of the mockup.

c. **Terms.** The following terms are important in mock door training and are explained during the initial phase of instruction.

(1) ***Left and Right Door.*** When the parachutist is facing the pilot's compartment, the door on his right is the right door; the door to his left is the left door.

(2) ***Anchor Line Cable.*** A cable is normally extended along the long axis of the cargo/troop compartment and secured at both ends. The cable is designed to accommodate the static line snap hook and to initiate parachute deployment.

(3) ***Stick of Parachutists.*** This is a group of parachutists exiting from the same door (or from one side of a ramp) during one pass over the DZ.

(4) ***Drop Zone.*** This is a designated area where personnel or equipment are delivered by means of parachute or free drop. The GUC designates the DZ location.

(5) ***Shuffle Position.*** This is a method of moving toward the jump door, used to avoid losing balance or tripping (Figure 4-6, page 4-10).

(a) The guide hand is extended down and out to maintain balance; the other hand grasps the static line. When jumping the left door, the parachutist has the static line over the left shoulder; when jumping the right door, he has the static line over the right shoulder.

(b) Facing the rear of the aircraft, the parachutist keeps both feet directly beneath the body and staggered with the outboard foot forward; this is the shuffle foot. The inboard foot is the trail foot.

(c) The parachutist moves by stepping forward 6 to 8 inches with his shuffle foot and then his trail foot. Both feet are staggered in the same heel-and-toe position.

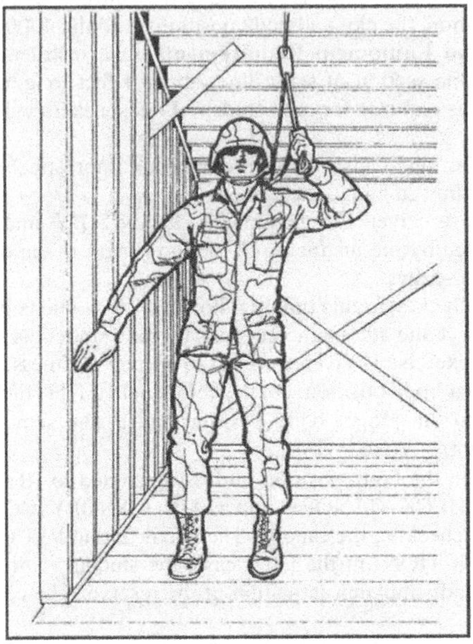

Figure 4-6. Shuffle position.

(6) ***Bight.*** The parachutist forms a bight of about 6 inches in the static line by making one fold and grasping the loop at eye level about 6 inches to the front (Figure 4-7). The remainder of the static line is routed over the shoulder. The free hand is used to steady the parachutist while moving toward the door.

Figure 4-7. Static line bight.

(7) ***Jump Commands.*** The last two jump commands, STAND BY and GO, are used for each student when practicing exits.

(a) On the command STAND BY, the parachutists shuffle toward the jump door.

- When the first jumper is perpendicular to the jump door, he takes one more shuffle step and halts his movement about 2 feet from the center of the jump platform. He keeps his feet spread and legs slightly flexed so that his weight is equally distributed over both feet to maintain balance.
- He makes eye-to-eye contact with the safety and hands the static line to the safety.
- He executes a left or right turn to face the open jump door, ensures his arm is not entangled with the static line, holds his elbows firmly into his sides, and places the palms of his hands (fingers spread) on the ends of the reserve parachute with his right hand protecting the rip cord grip.
- At the command STAND BY, the number 2 jumper will be positioned about even with the leading edge of the jump door, 2 feet from the skin of the aircraft and facing to the rear, in the shuffle position with his feet spread and legs slightly flexed so that the weight is equally distributed over both feet to maintain balance.
- Follow-on jumpers close up behind the preceding jumper and keep the shuffle position with the feet spread and legs slightly flexed so that their weight is equally distributed over both feet to maintain balance.

(b) At the command GO, the number 1 jumper walks toward the door and onto the jump platform, focusing on the horizon. He pushes off with either foot and **vigorously** jumps **up** and **out** away from the jump platform, immediately snapping into a good tight body position.

(c) The number 2 jumper (and all following jumpers) performs the following:

- He shuffles toward the jump door, ensuring he is about 2 feet from the skin of the aircraft.
- As he begins to shuffle, he assumes an elbow-locked position with the arm that is controlling his static line. He places his static line control hand so that it is nearly touching the back of the pack tray of the jumper in front of him, which establishes the proper interval between jumpers. He does **not** place his static line control hand in a position so that it extends past the pack tray of the jumper in front of him.
- When about perpendicular to the jump door, he takes one more shuffle step, makes eye-to-eye contact with the safety, and hands the static line to the safety.
- He ensures his arm is not entangled with the static line, and he holds his elbows firmly into his sides.
- He places the palms of his hands (fingers spread) on the ends of the reserve parachute, with his right hand protecting the rip cord grip.
- He executes a left or right turn to face the open jump door.
- He walks toward the door and onto the jump platform, focusing on the horizon. He pushes off with either foot and **vigorously** jumps **up** and **out** away from the jump platform, immediately snapping into a good tight body position.

(8) ***Body Position.*** The student remains in the body position, is critiqued by the instructor, and is told to RECOVER and wait for further instruction.

4-7. ADVANCED PHASE

The training provided during the advanced phase is presented in the same manner as in the basic phase except that the entire sequence of time warnings and jump commands is given, and mass exits are substituted for individual exits.

a. **Personnel.** Extra instructors may be needed to ensure that all students in the mock door apparatus react properly to each of the jump commands.

b. **Execution.** When the mass exit technique is taught, each stick receives the commands STAND BY and GO. Each succeeding student shuffles to the door and exits the aircraft. A one-second interval must be maintained between students.

Section III. SUSPENDED HARNESS

The suspended parachute harness apparatus is a modified troop parachute harness suspended from a spreader bar assembly by four web risers (Figure 4-8). The spreader bars react to riser (T-10-series) or toggle (MC1-series) manipulation much the same as the canopy. The suspended harness simulates the third and fourth points of performance: canopy control and prepare to land.

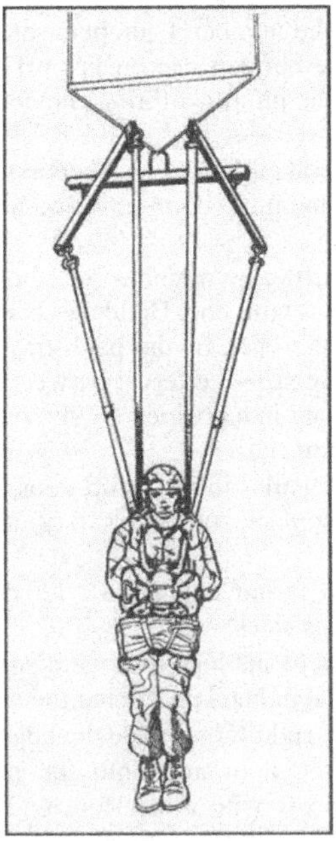

Figure 4-8. Suspended harness.

4-8. OBJECTIVES

The suspended harness apparatus is designed to teach the student to execute the following maneuvers:

- T-10-series slips (two-riser and one-riser).
- MC1-series turning, holding, running, and crabbing.
- T-10-/MC1-series landing attitude.
- T-10-/MC1-series emergency landings (tree, water, and wire).
- React to twists, collisions, and entanglements.

4-9. PERSONNEL AND EQUIPMENT REQUIREMENTS

The following personnel and equipment are needed to conduct training on the suspended harness.

a. One instructor is needed to give commands, and at least one assistant instructor is needed for each platform to control and supervise student performance. One troop harness is required for every two students. A modified MC1-series riser assembly is used, which facilitates training on a simulated T-10-series or MC1-series parachute.

b. A platoon is divided into teams of two students each. Number 1 is the parachutist, and number 2 is the coach.

(1) The parachutist obtains a troop parachute harness and moves to a specific point at the suspended harness apparatus.

(2) The coach mounts the platform and adjusts the risers so that the male fitting of the canopy release assembly is level with the parachutist's shoulders below. The coach dismounts the platform and helps the parachutist don the harness and attach it to the risers. The coach remounts the platform and takes up all the slack in the parachutist's risers.

(3) All parachutists face the instructor and await commands.

4-10. SEQUENCE OF COMMANDS

The instructor uses the following sequence of commands. (The let-up position is the starting position for all training on the suspended harness.)

a. **LET UP**. On this command, each parachutist:

- Has his head erect and his eyes on the horizon.
- Grasps all four risers (T-10-series) with his hands and locks his elbows.
- Grasps the steering toggles (MC1-series) with his hands (palms outward, eye level, elbows back) and applies moderate tension (enough to take out any slack in the control lines).
- Has his back straight.
- Has his feet and knees together. (The parachutist is standing flat-footed on the ground.)

b. **AT EASE IN THE HARNESS**. On this command, each parachutist comes to a modified position of parade rest in the harness. From the let-up position, the instructor directs the parachutist to practice appropriate parachute maneuvers.

c. **CHANGE OVER**. Number 2 switches places with number 1 and becomes the parachutist; number 1 becomes the coach. Instructors may give additional commands and instruction to ensure proper fitting, wearing, or removal of equipment and proper student performance on the apparatus.

Section IV. THE 34-FOOT TOWER

The 34-foot tower supports a replica of a section of a troop carrier aircraft (Figure 4-9). A jump door is on each side of the replica. Four steel cables are suspended parallel to the ground and slightly above each door. A trolley, which supports tow trolley risers, is attached to each cable. Each trolley riser has a ring attached to its free end. This ring is connected to a modified harness worn by students during training on the tower. The tower is a primary training apparatus to help teach basic jump techniques and points of performance. For ease in training, the instruction is divided into two phases: The **basic** training phase continues instruction presented on the mock door apparatus. The **advanced** training phase continues instruction on the mass exit technique, simulates parachute malfunctions, and familiarizes the students with jumping combat equipment.

Figure 4-9. The 34-foot tower.

4-11. BASIC TRAINING OBJECTIVES

Students practice the exit technique, the proper body position, and the 4000-count. The tower also gives each student experience to overcome his fear of height, and it simulates the opening shock.

a. **Training**. The initial period of instruction on the tower apparatus includes an orientation and demonstration of the apparatus, duties of instructor personnel (graders, JMs, harness checker), and duties of all detail personnel. Each student is given the commands STAND BY and GO by the JM in the tower. The student's exit, body position, 4000-count, and simulated canopy checks are critiqued and graded by the grader on the ground.

b. **Safety**. To prevent safety problems, all personnel must be safety conscious. The following measures are taken to prevent unsafe conditions:

(1) Tower JMs must wear safety belts.

(2) Students must have equipment adjusted properly, and instructors must check them to ensure a snug fit, the absence of frays or tears, the serviceability of snaps, and a properly fitted helmet.

(3) Qualified maintenance personnel must perform and record thorough inspections of the tower weekly. Inspections must include trolley risers, cables, trolleys, and JM safety belts.

(4) Students must lift their legs as they approach the mound at the end of the cable run to prevent injury.

4-12. PERSONNEL AND EQUIPMENT REQUIREMENTS

The following personnel manage, operate, and carry out 34-foot tower training.

a. **Instructors**. A minimum of seven instructors are required to operate this apparatus. Their positions are indicated in Figure 4-10 (page 4-16).

(1) Two JMs hook up students in the tower and issue the last two jump commands.

(2) One harness checker inspects all harnesses before students exit the tower.

(3) Two graders on the ground critique and grade the students' performance on the apparatus.

(4) Two mock door instructors correct students on problem areas and give them additional practice.

b. **Detail Personnel**. A minimum of 38 detail personnel are required.

(1) ***Mound Safety Officers/NCOs.*** A student stands at parade rest on top of the mound between and slightly to the rear of the moundmen assigned to his side of the tower. He is responsible for the safety and conduct of the moundmen. He should be the most senior officer or NCO on the mound on each side (2 total).

(2) ***Moundmen.*** Two students are assigned to each cable on the tower (16 total). They stand at parade rest on top of the mound, facing the tower. After the parachutists reach the mound and come into contact with the stop cable, the moundmen unhook the parachutist and then hook the rope to the trolley risers.

(3) ***Ropemen.*** One student at each point along the beaded path returns the trolley risers (with rope attached) to the tower after the parachutist is detached. The ropemen take commands from the ropeman safety and rope line safety/relay. The ropemen for each

two points rotate, and there are always two additional ropemen standing at the base of the pole supporting the tower (12 total).

(4) ***Riser Safeties.*** One student stands at parade rest off each side of the mound and outside of the path leading to the grader's booth (2 total). He is responsible for attaching the butterfly snaps on the risers to the D-rings of the harness.

(5) ***Ropeman Safeties.*** One ropeman safety is located between points 2 and 3 and the other between points 6 and 7 on the beaded path (2 total). They are responsible for controlling the flow of the ropemen from the mound to the tower.

(6) ***Rope Line Safeties/Relays.*** One student (officer or NCO) for each side of the tower (2 total) stands at parade rest near the base of the pole supporting the cables facing the rope line. He relays the commands of the jumpmaster to the ropemen and ensures students do not pass under the safety ropes and anchor cables.

(7) ***Base Safety Officers/NCOs.*** One student (officer or NCO) for each side of the tower (2 total) stands inside the base of the tower next to the stairs. He places the risers over the proper shoulder for the door (right or left) each parachutist will jump.

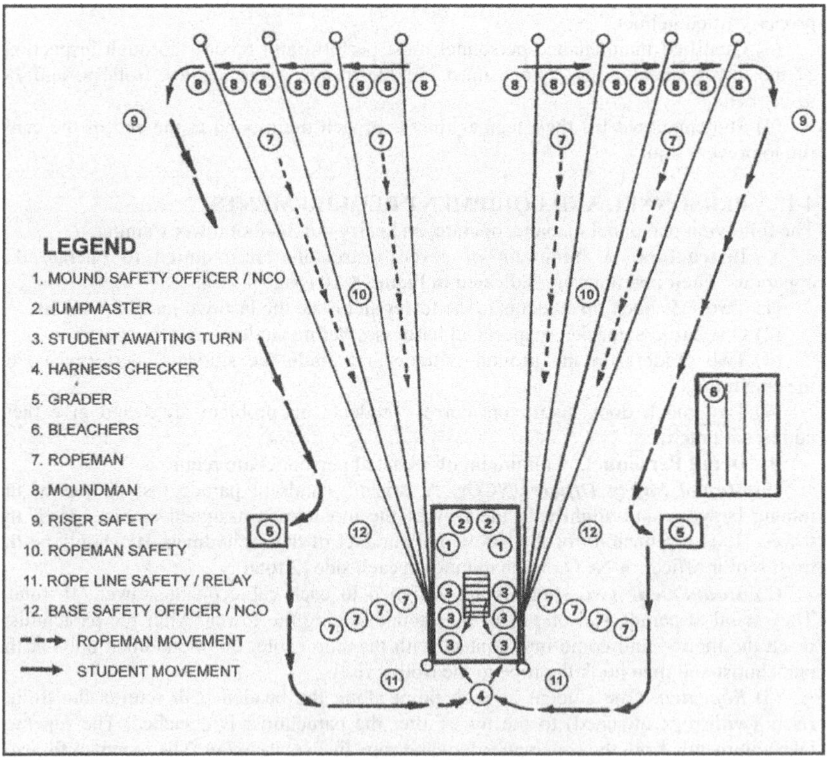

Figure 4-10. Tower personnel positions.

c. **Equipment**. The equipment includes one troop parachute harness assembly (with four risers) and a training reserve parachute for each student.

4-13. ADVANCED TRAINING OBJECTIVES

These objectives have the student practice exits using the mass exit technique with and without combat equipment. If exiting with combat equipment, he must also practice lowering the equipment. The student must practice simulating activation of the reserve parachute. He must do this instantly when told a malfunction has occurred.

a. **Training**. The initial period of instruction during this phase includes a lecture and demonstration on the mass exit technique.

b. **Common Student Errors**. The following errors are often made by students.

- Failing to maintain a 1-second interval between jumpers.
- Using improper body position (caused by rushing).
- Improper exit.
- Falling out of the door.
- Failing to count.

c. **Safety**. Safety considerations for the advanced phase are the same as in the basic phase.

d. **Personnel**. Personnel requirements are reduced slightly from the basic phase to a minimum of 29: 16 moundmen, 8 ropemen, 2 mound safety officers, 2 rope line safety officers, and a harness checker.

e. **Equipment**. Combat equipment is required for the students. (Actual packed combat equipment containers, or combat equipment containers which approximate packed loads, can be used to train students in the techniques of exiting with this type of load.)

Section V. METHODS OF RECOVERY

During methods of recovery training, the jumper learns the steps he must take to consolidate with his unit as a member of the fighting force. Once the jumper lands on the drop zone, he must quickly recover, correctly stow his air items, and quickly proceed to the unit assembly area.

4-14. TRAINING OBJECTIVES

This phase of training teaches the student to properly activate the canopy release assemblies on the parachute harnesses. He learns to perform the buddy-assist method of recovery and to react properly when using the various recovery training apparatuses. He is also taught how to recover the parachute and individual equipment from the DZ and to rapidly clear the DZ.

4-15. PERSONNEL AND EQUIPMENT REQUIREMENTS

Personnel required are one instructor, one assistant for every two dismount points, and four detail personnel for each dismount point. Sufficient harnesses and combat gear properly rigged to accommodate each student are also needed.

4-16. TRAINING APPARATUS

The hand-towed drag pad training apparatus is used in teaching students how to activate the canopy release assemblies. The hand-towed drag pad is made of two pieces of webbing attached to a metal bar with a loop on each end. Attached to the bar are two risers with the male fitting of the canopy release assemblies attached to the opposite ends (Figure 4-11). Students train in three-man teams as follows:

- Number 1 is the parachutist and wears the drag pad.
- Number 2 and number 3 pull the drag pad.

Figure 4-11. Hand-towed drag pad.

- Number 1 puts on the harness and passes the risers to numbers 2 and 3. Numbers 2 and 3 then attach the risers to the canopy release assemblies, and number 1 lies on his back.
- On the command PREPARE TO DRAG, number 1 places his chin on his chest, raises his feet 6 inches off the ground, and places his hands on the canopy release assemblies.
- On the command DRAG, numbers 2 and 3 drag number 1 across the ground.
- On the command RELEASE, number 1 reaches down and activates the canopy release assemblies using the prescribed method.
- The three students immediately return to the starting line.
- Numbers 2 and 3 hook up the parachutist to the drag pad again.
- At the command CHANGE OVER, number 2 becomes the parachutist (number 1), number 3 becomes number 2, and number 1 becomes number 3. Students rotate numbered positions until each student becomes proficient in the procedures.

4-17. CANOPY RELEASE ASSEMBLIES

Before the canopy release assemblies can be activated, the safety clip must be pulled down to expose the cable loop. There are two ways to activate the canopy release assembly. One is the hand-to-shoulder method and the other is the hand-assist method.

4-18. CANOPY RELEASE ASSEMBLY ACTIVATION

The instructor tells the student to pull the safety clip out and away from the body. The cable loop release does not require a great deal of strength to activate, and, if the parachutist is injured, the cable loop release can be easily activated by the thumb or fingers of either hand (Figure 4-12).

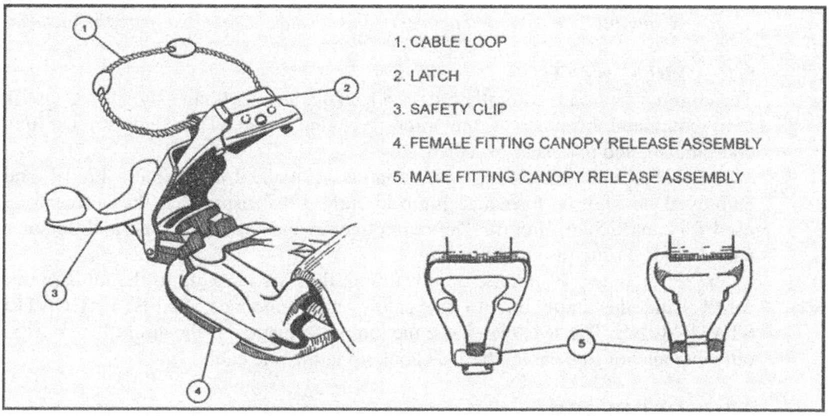

Figure 4-12. Cable loop release.

4-19. JUMP REFRESHER TRAINING

Before making a parachute jump, students make several refresher jumps from platforms. They are required to make a satisfactory PLF in each of the principal directions before they leave the platform area. (A detailed description of refresher training is in Appendix A.)

CHAPTER 5

JUMP COMMAND SEQUENCE AND JUMPER ACTIONS

The JM gives a sequence of nine jump commands to ensure positive control of parachutists inside the aircraft and immediately before exiting. Every command requires specific actions by each parachutist. When commands are executed properly, they ensure a safe exit from the aircraft.

5-1. PRESENTATION

The commands are given orally but, as a backup, arm-and-hand signals are also used with each command because of the aircraft engine noise. The signals must be smooth, coordinated, and precisely executed.

a. The commands listed, with variations explained in Chapters 17, 18, and 19, are employed on high-performance jump aircraft. JMs ensure that the correct sequence is used for a particular aircraft. The correct commands are explained and demonstrated to parachutists during prejump briefing.

b. Prior to the 10-minute time warning, the JMs hook up to the inboard anchor line cable, hand the static line to the safety, and announce, SAFETY, CONTROL MY STATIC LINE. The JMs then issue the jump commands. If the aircraft is configured with only one anchor line cable, the JMs hook up to the one cable.

5-2. GET READY

GET READY is the first jump command.

a. **Command**. This jump command alerts the parachutists seated in the aircraft and directs their complete attention to the JM.

(1) The JM starts the command with his arms at his sides and gives the arm-and-hand signal by extending both arms to the front at shoulder level with his palms facing the parachutists.

(2) He begins at shoulder level, fingers and thumbs extended and joined, palms facing toward the parachutists. He extends both arms forward until the elbows lock, with the palms toward the parachutists. He gives the oral command GET READY, then returns to the start position with arms at the sides (Figure 5-1).

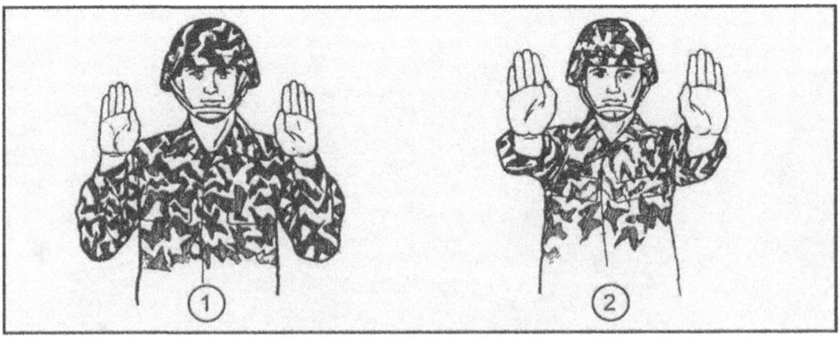

Figure 5-1. GET READY.

b. **Static Line**. The static line is over the appropriate shoulder and fastened to the top carrying handle of the reserve parachute. Parachutists do not remove the static line snap hook from the reserve parachute after the JM inspection or anytime before the command HOOK UP.

c. **Jumper Actions**. Each parachutist signifies alertness by leaning forward and placing both hands on his knees. Each parachutist positions his foot nearest the jump door under the seat and places his foot nearest the pilot's compartment in the aisle.

5-3. OUTBOARD PERSONNEL, STAND UP

OUTBOARD PERSONNEL, STAND UP is the second jump command. For this command, the arm-and-hand signal has two parts.

Part One. The JM starts at the shoulders, index and middle fingers extended and joined, with remaining fingers and thumbs curled to the palms. He gives the command OUTBOARD PERSONNEL, lowers the arms down to the sides at a 45-degree angle, and locks the elbows.

Part Two. The JM gives the command STAND UP. He extends and joins the fingers and thumb of each hand, rotates the hands so the palms face up, and then raises the arms straight overhead, keeping the elbows locked (Figure 5-2). At this command, parachutists sitting nearest the outboard side of the aircraft stand up, raise and secure the seats, face the jump doors, and assume the shuffle position.

NOTE: The method of releasing the seats from the floor varies, depending on the model and year of aircraft. Before takeoff, these devices are inspected and the method of release explained.

Figure 5-2. OUTBOARD PERSONNEL, STAND UP.

5-4. INBOARD PERSONNEL, STAND UP

INBOARD PERSONNEL, STAND UP is the third jump command. The arm-and-hand signal has two parts.

Part One. The JM starts with the hands centered on the chest at shoulder level, index and middle fingers extended and joined, remaining fingers and thumbs curled to the palms. He gives the command INBOARD PERSONNEL, extends the arms forward at a 45-degree angle, toward the inboard seats, and locks the elbows.

Part Two. The JM gives the command STAND UP. He first rotates his arms to the sides and down at a 45-degree angle. Then he extends and joins the fingers and thumb of each hand, rotates his hands so the palms face up, and raises his arms straight overhead, keeping the elbows locked (Figure 5-3).

The parachutists seated inboard react in the same manner as the outboard personnel described in the previous paragraph.

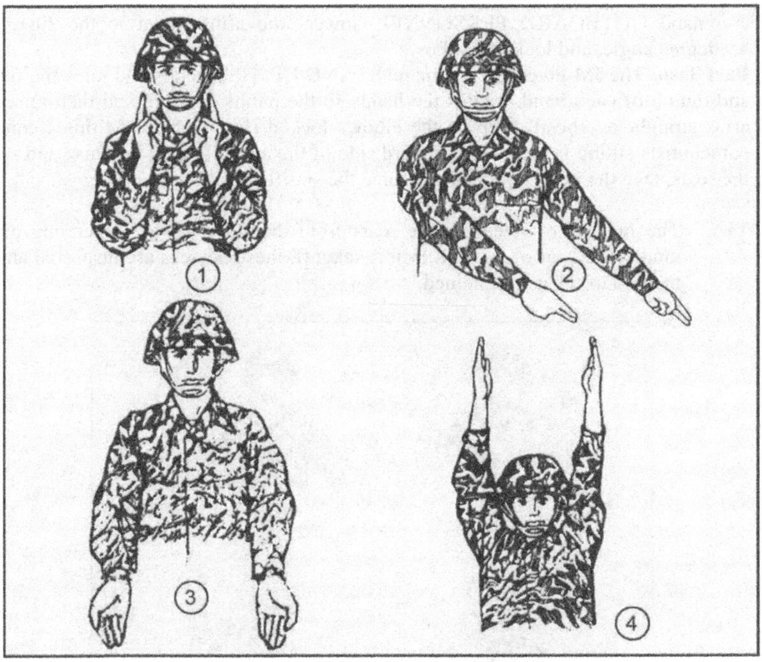

Figure 5-3. INBOARD PERSONNEL, STAND UP.

5-5. HOOK UP

HOOK UP is the fourth jump command.

a. **Command.**

(1) The JM begins with his arms either extended directly overhead with elbows locked or with arms bent, hands at shoulder level.

(2) He forms a hook with the index finger of each hand. He forms fists with the remaining fingers and thumb of each hand (Figure 5-4).

(3) As he gives the oral command, he moves his arms down and up in a pumping motion. He repeats the arm-and-hand signal at least three times.

Figure 5-4. HOOK UP.

b. **Jumper Actions.**

(1) At this command, each parachutist detaches the static line snap hook from the top carrying handle of the reserve parachute and hooks up to the appropriate anchor line cable, with the open portion of the snap hook toward the outboard side of the aircraft. Each parachutist must ensure that the snap hook locks properly.

(2) The safety wire is inserted in the hole and folded down. To protect the eyes, the wire is inserted by pointing it toward the rear of the aircraft. Then a bight is formed in the static line and held at eye level. The bight is not released until the parachutist moves into the door.

(3) Personnel jumping the left (right) door have the static line over the left (right) shoulder.

5-6. CHECK STATIC LINES

CHECK STATIC LINES is the fifth jump command.

a. **Command.**

(1) This is a plural command since there are several static lines attached to the anchor line cable. It begins at eye level, with the thumb and index finger of each hand forming an "O."

(2) The JM extends and joins his remaining fingers with the palms facing in. As he gives the oral command, he extends his arms to the front until the elbows are nearly locked, then returns to the starting position.

(3) He repeats the arm-and-hand signal at least three times, ensuring the knife edge of his hands are toward the parachutists and the palms face each other (Figure 5-5).

b. **Jumper Actions.**

(1) Upon receiving this command, each parachutist checks his static line and the static line of the parachutist to the front.

(2) Each parachutist checks visually and by feeling with his free hand. He does not release the bight for checks. He verifies the following items:

- Static line snap hook is properly attached to the anchor line cable with the safety wire properly inserted. Static line is free of frays and tears.
- Static line is not misrouted and is properly stowed on pack tray.
- All excess slack in the static line is taken up and stowed in the static line slack retainer.
- Pack closing tie is routed through the pack opening loop.
- Pack tray is intact.

NOTE: The last two jumpers in each stick face about. The next to last jumper inspects the last jumper's static line and gives him a sharp tap to indicate that the static line and pack tray have been checked and are safe for jumping.

(3) Each parachutist gives the parachutist to the front a sharp tap signifying that the static line and pack tray have been checked and are safe for jumping.

Figure 5-5. CHECK STATIC LINES.

5-7. CHECK EQUIPMENT

CHECK EQUIPMENT is the sixth jump command.

a. **Command.**

(1) The JM starts this arm-and-hand signal with the fingertips centered on his chest, palms facing the chest, and fingers and thumb of each hand extended and joined; or with his arms extended to the sides at shoulder level, fingers and thumbs extended and joined, and palms facing toward the parachutist.

(2) He gives the oral command, extends his arms to the sides at shoulder level, and then returns them to the chest; or bends his arms at the elbows, bringing the fingertips to the center of the chest, and then returns to the extended position (Figure 5-6).

(3) He repeats the arm-and-hand signal at least three times. (The JM must check his own equipment.)

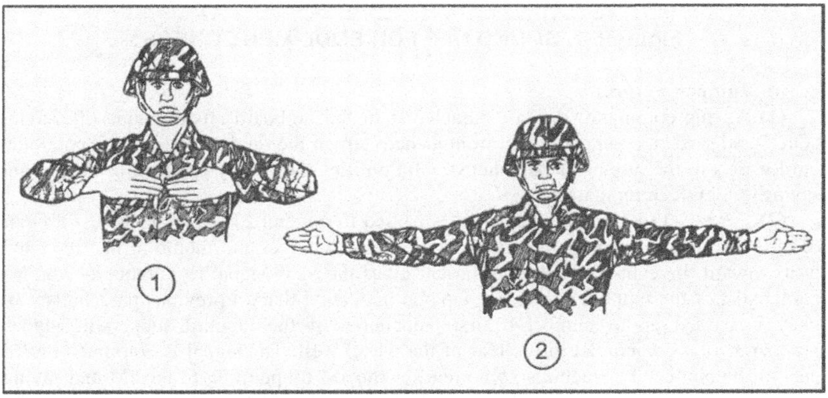

Figure 5-6. CHECK EQUIPMENT.

b. **Jumper Actions.**

(1) At this command, each parachutist checks his equipment, starting at the helmet, and ensures there are no sharp edges on the rim of the ballistic helmet and that the chin strap and parachutist retention straps are properly routed and secured. The parachutist then physically seats the activating lever of the chest strap ejector snap and the leg strap ejector snaps. If jumping combat equipment, the parachutist also ensures the ejector snap of the HPT lowering line is properly attached and seated.

(2) The parachutist completes these actions with the free hand while maintaining a firm grip on the static line bight with the other hand.

5-8. SOUND OFF FOR EQUIPMENT CHECK

This is the seventh jump command.

a. **Command.**

(1) The JM cups his hands and places the thumbs behind the ears.

(2) He gives the oral command SOUND OFF FOR EQUIPMENT CHECK (Figure 5-7).

Figure 5-7. SOUND OFF FOR EQUIPMENT CHECK.

b. **Jumper Actions**.

(1) At this command, the last parachutist in the outboard stick sounds off, saying "OK," and gives the parachutist in front a sharp tap on the thigh. The signal is continued until it gets to the number 1 parachutist, who notifies the JM by pointing to the JM and saying, "All OK, jumpmaster."

(2) For a C-130 aircraft, this signal is passed to number 25 parachutist (just forward of the wheel well), who forms a circle with his index finger and thumb of his free hand, turns toward the center of the aircraft, and gives the okay signal to number 24 (the last parachutist of the inboard stick). The tap and indication that all previous parachutists are okay is passed up to number 4, first parachutist of the inboard stick, who signals number 3, first parachutist to the rear of the wheel well. The signal is continued until it gets to the number 1 parachutist, who notifies the JM by pointing to the JM and saying, "All OK."

(3) A parachutist who has an equipment problem notifies the JM, AJM, or safety personnel by raising his outboard hand high above the anchor line cable, palm facing the JM. The parachutists do not pass this signal. The JM, AJM, or safety either corrects the deficiency or removes the parachutist from the stick.

NOTE: After the JM receives "All OK, jumpmaster," he regains control of his static line from the safety and takes the number 1 parachutist position.

5-9. STAND BY

STAND BY is the eighth jump command. This command (Figure 5-8, page 5-8) is given about 10 seconds before the aircraft reaches the release point and only after the aircraft has cleared all obstacles near the DZ.

a. **Command**.

(1) Starting at the shoulders, the JM extends and joins his index and middle fingers, curling the remaining fingers and thumb of each hand toward the palm.

(2) He extends his arms down to the sides at a 45-degree angle by locking the elbows, and points to both doors at the same time.

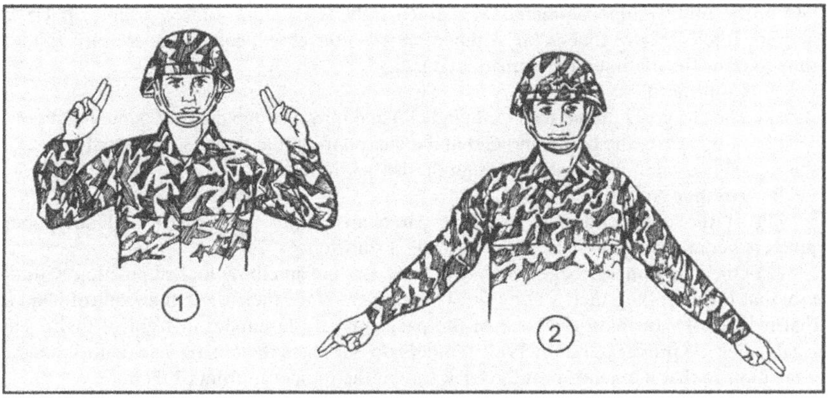

Figure 5-8. STAND BY.

b. **Jumper Actions.**

(1) At this command, parachutist number 1 shuffles toward the door, establishes eye-to-eye contact with the safety, hands the safety his static line, holds his elbows firmly into his sides with his palms on the end of the reserve, turns and centers himself in the open jump door, and awaits the command GO.

(2) All following parachutists maintain the static line bight and close up behind the preceding parachutist (Figure 5-9).

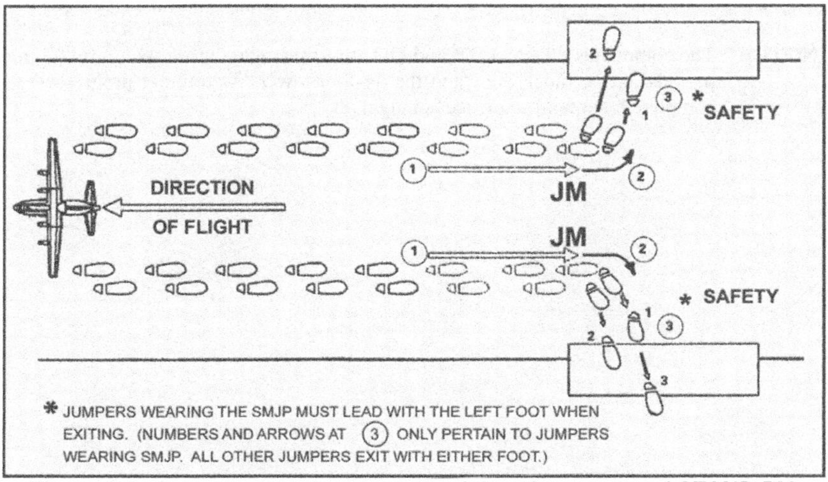

Figure 5-9. Jumper execution upon receiving the command STAND BY.

5-10. GO

GO is the ninth jump command. The green light is the final time warning on USAF aircraft. It tells the JM that as far as the aircrew is concerned, conditions are safe and it is time to issue the ninth jump command, GO.

a. **Command**.

(1) The JM gives the verbal command GO and may also tap the first parachutist out.

(2) In this case, the command GO and a sharp tap on the thigh is the signal to exit. If this signal is used, it is explained during the JM's briefing.

b. **Jumper Actions**.

(1) At the command GO, the first parachutist exits the aircraft. All subsequent jumpers begin moving toward the door using a shuffle.

(2) Once the jumpers begin to shuffle, they assume an elbow-locked position with the arm that is controlling their static line. The jumpers place their static line control hand so that it is nearly touching the back of the pack tray of the jumper in front of them. This establishes the proper jump interval. Jumpers do not place their static line control hand in a position so that it extends past the pack tray of the jumper in front of them.

(3) As each jumper approaches the door, he establishes eye-to-eye contact with the safety and hands his static line to the safety. Once the safety has control of the jumper's static line, the jumper returns his hand to the end of the reserve parachute with his fingers spread.

(4) After handing his static line to the safety (vicinity of the lead edge of the door), the jumper executes a left or right turn (as appropriate) and faces directly toward and centered on the door with both hands over the ends of the reserve parachute, fingers spread. He continues the momentum of his movement by walking toward the door, focusing on the horizon, and stepping on the jump platform. He pushes off with either foot and **vigorously** jumps up and **out** away from the aircraft. He immediately snaps into a good tight body position and exaggerates the bend into his hips to form an "L" shape.

NOTE: The commands STAND BY and GO are first taught during the initial training periods on the mock door and the 34-foot tower. As training progresses, the complete command sequence is taught.

CHAPTER 6

MAIN PARACHUTE MALFUNCTIONS AND EMPLOYMENT OF THE RESERVE PARACHUTE

A malfunction is any failure in the deployment or inflation of a parachute; or it is canopy damage, which can create a faulty, irregular, or abnormal condition that increases the jumper's rate of descent. The two classes of main parachute malfunction (total and partial) demand the jumper's immediate attention. Jumpers must be trained to identify the malfunction and take the appropriate action. Thorough training in what actions to take in the case of a malfunction is essential for parachutists. Practical exercises involving the activation of the reserve parachute are incorporated into all phases of training. Each type of malfunction is demonstrated so that jumpers can see exactly how each type of malfunction looks.

6-1. PULL-DROP METHOD (FOR MIRPS AND T-10 RESERVE)

When a total malfunction occurs, or the parachute provides no lift, the jumper must activate his reserve using the pull-drop method. Also, at the end of the 4000-count, if the jumper feels no opening shock, he should immediately activate his reserve using the pull-drop method. The jumper—

- Keeps a tight body position.
- Keeps his feet and knees together.
- Grasps the left carrying handle of the reserve parachute with his left hand.
- Turns his head left or right.
- Pulls the rip cord grip with his right hand and drops it to the ground.

When descending with only the reserve parachute inflated, the parachutist controls directional movement by slipping. The proper landing attitude is obtained by reaching up and grasping as many suspension lines in the opposite direction of drift as possible and slipping (as in the prepare-to-land attitude with the T-10-series parachute). Upon landing, the parachutist makes a quick recovery and collapses the canopy. In strong winds, if a quick recovery is impossible, the parachutist grasps one suspension line from either suspension line group and pulls the suspension line toward himself until the canopy collapses.

NOTE: If the MIRPS does not activate immediately, the jumper remains in a good tight body position while still grasping the left carrying handle. With the right hand forming a knife cutting edge, he sweeps the top panel of the reserve up and away, ensuring that the right hand does not interfere with the spring-loaded deployment assistance device.

6-2. DOWN-AND-AWAY METHOD (FOR T-10 RESERVE ONLY)

To activate the reserve parachute using the down-and-away method, the jumper—

- Returns to a tight modified body position with his feet and knees together.

- Places his hand on the middle of the reserve (over the rip cord protector flap). He exerts pressure on the reserve and grasps the rip cord grip with his right hand.

NOTE: Strong pressure must be maintained with the left hand to prevent the pilot chute and reserve canopy from springing out.

- Pulls the rip cord grip and drops it.
- With the right hand forming a knife edge, palm facing out, reaches between the pack assembly and canopy, and grabs as much canopy and reserve parachute suspension lines as possible.
- Throws the reserve parachute down and to the right (or left) side at about a 45-degree angle. (If the jumper is spinning, the canopy is thrown in the direction of the spin.) (If the reserve does not inflate, the jumper must retrieve the canopy and continue to throw it down and away until it inflates.)
- Uses the thumbs to clear all remaining suspension lines from the pack tray.

When the reserve parachute has activated, the jumper may have two inflated canopies. When descending with two inflated canopies, he has no direction control over the parachutes; all other jumpers remain clear.

6-3. TOTAL MALFUNCTION

A total malfunction is the failure of the parachute to open or to deploy.

a. **Causes of a Total Malfunction**. The failure of the parachute to deploy can be caused by a severed static line, a broken snap hook, or a broken anchor line cable. The jumper's failure to hook up also results in the failure of the parachute to deploy. Malfunctions of these types are rare.

b. **Streamer**. Although not defined as total malfunction, a deployed parachute with a "cigarette roll" or "streamer" provides little or no lift for the jumper. This malfunction must be treated as a total malfunction. This malfunction occurs when a portion of the skirt blows between two suspension lines and begins to roll with the opposite fabric. The heat generated by the friction of the fabric being rolled causes the nylon to fuze and blocks the air channel in the canopy. The jumper immediately activates his reserve using the pull-drop method.

c. **Towed Jumper**. Although not classed as a parachute malfunction, a parachutist can be towed behind the aircraft by a misrouted static line or by a piece of equipment that has snagged the aircraft during the jumper's exit. During the 4000-count, the jumper feels an excessive opening shock and then feels himself being towed by the aircraft. The jumper remains in a tight body position, protecting his rip cord grip until he is either retrieved inside the aircraft or is cut free by the loadmaster on the pilot's order. If the jumper is being towed by the static line and is cut free, the main parachute will not deploy, and the jumper will have to activate his reserve using the pull-drop method. The jumper's actions are as follows:

(1) ***Retrieving the Jumper Inside the Aircraft.*** The jumper remains in a tight body position until he is completely inside the aircraft. A towed jumper must not use his hands to assist the retrieving personnel. The most important action of a towed jumper is to protect his rip cord grip.

(2) ***Cutting the Jumper Away.*** Once the jumper is cut free of the aircraft, the main parachute may, or may not, deploy. If the jumper was towed by something other than the static line, the main parachute will deploy and inflate. There is no need to activate the reserve parachute. If the jumper was towed by the static line and is cut free, the jumper must immediately deploy his reserve using the pull-drop method.

WARNING

The towed jumper must remain in tight body position and protect the rip cord grip with his right hand. Accidental activation of the reserve while being towed may be fatal.

6-4. PARTIAL MALFUNCTION

The four types of partial malfunctions are **complete inversion**, **semi-inversion**, **blown section or gore**, and **broken suspension lines**.

a. **Complete Inversion**. This malfunction may occur when a portion of the skirt blows inward between a pair of suspension lines on the opposite side of the parachute. This portion of the skirt forms a secondary lobe that fills with air and enlarges at the expense of the rest of the canopy. The portion of the canopy forming the secondary lobe is inverted. The canopy turns inside out with no decrease in its lifting surface.

(1) It is difficult to detect if a complete inversion occurs during the initial deployment of the canopy. With a T-10-series, the rear risers control the front of the canopy and the front risers control the rear of the canopy. With an MC1-series, the orifice and control toggles are to the parachutist's front and maneuvering techniques are reversed.

(2) There may be no need for the parachutist to activate the reserve parachute unless the canopy was damaged during inversion. The parachute failed to function properly, but will support the parachutist. The inversion may increase the jumper's rate of descent. If the jumper's rate of descent is greater than that of other jumpers, he deploys his reserve. If he is using the T-10 reserve and his rate of descent is only slightly greater than other jumpers, he uses the down-and-away method. Otherwise, he uses the pull-drop method (the only method of activating the MIRPS).

b. **Semi-Inversion**. This malfunction may occur if development of the secondary lobe stops before completely inverting. This malfunction may remain stable, become completely inverted, or revert to normal during descent. The total lifting capability of the canopy is decreased by the formation of a secondary lobe. The fabric can be burned by friction and weakened during descent. With this malfunction, the parachutist must deploy his reserve. (If he is using the T-10 reserve, he uses the down-and-away method.) The T-10-series and MC1-series parachutes have anti-inversion nets that reduce the chance of this type of malfunction.

c. **Blown Section or Gore**. This malfunction occurs when the strain placed on the canopy during inflation is great and a panel, section, or gore is ripped or torn out, resulting in a hole(s) in the canopy. The jumper compares his rate of descent with that of other jumpers. If the jumper is falling faster than other jumpers around him, he must deploy the reserve parachute. (If he is using the T-10 reserve, he uses the down-and-away method.) Large holes in the canopy should be treated like a blown section or gore.

d. **Broken Suspension Lines**. This malfunction occurs when enough suspension lines break causing the jumper's rate of descent to be greater than fellow jumpers; the parachutist must activate his reserve using the down and away method. If control lines on an MC1-series canopy break, the parachutist controls the canopy by pulling only one of the rear risers in the direction that he wishes to turn. **He must use a rear riser**.

PART TWO
Duties and Functions of Key Personnel, Advanced Airborne Techniques and Training

CHAPTER 7
RESPONSIBILITIES AND QUALIFICATIONS OF KEY PERSONNEL

The initial training and follow-on refresher training of key personnel are of major concern to commanders. The proper training and supervision of key personnel ensure that correct procedures and operational safety measures are followed during airborne operations.

7-1. COMMANDER'S RESPONSIBILITIES

The airborne commander designates the key personnel for each airborne operation. These key personnel are the primary jumpmaster (JM), assistant jumpmasters (AJMs), safety personnel, departure airfield control officer (DACO), drop zone support team leader (DZSTL), drop zone safety officer (DZSO), and malfunction officer (MO).

a. Each aircraft has designated JM, AJM, and safety personnel. The airborne commander gives the designated JM command authority over, and responsibility for, all personnel onboard a jump aircraft. JM, AJM, and safety duties are described in Chapters 8, 9, and 10.

b. The DACO is located at the departure airfield and has coordination responsibility with the GLO and TALCE/aircrew for the loading of personnel, equipment, and supplies into the aircraft. Also, the DACO provides the JM with changes to station time and the overall operational plan, current DZ weather, airfield crossing procedures, and the aircraft parking plan. Complete DACO duties are discussed in Chapter 11.

c. Each DZ has a DZSO or DZSTL. The DZSO or DZSTL has command authority over the actions and safety of all personnel on the drop zone. DZSO and DZSTL procedures for DZ operations are described in Chapters 20, 21, and 22.

d. The MO, as a member of the DZST, is located on the DZ. Detailed MO duties are described in Chapter 23.

The JM, DACO, DZSTL, or DZSO can delegate their authority to subordinates, but they **cannot** delegate their responsibilities.

7-2. KEY PERSONNEL PREREQUISITES

The following minimum standards must be met before personnel will be allowed to perform JM, AJM, safety, DACO, DZSO, DZSTL, or MO duties for personnel and heavy equipment airdrop operations. Tables 7-1 and 7-2 (pages 7-3 and 7-4) indicate the duties that JM, DZSO, and DZSTL qualified and current personnel may perform, the airdrop method that will be used, and the type of airdrop mission that may be flown.

a. **JM, AJM, and Safety Personnel**. To be appointed as JM, AJM, or safety personnel, individuals must meet the following prerequisites:

(1) ***Primary Jumpmaster.***

(a) Commissioned officer, warrant officer, or NCO (US Army and US Navy: E5 or above; USMC and USAF: E4 or above).

(b) JM qualified. To be JM qualified, the JM must be a graduate from an authorized JM course (see Appendix B) at Fort Benning, GA; at a Fort Benning MTT; at Fort Bragg, NC; at a Fort Bragg MTT; or from an SOF JM course. (JMs qualified through SOF JM course must be JM refreshed prior to assuming JM duties outside SOF units.)

(c) JM current. To be JM current, the JM must have either performed JM or AJM duties within the preceding 180 days; or, if a senior or master rated parachutist, performed duty as a safety on military fixed-wing aircraft, utilizing door or ramp exit within the preceding 180 days; or completed a JM refresher course (see Appendix C) in the preceding 180 days and be a current jumper. (JM or safety duties performed on rotary-wing aircraft will apply for JM currency.)

(d) Prior experience. Perform duties twice as the AJM and perform duties twice as a safety. USAF personnel must complete the job qualification requirements for their Air Force specialty code (AFSC).

(2) ***Assistant Jumpmaster.***

(a) Commissioned officer, warrant officer, or NCO (US Army and US Navy: E5 or above; USMC and USAF: E4 or above).

(b) JM qualified and current.

(c) Prior experience. Perform duties twice as a safety.

(3) ***Safety Personnel.***

(a) Commissioned officer, warrant officer, or NCO (US Army: E5 or above; USMC, USN, USAF: E4 or above).

(b) JM qualified and current.

b. **DACO, DZSO, DZSTL, and MO.** To be appointed a DACO, DZSO, DZSTL, or MO, individuals must meet the following prerequisites:

(1) ***DACO.***

(a) Commissioned officer, warrant officer, or NCO (US Army: E5 or above; USMC, USAF: E4 or above).

(b) JM qualified and current.

(c) Prior experience. Perform duties as assistant DACO at least once.

(2) ***DZSO.***

(a) Commissioned officer, warrant officer, or NCO (US Army and US Navy: E5 or above; USMC, USAF: E4 or above).

(b) JM qualified and current or CCT/STS certified.

(c) Prior experience. Perform the duties of assistant DZSO in support of an airborne operation involving personnel or heavy equipment at least once.

NOTE: For combination airdrop operations the DZSO/DZSTL must follow the procedures for heavy drop operations, but observe the jumpers as they exit the aircraft.

(3) ***DZSTL.***

(a) Commissioned officer, warrant officer, or NCO (US Army: E5 or above; USMC, USAF: E4 or above), or civilian equivalent. (This position is not required for USN.)

(b) JM qualified and current for PE or HE.

(c) Certified as a DZSTL by having attended one of the following since 1988:

- USAIS Pathfinder Course.
- USAIS Jumpmaster Course (for CARP DZs only).
- USAIS DZSTL MTT.
- 82d Airborne Division Advanced Airborne School DZSTL Course.
- 82d Airborne Division Advanced Airborne School JM Course (for CARP DZs only).
- US Army Special Operations Command (USASOC) Jumpmaster Course.

(d) Prior experience. Observed and assisted a qualified and current DZSTL while performing his duties during an airdrop operation involving personnel or heavy equipment.

(e) DZSTL current. To be DZSTL current, the DZSTL must have performed the duties of DZSTL or assistant DZSTL within the preceding 180 days or completed a DZSTL refresher course taught by a current DZSTL within the preceding 180 days.

NOTE: DZSTLs in support of CDS airdrops are **not** required to be airborne qualified, on jump status, or JM qualified and current, but they must have attended an authorized Pathfinder or DZST course. For combination airdrop operations the DZSO/DZSTL must follow the procedures for heavy drop operations, but observe the jumpers as they exit the aircraft.

(4) ***Malfunction Officer.***

(a) Commissioned officer, warrant officer, or NCO (US Army and US Navy: E5 or above; USMC, USAF: E4 or above), or civilian equivalent. USAF: DZC/DCSO will fill the duties as the MO and be designated in writing as an MO. He will have a thorough understanding of the parachute equipment used for the operation. USAF combat control personnel are authorized to perform the duties of MO during unilateral operations.

(b) Qualified parachute rigger from the unit providing the air items used during the operation (IAW AR 59-4/MCO 13480.1B). (For the USMC only, the malfunction officer does not have to be from the organization providing the air items.)

	DUTY TO PERFORM	AIRDROP METHOD	TYPE AIRDROP
1. JM SCHOOL GRADUATE BEFORE SEP 1988:			
(A) JM (C)	JM, DZSO, DZSTL	CARP, VIRS, WSVC, GMRS	CDS, HE, PERS
(B) JM (NC)	NONE		
2. JM SCHOOL GRADUATE AFTER SEP 1988:			
(A) JM (C)	JM, DZSO	CARP	CDS, HE, PERS
(B) JM (NC)	DZSO	CARP	CDS
C = CURRENT NC = NOT CURRENT			

Table 7-1. Duties that Jumpmaster School certified personnel may perform.

	DUTY TO PERFORM	AIRDROP METHOD	TYPE AIRDROP
1. USAIS PATHFINDER SCHOOL GRADUATE AFTER SEP 1988:			
(A) JM (C) DZST (C)	JM, DZSTL	CARP, *GMRS, VIRS	CDS, HE, PERS
(B) JM (NC) DZST (C)	DZSTL	CARP	CDS
(C) JM (C) DZST (NC)	JM		
(D) NONAIRBORNE DZST (C)	DZSTL	CARP	CDS
C = CURRENT NC = NOT CURRENT * GMRS DZ NORMALLY RESERVED FOR SOCOM UNITS, AND REQUIRES DOCUMENTED TRAINING FROM DZSTL MTT			

Table 7-2. Duties that USAIS Pathfinder School graduates may perform.

CHAPTER 8

JUMPMASTER DUTIES AT THE UNIT AREA

The success of airborne operations depends mainly on how well the JM executes his duties. He must receive mission briefings, conduct prejump training, supervise rigging of equipment, and move to the departure airfield, all within a rigid time schedule.

Section I. ESSENTIAL INFORMATION

A key factor in the JM duties is the mission briefing. H-hour (TOT) is established at this time and the backward planning process begins.

NOTE: Jumpmaster duties begin immediately upon notification.

8-1. DESIGNATION NOTIFICATION

Upon notification that he has been designated as JM, the individual obtains or is provided the following information:

- Mission and ground tactical plan.
- Air movement plan to include time of flight, formations, route, direction of flight over drop zone, drop altitude, location and design of code letters, racetracks, and emergency call signs/frequencies.
- Names of AJM(s) and safety personnel, and time and place to brief them.
- Time and place of initial manifest call.
- Time and place of final manifest call.
- Time and place to conduct operations briefing.
- Time and place to conduct prejump training.
- Time and place to check and inspect parachutists' uniforms and equipment.
- Transportation (movement to marshaling area and departure airfield plan and times).
- Tactical cross-load plan.
- Time and place of parachute issue, including types of parachutes.
- Weather decision time(s).
- Time and place of troop safety briefing.
- Type of aircraft for the operation and special items of equipment being worn by jumpers (DMJP, AIRPAC, AT4JP, SMJP, or CWIE) or A-series containers aboard aircraft (door bundles or wedge).
- Aircraft tail numbers, chalk numbers, and parking spots.
- Load time.
- Time and place of aircrew/JM briefing.
- Station time.
- Takeoff time.
- Time on target.

NOTE: If, during the joint planning and preparation phase for airborne operations, it is decided host nation aircraft are to be used without navigational equipment, a detailed pilot, loadmaster, and JM brief must take place.

- Landing plan to include drop zones, drop times, delivery sequence, number/type of loads (PP, HD, CDS, LAPES), and types of drops (CARP, GMRS, VIRS, or WSVC).
- Mission and ground tactical plan.
- Air item turn-in plan.
- Medical support plan.

8-2. ASSISTANT'S BRIEFING

After he receives the initial operation briefing, the JM returns to the unit and briefs his AJM and safety personnel. The JM assigns them duties for the remainder of the operation. At this time, the JM determines who assumes responsibility for parachutists remaining onboard. The manifest of personnel scheduled to jump is prepared. The JM schedules a full rehearsal with the JM team before the JM team assembles the chalk (planeload).

a. Upon completion of the briefing, the JM organizes the chalk IAW the cross-loading plan and conducts the initial manifest call. AJM(s) and safety personnel check the identification card and identification tags of each parachutist.

b. Items for cross-loading include door bundles and large, bulky equipment carried by individual parachutists (CWIE, DMJP). The JM determines which chalk and informs parachutists which position and door they will jump.

c. The JM, aided by his AJM(s) and safety personnel, inspects each parachutist's equipment to ensure proper rigging. Parachutists pack and rig their equipment and containers before airborne operations.

NOTE: Primary jumpmaster can delegate authority, but not responsibility.

8-3. JUMPMASTER/SAFETY KIT BAG

The JM ensures aviator kit bags for use onboard the aircraft have been prepared to contain extra items that may be needed during any phase of the airborne operation. This is referred to as the JM/safety kit bag and is used by the JM, AJM, and safety personnel. Items to consider for use onboard the aircraft (depending on the type of airborne operation) are—

- Flashlight (night operations).
- Masking tape/cloth.
- Roll of 1/4-inch cotton webbing.
- Safety wires (with lanyards).
- Harness single point release (HSPR) complete, or H-harness complete.
- Hook-pile tape (HPT) lowering lines.
- Foam impact pads, chin straps, pull-the-dot fasteners with tabs, headbands with attaching clips, and parachutist retention straps.
- Quick-release snaps.
- Upper tie-downs for M1950 weapons cases.

- Retainer bands.
- Trash bags.
- Earplugs and airsickness bags.
- Two extra reserves and extra aviator kit bags (for static lines and deployment) (1 bag for every 15 deployment bags).
- A knife will be carried by the JM, AJM, and safety personnel. (Knife should not be carried on a point of contact.)

8-4. OPERATION BRIEFING

As soon as practical after the initial manifest call, the JM briefs personnel on the details of the operation. Following the operation briefing, prejump training is conducted in the unit area or at the departure airfield. It should be scheduled and conducted within 24 hours before takeoff and include the following:

- Drop zone.
- Type of aircraft.
- Chalk number(s).
- Type of parachute(s).
- Briefing on serials, container delivery system, heavy drop, and type of aircraft, if a part of a larger airborne operation.
- Weather decision time (for GO, NO-GO decision).
- Type of individual equipment and separate equipment that troops will be jumping (CWIE, AIRPAC, DMJP, ALICE pack, SMJP, M1950 weapons case).
- Time and place of parachute issue.
- Load time.
- Station time.
- Takeoff time.
- Length of flight.
- Time on target.
- Direction of flight over DZ.
- Drop altitude.
- Predicted winds on the DZ and direction.
- Route checkpoints.
- DZ assembly aids and area.
- Parachute turn-in point(s).
- Time and place of final manifest call.
- Medical support plan.
- Obstacles on or near the DZ.

NOTE: If the unit jumps nonstandard equipment containers, the rigging on these containers must be approved IAW individual service regulations. These rigging procedures must be forwarded for testing to US Army Airborne Special Operations Test Directorate (USAABNSOTD), Fort Bragg, NC, and then for approval to Commander, 1st Battalion (Abn), 507th Infantry, ATTN: Jumpmaster School, Fort Benning, GA 31905.

Section II. SUSTAINED AIRBORNE TRAINING

All personnel require sustained airborne training. The JM usually does not know the proficiency of all parachutists he is responsible for; therefore, basic airborne jump techniques are rehearsed so each parachutist can demonstrate his ability to perform them. Jumpmasters, safeties, and key leaders will make on-the-spot corrections of any jumper not properly performing the required training. (See paragraph 8-15 for a sample prejump training narrative.)

8-5. MINIMUM TRAINING

Sustained airborne training will consist of, at a minimum, prejump training to include PLFs and mock door training. Prejump training will include a review of the five points of performance, towed jumpers, collisions and entanglements, malfunctions, activation of the reserve, and emergency landings. Mock door training will include rehearsal of every detail involved with the airborne operation to include accidental activation of the reserve on board the aircraft. Jumpers must make a minimum of two exits from the mock door.

a. **Training Apparatuses**. Army personnel will use aircraft fuselage mock-ups to rehearse preflight and in-flight action. (Aircraft fuselage mock-ups can be complete airframes or may be field-expedient training devices as simple as rope strung to simulate an anchor line cable and ballistic helmets on the ground to mark the jump door. The use of field-expedient training devices is not the preferred method for conducting sustained airborne training.) Also, actions-in-the-aircraft training reminds parachutists what occurs in flight before jumping. The JM can use the mock door apparatus to show parachutists where their relative positions will be in the aircraft. If in-flight rigging is to be performed, the rigging station locations can be indicated also. The JM reviews and leads a rehearsal of all actions related to in-flight procedures so the jump mission will be smooth and safe.

b. **Execution**. Each parachutist should be seen by the JM and should hear him (a bullhorn should be used, if necessary). Performance-oriented training is conducted for emergency landings. AJM and safety personnel must make aggressive and positive on-the-spot corrections. Prejump training must be taught proficiently.

8-6. PREJUMP TRAINING

All personnel must attend prejump training. Jumpers must be positioned so that their actions can be viewed by the JM and so they can hear him. A bullhorn can be used, if necessary. Prejump training is performance-oriented and should be tailored to fit the mission. AJM and safety personnel must make aggressive and positive on-the-spot corrections. Prejump training must be taught proficiently.

8-7. FIVE POINTS OF PERFORMANCE

Training on the five points of performance must be attended by all parachutists and JMs.

a. **Proper Exit, Check Body Position and Count**.

(1) Keep chin on chest.

(2) Keep eyes open.

(3) Keep elbows into sides.

(4) Hands over the ends of the reserve parachute, fingers spread, right palm protecting the rip cord grip.

(5) Bend body forward at the waist.

(6) Keep feet and knees together.

(7) Lock knees.

(8) Count to four thousand (fixed-wing aircraft).

(9) Count to six thousand (helicopters).

(10) Immediately activate the reserve parachute using the proper method if opening shock is not felt.

b. **Check Canopy and Gain Canopy Control.**

(1) Reach up and grasp toggles (MC1-series); or grasp risers (T-10-series).

(2) Make a 360-degree check of the canopy.

(3) Remove twists, if any: grasp a set of risers in each hand, thumbs down with knuckles to the rear, and apply outward pressure while bicycling the legs in a vigorous motion.

(4) Immediately recheck canopy and gain canopy control.

c. **Keep a Sharp Lookout for All Jumpers During Descent.**

(1) Remember the three rules of the air.

(2) Always look before turning/slipping.

(3) Always slip/turn in the opposite direction from other jumpers.

(4) Always give lower parachutist the right-of-way.

(5) Avoid other parachutists all the way to the ground; maintain 25 feet of separation (T-10-series) or 50 feet of separation (MC1-series) between jumpers in the air.

(6) At the end of the third point of performance, release all appropriate equipment tiedowns.

d. **Slip/Turn into the Wind and Prepare to Land.**

(1) Check to ensure the air below and around is clear of other parachutists. At no higher than 200 feet, lower equipment. Immediately regain canopy control and continue to keep a sharp lookout for other parachutists.

(2) When 100 feet above ground level, slip into the wind (T-10-series), or turn and hold into the wind (MC1-series).

(3) Keep feet and knees together.

(4) Keep knees slightly bent and unlocked.

(5) Point balls of the feet toward the ground.

(6) Keep head and eyes toward the horizon. Before making contact with the ground, turn the lower portion of the body (below the waist) to a 45-degree angle (front or rear PLF), exposing the portion of the body that will come in contact with the ground.

e. **Land.**

(1) Make a PLF using the five points of contact.

(2) Make no stand-up landings.

(3) Remain on your back and activate one canopy release assembly using either the hand-assist or hand-to-shoulder method.

(4) Recover and turn in equipment.

- Remain in prone position; place weapon into operation and remove harness.
- Remove air items from D-rings on parachute harness.
- Elongate canopy into the wind, remove debris, and figure-eight roll.
- Insert canopy into aviator kit bag with bridle loop on top. Insert waistband through bridle loop.

- Snap aviator kit bag closed; do not zip.
- Attach reserve parachute to the carrying handle of the aviator's kit bag.
- Maintain noise and light discipline. Move rapidly to the nearest turn-in point.

8-8. FIVE POINTS OF CONTACT

Training on the five points of contact includes the following:

a. **Review of Five Points of Contact**.

- Balls of the feet.
- Calves.
- Thighs.
- Buttocks.
- Pull-up muscle (right or left side).

b. **Parachute Landing Falls**. All manifested jumpers must perform one satisfactory PLF in each of the four directions.

- Left side.
- Right side.
- Front.
- Rear.

8-9. TOTAL MALFUNCTIONS (NO LIFT CAPABILITY)

Parachutists must receive training on total malfunctions and control of the reserve parachute.

a. **Reserve Activation**. Activate the reserve using the pull-drop method.

- Remain in a tight body position.
- Keep feet and knees together.
- Grasp the left carrying handle of the reserve parachute with the left hand.
- T-10: Turn head to either side. With the right hand, pull the rip cord grip and drop it.
- MIRPS: Sweep up and away on rip cord protector flap.

b. **Reserve Parachute Control**. While descending under the reserve, control the canopy by slipping and assume the landing attitude by reaching up and grasping as many suspension lines in the opposite direction of drift as possible and slipping with both hands (as in the prepare-to-land attitude with the T-10-series parachute).

c. **MIRPS Sweep Method**. If the reserve parachute does not activate immediately, the jumper remains in a good tight body position while still grasping the left carrying handle. With the right hand forming a knife cutting edge, he sweeps up on the rip cord protector flap, ensuring that the right hand does not interfere with the spring-loaded deployment assistance device.

8-10. PARTIAL MALFUNCTIONS

Parachutists must receive training on partial malfunctions.

a. **Partial Malfunction Indicators**.

- Complete inversion.
- Semi-inversion.
- Blown section or gore.

- More than six broken suspension lines.
- Holes.

Should any of these malfunctions occur and cause the parachutist's rate of descent to increase in comparison to other parachutists, the parachutist activates his reserve parachute as follows:

(1) ***T-10 Reserve.***

(a) Snaps back into a tight body position.

(b) Places the left hand over the rip cord protector flap, fingers extended and spread. Applies pressure to prevent canopy from deploying.

(c) Pulls the rip cord grip with the right hand and drops it.

(d) Forms a knife cutting edge with the right hand to reach into the pack tray, at the same time releasing pressure with the right hand, and grasps as much canopy as possible.

(e) Keeps feet and knees together.

(f) Throws canopy down and away from the body in the same direction as the spin. (Spin is to right, throw right; spin is to left, throw left.) If the reserve does not inflate, pulls it back in and throws it in the opposite direction.

(g) Immediately frees all suspension lines using the thumbs in a downward raking motion.

(2) ***MIRPS.***

(a) Snaps back into a tight body position.

(b) With the left hand, grasps the left carrying handle.

(c) Pulls the rip cord grip with the right hand and drops it.

(d) If the reserve does not deploy, forms a knife cutting edge with the right hand and sweeps the rip cord protector flap up.

b. **Reserve Inflates**. When the reserve parachute inflates, and there are two inflated canopies, the parachutist has no directional control over his parachute. All other parachutists slip or steer clear. To assume the proper landing attitude, the parachutist reaches high on all four risers of the main parachute and maintains this attitude until making ground contact. Immediately upon landing, he releases the main parachute (using the canopy release assemblies) and collapses the reserve by using either the quick-recovery method or by detaching the connector snaps from the D-rings of the main lift web.

8-11. COLLISIONS

The parachutist must always attempt to slip or turn away. If unable to avoid a collision, he uses the spread-eagle method to bounce off another canopy or suspension lines. If a parachutist enters another parachutist's suspension lines, the entering parachutist assumes the modified position of attention with the right hand protecting (but not grasping) the rip cord grip, in hope that he will exit the same location without becoming entangled. If not, the entering parachutist may use his left hand to assist in exiting the other jumper's canopy and suspension lines the same way he entered.

8-12. ENTANGLEMENTS

If parachutists become entangled, their actions required to correct the problem depend upon the type parachute used. Reaction techniques are as follows:

a. **T-10-Series Parachute**. The higher parachutist moves hand under hand down to the lower parachutist. They attempt to establish eye-to-eye contact and hold onto each other by the left main lift web(s). They must **NOT** touch the other jumper's canopy release assemblies. They decide which parachute landing fall to execute upon contact with the ground and both parachutists execute the same PLF. If they are face-to-face, they will **NOT** execute a front PLF. If they are back-to-back, they will **NOT** execute a rear PLF. If one parachutist has a completely inflated canopy, neither parachutist activates the reserve parachute. If both parachutes lose lift capabilities, parachutists use the pull-drop method to activate their reserve parachutes.

b. **MC1-Series Parachute**. Both parachutists immediately activate their reserve parachutes for a partial malfunction. Neither parachutist attempts to climb to the other parachutist. The higher parachutist avoids the lower parachutist when landing.

8-13. EMERGENCY LANDINGS

Depending upon the type of emergency landing initiated, the parachutist performs specific actions.

a. **Tree Landing**. The parachutist—

(1) Attempts to avoid the trees.

(2) Retains combat equipment (if not already lowered).

(3) Checks below, then jettisons combat equipment (if already lowered). (Ballistic helmet remains on the head.)

(4) Maintains canopy control until making contact with trees.

(5) Rotates forearms in front of his face and chest when making contact with trees.

(6) Prepares to execute PLF if passing through trees.

(7) Considers the possibility of activating the reserve parachute and climbing down the outside of it if hung up in trees.

b. **Wire Landing**. The parachutist—

(1) Attempts to avoid the wires.

(2) Lowers combat equipment, checks below, then jettisons equipment.

(3) Raises both arms to the elbow-locked position and places the palms of the hands on the inside of the front set of risers before contact. The feet and knees are together (bend in knees is exaggerated), and chin is on the chest.

(4) Upon making contact with the wires, the jumper pushes forward on the front set of risers, arches his back, and kicks vigorously with the legs, initiating a rocking motion, and attempts to work through the wires.

(5) Prepares to do a PLF if he should pass through the wires.

c. **Water Landing without Life Preserver**. The parachutist—

(1) Attempts to avoid the water (lake or river).

(2) Checks below and then, if clear, jettisons headgear.

(3) Releases all equipment tie-downs and lowers equipment, but does **not** jettison it.

(4) Activates quick release on the waistband.

(5) Unsnaps the left connector snap on the reserve parachute.

(6) Rotates the reserve to the right side of the parachute harness.

(7) Seats himself well into the saddle.

(8) Activates the ejector snap on the chest snap.

(9) Places hands on the ejector snaps on the leg straps.

(10) Activates the ejector snaps on the leg straps, throws arms up, and arches out of the harness when entering the water.

(11) Prepares to execute a PLF if the water is shallow.

(12) Swims upstream, upwind, or away from the parachute to avoid becoming entangled with it.

d. **Water Landing With Life Preserver**.

(1) Emergency water landings could occur on a tactical training mission where the route to the DZ is over a large body of water. On such flights, life preservers are issued to parachutists. If the aircraft malfunctions, the parachutists may need to jump over the water.

(2) Emergency water landings require parachutists to leave any combat equipment onboard the aircraft so they do not become entangled with the equipment in the water.

(3) A parachutist may find himself over water with attached combat equipment due to his drifting off a DZ bordered by water, or during emergency bailout after being hooked up onboard the aircraft. If this occurs, the jumper must lower his combat equipment but **not** jettison it.

(4) Deliberate water landings are executed for training in selected water drop zones. Parachutists may wear combat equipment only after is has been waterproofed and float checked.

(5) The parachutist wears the inflatable life preserver under his harness with the inflatable portions under his armpits. Parachutists wearing a UDT vest will route the chest strap under the UDT vest to prevent crushing the chest if inflated.

(6) During the third point of performance, the parachutist inflates the life preserver by discharging the attached CO2 cartridges. If necessary, the life preserver can be inflated by blowing air into the inflation valve hose.

(7) Upon entering the water, the parachutist activates both canopy release assemblies and swims upstream or away from the canopy.

8-14. RESERVE ACTIVATION INSIDE AIRCRAFT

If the reserve parachute is activated inside an aircraft, the aircrew must follow certain procedures.

a. **Fixed-Wing Aircraft**. If the parachutist is aft of the wheel well, and the jump doors (or ramp) are open and the reserve canopy is in or going out the door (or ramp), the JM and safety personnel do not attempt to retain the parachutist inside the aircraft.

(1) The area is cleared, if possible, and the parachutist exits immediately. If the canopy is not in the door (or ramp), the deployed reserve canopy is secured as quickly as possible by anyone nearby. Then, the parachutist is moved to the forward section of the aircraft, the open reserve is removed, another reserve is attached, and the parachutist is returned to the stick to jump.

(2) If the parachutist is forward of the wheel well, and the jump doors (or ramp) are open, the parachutist either steps on or grasps the reserve canopy and traps it so it cannot inflate. Safety personnel move the parachutist to the forward section of the aircraft. They remove the reserve and attach another, and the parachutist jumps on the next pass over the DZ.

(3) If the jump doors (or ramp) are closed and a reserve parachute deploys, the JM or safety personnel move the parachutist to the forward section of the aircraft. The deployed

reserve is removed, and the loadmaster is told not to open the jump doors (or ramp). Another reserve is attached, and the parachutist is returned to the stick to jump.

b. **Rotary-Wing Aircraft**. If the reserve parachute is activated in an aircraft that requires the parachutist to sit in the door, no attempt is made to stop the parachutist from exiting the aircraft. The parachutist immediately exits. In an aircraft that requires parachutists to exit over the ramp, the procedures described for fixed-wing aircraft are followed.

8-15. TOWED PARACHUTIST PROCEDURES

A towed parachutist is retrieved or cut free as follows:

a. If the parachutist is being towed by the static line, he is retrieved back inside the aircraft. Once he nears the jump door, he does not reach for the JM but continues to protect the rip cord grip; the JM or safety reaches for him. If the jumper cannot be retrieved, he is cut free over the drop zone. Once he feels himself fall free from the aircraft, he must immediately activate his reserve parachute for a total malfunction.

b. If the parachutist is being towed by anything other than the static line, the JM or safety tries to jog him free from the aircraft. If the parachutist cannot be freed, he is retrieved. If he is freed from the aircraft, he does not need to activate his reserve since his main parachute will deploy.

c. Only the aircraft loadmaster and the JM (with assistance from the safety) will perform towed parachutist retrieval.

8-16. SAMPLE PREJUMP TRAINING NARRATIVE

Prior to beginning prejump training, jumpmaster personnel check ID card, ID tags, and ballistic helmets for serviceability and proper routing of parachutist helmet retention strap. Then, the JM organizes jumpers into either a half-moon or extended rectangular formation. The JM uses a half-moon formation for 30 or fewer jumpers and an extended rectangular formation for more than 30 jumpers. When in a half-moon formation, jumpers must be positioned so the JM can see them easily. The extended rectangular formation provides better control for on-the-spot corrections than does the half-moon formation. Once the jumpers are in the chosen formation, safety personnel check identification tags and cards. When finished, they make on-the-spot corrections on the jumpers as they conduct prejump training. Figure 8-1 shows a sample prejump training briefing that will help the JM conduct an effective prejump training session for his jumpers. The jumpmaster should give the prejump training narrative outlined in Figure 8-1 verbatim. The jumpmaster will not read the narrative to the jumpers.

NOTES: 1. The following items are to be covered during prejump training:

- Five points of performance.
- Recovery of equipment.
- Towed parachutist procedures.
- Malfunctions.
- Collisions and entanglements.
- Emergency landings (tree, wire, and water).
- Parachute landing falls.

2. Although the prejump briefing can be given by anyone in the jumpmaster team (JM, AJM, safety), the primary jumpmaster can delegate his authority but cannot delegate his responsibility.
3. Prejump training is performance-oriented training and should be tailored to each mission. The jumpmaster should refer to his unit SOP for additional guidance. During prejump training, the JM uses the command HIT IT as often as needed to keep the jumpers actively involved

Prior to prejump training, place your jumpers into a formation that allows you to easily control them and make on-the-spot corrections. The extended rectangular formation and the horseshoe formation are two preferred formations.

Prior to placing them in formation, ensure the jumpmaster team inspects ballistic helmets, ID tags, and ID cards. The jumpmasters or the safeties can accomplish this inspection.

Although prejump can be given by anyone on the jumpmaster team, the primary jumpmaster can delegate authority but not responsibility.

Holding, running, one riser slips, and other information can be inserted into the prejump if the airborne commander deems it necessary.

Although prejump training should be tailored to fit the mission, emergency landings should always be covered due to the many variables involved with emergency situations; for example, if jumpers have to conduct an emergency bailout over unfamiliar terrain.

Prejump training is performance-oriented training and the jumpmaster team must ensure that the jumpers are performing the actions as they are being covered. During prejump training, use the "HIT IT" exercise as often as needed to keep the jumpers actively involved. Jumpmasters will refer to their unit ASOPs for additional guidance.

When jumping rotary-wing aircraft, jumpers will extend their count to six thousand.

The first point of performance is PROPER EXIT, CHECK BODY POSITION, AND COUNT. Jumpers "HIT IT." Upon exiting the aircraft, snap into a good tight body position. Keep your eyes open, chin on your chest, elbows tight into your sides, hands on the end of the reserve with your fingers spread, right hand covering the rip cord grip. Bend forward at the waist keeping your feet and knees together, knees locked to the rear, and count to four thousand.

At the end of your four thousand count, immediately go into your second point of performance—CHECK CANOPY AND IMMEDIATELY GAIN CANOPY CONTROL. When jumping the T-10-series parachute, reach up to the elbow locked position and secure a set of risers in each hand; simultaneously conduct a 360-degree check of your canopy. When jumping the MC1-series parachute, secure a toggle in each hand and pull them down to eye level, simultaneously conducting a 360-degree check of your canopy. If, during your second point of performance, you find that you have twists, reach up with both hands and grab a set of risers, thumbs down, knuckles to the rear. Pull the risers apart and begin a vigorous bicycling motion. When the twist comes out, check canopy and immediately gain canopy control.

Figure 8-1. Sample prejump training briefing.

Your third point of performance is KEEP A SHARP LOOKOUT DURING YOUR ENTIRE DESCENT. Remember the three rules of the air and repeat them after me. Look before you slip/turn; slip/turn iin the opposite direction to avoid collisions, and the lower jumper has the right of way. Avoid all jumpers all the way to the ground and maintain a 25-foot separation when jumping the T-10-series parachute and a 50-foot separation when jumping the MC1-series parachute. Some time during your third point of performance, release all appropriate equipment tie downs.

This brings you to your fourth point of performance, which is PREPARE TO LAND. At 200 to 100 feet AGL, look below you to ensure there are no fellow jumpers, and lower your equipment. Regain canopy control. At approximately 100 feet AGL, slip/turn into the wind and assume a landing attitude. When jumping the T-10-series parachute and the wind is blowing from your left, reach up on the left set of risers and pull them deep into your chest. If the wind is blowing from your front, reach up on the front set of risers and pull them deep into your chest. If the wind is blowing from your right, reach up on your right set of risers and pull them deep into your chest. If the wind is blowing from your rear, reach up on your rear set of risers and pull them deep into your chest. When jumping the MC1-series parachute and the wind is blowing from your left, pull your left toggle down to the elbow locked position. When you are facing into the wind, let up slowly to prevent oscillation. If the wind is blowing from your rear, pull either toggle down to the elbow locked position. When you are facing into the wind, let up slowly to prevent oscillation. If the wind is blowing from your front, make minor corrections to remain facing into the wind. Once you are facing into the wind, assume a landing attitude by keeping your feet and knees together, knees slightly bent, with your head and eyes on the horizon.

When the balls of your feet make contact with the ground, you will go into your fifth point of performance, LAND. You will make a proper PLF by hitting all five points of contact. Touch them and repeat them after me. 1) BALLS OF THE FEET. 2) CALF. 3) THIGH. 4) BUTTOCKS. 5) PULL UP MUSCLE. You will never attempt the make a stand up landing. Remain on your back and activate one of your canopy release assemblies using either the hand-to-shoulder method or the hand assist method. To activate your canopy release assembly using the hand-to-shoulder method, with either hand reach up and secure a safety clip and pull it out and down exposing the cable loop. Insert the thumb from bottom to top through the cable loop, turn your head in the opposite direction, and pull out and down on the cable loop. To activate the canopy release using the hand assist method, with either hand reach up and secure a safety clip and pull it out and down exposing the cable loop. Insert the thumb from bottom to top. Reinforce that hand with the other hand, turn your head in the opposite direction, and pull out and down on the cable loop. If your canopy fails to deflate, activate the other canopy release assembly. Place your weapon into operation and remove the parachute harness.

I will now cover RECOVERY OF EQUIPMENT. Once out of the parachute harness, remove all air items from the parachute harness. Place the parachute harness inside the aviator's kit bag smooth side facing up, leaving the waistband exposed. Elongate the canopy and suspension lines removing all debris as you go. Once you reach the apex of the canopy, insert your thumb into the bridle loop and figure-eight roll the canopy and suspension lines all the way down the aviator's kit bag. Place the canopy and suspension lines into the aviator's kit bag. Route the waistband through the bridle loop leaving 6 to 8 inches of the waistband exposed. Snap the aviator's kit bag closed. Do not zip. Secure the reserve parachute to the aviator's kit bag, place it over your head, conduct a 360-degree police of your area, and move to your assembly area.

Figure 8-1. Sample prejump training briefing (continued).

I will now cover TOWED JUMPER PROCEDURES. "JUMPERS, HIT IT." If you become a towed jumper and are being towed by your static line and are unconscious, you will be retrieved inside the aircraft. If you are conscious, maintain a good tight body position with your right hand protecting your rip cord grip, and an attempt will made to retrieve you inside the aircraft. As you near the jump door, DO NOT REACH FOR US, continue to protect your rip cord grip. If you cannot be retrieved, you will be cut free. Once you feel yourself falling free from the aircraft, immediately activate your reserve parachute for a total malfunction.

If you are being towed by your equipment, regardless of whether you are conscious or unconscious, we will cut or jog your equipment free and your main parachute will deploy.

NOTE: If you are being towed from a rotary-wing aircraft, maintain a good tight body position and protect your rip cord grip. The aircraft will slowly descend to the DZ, come to a hove, and the jumpmaster will free you from the aircraft.

The next item I will cover is MALFUNCTIONS. There are two types of malfunctions: total and partial. A total malfunction provides no lift capability what-so-ever; therefore, you must activate your reserve using the PULL DROP METHOD.

There are several types of partial malfunctions and actions for each. If you have broken suspension lines or blown sections or gores, compare your rate of descent with fellow jumpers. If you are falling faster than fellow jumpers, activate your reserve for a partial malfunction. If you have a squid, semi-inversion, or a complete inversion with damage to the canopy or suspension lines, you must immediately activate your reserve for a partial malfunction. If you have a complete inversion with no damage to the canopy or suspension lines, do not activate your reserve.

I will not cover ACTIVATION OF THE MODIFIED IMPROVED RESERVE PARACHUTE SYSTEM. To activate the MIRPS, you will use the "PULL DROP METHOD." "JUMPERS, HIT IT." Maintain a good tight body position. Grasp the left carrying handle with your left hand; with right hand grasp the rip cord grip. Pull out on the rip cord grip and drop it. Your reserve will activate.

NOTE: If your reserve does not activate, maintain a good tight body position, and with your right hand form a knife cutting edge and sweep the rip cord protector flap up and away allowing the DADS to properly deploy.

To activate the T-10 reserve for a total malfunction, use the "PULL DROP METHOD." "JUMPERS, HIT IT." If you do not feel an opening shock at the end of your count, maintain a good tight body position. With your left hand, grasp the left carrying handle; with your right grasp the rip cord grip. Turn your head to the left or right; pull the rip cord grip and drop it. Your reserve will activate.

To activate the T-10 reserve for a partial malfunction, use the down-and-away method. "JUMPERS, HIT IT. CHECK CANOPY." Snap back into a good tight body position. With the left hand, cover the rip cord protector flap and with the right hand, grasp the rip cord grip. Apply inward pressure with the left hand and with the right hand pull the rip cord grip and drop it. Form a knife cutting edge with the right hand and insert it into the upper right hand corner of the reserve. Grasp as much canopy and suspension lines as possible and pull out and up over either shoulder and throw it down and away in the direction you are

Figure 8-1. Sample prejump training briefing (continued).

spinning. If the canopy fails to inflate, pull it back into your body and throw it down and away in the opposite direction. Once your reserve activates, with each hand form a fist, thumbs exposed, and with a sweeping motion, clear all of the suspension lines from the pack tray ensuring they are deployed.

NOTE: If you have to activate the reserve for a partial malfunction, any attempt to control either canopy will be useless as one canopy will act as a brake for the other.

The next items I will cover are COLLISIONS AND ENTANGLEMENTS. "JUMPERS, HIT IT. CHECK CANOPY." If you see another jumper approaching, immediately attempt to slip/turn away. If you cannot avoid the collision, assume a spread eagle position and attempt to bounce off the jumper's canopy and suspension lines and then slip/turn away.

If you become entangled and are jumping the T-10-series parachute, the higher jumper will climb down to the lower jumper using the hand-under-hand method. Once both jumpers are even, you will face each other and grasp each other's main lift web. Both jumpers will discuss which PLF to execute. Both jumpers will conduct a PLF with the direction of drift. Neither jumper will fall into the other jumper. Both jumpers will continue to observe their canopies. If one canopy collapses, neither jumper will activate their reserve as one T-10-series parachute can safely deliver two combat-equipped jumpers to the ground. If both canopies collapse, the jumpers will pull towards each other to create a clear path for the activation of their reserve parachutes, then activate their reserves using the PULL DROP METHOD.

If you are jumping the MC1-series parachute, both jumpers will remain where they are, obtain a clear path, and immediately activate their reserve using the PULL DROP METHOD.

The next items I will cover are EMERGENCY LANDINGS. The first emergency landing I will cover is the tree landing. If you are drifting towards the trees, immediately slip/turn away. If you cannot avoid the trees and have lowered your equipment, look below you to ensure there are no fellow jumpers and then jettison your equipment making a mental note of where it lands. If you have not lowered your equipment, keep it on you to provide extra protection while passing through the trees. At approximately 100 feet AGL, assume a landing attitude by keeping your feet and knees together, knees slightly bent, with your head and eyes on the horizon. When the balls of your feet make contact with the trees, rotate your hands in front of your face with your elbows high. Be prepared to execute a PLF if you pass through the trees.

If you get hung up in the trees, keep your ballistic helmet on and lower and jettison all unneeded equipment. Activate the chest strap ejector strap and activate the quick release in your waistband. Place your left hand over the rip cord protector flap and apply pressure. Grasp the rip cord grip with your right hand and pull it and drop it. Control the activation of the reserve parachute toward the ground ensuring that all suspension lines are completely deployed. Disconnect the left connector snap and rotate the reserve to the right. Grasp the main lift web with either hand below the canopy release assembly and with the other hand activate the leg strap ejector snaps and climb down the outside of the reserve.

Remember, when in doubt, stay where you are and wait for assistance.

Figure 8-1. Sample prejump training briefing (continued).

The next emergency landing I will cover is the wire landing. If you are drifting toward wires, immediately slip/turn away. If you cannot avoid the wires, look below you to ensure there are no fellow jumpers and then lower and jettison your equipment making a mental note of where it lands. Assume a landing attitude by placing your hands, fingers extended and joined, high on the inside of the front set of risers with the elbows locked. Place your chin on your chest, keep your feet and knees together, and exaggerate the bend in your knees. When the balls of your feet make contact with the wires, begin a vigorous rocking motion in an attempt to pass all the way through the wires. Be prepared to execute a PLF if you pass all the way through the wires. If you get hung up in the wires, stay where you are and wait for assistance.

The last emergency landing I will cover is the water landing. The water landing is the most dangerous emergency landing because it takes the most time to prepare for. If you are drifting towards a body of water, immediately slip/turn away. If you cannot avoid the water, look below you to ensure there are no fellow jumpers, and then lower, DO NOT JETTISON, your equipment. Next, jettison your ballistic helmet. Activate the quick release in your waistband, disconnect the left connector snap and rotate the reserve to the right. Seat yourself well into the saddle and activate the chest strap ejector snap. Regain canopy control. Before entering the water, assume a landing attitude by keeping your feet and knees together, knees slight bent, and place your hands on the leg strap ejector snaps. When the balls of your feet make contact with the water, activate the leg strap ejector snaps, arch your back, throw your arms above your head, and slide out of the parachute harness. Swim upwind or upstream away from the canopy. Be prepared to execute a PLF if the water is shallow. If the canopy comes down on top of you, locate a radial seam, follow it to the skirt of the canopy and swim upstream or upwind away from the canopy.

NOTE: If you are jumping the B-7 life preserver, activate it in the air. If you are jumping the B-5 life preserver, activate it in the water after you remove the parachute harness.

NIGHT JUMPS: When conducting night jumps, be sure to give your canopy an extra look, maintain noise and light discipline, and be prepared to execute a PLF because you will hit the ground 5 to 10 seconds before you think you will.

AWADS: When jumping under AWADS conditions, do not lower your equipment until you have passed through the clouds. Do not slip/turn unless you must to avoid a collision. If you have any type of malfunction, immediately activate your reserve using the pull drop method, because you cannot compare your rate of descent. Ensure you recheck your canopy once you pass through the clouds.

PLFs: We will now move to the PLF platform and conduct one satisfactory PLF in each of the four directions, ensuring you conduct a proper PLF.

Figure 8-1. Sample prejump training briefing (continued).

CHAPTER 9

JUMPMASTER AND SAFETY DUTIES AT THE DEPARTURE AIRFIELD

Time is a critical factor at the departure airfield. The following events occur at the same time to allow the unit to meet station time:

- *DACO/PJM update briefing.*
- *PJM aircraft inspection and coordination with aircrew.*
- *Control of parachute issue by AJM/safeties.*
- *Rigging/inspection of parachutists.*
- *Loading of aircraft.*

The PJM usually turns control of the chalk (planeload) over to the AJM and safeties while he accomplishes update briefings, aircraft inspection, and aircrew coordination. The AJM and safeties control parachute issue and prepare for rigging/inspection of the chalk. All JM team personnel must draw their own parachutes for wear during airdrop operations; the aircrew cannot provide parachutes for the JM team.

Section I. KEY PERSONNEL

The commander selects the best-qualified jumpers to perform the JM, AJM, DACO, and DZSO/DZSTL duties for a unit airdrop operation. Those selected key personnel must correctly perform their assigned duties to ensure mission success and jumper safety. This responsibility may be delegated to qualified personnel in accordance with individual service component regulations.

9-1. PRIMARY JUMPMASTER/ASSISTANT JUMPMASTER DUTIES

PJM/AJM duties are as follows:

a. **PJM/DACO Briefing**. Upon arrival at the airfield, the PJM reports to the DACO for an update briefing to include:

- Change in the station time.
- Change in the overall operation plan.
- Current weather and winds.
- Parking plan of aircraft (location and tail number of the assigned aircraft).
- Coordination with the USAF guide if wheeled vehicles are used for transport to aircraft.

b. **Manifest Distribution**. Normally the manifest (DA Form 1306) is distributed as follows:

- DACO—two copies (original plus one copy).
- PJM—one copy.
- Pilot or his representative—one copy.
- Parachute issue facility—one copy.
- Unit suspense file—one copy.

c. **PJM/Aircrew Initial Coordination**. After DACO coordination, the PJM should proceed to the aircraft for initial coordination. Normally, the aircraft is open with a crewmember onboard one hour before station time. Items to be discussed, verified, or agreed to include:

(1) Aircraft configuration IAW the unit mission. If the aircraft is incorrectly configured, the requesting unit has the option to accept or reject it. If the mission request asks for both doors to be open for a mass troop drop but one door is inoperative (in flight or on the ground), the requesting unit has the option to continue the mission using one door or to abort the mission. For single door operations, one door may be used for personnel airdrop. However, if the other paratroop door is open, it must be manned by a current JM. If only one door is being used, it must be manned by a current JM/safety team. A single-door aircraft operation must be conducted in a military controlled air space environment.

(2) Control of the jump doors.

(3) Drop altitude, speed, and heading.

(4) Racetracks.

(5) Towed parachutist procedures (in detail).

(6) Emergency actions onboard.

(7) Time warnings and checkpoints.

(8) Type of drop, for example, CARP, GMRS, VIRS.

(9) Load time.

(10) Station time.

(11) Takeoff time.

(12) Initial contact time with CCT or DZST for update on DZ conditions (if communications are being used).

(13) Drop time.

(14) Additional details:

(a) If a ground abort occurs, designate which key personnel onboard must be advised.

(b) If the PJM is not the last parachutist, designate who is in command of the troops on board in an emergency.

(c) Emphasize to the aircrew the importance of accurate direction and velocity of DZ winds (before the 1-minute time warning) and accurate time warnings.

(d) Ensure seats are stowed by releasing the support leg from the aircraft floor, rotating the support leg 270 degrees until it rests on the seat, folding the seat up until it is vertical to the aircraft floor, and securing it there by attaching the clip provided on the stowage strap to the upper seat back support.

d. **Aircraft Inspection**. The PJM, accompanied by a crew member (usually the loadmaster), inspects the aircraft and coordinates any activities related to the airborne operation. The PJM must check the exterior and interior portions of the aircraft directly related to the airborne operation.

9-2. SAFETY PERSONNEL

While the PJM inspects the aircraft, safety personnel control the chalk, making sure personnel remain in assigned stick at all times. Personnel must also be accounted for at all times.

a. **Parachute Issue**. AJM/safety personnel supervise the chalk during parachute and air item issue. They draw extra aviator kit bags (1 for every 15 jumpers) and at least two extra reserve parachutes. The extra kit bags are used to store the static lines and deployment bags after the jump. (The extra kit bags are placed in or with the PJM kit bag.)

b. **Departure Airfield Layout Inspection**. All air items and combat equipment should be displayed for inspection by AJM/safety personnel before donning or loading on the aircraft.

c. **Parachutes and Equipment**. AJM/safety personnel ensure that all parachutists use the buddy system when donning parachutes and equipment. (Personnel should not start donning parachutes and equipment earlier than one hour before load time to avoid unnecessary time in the harness.)

d. **Final DACO Coordination**. If directed by the JM, AJM/safety personnel report to the DACO for any special or last-minute instructions that must be passed to the JM.

e. **JMPI**. AJM/safety personnel assist in rigging, inspecting, and correcting deficiencies as directed by the PJM. The PJM's role during JMPI is to observe and supervise. The PJM should only perform JMPI to facilitate meeting station time.

NOTE: All current and qualified JMs assist during the JMPI.

Section II. JUMPMASTER PERSONNEL INSPECTION

The PJM is responsible for the inspection of his parachutists before an airborne operation. Only by a complete and systematic equipment inspection of each parachutist can the PJM ensure that personnel aboard his aircraft are safe to jump.

9-3. HANDS-ON INSPECTION

During JMPI, the PJM uses a control hand and a working hand. During the course of the inspection, the control hand and working hand may change. The control hand remains stationary, while the eyes follow the working hand. The principle is to look at what is touched by the working hand. When locations are described in this sequence (top, right, left), it is in relation to the parachutist—**not** the JM. The word "trace" describes the working hand moving along the item being inspected to ensure that it is not twisted, cut, frayed, or misrouted. When tracing metal items, the JM inspects for sharp edges and proper assembly. The inspection sequence described in this section is followed for a typical combat parachutist rigged with LBE (exposed), combat pack (ALICE pack/FPLIF), M1950 weapons case, and main and reserve parachutes. (Other items and combinations of equipment are prescribed elsewhere in this manual.) When performing JMPI on these items, the JM starts at the point of attachment to the parachute harness and completely inspects that item before proceeding with the remainder of the inspection sequence.

9-4. BALLISTIC HELMET (FRONT)

The jumper moves to the JM and the JM tells him, "Open your rip cord protector flap." The JM then proceeds as follows:

a. Place both hands on the rim of the ballistic helmet, on the extreme right side, fingers extended and joined, palms facing the parachutist. Using the right hand as the working hand and the left hand as the control hand, trace across the rim of the ballistic helmet with the working hand to the opposite side of the ballistic helmet, inspecting for any sharp or protruding edges that may damage or cut the static line during aircraft exit. Once the hands are parallel, insert the thumb of each hand under the rim and feel for the locking nuts to ensure they are present and secured.

b. Gently tilt the parachutist's head back and look at the headband. Ensure that the smooth leather portion is toward the parachutist's head and the openings of the attaching clips are down, visible, and secured. Place the right index finger on the pull-the-dot fastener with tab. Ensure that it is a serviceable pull-the-dot fastener in that there are four plies of nylon, three of which must run through the snap portion, and it is secured.

c. Bypass the pull-the-dot fastener with tab and trace down to the point of attachment for the chin strap. Ensure the chin strap is properly routed through the adjusting buckle and that the parachutist retention strap is routed around the long continuous portion of the chin strap and under the adjusting buckle, with the pile portion of the parachutist retention strap away from the jumper's face.

d. Continue to trace the long continuous portion of the chin strap down under the parachutist's chin and back up to the point of attachment on the opposite side, and conduct the same inspection.

e. Finally, trace the inside of the nylon portion of the adjusting buckle until you make contact with the left thumb, which is still on the locking nut. Trace the short, sewn portion of the chin strap across the front of the parachutist's chin. Drop both hands.

9-5. ADVANCED COMBAT HELMET (FRONT)

The ACH is the programmed replacement for the ballistic helmet. It is compatible with NVDs and chemical defense equipment, allows maximum sensory and situational awareness, provides an unobstructed field of view, and increases ambient hearing capabilities. With proper sizing, the ACH retention and suspension system provides balance, stability, comfort, fit, ventilation, and superior impact protection throughout all static line airborne operational scenarios. The JM inspects the ACH as follows:

a. Move to the jumper's front and issue the command "OPEN YOUR RIP CORD PROTECTOR FLAP." Place your hands on the extreme right hand side of the helmet, fingers and thumb extended and joined, palms facing the jumper. The left hand is the control hand and the right hand is the working hand. With the working hand, trace across the rim of the ACH and feel for any sharp or protruding edges.

b. Place the right index finger on the ladder locks to ensure the chinstrap is properly routed through the ladder locks and free of all cracked components and is serviceable. Trace the chinstrap down to the fastener and ensure that it is free of all cracked components and is properly secured. Continue to trace the long continuous portion of the chinstrap under the jumper's chin to its point of attachment on the opposite side and conduct the same inspection. Place your index finger on the short sewn portion of the chinstrap and trace it across the jumper's chin.

9-6. CANOPY RELEASE ASSEMBLIES

The JM inspects the right canopy release assembly.

a. Look at the right canopy release assembly. Tap the canopy release assembly with the knuckles of the left hand and ensure that it sounds solid. Turn the right canopy release assembly one-quarter turn out and look at it closely. Ensure that it is properly assembled, has no cracked components, and is free of all foreign matter. Look at the left canopy release assembly and, with the right hand, conduct the same inspection. If the static line is resting against either canopy release assembly, move it so that it does not obstruct the view of the assembly.

b. When inspecting a parachute harness equipped with the modified D-rings, use the thumb of the left hand to pull down on the right upper D-ring to ensure it is secured to the jumper's right main lift web with gutted Type II or Type III nylon cord, and that it is routed under the jumper's chest strap. With the right hand, conduct the same inspection on the jumper's left upper D-ring.

9-7. CHEST STRAP

The JM inspects the chest strap next.

a. Simultaneously slide both hands down the main lift web until the little fingers make contact with either the D-rings or the L-bar connector link.

b. When two sets of D-rings are present, look at the left existing D-ring verifying it is rotated in the up position and safetied behind the main lift web using either Type II or Type III nylon cord (core threads removed).

c. Move the left hand below the "new" D-ring and rotate the main lift web inward, verifying the D-ring is positioned immediately below the existing D-ring and directly above the confluence wrap.

d. Visually verify the screw pin is fully seated. It will be slightly recessed below the screw pinhead housing and staked. (Any sign of screw pin misalignment immediately renders the harness unserviceable. It will not be jumped and must be removed from service.)

e. Repeat this inspection procedure for the right main lift web.

f. Look at the chest strap and ensure that it is not misrouted around the main lift web.

g. Insert the right hand (fingers and thumb extended and joined, fingers pointed skyward, palm facing the jumpmaster) from bottom to top behind the chest strap (next to where it is sewn into the main lift web)

h. Trace the chest strap until the right hand is behind the ejector snap, ensuring it is not twisted, cut, or frayed. The ejector snap pad will not come between the right hand and the ejector snap.

i. With the thumb of the right hand, press in on the activating lever for the ejector snap to ensure that it is properly seated over the ball detents and is free of all foreign matter. Leave the right hand and thumb in place and move to the right side of the jumper.

9-8. WAISTBAND

The JM begins the inspection by moving to the right side of the parachutist.

a. Insert the left hand (fingers and thumb extended and joined, fingers pointed skyward, palm facing the jumpmaster) from the bottom to the top behind the waistband (next to where it is sewn into the pack tray). Look at the waistband where it is sewn into the pack tray and ensure that at least 50 percent of one row of stitching is present.

b. Trace the waistband forward to ensure that it is not twisted, cut, or frayed and that it has not been misrouted behind the horizontal back strap. Trace the waistband forward until the left hand makes contact with the right D-ring. Look at the waistband to ensure that it is routed over the right main lift web and under the right D-ring.

c. Rotate the right hand down and grasp the top carrying handle of the reserve parachute, palm facing the reserve. Simultaneously lift up and out on the reserve parachute and place the left hand, palm facing the jumper, into the jumper's chest. Look at the waistband where it is routed behind the reserve parachute to ensure that it is routed through both waistband retainers and is not twisted, cut, or frayed.

d. Withdraw the left hand from the jumper's chest, reach under the right forearm, and grasp the left carrying handle of the reserve parachute (palm facing away from the reserve, fingers spread). Release the top carrying handle of the reserve with the right hand and move to the left side of the parachutist. Insert the right hand (fingers and thumb extended and joined, pointing up, and the palm facing away from the parachutist) from bottom to top behind the waistband as close to the left D-ring as possible. Look at the waistband to ensure that it is routed over the left main lift web and under the left D-ring.

e. Trace waistband back to the metal adjuster, ensuring that it is not twisted, cut, or frayed. Leave the right hand in place behind the metal adjuster. Remove the left hand from the left carrying handle and insert the index and middle fingers of the left hand into the quick release of the waistband, ensuring that it is a two- to three-finger quick release, and not a false quick release. Remove the index and middle fingers from the quick release and, with the index finger and thumb of the left hand, pinch off the free-running end of the waistband where it comes out of the metal adjuster. Trace the free-running end of the waistband until the fingers fall off the end. Ensure the free-running end is not cut, torn, or frayed and that it is easily accessible to the jumper.

f. Reinsert the left hand into the left carrying handle of the reserve, with the palm facing away from the reserve and the fingers spread. Look at the right hand and trace the waistband adjuster panel back to the pack tray, ensuring it is not twisted, cut, frayed, or misrouted under the horizontal back strap. Look at the waistband adjuster panel where it is sewn to the pack tray and ensure that at least 50 percent of one row of stitching is present. Drop both hands and move to the jumper's front.

9-9. RESERVE PARACHUTE

The JM begins by grasping the top carrying handle with the left hand (with the palm facing the reserve) and pulling up and out slightly. He looks at the left connector snap and, with the index finger or thumb of the right hand, checks the left connector snap to ensure that it is secured to the left D-ring, that it has spring tension, and is not safetied.

a. With the left hand, grasp the top carrying handle of the reserve, palm facing the reserve parachute, and lift up and out. Look at the left connector snap and, with the index finger of the right hand, finger the left connector snap one time to ensure that it is properly secured to the left D-ring, has spring tension, and has not been safetied. Grasp the top carrying handle of the reserve parachute with the right hand, palm facing the reserve, and lift up and out. Look at the right connector snap and, with the index finger of the left hand, finger the right connector snap one time to ensure that it is properly secured to the D-ring, does not have spring tension, and has been safetied. To further ensure that the right connector snap has been safetied, with the left index finger, pull down and out on the safety wire and lanyard to ensure that the safety wire has been routed through the drilled hole in the right connector snap. Insert the left index finger from the top to the bottom on the inside of the right connector snap to ensure the safety wire has been bent down toward the ground.

b. Place the right hand on the left end panel of the reserve parachute. Form a knife edge with the left hand, palm facing the jumpmaster, and trace from top to bottom between the rip cord grip and the right end panel of the reserve parachute to ensure that the rip cord grip is not winterized (unless the mission calls for it) and that the right pack opening spring band is not misrouted over the rip cord grip. Insert the index finger of the left hand into the rip cord grip stow pocket and feel for the steel swaged ball. With the left index finger and thumb, pinch off the first locking pin where it comes out of the rip cord grip stow pocket. Apply inward pressure on the left end panel with the right hand and physically seat the first locking pin from the jumper's right to the jumper's left through its cone. Trace the cable over to the second locking pin and continue tracing until the fingers fall off end, ensuring that the locking pins are not bent, cracked, or corroded and that the cable is not kinked or frayed.

c. Grasp the right end panel of the reserve parachute with the left hand and look at the first set of locking pins, cones, and grommets. Rotate the reserve parachute 360 degrees ensuring there is no exposed canopy or suspension lines and no rust or corrosion, and that the locking pins are properly assembled. Look at the second set of locking pins, cones, and grommets and repeat the same inspection. With the index finger and thumb of either hand, pinch off the rip cord protector flap, visually and physically ensure the log record book is present, then close the rip cord protector flap. When inspecting the MIRPS, ensure that a 1/4-inch strip of yellow binding tape has been permanently sewn across the top of the rip cord protector flap. The jumpmaster then places either hand on the bulge created by the deployment assistance device to ensure it is centered behind the rip cord protector flap.

d. Inspect the pack opening spring bands for exposed metal, spring tension, and proper routing. Form a knife edge with the left hand (fingers and thumb extended and joined, palm facing the jumpmaster) and sweep the top carrying handle and universal static line snap hook back toward the jumper (this will be the control hand). Begin the inspection with the top right pack opening spring band.

(1) With the index finger and thumb of the right hand, pinch off the tab portion of the top right pack opening spring band and pull it down toward the rip cord protector flap. Look at the pack opening spring band to ensure that it is routed through the reinforced nylon webbing at the back of the reserve, it is properly routed under the top carrying handle, and that there is no exposed metal on the pack opening spring band. When the tab portion of the pack opening spring band is released, the pack opening spring band should snap back into place. Repeat the same inspection for the top left pack opening spring bands. Place the control hand on the bottom right corner of the reserve parachute, lifting it up and out, so that the bottom left and bottom right pack opening spring bands can be inspected in the same manner.

(2) Lower the reserve parachute back to its original position, leaving the control hand on the bottom right corner of the reserve, and inspect the right pack opening spring band.

e. Conduct an overall inspection of the reserve parachute to ensure that it is free of grease, oil, dirt, mud, tears, and exposed canopy. Place both hands on the top right corner of the reserve parachute, palms facing the reserve. The left hand is the control hand and the right hand is the working hand. With the head and eyes 6 to 8 inches away from the working hand, trace across the top panel and down the left end panel. Move the control hand down to the bottom right corner of the reserve and lift the reserve parachute up and out, exposing the bottom panel. Trace the bottom panel across until contact is made with the control hand. Lower the reserve parachute back to its original position, returning the control hand to the top right corner. Flip the right hand over and trace up the right end panel until contact is made with the control hand. Raise the control hand up out of the way and trace across the top right corner where the control hand had been. Raise the reserve parachute to the jumper and command, HOLD, SQUAT.

9-10. LEG STRAPS

Insert the index and middle fingers of both hands, from outside to inside, under the leg straps right below the aviator's kit bag where the natural pocket is formed. Simultaneously slide both hands rearward on the leg straps back to the saddle, ensuring that the leg straps are not crossed.

a. With the left hand, trace the right leg strap forward to the quick-fit V-ring. Ensure that the leg strap is not twisted, cut, or frayed. With the left thumb, push in on the activating lever of the ejector snap to ensure it is properly seated over the ball detents and is free of all foreign matter.

b. Leave the left hand in place and look at the left leg strap. With the right hand, trace the left leg strap up to the quick-fit V-ring, ensuring that it is not twisted, cut, or frayed, and is properly routed through the exposed carrying handle of the aviator's kit bag, over the bottom and under the top. With the right thumb, push in on the activating lever of the ejector snap to ensure it is properly seated over the ball detents and is free of all foreign matter. Look at the aviator's kit bag and ensure that it has not been reversed and that the sewn, reinforced portion is facing away from the jumper. Tap the jumper on the thigh and command, RECOVER.

9-11. STATIC LINES

If the static line has not been routed over the parachutist's shoulder, the JM routes it at this time. The following explanation is for the static line when routed over the right shoulder. For the left shoulder, the hands are reversed.

a. **Static Line.**

(1) With the right hand, secure the static line snap hook. Pull up on the static line snap hook and ensure that it is attached to the top carrying handle of the reserve parachute. Open the palm and look at the static line snap hook. Place the index finger of the left hand on the static line snap hook next to the drilled hole and ensure the hole for the safety wire is present and free of all obstructions.

(2) Regrasp the static line snap hook and, with the left thumb and index finger, pinch the safety wire and lanyard at its point of attachment; look at the safety wire and lanyard. Ensure that the lanyard is attached to the upper loop portion of the static line and not to the cut-away portion of the static line snap hook. Trace the safety wire and lanyard to ensure it is not too long, too short, or excessively bent. Drop the safety wire and lanyard.

(3) Grasp the static line above the static line snap hook with the left hand, palm facing the jumper, thumb down just above the upper loop portion. Rotate it to the parachutist's right and push it toward the static line snap hook, exposing the inside of the looped portion of the static line. Look inside the looped portion of the static line to ensure that it is not cut, frayed, or burned. Rotate the static line snap hook 180 degrees and inspect it again in the same manner. Drop the left hand.

(4) Use either hand to form a bight in the static line, and inspect the static line retainer to ensure it is not cut, torn, or frayed more than 50 percent. Insert the static line bight from top to bottom through the static line slack retainer and pull the excess through. Flip the bight on top of the pack tray and place either hand on it. The hand that controls the bite becomes the control hand. With the index finger and thumb of working hand pinch off the first stow and pull it one or two inches toward the center of the pack tray. Look behind the stow to ensure that the static line has not been misrouted around the static line stow bar and that it is free of cuts, frays, or burns. Release the first stow and let it pop back into place.

(5) Insert the index finger of the working hand from bottom to top behind the first strand of static line as close as possible to the first stow. Trace the first strand static line over to the second stow to ensure that it is free of all cuts, frays, and burns. Once contact is made with the second stow, pinch it off with the index finger and thumb of the working hand, pull it one to two inches toward the center of the pack tray, and conduct the same inspection. Place the index finger or thumb of the working hand behind the second strand of static line and trace it away from you ensuring it is not cut, frayed, or burned. Continue to inspect the static line in the same manner down to the pack opening loop ensuring that you inspect the last strand of static line with the index finger only. Ensure that the last strand of universal static line is routed from the right outer static line stow bar.

(6) Once contact is made with the pack opening loop ensure that it is situated between the pack closing loops at the 6 and 9 o'clock position. Insert the index finger of the working hand from bottom to top into the pack opening loop. Pull down and out on the pack opening loop, look inside to ensure the pack closing tie has been routed through the pack opening loop, and that the pack opening loop is not torn or frayed. Let the pack opening loop pop off your finger.

(a) Place the index finer of the working hand on the pack closing loop at the 6 o'clock position. Look at the pack closing loop to ensure the pack closing tie is routed through the pack closing loop and the pack closing loop is not torn or frayed more than 50 percent. Inspect the remaining pack closing loops in the same manner using a clockwise motion (9 o'clock, 12 o'clock, 3 o'clock).

(b) Look at the pack closing tie and the surgeon's knot locking knot. Ensure the surgeon's knot locking knot is properly positioned between the pack closing loops at the 3 o'clock and 6 o'clock position. Insert the index finger of the working hand from bottom to top behind the surgeon's knot locking knot. Pull down and out to ensure it is secure and that the pack closing tie has been properly constructed of one turn **only** of 1/4-inch cotton webbing. Let the pack closing tie pop off the end of your finger, stand up directly behind the jumper, and issue the command TILT YOUR HEAD FORWARD.

b. **Universal Static Line.**

(1) With the right hand, grasp the USL snap hook and give it a slight tug, ensuring that the USL snap hook is secured to the top carrying handle of the reserve parachute. Let the USL snap hook rest in the palm of the right hand. Place the index finger of the left hand on the girth hitch to ensure that it is routed around the narrow portion of the static line snap hook and that the green marking stitching is present on the static line. Place your index finger on the rivet ensuring that it is present, then place the index finger on the opening gate of the USL snap hook and finger it one time ensuring that it has spring tension and that the opening gate is facing toward the jumper. Regrasp the USL snap

hook and position it perpendicular to the reserve. With the left hand palm facing the jumper, thumb pointing down, grasp the USL just above the girth hitch and rotate it down into the jumper's right. Push the static line toward the girth hitch and inspect the inside upper loop portion of the static line ensuring it is free of all cuts, frays, or burns. With the right thumb or index finger, push in on the upper loop portion of the static line and conduct the same inspection. Redress the static line around the narrow portion of the USL snap hook.

(2) Since the static line is routed over the jumper's right shoulder, form an "O" with the index finger and thumb of the right hand around the USL snap hook. The right hand now becomes the working hand. Raise the hand, simultaneously inspecting the static line as it runs through the "O" ensuring the static line is free of all cuts, frays, or burns. When the working hand is as high as it will go, issue the command TURN. Once the jumper has completed the turn, the right hand should have been raised high enough to pull all of the slack from the static line slack retainer. Keep the USL tight between the control hand and the first stow and place the index finger (or index finger and middle finger) of the working hand behind the USL below the control hand so there is skin-to-skin contact. Trace the USL down to the first stow ensuring that it is free of all cuts, frays, and burns and it has not been misrouted under or through either riser assembly.

(3) The JM will then use either hand to form a bight in the static line. The JM will now inspect the static line retainer to ensure it is not cut, torn, or frayed more than 50 percent. Insert the static line bight from top to bottom through the static line slack retainer and pull the excess through. Flip the bight on top of the pack tray and place either hand on it. The hand that controls the bite becomes the control hand. With the index finger and thumb of the working hand, pinch off the first stow and pull it one or two inches toward the center of the pack tray. Look behind the stow to ensure that the USL has not bee misrouted around the static line stow bar and it is free of cuts, frays, or burns. Release the first stow and let it pop back into place.

(4) Insert the index finger of the working hand from bottom to top behind the first strand of USL as close as possible to the first stow. Trace the first strand of USL over to the second stow to ensure that it is free of all cuts, frays, and burns. Once contact is made with the second stow, pinch it off with the index finger and thumb of the working hand, pull it one to two inches toward the center of the pack tray, and conduct the same inspection. Place the index finger or thumb of the working hand behind the second strand of static line and trace it away from you ensuring it is not cut, frayed, or burned. Continue to inspect the USL in the same manner down to the pack opening loop ensuring that you inspect the last strand of static line with the index finger only, and the last strand of USL is routed from the right outer static line stow bar.

(5) Once contact is made with the pack opening loop ensure that it is situated between the pack closing loops at the 6 and 9 o'clock position. Insert the index finger of the working hand from bottom to top into the pack opening loop. Pull down and out on the pack opening loop and look inside to ensure the pack closing tie has been routed through the pack opening loop and that the pack opening loop is not torn or frayed. Let the pack opening loop pop off your finger. Place the index finger of the working hand on the pack closing loop at the 6 o'clock position.

(a) Look at the pack closing loop to ensure the pack closing tie is routed through the pack closing loop and the pack closing loop is not torn or frayed more than 50 percent.

Inspect the remaining pack closing loops in the same manner using a clockwise motion (9 o'clock, 12 o'clock, 3 o'clock).

(b) Look at the pack closing tie and the surgeon's knot locking knot. Ensure the surgeon's knot locking knot is properly positioned between the pack closing loops at the 3 o'clock and 6 o'clock position. Insert the index finger of the working hand from bottom to top behind the surgeon's knot locking knot and pull down and out. Ensure it is secure and that the pack closing tie has been properly constructed of one turn **only** of 1/4-inch cotton webbing. Let the pack closing tie pop off the end of your finger, stand up directly behind the jumper, and issue the command TILT YOUR HEAD FORWARD.

9-12. BALLISTIC HELMET (BACK)

Place both hands on the rim of the ballistic helmet on the extreme left-hand side, fingers and thumb extended and joined, fingers pointing skyward, and palms facing the jumper.

a. The left hand is the control hand and the right hand is the working hand. With the working hand, trace the rim of the ballistic helmet across to the opposite side, ensuring there are no sharp or protruding edges that could cut or damage the jumper's static line upon exiting the aircraft.

b. Once the hands are parallel, place the thumb of each hand on the rim of the ballistic helmet and tilt the jumper's head forward. Look at the parachutist's retention strap to ensure that it is properly routed between the helmet shell and the foam impact pad and that it is not twisted.

c. With the index finger and thumb of either hand, pinch off the foam impact pad and pull down slightly to ensure that it is secured to the shell of the ballistic helmet and that it is serviceable.

9-13. ADVANCED COMBAT HELMET (BACK)

Place both hands on the left side of the ACH with the fingers extended and joined, fingers pointing skyward, with the palms facing the jumper. The left hand is the control hand and the right hand is the working hand.

a. With the working hand, trace the rear of the ACH checking for sharp or protruding edges that could cut or fray the static line. Once both hands are parallel with each other, tilt the jumper's head forward.

b. Place the right index finger on the ladder lock to ensure the back adjustment strap is properly routed through the ladder lock, is free of all cracked components, and is serviceable. Trace the back adjustment strap forward until contact is made with the chin strap to ensure the back strap is not twisted, cut, or frayed. Place the left index finger on the left ladder lock and conduct the same inspection.

c. Conduct a visual inspection of the nape pad to ensure it is present, secure, and serviceable.

NOTES:
1. There are four different configurations for the individual pads inside the ACH.
2. Ensure there is a minimum of seven pads inside the ACH and four pads are covering the locking nuts.
3. The pads in the rear should extend slightly past the bottom of the ACH.

9-14. RISER ASSEMBLIES

Reach as far forward over the jumper's shoulders as possible and, with each hand, grasp a riser assembly just above the canopy release assemblies.

a. Give the left riser assembly a sharp tug to the rear. Open the hand to form an "L." Apply upward pressure with the left thumb and trace the riser assembly rearward to where it disappears into the main pack tray, ensuring that it is not twisted, cut, torn, or frayed. Leave your left hand in place and repeat the same procedure for the right riser assembly.

b. Ensure a log record book is present in either riser assembly.

9-15. PACK TRAY

An overall inspection of the pack tray is conducted to ensure the pack tray is free of grease, oil, dirt, mud, or tears. Place both hands on the top left corner of the pack tray, palms facing the pack tray. The left hand is the control hand and the right hand is the working hand.

a. With the head and eyes 6 to 8 inches away from the working hand, trace across the top pack closing flap, down the right pack closing flap, across the bottom pack closing flap, flip the right hand over, and trace up the left pack closing flap.

b. When the working hand makes contact with the control, raise the control hand out of the way and trace across the top left corner of the pack tray where the control hand had been. Form knife edges with both hands, palms facing the JM, and issue the command ARCH YOUR BACK.

9-16. DIAGONAL BACK STRAPS

Insert each hand under the "X" formed by the diagonal back strap.

a. Look at the diagonal back straps to ensure they have been properly routed over the appropriate shoulder and that the top diagonal back strap has one more row of exposed stitching than the one on the bottom

b. Look at the diagonal back strap retainers to ensure that they are routed through the sizing channels in the diagonal back straps. The diagonal back strap retainers are routed around the diagonal back strap keepers and the pull-the-dot fasteners are secured.

c. With both thumbs, apply upward pressure to the pull-the-dot fasteners to further ensure that they are secure. Focus attention on the left hand and the left side of the jumper.

d. With the left hand, trace down the diagonal back strap to the back strap adjuster, ensuring it is not twisted, cut, or frayed. Grasp the back strap adjuster with the left hand and focus attention on the jumper's right side.

e. With the right hand, trace the diagonal back strap, ensuring that it is not twisted, cut, or frayed; bypass the back strap adjuster; and continue the inspection of the horizontal back strap.

9-17. HORIZONTAL BACK STRAP

The JM picks up the horizontal back strap.

a. Trace the horizontal back strap down to where it disappears into the main lift web. Withdraw the right hand from under the horizontal back strap and reinsert it (fingers and thumb extended and joined, fingers pointed skyward, palms facing the JM) from bottom to top behind the horizontal back strap where it reemerges from the main left web. Issue the jumper the command, BEND FORWARD AT THE WAIST.

b. Release the back strap adjuster and, with the left hand, reach across the pack tray and grasp the center of the bottom pack closing flap. With the head and eyes 6 to 8 inches from the working hand, trace the horizontal back strap across the jumper's back, ensuring that it is not twisted, cut, or frayed. Ensure that the horizontal back strap is properly routed through both horizontal back strap retainers, the horizontal back strap retainers are routed around the horizontal back strap keepers, and the pull-the-dot fasteners are secured.

c. Continue tracing the horizontal back strap to where it disappears into the main lift web on the left side of the jumper. With the left hand, palm facing skyward, regrasp the back strap adjuster on the jumper's left side.

d. Withdraw the right hand from behind the horizontal back strap and reinsert it (fingers and thumb extended and joined, palm facing the JM) from top to bottom or bottom to top behind the horizontal back strap where it reemerges from the main lift web. Trace the horizontal back strap up to where contact is made with the control hand still in place on the back strap adjuster. Withdraw the right hand from behind the horizontal back strap and get left hip to left hip with the jumper.

9-18. SADDLE

Place the fingertips of the right hand (fingers and thumb extended and joined, fingers pointed down, palm facing the jumper) on the accessory attaching ring of the lowering line adapter web or the single "X" boxed stitch on the left main left web.

a. Trace the saddle across the jumper's buttocks, ensuring that the saddle is not inverted, twisted, cut, or frayed, and that neither leg strap has been misrouted around the saddle. Trace the saddle until contact is made with the single "X" boxed stitching on the right main lift web.

b. Give the jumper the good seal of approval by lightly tapping him on the buttocks and issuing the command RECOVER.

9-19. WEAPONS CASE, M1950

The M1950 weapons case will be inspected in its entirety before inspecting the reserve parachute.

a. The inspection of the M1950 weapons case begins with its point of attachment on the left D-ring. Look at the opening gate of the quick-release snap to ensure that the opening gate is facing the jumper's body and is the outermost item on the left D-ring (unless the harness is not equipped with either the lowering line adapter web or triangle links). With the right index finger, finger the opening gate one time to ensure that it is properly attached to the left D-ring, it has spring tension, and it has not been safetied. With the heel of the right hand, press up on the activating arm of the quick-release snap to ensure that it is seated between the ball detents. With the index finger of the right hand, trace down until contact is made with the V-ring. Ensure the quick-release link is routed through the V-ring and is secured by the rotating claw.

b. Continue to trace down the inside of the M1950 weapons case until contact is made with the adjusting strap. Ensure the adjusting strap is routed through the appropriate set of adjusting strap connectors, secured by means of a half hitch, and is not twisted, cut or frayed. Continue tracing down the inside of the M1950 weapons case until the fingers fall off the bottom.

c. Form a knife-edge with the right hand, palm facing skyward, and trace from front to rear along the bottom of the M1950 weapons case to ensure the muzzle of the weapon is not protruding.

d. Place the index finger of the right hand on the slide fastener at the bottom of the closing flap. Ensure the slide fastener is secure by tracing up the outside of the M1950 weapons case to the vicinity of the lift fastener. With the index finger and thumb of the right hand, pinch off the slide fastener and tab thong. Pull down and out to ensure the slide fastener and tab thong are secured by the upper tie-down tape or have been separated over the lift fastener.

e. Drop the right hand down 10 to 12 inches from the top of the M1950 weapons case and give it a sharp slap, feeling for the forward assist of the M16-series rifle or the charging handle of the M249.

f. With the index finger and thumb of the right hand, pinch off the bow knot of the upper tie-down tape on the front of the M1950 weapons case. Visually inspect the upper tie-down tape to ensure it is properly routed behind the M1950 weapons case, around the main lift web, above the chest strap, and secured by a bow knot. This concludes the inspection of the M1950 weapons case.

g. With the left hand, grasp the top carrying handle of the reserve parachute, palm facing the reserve parachute, and lift up and out. Place the index finger of the right hand on the inner mounting screw of the L-bar connector link to ensure that it is present and tightly secured. Rotate the right index finger around to the outer mounting screw of the L-bar connector link and conduct the same inspection.

h. Place the right index finger behind the main lift web from the outside to the inside and feel for the exposed metal of the L-bar connector link. Inspect the reserve parachute in the same manner as if it were on a Hollywood jumper and issue the jumper the command HOLD.

9-20. ALICE PACK WITH HARNESS, SINGLE-POINT RELEASE, AND HOOK-PILE TAPE LOWERING LINE

After the JM has completed the overall inspection of the reserve parachute, he then inspects the ALICE pack.

a. With the left hand, secure the snap hook for the right adjustable D-ring attaching strap. With the right hand, secure the snap hook for the left adjustable D-ring attaching strap. Begin the inspection with the right adjustable D-ring attaching strap. Look at the snap hook to ensure that the opening gate is facing the jumper's body and it is attached to the outside of the right connector snap. With the index finger of the left hand, finger the opening gate one time to ensure that it is properly secured to the right D-ring and has spring tension. With the left thumb, flip the free-running end of the right adjustable D-ring attaching strap out of the way.

b. Place the index of the left hand on the front of the right adjustable D-ring attaching strap just below the snap hook. Trace down the right adjustable D-ring attaching strap until contact is made with the white attaching loop, ensuring that it is not twisted, cut, or frayed. Once contact is made with the white attaching loop, bypass the triangle link and pick up the inspection of the white attaching loop in front of the triangle link.

c. With the left index finger, trace down the attaching loops to ensure that the white attaching loop is routed from bottom to top through the triangle link, the green attaching loop is routed from bottom to top through the white attaching loop, the red attaching loop is routed from bottom to top through the green attaching loop, and all are routed from bottom to top through the grommet in the female portion leg strap release assembly. Place the index finger of the left hand on the single "X" boxed stitch on the release handle cross strap.

d. Look at the release handle cable where it emerges from the release handle cross strap. Ensure the release handle cable is properly routed through the red attaching loop and secured by the cable loop retainer. Leave the left index finger in place and, with the right hand; conduct the same inspection on the left adjustable D-ring attaching strap. After inspecting the left adjustable D-ring attaching strap, focus on the release handle.

e. With the right hand, lift up on the release handle. Ensure the release handle is properly routed between the two plies of the release handle cross strap and secured by the hook-pile tabs.

f. Form knife edges with each hand, palms facing each other, and place them on the equipment retainer straps. Trace the equipment retainer straps down between the external cargo compartments of the ALICE pack until you make contact with the adjustable cross strap.

g. Leave the left hand in place and with the right hand grasp the free-running end of the adjustable cross strap and give it a tug to the jumper's left, ensuring that all the slack has been removed from the adjustable cross strap. Raise the ALICE pack to the jumper and issue the command HOLD.

h. Reform the knife edges with each hand, palms facing each other, and place them on the equipment retainer straps where they come out of the cushioned envelope at the top of the ALICE pack. Trace the equipment retainer straps down, bypassing the girth hitch formed by the hook-pile tape lowering line, until contact is made with the friction adapters. Look at the friction adapters to ensure the equipment retainer straps are properly routed through the friction adapters.

i. Simultaneously, place the index finger and middle finger of each hand on top of the quick releases to ensure they are no more than three fingers and no less than two. With the thumbs of each hand, simultaneously pick up the free-running ends of the equipment retainer straps. Look at them to ensure that they are "S"-folded or accordion folded and not rolled and are secured by two turns of retainer band.

j. With the left hand, palm facing the ALICE pack, grasp the hook-pile tape lowering line in the vicinity of the girth hitch. Look at and pull up on the girth hitch to ensure it is properly routed from north to south and not east to west to the "X" formed by the equipment retainer straps.

k. Place the index finger of the right hand on the hook-pile tape lowering line so there is skin-to-skin contact with the control hand. Begin tracing the hook-pile tape lowering line ensuring that it is routed over the left adjustable shoulder-carrying strap. Continue to trace the hook-pile tape lowering line until contact is made with the first set of hook-pile tabs. Ensure the hook-pile tabs are secured and no folds of the hook-pile tape lowering line are protruding from the retainer flap.

l. Continue to trace down the hook-pile tape lowering line ensuring that it is secured to the tubular portion of the frame by two retainer bands. Continue tracing the hook-pile tape lowering line until contact is made with the second set of hook-pile tabs. Ensure they are properly secured and no folds of the hook-pile tape lowering line are protruding from the retainer flap. Continue to trace the hook-pile tape lowering line until it disappears behind the nylon chaff of the M1950 weapons case. Leave the right hand in place and, with the left hand, release the hook-pile tape lowering line.

m. With the left hand, grasp the trail edge of the M1950 weapons case and pull it forward. With the right index finger, pick up the hook-pile tape lowering line on the backside of the nylon chaff and trace it up to its point of attachment. Once the hook-pile tape lowering line has been traced to its point of attachment, look at the ejector snap to ensure the opening gate is facing the jumper's body.

n. With the right thumb, press in on the activating lever to ensure that it is properly seated over the ball detent and free of all foreign matter. Turn the ejector snap 1/4-turn out to ensure the small tooth is present.

o. Visually inspect the yellow safety lanyard to ensure that it is serviceable and has not been wired, tied, or taped down.

p. Place the index finger of the right hand on the surgeon's knot locking knot to ensure that the lowering line adapter web has been tied off to the main lift web.

q. With the right index finger, trace up the lowering line adapter web until it disappears behind the waistband, ensuring that it is not twisted, cut, or frayed, or misrouted in any manner.

r. Move back to the front of the jumper and issue the command SQUAT. Continue to inspect the remainder of the jumper in the same manner as a Hollywood jumper.

9-21. JMPI OPTIONS WITH COMBAT EQUIPMENT

The jumpmaster is responsible for inspecting each jumper and item of equipment as outlined in paragraphs 9-3 through 9-20. Long flights, heavy loads, extreme heat, tactical updates, and special items of equipment (DMJP, STINGER, and so on) may warrant delaying the attachment of equipment until a predesignated time later in the mission timeline. Under such conditions, the airborne commander may decide to conduct the JMPI without equipment. The commander **must** consider individual and unit proficiency when conducting the command risk assessment. When equipment is attached after the JMPI, a jumpmaster should attach and inspect the equipment.

9-22. MOLLE

Rigging and JMPI procedures for the MOLLE are contained in this paragraph.

a. **Rigging Procedures.** Rigging procedures include preparing the MOLLE for rigging and securing the HSPR to the MOLLE.

(1) ***Preparation.*** Prepare the MOLLE for rigging as follows:

(a) Remove the excess slack from both adjustable shoulder straps.

(b) Secure all excess webbing (roll and tape).

(c) Attach two retainer bands to the left side of the frame.

(d) When jumping special items of equipment (60-mm modified M1950 weapons case, AT4JP, DMJP, SMJP, or side-mount AIRPAC) or any other equipment that exceeds

30 inches in width, either empty the outer pockets or remove them and place them inside the MOLLE.

(2) ***Secure the HSPR to the MOLLE.*** Both sides of the HSPR are prepared in the same manner. Secure the HSPR to the MOLLE as follows:

(a) Place the HSPR on a smooth flat surface with the attaching loops facing up and the equipment retainer straps toward the jumper. Remove all twists or turns from the equipment retainer straps and the male portion leg strap release assembly.

(b) Route the release handle assembly between the two plies of the release handle cross strap from bottom to top (release handle cable first). Secure the release handle assembly in place with the hook-pile tape.

(c) Place the adjustable D-ring attaching strap alongside the HSPR so the opening gate of the snap hook is facing downward and the triangle link is over the white attaching loop.

(d) Route the white attaching loop from bottom to top through the triangle link; route the green attaching loop from bottom to top through the white attaching loop; route the red attaching loop from bottom to top through the green attaching loop and then through the grommet on the female portion of the leg strap release assembly. Ensure the cable loop retainer is facing upward.

(e) Route the release handle cable through the red attaching loop and the cable loop retainer.

(f) Rotate the HSPR over (so the bottom is on top) and remove all twists and turns. Ensure all running ends of webbing on the MOLLE are properly secured with an elastic keeper or retainer band.

(g) With the frame side facing up, place the MOLLE upside down with the bottom of the pack toward the adjustable D-ring attaching straps on the HSPR.

(h) Route the equipment retainer straps through the carrying strap on the top of the pack (bottom, as rigged), under the top horizontal support of the frame (between the shoulder straps), and over the back pad. Cross the retainer straps forming an "X" pattern at the back of the MOLLE.

(i) From the bottom of the MOLLE (top, as rigged), route the two friction adapters between the frame and the waist belt. Secure the retainer straps to their appropriate friction adapter. Form a quick release in each equipment retainer strap.

(j) Adjust the HSPR before tightening. On the front of the MOLLE, ensure the equipment retainer straps are routed on each side of the pouch. Ensure the white attaching loops are in line with the center of the bottom portion of the MOLLE (top, as rigged).

(k) Tighten the equipment retainer straps by pulling on the lower looped portion of the quick releases using a see-saw method. After tightening, ensure the quick releases are accessible for the JM to inspect during JMPI.

(l) Adjust the length of the quick releases so they are no shorter than two fingers width and no longer than three fingers width.

(m)S-fold and secure the equipment retainer straps with masking tape **or** retainer bands—always one, never both. Do not secure the quick releases to the S-folds.

(n) Route the male portion of the leg strap release assembly from the point where it is sewn to the equipment retainer strap by the most direct route down the side of the MOLLE and attach it to the female portion of the leg strap release assembly forming the adjustable leg strap.

(o) Tighten and S-fold, or roll, the excess webbing and secure in the webbing retainer. Tighten both adjustable shoulder-carrying straps and secure excess webbing.

(p) Secure any exposed items with 1/4-inch cotton webbing, or tape.

(r) Bend each side of the waist belt toward the frame and secure to the frame with a retainer band or 1/4-inch cotton webbing.

b. **JMPI Procedures.** After the JM has completed the inspection of the reserve parachute, he lifts up on the reserve parachute and issues the command HOLD. He then inspects the MOLLE as follows:

(1) With the left hand, secure the snap hook on the right adjustable D-ring attaching strap. With the right hand, secure the snap hook on the left adjustable D-ring attaching strap. Begin the inspection with the right adjustable D-ring attaching strap.

(a) Look at the snap hook to ensure that the opening gate is facing the jumper's body and is the outermost item on the right D-ring. With the left index finger, finger the opening gate to ensure that it is properly secured to the right D-ring, and it has spring tension.

(b) With the left thumb, flip the free-running end of the right adjustable D-ring attaching strap out of the way.

(c) With the index and middle finger of the left hand behind the right adjustable D-ring attaching strap, trace down the right adjustable D-ring attaching strap until contact is made with the white attaching loop, ensuring that strap is not twisted, cut, or frayed.

(d) Once contact is made with the white attaching loop, bypass the triangle link and pick up the inspection of the white attaching loop in front of the triangle link. With the left index finger, trace down the attaching loops to ensure that the white attaching loop is routed from bottom to top through the triangle link, the green attaching loop is routed from bottom to top through the white attaching loop, and the red attaching loop is routed from the bottom to the top through the green attaching loop and then through the grommet in the female portion of the leg strap release assembly. The only way to ensure that the red attaching loop has been routed properly is to pull back on the female portion of the leg strap release assembly and look behind it.

(e) Place the left index finger on the single "X" boxed stitch on the release handle cross strap just below the female portion of the leg strap release assembly. Look at the release handle cable where it comes out of the release handle cross strap to ensure that it is routed through the red attaching loop and is secured in the cable loop retainer.

(f) Leave the left hand in place and focus attention on the right hand and left adjustable D-ring attaching strap. Inspect it in the same manner.

(2) Once the inspection of the left adjustable D-ring attaching strap has been completed, focus attention on the release handle. With the right hand, gently pick up on the release handle assembly and ensure that it has been routed through the two plies of the release handle cross strap and is secured by the hook-pile tab provided.

(a) Form a knife edge with each hand, palms facing each other, and place them on the single "X" boxed stitch. Simultaneously trace down the equipment retainer straps to the outside of the pouch of the MOLLE pack until contact is made with the adjustable cross strap.

(b) Leave the left hand in place. With the right hand grasp the free-running end of the adjustable cross strap and give it a sharp tug to the jumper's left, ensuring that all of the slack has been taken out of the adjustable cross strap.

(c) Continue to trace down the equipment retainer straps until your fingers fall off the bottom.

(3) Raise the MOLLE pack up and visually inspect the equipment retainer straps to ensure they are routed through the top carrying strap, under the top horizontal support frame, and over the back pad. Then issue the next command of HOLD.

(a) Reform knife edges with each hand, palms facing each other, and place them on the equipment retainer straps where they reemerge under the support frame on top of the back pad. Simultaneously trace down the equipment retainer straps bypassing the girth hitch formed by the hook-pile tape lowering line, and trace all the way down to the friction adapters. Look at the friction adapters to ensure that the equipment retainer straps have been properly routed through them.

(b) Simultaneously place the index and middle finger of each hand on top of the two- to three-finger width quick releases in the equipment retainer straps to ensure that they are no more than three fingers width and no less than two fingers width.

(c) Simultaneously pick up the free-running ends of the equipment retainer straps with your thumbs. Look at them to ensure that they are S-folded (or accordion folded) and secured by at least two turns of a retainer band **or** masking tape—always one, never both.

(4) Next, secure the free-running ends of the adjustable shoulder-carrying straps and conduct a visual inspection ensuring that they are S-folded (or accordion folded) and secured with a retainer band **or** masking tape—always one, never both.

c. **Hook-Pile Tape Lowering Line.** Inspect the hook-pile tape lowering line as follows:

(1) With the left hand (palm facing the MOLLE pack, thumb down), grasp the hook-pile tape lowering line just above the girth hitch. This is the control hand. Look at and pull up on the girth hitch to ensure it is properly secured to the "X" formed by the equipment retainer straps, from north to south, and is centered on the back pad.

(2) Place the index finger of the right hand on the hook-pile tape lowering line directly beside the control hand making skin-to-skin contact. Begin tracing the hook-pile tape lowering line to ensure that it is routed over the left adjustable shoulder-carrying strap. Continue to trace the hook-pile tape lowering line until contact is made with the first hook tab.

(3) Ensure that the first hook tab is properly secured and no folds of the hook-pile tape lowering line are protruding out of the retainer flaps. Continue to trace down the hook-pile tape lowering line ensuring that it is secured to the frame of the MOLLE by a retainer band on each end of the retainer flap. Continue tracing the hook-pile tape lowering line until contact is made with the second hook tab to ensure that it is properly secured and no folds of the hook-pile tape lowering line are protruding out of the retainer flaps. Continue to trace the hook-pile tape lowering line until it disappears behind the nylon chafe on the M1950 weapons case.

(4) Release the hook-pile tape lowering line with the left hand and grasp the trail edge of the M1950 weapons case and pull it forward.

(5) With the right index finger, continue the inspection of the hook-pile tape lowering line on the backside of the nylon chafe and trace it up to its point of attachment.

(6) Once the hook-pile tape lowering line has been traced to its point of attachment, look at the ejector snap to ensure the opening gate is facing the jumper's body. With the

right thumb, press in on the activating lever to ensure that it is properly seated over the ball detent and free of all foreign matter. With the right index finger and thumb, pinch off the yellow safety lanyard and trace down to ensure that it is serviceable and has not been wired, tied, or taped down.

(7) With the right index finger, pull out on the accessory attaching ring of the lowering line adapter web to ensure that it has been tied off to the main lift web. With the right index finger, trace up the lowering line adapter web until it disappears behind the waistband ensuring that it is not twisted, cut, frayed, or misrouted in any manner.

(8) Move back to the front of the jumper, issue the command SQUAT, and inspect the remainder of the jumper in the same manner as a Hollywood jumper.

9-23. JMPI SEQUENCE FOR AIRPAC

The jumpmaster inspects the AIRPAC as follows:

NOTE: This JMPI sequence is for the front-mounted AIRPAC when jumped with a M1950 weapons case.

a. Place the left hand on the snap hook for the right D-ring attaching strap and the right hand on the snap hook for the left D-ring attaching strap. Leave the right hand in place and focus attention on the left hand and the right D-ring attaching strap.

(1) With the index finger of the left hand, finger the opening gate to ensure it has spring tension and is properly attached to the right D-ring to the outside of the connector snap. Place the index and middle finger of the left hand from outside to inside behind the nylon portion of the D-ring attaching strap and trace down until the middle finger makes contact with the circular ring, ensuring the strap is not twisted, cut, or frayed.

(2) Bypass the circular ring and place the index finger on the red attaching loop. Visually inspect the circular release ring to ensure it has been routed through the circular ring and folded back against the nylon portion of the D-ring attaching strap. Ensure the red attaching loop has been routed bottom to top through the circular release ring and then through the grommet on the female portion of the leg strap release assembly. With the thumb and index finger, peel back on either the top or bottom corner of the female portion of the leg strap release assembly to further ensure the red attaching loop has been properly routed.

(3) Place the index finger of the left hand on the single "X" boxed stitch just below the circular ring. Conduct a visual inspection of the release handle cable to ensure it is properly routed under both plies of the release handle cross strap and through the red attaching loop. Conduct the same inspection on the left D-ring attaching strap.

b. Leave the left index finger in place and, with the right hand, gently lift up on the release handle to ensure it has been properly routed under both plies of the release handle cross strap and has been secured by the hook-pile tabs.

c. Form a knife edge with both hands, palms facing each other, and place them on the outer edge of the equipment retainer straps. Simultaneously trace down the equipment retainer straps until contact is made with the friction adapters. Leaving both hands in place, look at the friction adapters, ensuring the equipment retainer straps have been properly routed through them and the free-running ends are secured by means of the webbing retainers.

d. With either hand, carefully open the circular closing flap and hold it up and out of the way. This hand is now the control hand. Place the index finger of the working hand on the single "X" boxed stitch on the 6-o'clock set of equipment retainer straps. Visually inspect both friction adapters to ensure the free-running ends of the equipment retainer straps are properly routed through them and are secured by means of the webbing retainers.

e. Conduct the same inspection for the 9-o'clock and 3-o'clock set of friction adapters. Place the index finger of the working hand on the eyelet of the cotter pin.

f. Ensure that the white grommet securing loop has been routed through all three grommets, is not cut or frayed, and has been secured by the cotter pin.

g. Close the circular closing flap, lift the AIRPAC to the jumper, and issue the command HOLD.

h. With the left hand, grasp the hook-pile tape lowering line, palm facing the jumpmaster. Pull up if the short bridal is attached horizontally, or to the jumper's left if it is attached vertically.

i. Insert the index finger and middle finger of the right hand, from the jumper's left to right, behind the locking D-ring. Ensure that it is properly attached to the triangle link and is locked into place.

j. Trace up the short bridal, ensuring it is not twisted, cut, or frayed, until contact is made with the girth hitch of the hook-pile tape lowering line.

k. Place the index finger of the right hand on the girth hitch, ensuring the cotton buffer is present.

l. Place the index finger and middle finger of the right hand from the jumper's left to right behind the short bridal on the opposite side of the girth hitch. Trace the short bridal until contact is made with the triangle link, ensuring it is not twisted, cut, or frayed, paying special attention to the point of attachment at the triangle link.

m. Leave the left hand in place and put the index finger of the right hand on the hook-pile tape lowering line where it emerges from the lowering line stow pocket. Ensure that no excess of the hook-pile tape lowering line is protruding from the retainer flap. Trace the hook-pile tape lowering line until the index finger disappears behind the nylon chaff portion of the M1950 weapons case ensuring it is not cut, frayed, or burned.

n. Leave the right index finger in place and, with the left hand, grasp the trail edge of the M1950 weapons case and pull it toward him. Remove the right index finger and rotate it around to the backside of the M1950 weapons case and continue the inspection of the hook-pile tape lowering line where it reemerges from the nylon chaffing.

o. Trace the hook-pile tape lowering line to its point of attachment.

p. With the thumb of the right hand, push in on the activating lever of the ejector snap to ensure it is properly seated over the ball detent. Ensure it is free of any foreign matter, and the opening gate is facing the jumper's body.

q. With the thumb and index finger of the right hand, pinch off the yellow safety lanyard and trace it until his fingers fall off the end, ensuring it is not wired, tied, or taped.

r. Move to the front of the jumper and issue the command SQUAT.

9-24. TECHNICAL INSPECTION FOR SIDE-MOUNT AIRPAC

The side-mount container is technically inspected and attached at the 20-minute time warning.

a. Ensure all visible container retaining straps are properly secured to their appropriate friction adapter. (All visible container retaining straps have a five- to six-finger quick release.)

NOTE: Five- to six-finger quick releases will depend on the item of equipment inside the side-mount AIRPAC.

b. Ensure the free-running ends of the container retaining straps are S-folded and secured to the quick-release loop by two turns of a retainer band.

c. Ensure the slide fastener is secure.

d. Ensure there is no excess in the fabric flap.

e. Ensure packing material is present, if required for particular weapon system.

9-25. M82, MEDIC JUMP PACK

Inspect the M82, medic jump pack (MJP) as follows:

a. With the left hand, secure the snap hook for the right adjustable D-ring attaching strap. With the right hand, secure the snap hook for the left adjustable D-ring attaching strap.

b. Begin the inspection with the right adjustable D-ring attaching strap. Look at the snap hook to ensure that the opening gate is facing the jumper's body and is the outermost item on the right D-ring. With the left index finger, finger the opening gate to ensure it is properly secured to the right D-ring and has spring tension. With the left thumb, flip the free-running end of the right adjustable D-ring attaching strap out of the way.

(1) With the index and middle finger of the left hand behind the right adjustable D-ring attaching strap, trace down the right adjustable D-ring attaching strap until contact is made with the white attaching loop, ensuring it is not twisted, cut, or frayed.

(2) Once contact is made with the white attaching loop, bypass the triangle link and continue the inspection of the white attaching loop in front of the triangle link. With the left index finger, trace down the attaching loops to ensure the white attaching loop is routed from bottom to top through the triangle link; the green attaching loop is routed from bottom to top through the white attaching loop; and the red attaching loop is routed from the bottom to top through the green attaching loop and then through the grommet in the female portion of the leg strap release assembly.

(3) Pull back on the female portion of the leg strap release assembly and look behind it to further ensure the red attaching loop is routed properly. Place the index finger of the left hand on the single "X" boxed stitch on the release handle cross strap just below the female portion of the leg strap release assembly.

(4) Look at the release handle cable where it comes out of the release handle cross strap to ensure it is routed through the red attaching loop protruding out of the grommet in the female portion of the leg strap release assembly and is secured in the cable loop retainer.

(5) Leave the left hand in place and focus attention on the right hand and the left adjustable D-ring attaching strap and inspect it in the same manner.

c. After inspecting the left adjustable D-ring attaching strap, focus attention on the release handle. With the right hand, gently pick up on the release handle and ensure that it has been routed through the two plies of the release handle cross strap and is secured by the hook-pile tab provided.

d. Form a knife edge with each hand, palms facing each other, and place them on the equipment retainer straps. Simultaneously trace down the equipment retainer straps until contact is made with the adjustable cross strap. Once contact has been made with the adjustable cross strap, leave the left hand in place.

e. With the right hand, grasp the free-running end of the adjustable cross strap and give it a sharp tug to the jumper's left ensuring that all slack has been taken out the adjustable cross strap.

f. Raise the M82 to the jumper and issue the command HOLD.

g. Re-form knife edges with each hand, palms facing each other, and place them on the equipment retainer straps where they come out of the cushioned envelope on the top of the M82. Simultaneously, with both hands, trace down the equipment retainer straps, bypassing the girth hitch formed by the hook-pile tape lowering line and trace all the way down to the friction adapters.

h. Look at the friction adapters to ensure the equipment retainer straps have been properly routed through them. Place the index and middle fingers of each hand on top of the two- to three-finger quick releases in the equipment retainer straps to ensure they are not more than three fingers and no less than two. Simultaneously pick up the free-running ends of the equipment retainer straps and ensure they are S-folded (or accordion folded) and have not been rolled. Ensure they are secured by two turns of a retainer band.

i. With the left hand (palm facing the M82, thumb down), grasp the hook-pile tape lowering line just above the girth hitch.

j. Look at, and pull up on, the girth hitch to ensure it is properly routed south to north through the "X" formed by the equipment retainer straps.

k. Place the right hand index finger on the hook-pile tape lowering line directly beside the control hand. Begin tracing the hook-pile tape lowering line ensuring it is routed over the left adjustable shoulder-carrying strap. Trace the hook-pile tape lowering line until contact is made with the first hook tab. Ensure the first hook tab is properly secured and no folds of the hook-pile tape lowering line protrude from the retainer flap.

l. Continue to trace down the hook-pile tape lowering line ensuring it is secured to the tubular portion of the frame by retainer bands. Continue tracing the hook-pile tape lowering line until contact is made with the second hook tab. Ensure the second hook tab is properly secured and no folds of the hook-pile tape lowering line protrude from the retainer flap. Continue to trace the hook-pile tape lowering line until the right index finger disappears behind all three plies of nylon chaffing on the M1950 weapons case.

m. Release the hook-pile tape lowering line with the left hand and grasp the trailing edge of the M1950 weapons case and pull it forward.

n. With the right index finger, pick up the hook-pile tape lowering line on the back side of the three plies of nylon chaffing and trace it to its point of attachment. Look at the ejector snap to ensure the opening gate faces the jumper's body.

o. With the right thumb, press in on the activating lever to ensure it is properly seated over the ball detent and free of foreign matter.

p. With the right index finger and thumb, pinch off the yellow safety lanyard and trace down to ensure it is serviceable and has not been wired, tied, or taped down.

q. Move back to the front of the jumper and inspect the remainder of the jumper in the same manner as a Hollywood jumper.

Section III. MOVEMENT ON THE AIRFIELD

After personnel inspection, safety personnel load the parachutists aboard the aircraft. Load time is the time agreed on by the Army and Air Force for loading the aircraft. Station time is the time the aircrew, parachutists, and equipment are inside the aircraft and are prepared for takeoff (everyone seated with seat belts fastened and helmets on). (See Table 9-1.)

9-26. AIRFIELD MOVEMENT PROCEDURES

These procedures must be followed when moving parachutists on or across an active airfield:

a. Coordinate with DACO for permission to cross the airfield.

b. Keep parachutists in closed formation.

c. Cross active runways at authorized crossing points.

d. Cross only on light signals from airfield control tower:

- Green (GO).
- Red (STOP, DO NOT PROCEED).
- Flashing red (CLEAR TAXI STRIPS AND RUNWAYS).
- Flashing red and green (EMERGENCY WARNING, BE ALERT).
- Flashing white (RETURN TO STARTING POINT).

e. Display checkered flags on the first and last escort vehicles.

f. Keep vehicles in low gear while crossing runways.

g. Do not raise radio antennas within 50 feet of any aircraft.

h. Do not smoke in the vicinity of any aircraft.

i. Avoid aircraft propellers.

j. Avoid jet engine intakes/exhausts. Stay about 50 feet from intakes and 200 feet from exhausts.

9-27. LOADING AIRCRAFT

Parachutists are loaded in the aircraft in reverse order. During loading, safety personnel move forward in the aircraft ahead of the chalk and supervise seating of the chalk to ensure that all seats are filled, seat belts are fastened, and that personnel are in proper stick order. They also assist in loading equipment aboard the aircraft. The aircrew briefing may be given before or after loading the aircraft but must be completed before takeoff.

NOTE: The actions taken are according to type of aircraft used during the airdrop.

9-28. IN-FLIGHT EMERGENCY PROCEDURES

Jumpers must be briefed on in-flight emergency procedures described in Table 9-1.

NOTE: Rotary-wing aircraft in-flight emergency procedures are different from those shown in Table 9-1. They are explained by either the PJM or pilot before boarding the aircraft.

SITUATION	SIGNAL	ACTIONS IN USAF AIRCRAFT	ACTIONS IN ARMY AIRCRAFT
Crash landing during takeoff	Continuous ringing of alarm bell or oral warnings	Remain seated until aircraft stops, then exit.	Pull legs inside aircraft; remain in position; cover head with arms.
Crash landing during flight	Six short rings of alarm bell or oral warning	Jump if time permits. If not, secure seat belts; brace for impact on continuous ring; remain seated until aircraft has stopped.	As directed by pilot.
Emergency bailout	Three short bells of alarm bell or oral warning	Stand up and hook up on continuous ring; exit aircraft under direction of PJM.	Exit aircraft under direction of PJM.
Ditching over water with insufficient drop attitude	Six short rings and oral warning	Use available padding, remain seated, and secure seat belts.	Pull legs inside aircraft; remain in position; cover head with arms.
Lighten load	Oral warning	Designated parachutist may assist PJM or loadmaster in jettisoning cargo.	As directed by pilot.
Fire in flight	Oral warning	Remove parachutists from fire area and extinguish fire.	As directed by pilot.

Table 9-1. In-flight emergency procedures.

CHAPTER 10

JUMPMASTER AND SAFETY DUTIES IN FLIGHT

This chapter provides an overview of the duties of the jumpmaster, assistant jumpmaster, and safety personnel during flight. For aircraft-specific personnel duties, see Part Four, Aircraft Used in Airborne Operations.

Section I. PRIMARY JUMPMASTER, SAFETY PERSONNEL, AND ASSISTANT JUMPMASTER

After aircraft takeoff, the JM must remain oriented to aircraft position at all times and keep the parachutists informed of any deviations from flight plan. He may coordinate with the navigator or use strip maps and checkpoints. He also remains in communication with the pilot. This is performed by relay through the loadmaster over the interphone. On Army aircraft, the JM or safety personnel should wear a flight helmet or headset for direct communication with the pilot and to monitor the ground control element. If the JM/safety cannot wear a flight helmet or headset, communication can be made through the crew chief.

10-1. PRIMARY JUMPMASTER

General rules stress that the JM must—

- Not sacrifice safety for any reason.
- Rehearse JM procedures on the ground.
- Hook up before opening jump door(s) or ramp.
- Face the open jump door or tailgate when in flight.
- Maintain a firm handhold on the aircraft when working in or close to the open jump door or ramp.
- Not allow anyone in or near the open jump door without a helmet (or equivalent) and safety harness or parachute.

The JM must remember that jumpers' eyes require 20 minutes to adjust to lower light levels before night operations. Therefore, only red compartment lights or night vision imaging systems (NVIS) lighting will be used 20 minutes before drop time. This allows for physical adjustment and minimizes the impairment of a jumper's night vision during the critical first seconds of a night drop. If the white light is substituted for the red light before drop time, the jumpers' eyes must readjust by turning on the red compartment lights for 20 minutes.

10-2. SAFETY PERSONNEL

During flight, safety personnel constantly check on the condition of all parachutists and distribute airsickness bags where needed. They also assist the JM in relocating personnel who are too sick to jump, jump refusals, and other types of no-jumps. If a jump refusal occurs, safety personnel move the parachutist forward in the cargo compartment to be seated. During in-flight rigging missions, safety personnel assist in parachute issue. They operate rigging, JMPI, and correction stations as directed by the JM. The safety double checks to ensure the pilot and loadmaster have designated a specific individual to cut away a towed jumper.

10-3. PRIMARY JUMPMASTER/ASSISTANT JUMPMASTER DUTIES

JM/AJM duties are as follows:

a. Enforce flight rules and regulations.
b. Issue time warnings.
c. Issue jump commands.
d. Perform door safety checks.
e. Perform outside air safety checks.
f. Control exit of all parachutists.
g. Eject door bundles.
h. Perform in-flight rigging mission.

Section II. DOOR PROCEDURES AND DOOR BUNDLE EJECTION

Time warnings, door procedures, and door bundle ejection are events that commonly occur during each airdrop operation. Time warnings and door procedures are completed during each airdrop operation, but door bundles are not always ejected. The time warnings and door procedures are considered critical aspects of JM operations during all airdrop operations.

NOTE: The 20-minute time warning may be increased to 30 minutes aboard the C-141 or C-17 if the JMs need more time to accomplish their duties.

10-4. THE 20-MINUTE TIME WARNING

JMs unbuckle their seat belts and stand up. They move to the rear of the aircraft, turn, and face the parachutists. (All time warnings begin and end at the shoulders with closed fists.) They extend both arms straight forward, extending and spreading the fingers and thumb of each hand. This is repeated twice while the oral command is given.

a. All CWIEs, AT4JPs, and DMJPs are attached to designated parachutists and are inspected. Safety personnel assist the JMs with attaching equipment and ensuring the tie-downs are secure and the lower lines are attached and secure. JMs and safeties unlash door bundles (if any) and move them near the jump door(s). Once positioned, the static line of each cargo parachute is attached to the outboard anchor line cable and the following inspection sequence is completed:

- Static line and clevis (safety wire must be bent so that it has metal-to-metal contact).
- Pack closing tie.
- Drogue device (one for C-130; two for C-141).
- Point of attachment to the bundles (risers).
- Tie-down tapes (one on each corner).
- Pack tray and bundle (for any loose or excess webbing).

b. When the inspection is completed, the JM slaps the smooth side of the bundle and ensures it is facing the trailing edge of the door(s).

NOTE: The 10-minute time warning may be increased to 15 minutes aboard the C-141 or C-17 if the JMs need more time to accomplish their duties.

10-5. THE 10-MINUTE TIME WARNING

When the loadmaster informs the JMs that the aircraft is 10 minutes from the DZ, the JMs unbuckle their seat belts and stand up (if seated). They hook up to the inboard anchor line cable, move to the rear of the aircraft near the jump door, and transfer control of their static lines to safety personnel. They announce, SAFETY, CONTROL MY STATIC LINE, and ensure the safety has positive control before turning to face the parachutists. The JMs begin jump commands at the 10-minute time warning. During this process, the safeties help the JMs hook up, watch for excess static line, and stow any excess static line in the static line slack retainer. After inspection of the JMs' static lines, safeties move out to inspect the sticks, as indicated in the following paragraph.

10-6. FIRST SEVEN JUMP COMMANDS

The JM is now ready to issue the first seven jump commands.

a. After the parachutists are standing, safety personnel inspect the following items on each parachutist while moving forward in the aircraft:

- Waistband for proper quick release.
- Ejector snap on the HPT lowering line for proper attachment.
- Quick-release assembly on the weapons case for proper attachment.
- Adjustable leg straps on harness, single-point release.

NOTE: Safeties must be alert for and correct any excess webbing or loose lowering line.

b. Once the safeties have checked the last parachutist, and after the command HOOK UP, they return to the aft end of the aircraft. While moving to the aft end, safeties check the full length of each parachutist's static line to ensure proper routing and attachment to the anchor line cable. They check with both eyes and hands and stow any excess static line in the static line slack retainer so that each static line is taut from the anchor line cable to the main pack tray. As the safety checks each static line, he reminds each jumper to "make eye-to-eye contact and hand the safety your static line." After he checks the first parachutist and the JM's static line, the safety resumes control of the JM's static line.

c. After the JM receives "All okay, JM," he regains control of his static line from the safety, takes the number 1 parachutist position, and waits for slow-down procedures.

d. Once the aircrew has completed their slow-down checklist (slow aircraft to drop speed, open jump doors, deploy air deflectors, and position jump platforms), the loadmaster relinquishes control of the jump door to the JM by giving the JM a thumbs-up signal. He then moves to the ramp area.

10-7. DOOR SAFETY CHECK

Safety personnel position themselves near the trail edge of the jump door and control the static line for the JM as he performs the door safety check and outside air safety check. He grasps the lead edge of the door with his lead hand and transfers control of his static line back to the safety (Figure 10-1). Then, he grasps the trail edge of the door with his trail hand. Using only the lead hand, he ensures the pip pin is in place, securing the jump door in the open position on the C-130, or that the door-latch catch operates properly on the C-141. (See Chapter 16 for door procedures for the C-17A.)

a. **Jump Platform**. The JM checks the jump platform by—

(1) Tapping the lead down lock with the lead foot and then the trail down lock with the trail foot and visually checking to ensure they are engaged.

(2) Placing his trail foot in the center of the jump platform and shifting his weight to the platform, ensuring the platform can support his body weight. The trail foot remains in place. (If the JM removes his foot from the platform, he must again check the down locks and ensure the platform can support his body weight.) From this rest position, the JM performs all remaining duties in the door until it is time to issue the eighth jump command (or place bundles in the door).

b. **Trail Edge of Door**. With the trail hand, the JM starts at the top of the trail edge of the door, tracing all the way down to the trail down lock, then back to the top to ensure no sharp or protruding edges exist.

c. **Air Deflector**. The JM looks at the air deflector, ensuring it is fully deployed.

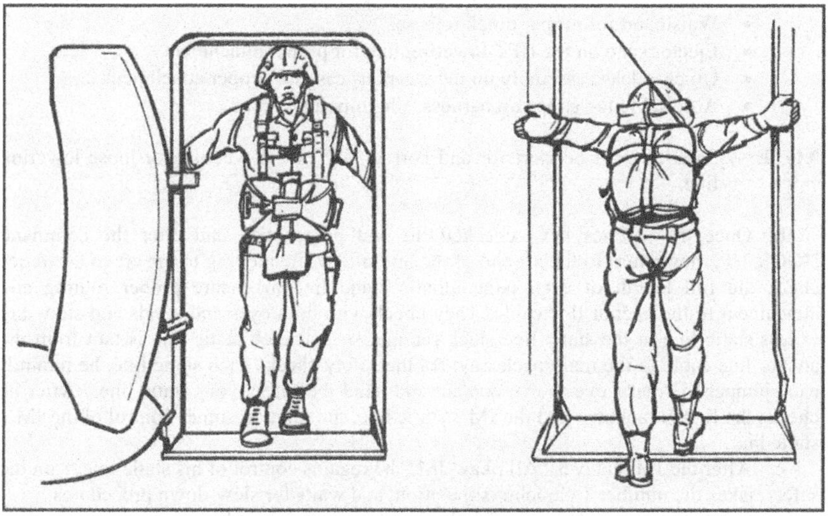

Figure 10-1. JM door position.

10-8. INITIAL OUTSIDE AIR SAFETY CHECK AND CHECKPOINTS

The JM must make outside air safety checks to ensure there are no unsafe conditions outside the aircraft (aircraft in the formation to the rear that are below drop altitude or other low-flying aircraft). From the same rest position, the JM leans far enough outside the aircraft to make a proper air safety check, and visually checks—

- Direction of flight.
- Overhead.

- Rear.
- Straight down.
- Straight to his front.
- Direction of flight.

The JM continues observing outside the aircraft and spotting for checkpoints en route to the DZ. In the absence of checkpoints, he uses additional time advisories from the air crew, which must have been identified during the pilot/JM briefing. The navigator and primary loadmaster should also be present for this briefing. The JM relays checkpoints or time advisories to the parachutists by leaning back inside the aircraft, keeping his foot centered on the platform, facing the parachutists, and issuing the checkpoint or time advisories.

NOTES: 1. When the JM leans back inside the aircraft to assume the rest position, he will not collapse his trail elbow with the lead arm locked. Allowing his trail elbow to collapse could cause accidental MIRPS activation resulting in JM extraction from the aircraft. Such an extraction can result in serious injury or death.

2. The JM may lock his elbows, if deemed necessary, to properly conduct outside air safety checks, or when checking for towed jumpers. However, the JM is not required to lock his elbows when performing those actions.

3. The loadmaster should relay time warnings to the jumpmaster or safety by using hand signals. The exact hand signals must be coordinated between the jumpmaster and loadmaster during the aircrew brief. However, suggested hand signals are as follows: 1-minute time warning--the index finger extended vertically; 30-second time advisory—the index finger and thumb held closely together.

10-9. THE 1-MINUTE TIME WARNING

Approximately one minute from the green light, the safety passes along loadmaster time warnings if the jumpmaster fails to hear them. The JM relays the 1-minute warning to the parachutists by leaning back, keeping his foot centered on the platform, facing the jumpers, and extending the index finger of his lead hand. He announces, ONE MINUTE. The jumpers relay the announcement to the rear. The JM continues observing outside for the 30-second checkpoint, or until he receives a 30-second time advisory from the loadmaster.

10-10. FINAL OUTSIDE AIR SAFETY CHECK

The JM announces, THIRTY SECONDS. The jumpers relay the announcement to the rear. After the JM has given the 30-second checkpoint or relayed a 30-second time advisory, he makes his final outside safety check and, based on his checkpoint or USAF/other service aircraft advisory, determines when the aircraft is about 10 seconds (20 seconds for bundle drops) from the release point. Once again, he leans out and visually checks—

- Direction of flight.

- Overhead.
- Rear (ensures no aircraft in the formation have dropped below jump altitude).
- Straight down.
- Straight to the front.
- Direction of flight (the JM may be able to see the DZ at this time, depending on the aircraft axis of approach).

The safety, positioned at the aft end of the door, controls the JM's static line during all outside safety checks.

NOTE: For a CARP/VIRS DZ, the JM has no responsibility to positively identify the DZ, code letter, or color of smoke. The responsibility to positively identify the DZ, code letter, or color of smoke rests with the USAF/other service aircraft.

a. **Unsafe Condition Outside Aircraft**. If the JM observes any unsafe conditions outside the aircraft, he notifies the loadmaster by a preplanned arm-and-hand signal (for example, a cutting motion across the throat) that a NO DROP situation exists for this pass.

b. **Bundle Drops, 20 Seconds from Release Point**. The JM moves back inside the aircraft and positions the first door bundle on the jump platform so that it is on its balance point with its longest dimension vertically in the door and the parachute on the top or inboard side of the bundle. Then the JM—

(1) Maintains a firm grasp on the bundle with the lead hand and a firm grasp on the trail edge of the jump door with the trail hand. He ensures that the cargo parachute static line is routed above the trail arm.

(2) Keeps his eyes on the jump caution light. When the light turns green, he ejects the door bundle(s), ensuring that it goes straight out and does not tumble.

(3) Maintains a firm grasp with his trail hand and turns toward the cargo area. The JM and AJM make eye-to-eye contact and give each other the thumbs-up signal, indicating that the door bundle has been ejected, that they are not aware of any unsafe conditions, and that they are ready to exit personnel. (Parachutists are not positioned in the door until this is accomplished.) This procedure is followed when ejecting door bundles from one door or both doors, as long as both doors are open.

(4) Moves on line with the lead edge of the jump door and issues the eighth jump command IAW the exit procedures.

c. **No Bundle Drops, 10 Seconds from Release Point**. When door bundles are **not** used on an airborne operation, the JM makes his final check to determine that the aircraft is about 10 seconds out from the RP, based on his checkpoint or USAF/other service aircraft advisory. Then he—

(1) Maintains a firm grasp on the trail edge of the door with the trail hand and turns as far as possible toward the cargo area. The JM and AJM make eye-to-eye contact and give each other the thumbs-up signal, indicating that there are no unsafe conditions outside the aircraft and that they are ready to exit personnel. This procedure is followed when exiting personnel from one or both doors, as long as both doors are open.

(2) Releases the trail edge of the door and issues the eight jump command (keeping his body centered on the trail edge of the door).

10-11. EIGHTH JUMP COMMAND

The JM issues the eighth jump command, STAND BY. The safety positions himself near the trail edge of the door, ensuring he does not impede the flow of jumpers.

a. **Computed Air Release Point**. After issuing the eighth jump command, the JM immediately regains control of his static line from the safety. He backs away from the door toward the middle of the aircraft to allow the safety room to control static lines. The safety receives the first parachutist's static line with his forward hand and secures it with his aft hand by pinning it against the aft anchor line support.

b. **Ground Marking Release System**. After issuing the eighth jump command, the JM continues to spot for the DZ and DZ markings. He backs away from the door toward the middle of the aircraft to allow the safety room to control static lines. When the line of the panels break the trailing edge of the jump door, the JM issues the command, GO. The safety receives the first parachutist's static line with his forward hand and secures it with his aft hand by pinning it against the aft anchor line support. For USAF operations, a safety is not used and either the JM or loadmaster fulfills the duty after coordination.

10-12. NINTH JUMP COMMAND

When the JM issues the ninth jump command, GO, he ensures that he has backed away from the door where he can best control his stick and away from the safety, to avoid confusion or congestion. Safety personnel take static lines while the JM controls the flow of parachutists. The safety grabs each static line with his forward hand, passes it back to his aft hand, and ensures the line is firmly seated against the aft anchor line cable support. After all parachutists have exited the aircraft, the JM and AJM hand off their static lines to the safeties and exit the aircraft. The safety immediately makes a visual check for toward jumpers. He then looks directly at the loadmaster, passes him a thumbs-up signal, and states YOUR DOOR, AIR FORCE.

a. **Red Light Procedures**. If a JM sees a red jump caution light, he sounds off with "RED LIGHT, RED LIGHT, RED LIGHT," and then moves forward to block the flow of the stick. If any parachutist tries to exit on the red light, he will be allowed to exit, except in the case of an unsafe parachutist. No one touches or physically tries to stop a parachutist who is past the leading edge of the door. Parachutists will be stopped only by oral command; this reduces the risk of accidentally activating the jumper's reserve.

b. **Jump Refusals**. When removing a jump refusal from the door, the JM tells the refusal, "GREEN LIGHT, GO; GREEN LIGHT, GO; GREEN LIGHT, GO." If the jumper does not exit, he tells the refusal, "YOU ARE A JUMP REFUSAL AND I AM TAKING YOU OUT OF THE DOOR." Then, the JM should approach the parachutist from the rear and grasp the sides of his parachute. The JM **never** puts his hand in front of the parachutist's face or grabs the back of his ballistic helmet.

(1) The safety warns the refusal that he is being brought back inside the aircraft. Under **NO** condition should the JM reach around in front of the refusal to pull him back into the aircraft; the JM could accidentally hit or grab the reserve rip cord grip.

(2) After the jump refusal is brought inside the aircraft, he is guided as far forward in the aircraft as is possible, seated, buckled up, unhooked, and directed to not touch his equipment. During training, a jump refusal will stop the jump for that pass.

c. **Unsafe Parachutist**. If a parachutist has passed the leading edge of the door and must be stopped because of a misrouted static line or other critical deficiency, the safety approaches the jumper from the back to grab the sides of his parachute. The safety never reaches around in front of the parachutist, puts his hands in front of the parachutist's face, or grabs the back of his helmet.

10-13. TOWED PARACHUTIST (FIXED-WING AIRCRAFT)

Actions to retrieve or cut free a towed parachutist are as follows:

a. **JM Actions**. If the JM observes a towed parachutist, he takes the following actions. The JM—

(1) Stops the stick of parachutists (if applicable).

(2) Notifies the loadmaster, who then notifies the pilot and requests that drop altitude be maintained.

(3) Identifies how the parachutist is being towed. If the parachutist is being towed by anything other than the lowering line or static line, the JM attempts to free him. If the parachutist is being towed by a lowering line, the JM immediately cuts the lowering line, thus freeing the parachutist. If the parachutist is being towed by the static line, the JM initiates recovery procedures.

b. **JM Recommendations**. The JM observes the towed parachutist and recommends whether to retrieve or cut the parachutist free. The recommendation is relayed by the loadmaster to the pilot. The pilot makes the decision. If the decision is to cut the jumper free, the loadmaster will cut the static line on command of the pilot. Priority is as follows:

(1) ***Door***. Safety personnel move the remainder of the stick toward the front of the aircraft. If the parachutist is to be retrieved, the loadmaster uses the Towed Parachutist Retrieval System (TPRS) or installs the retrieval bar on the C-141 or the CGU1-B cargo strap on the C-130, retracts or folds in the jump platform, and initiates retrieval. All personnel stay clear of the door and line of travel of the static line retriever cable. When the parachutist has been retrieved to the door, the JM and safety personnel gain physical control of the parachutist. The loadmaster relieves tension from the static line retriever so that the parachutist can be brought inside the aircraft. The retrieved parachutist is moved all the way forward and is seated. He does not jump again. If the retrieval is unsuccessful and the parachutist must be cut free, the loadmaster cuts the static line.

(2) ***Ramp***. If a parachutist is towed following a ramp exit and is to be cut free, the loadmaster partly retrieves the static lines to reach the towed parachutist's static line in order to cut it. If the parachutist is to be retrieved, the loadmaster uses the Towed Parachutist Retrieval System (TPRS) or installs a CGU1-B cargo strap (C-130) about 5 1/2 feet above the ramp. The static lines are retrieved over the CGU1-B strap. As the parachutist is retrieved to the ramp area, the JM and safety personnel gain physical control of him. The parachutist is pulled into the aircraft (under the strap) as the loadmaster relieves tension from the static line retriever cable. The parachutist is moved all the way forward and is seated; he does not jump again.

c. **Modifications to Towed Jumper Procedures**. Occasionally, the above towed jumper procedures must be modified. JMs are responsible for the safety of the parachutists' equipment checks and deployment safety as long as they are onboard the deployment aircraft. The loadmaster(s) is trained and responsible for retrieving towed parachutists. No additional JMs or safeties are required to remain onboard the aircraft unless specified in the jumping unit's ASOP. If all parachutists have exited, those actions described as JM responsibilities are accomplished by the loadmaster(s). The aircrew is responsible for all equipment left onboard the aircraft by the jump unit until it can be retrieved or turned over to the unit concerned.

NOTE: Towed jumper procedures for rotary-wing aircraft are addressed in Chapter 17.

d. **Safety Duties**. After all parachutists have exited (including JMs), the safety makes a visual check for towed parachutists to the rear of the jump door, gives the USAF loadmaster a thumbs-up signal, and an oral YOUR DOOR, AIR FORCE. This indicates that all parachutists are free and clear of the aircraft.

e. **Static Line and Deployment Bag Retrieval**. Safety personnel and the loadmaster retrieve the static lines and deployment bags. Once the static lines and deployment bags are inside the aircraft, safety personnel detach the static lines and store them with the deployment bags in extra aviator kit bags.

f. **During Return to Departure Airfield**. While en route to the departure airfield, safety personnel obtain the name, rank, social security number, unit, and reason (that is, sickness, equipment malfunction, jump refusal) for any manifested parachutist remaining onboard. They also check the aircraft for any Army equipment that was left onboard for turn-in to the DACO, collect trash and airsick bags, and reinstall seats and seat belts if subsequent lifts are planned.

g. **At Departure Airfield**. On return to the departure airfield, safety personnel turn in all air items (reserves, deployment bags, and kit bags) left onboard the aircraft to the storage facility (**obtain a receipt**). They also turn over any unit or personal equipment left aboard the aircraft to the DACO. All jump refusals and personnel left aboard the aircraft are immediately turned over to the DACO with a full account of the circumstances for each. Any jump refusal's personal equipment remains under DACO control until relieved of authority by competent unit personnel.

CHAPTER 11

DEPARTURE AIRFIELD CONTROL OFFICER

The DACO is appointed by the airborne commander and is responsible for coordination and control of the loading of personnel, equipment, and supplies into aircraft. The DACO is located at the departure airfield.

11-1. INITIAL COORDINATION

When advised of appointment, the DACO is furnished the following information:

- The unit(s) that is jumping.
- Type of jump and number of personnel.
- Type and number of aircraft.
- Number of lifts.
- Load time.
- Station time.
- Weather decision time.
- Drop zone.
- Names of DZSO, assistant DZSO, DZSTL, JM, AJM, and safety personnel.

11-2. TANKER/AIRLIFT CONTROL ELEMENT COORDINATION

At the departure airfield, the DACO makes contact with the tanker/airlift control element(TALCE)/aircrew to discuss, coordinate, and confirm the following:

- Aircraft parking plan.
- Aircraft tail number(s).
- Time of pilot/jumpmaster briefing.
- Time aircraft will be available for jumpmaster inspection.
- Weather data.
- Flight line safety measures.
- Appointment and identification of the GLO.
- Current safety regulations (TALCE/aircrew advises DACO).
- Aircraft loading procedure and time.
- Return of jump refusals (how they are to be managed).
- Manifest distribution to TALCE/aircrew.
- Inspection of aircraft for air items and other airborne unit equipment left onboard upon its return to departure airfield.

11-3. DROP ZONE SAFETY OFFICER/DROP ZONE SUPPORT TEAM COORDINATION

The DACO—

- Contacts the DZSO/DZSTL one hour before drop time.
- Informs the DZSO/DZSTL of aircraft aborts.
- Updates the DZSO/DZSTL on changes.

- Requests a closure report from the DZSO/DZSTL.
- Requests timely wind readings from the DZSO/DZSTL.

11-4. ADDITIONAL RESPONSIBILITIES OF THE DEPARTURE AIRFIELD CONTROL OFFICER

The DACO—

- Maintains radio and telephone contact with TALCE/aircrew and higher headquarters.
- Controls movement of all vehicles.
- Coordinates with TALCE/aircrew/GLO for airfield lighting to assist night airdrop departure airfield operations.
- Ensures all units police their areas before enplaning.
- Reports all serious incidents.
- Is physically present for all weather briefings and decisions.
- Ensures parachute maintenance personnel are present for issue and technical assistance.
- Becomes familiar with regulations and the SOP of the unit involved in the operation.

11-5. AIRFIELD AND RUNWAY SAFETY

The DACO—

- Is responsible for the movement in and around aircraft, taxiways, and runways.
- Briefs JM/AJM on airfield procedures.

PART THREE

Equipment

CHAPTER 12

INDIVIDUAL COMBAT EQUIPMENT JUMP LOADS

Individual equipment and weapons are attached or worn by the parachutist in several configurations--for example, exposed, placed inside containers, or a mix of the two. Unit airborne SOPs specify ways of packing equipment and other mission-essential items consistent with safety requirements and this manual. However, to ensure personnel safety, all small arms must be on "SAFE" when jumping.

Section I. LOAD PLACEMENT

Fragile items, such as weapon sights, are padded. Crushable items are not placed directly under the attaching harness. Exposed weapons or equipment, snap hooks, and projections on containers are potential safety hazards and must be taped.

12-1. LOAD DISTRIBUTION

Individual equipment attached to the equipment belt is placed on the front or sides of the body away from the PLF points of contact. The medium or large ALICE pack is attached to the front of the parachutist using an H-harness or HSPR. A lowering line is attached if the pack exceeds 35 pounds or has a frame attached. Hard, bulky, or irregularly shaped items are not placed to the rear of the parachutist or on the thighs. Jumpmasters and unit leaders must inspect parachutists to ensure weight and distribution of the load does not interfere with the parachutist's ability to perform a strong exit from the aircraft door.

12-2. CONSIDERATIONS

In accordance with AFI 11-2-C-130-Vol III, commanders will not load parachutists with equipment to exceed 350 pounds total weight and will reduce this weight by 12.5 pounds for every thousand feet of exit altitude above 3,000 feet MSL for fixed- and rotary-wing jumps. Total weight will not exceed 325 pounds for C-130 ramp jumps. Failure to comply could result in D-ring failure if the reserve is deployed at terminal velocity and or the JM is unable to retrieve the towed parachutist. The variety and weight of equipment or weapons attached to a jumper may exceed the parachute load limits and a safe descent rate. Also, the jumper's actions or time available to release tie-down straps and lower equipment may interfere with control of the parachute close to the ground (Tables 12-1 and 12-2, page 12-2). Commanders should use all the equipment methods of entry available to them (CDS, door bundles, wedge, and so on) to prevent overloading any individual parachutist.

TWO EQUIPMENT LOADS AND BASIC LOAD (POUNDS)									
Soldier Load Type	Load-Bearing Equipment with Two Canteens (Water)	ALICE Pack with Sleeping Bag (Winter)	Battle Dress Uniform Helmet, and Boots (1)	M1950 Weapons Container	Weapons Loaded with Ammunition (2)	Suspended Weight of T-10-/ MC1-Series Parachute and Reserve	Soldier Weight	Total Suspended Weight (Pounds) (3)
M16 Rifleman	11.5	32	15	7.3	31.0	23	205	324.8
M60 Machine Gunner	11.5	32	15	7.3	54.4	23	205	348.2
M203 Gunner	11.5	32	15	7.3	40.0	23	205	333.8
Dragon Gunner	11.5	32	15	6.5 (weapons only)	55.9	23	205	348.9
Radio Operator	11.5	32	15	7.3	71.6	23	205	365.4

(1) Weight of uniform does not include winter gear (for example, parka, liners, underwear).
(2) Includes basic load of ammunition, grenades, Claymore, bayonet, and cleaning kit.
(3) Does not include arctic gear.

Table 12-1. Weight of parachutist.

DESCRIPTION	WEIGHT (POUNDS)	REFERENCE	REMARKS
Maximum Load-Bearing Capacity T-10-Series/MC1-Series Main Canopy	360	Natick Research, Development, and Engineering Center	For every 1,000 feet above 3,000 feet MSL subtract 12.5 pounds from the jumper's total weight, which includes the parachute assembly. Failure to comply may result in D-ring or reserve failure if the reserve is deployed at terminal velocity.
Maximum Load-Bearing Capacity to Achieve 22 Feet Per Second or Less Descent Rate, T-10-Series/ MC1-Series Main Canopy	360	Natick Research, Development, and Engineering Center	Total suspended weight
Air Movement Planning Weight of Combat-Equipped Parachutist	260		Parachutist with one equipment container

Table 12-2. Parachute load limits.

Section II. LIFE PRESERVERS

Jumpers wear life preservers whenever a flight is conducted over water, a water obstacle is on the DZ, or the intended DZ is close to a water obstacle. (See Appendix D.) The B-7, the B-5, the LPU-10/P, and the life preserver inflatable (NSN 4220-0027-6892-9050) life preservers are currently certified for use by parachutists. Other life preservers may be used if approved by the individual service components.

12-3. B-7 LIFE PRESERVER

The B-7 is worn **under** the parachute harness. To fit the B-7 life preserver, the parachutist places one flotation packet under each arm so that the packet flaps are to the outside and the toggle cords are down and to the front. He runs the shoulder strap from front to rear over the left shoulder, under the back strap, then from rear to front over the right shoulder, and attaches it to the ring on the right flotation packet. He adjusts the shoulder strap so that the flotation packets fit snugly against his armpits. The parachutist then attaches the chest strap to the attachment rings on the left flotation packet, forming a quick release.

WARNING

Ensure that the B-7 life preserver is worn so that the inflatable portion is *not* between the parachute harness and the body. Serious injury may result if it is in this position when inflated.

12-4. B-5 LIFE PRESERVER

The B-5 is worn **under** the parachute harness. The parachutist inflates the vest by pulling two toggle cords (at the bottom of the vest) which activate CO2 cartridges that fill the vest with gas. An alternate provision for inflating the vest is by blowing into the manual inflation valve rubber hoses located on the upper right side of the vest. Manual inflation should be used only if the CO2 inflation valves fail to operate. The flotation vest is placed over the neck so that the inflatable vest is on the parachutist's chest. The manual inflating valves should be completely closed when donning the life vest. The back strap and leg straps are then adjusted.

WARNING

Do not inflate the B-5 life preserver until the parachute harness is removed. The B-5 can crush an individual's chest if inflated beneath a parachute harness.

12-5. LPU-10/P LIFE PRESERVER

The LPU-10/P is a standard USAF carbon-dioxide cartridge-activated life preserver assembly worn during flights over water or during airdrops when water obstacles are near or on the intended DZ. The LPU-10/P has an adjustable harness and underarm inflation bladders. It is designed to keep the wearer's head above water at weights up to 250 pounds for up to 10 minutes. The LPU-10/P is compatible with the USAF C-9, T-10, and MC-4 parachute harness assemblies. The LPU-10/P must be maintained IAW USAF TO 14S-1-102.

a. **Donning the LPU-10/P**. The LPU-10/P is worn **under** the parachute harness. The harness is worn so that the inflatable pockets are under the parachutist's arms. The manual inflating valves should be completely closed when donning the life vest. The shoulder and waist straps are then adjusted to ensure the inflation container is one hand width beneath the armpit and not constrained by the parachute harness.

WARNING

The inflation wings must be one hand-width beneath the jumper's armpit and clear of the harness straps. If the inflation pockets are too snug under the armpit, or if they are between the harness and the jumper's body, the jumper can experience severe pain or crushed ribs during inflation.

b. **Inflating the LPU-10/P.** The parachutist inflates the flotation bladders by pulling two toggle cords (at the bottom of the vest) which activate CO2 cartridges that fill the bladders with gas. An alternate provision for inflating the vest is by blowing into the manual inflation valve rubber hoses located on the bottom side of the wings. Manual inflation should be used only if the CO2 inflation valves fail to operate.

Section III. HARNESSES AND LOWERING LINE

All load carriers (ALICE packs and weapons cases) are attached to the parachutist by harnesses and, if lowered, rigged with the lowering line. Two types of harnesses are used--the older H-harness (modified) and the new Standard A, HSPR assembly. (The HSPR replaces the modified H-harness through attrition.)

12-6. H-HARNESS

The H-harness (modified) (Figure 12-1) consists of two equipment retainer straps.

a. These straps are connected by two cross straps. Each equipment retainer strap has two friction adapters 3 inches apart. Two D-ring attaching straps terminate in a free-running end on one end and a snap hook on the other end.

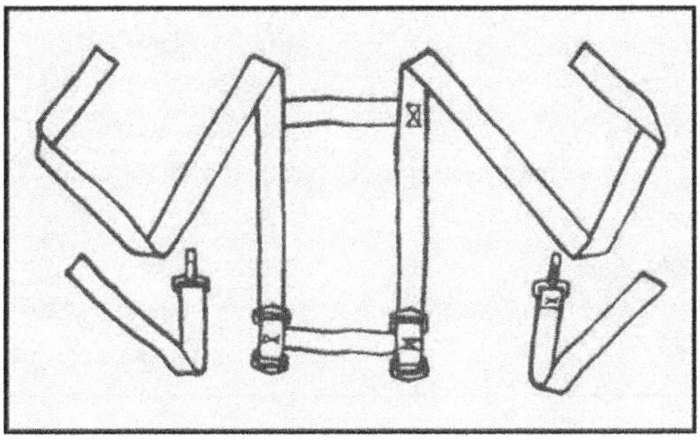

Figure 12-1. H-harness (modified).

b. The H-harness is used to rig the ALICE pack to the parachute harness. When rigging the H-harness, the parachutist ensures that equipment fits snugly under the reserve parachute and that the D-ring attaching strap snap hooks are spaced a four-finger distance from the H-harness friction adapters.

12-7. HARNESS, SINGLE-POINT RELEASE

The HSPR is an H-type design. It is made of nylon webbing with friction adapters to secure it around the load, and it has two adjustable D-ring attaching straps. To stabilize the pack to the parachutist during movement in the aircraft, exit, and main parachute deployment, two adjustable leg straps are provided to secure the pack to the parachutist's right and left legs. The leg straps are equipped with the male portion of the leg strap release assembly. The harness has a single-point release assembly that simultaneously releases the load and leg straps from the parachutist and parachute harness (Figure 12-2).

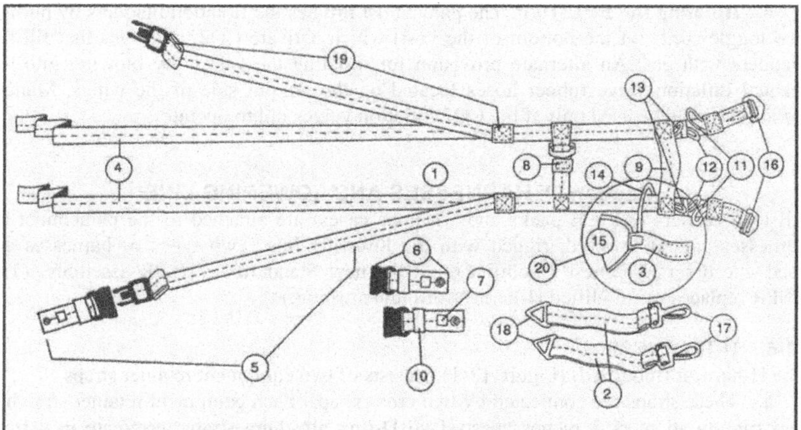

1. HARNESS, SINGLE-POINT RELEASE	11. WHITE ATTACHING LOOP
2. ADJUSTABLE D-RING ATTACHING STRAP	12. GREEN ATTACHING LOOP
3. RELEASE HANDLE	13. RED ATTACHING LOOP
4. EQUIPMENT RETAINER STRAP	14. RELEASE HANDLE LANYARD
5. ADJUSTABLE LEG STRAP	15. RELEASE HANDLE CABLE
6. FEMALE PORTION, LEG STRAP RELEASE ASSEMBLY	16. FRICTION ADAPTER
7. CABLE LOOP RETAINER	17. SNAP HOOK
8. ADJUSTABLE CROSS STRAP	18. TRIANGLE LINK
9. RELEASE HANDLE CROSS STRAP	19. MALE PORTION, LEG STRAP RELEASE ASSEMBLY
10. WEBBING RETAINER	20. RELEASE HANDLE CABLE ASSEMBLY

Figure 12-2. HSPR (NSN 1670-01-227-7992).

12-8. HOOK-PILE TAPE LOWERING LINE

The HPT lowering line (Figure 12-3) is used to lower all equipment attached to the parachutist. The HPT has been modified to accommodate the DMJP and AT4JP when lowered as a tandem load.

a. The standard 15-foot lowering line is made of tubular nylon (1 inch wide) with two retainer flaps sewn on. The retainer flaps have HPT sewn to the edges.

b. Two-inch tabs are sewn on the lowering line and, when the line is stowed, the tabs are secured to prevent line spillage. The ejector snap has a yellow safety lanyard (1 inch by 8 inches) attached.

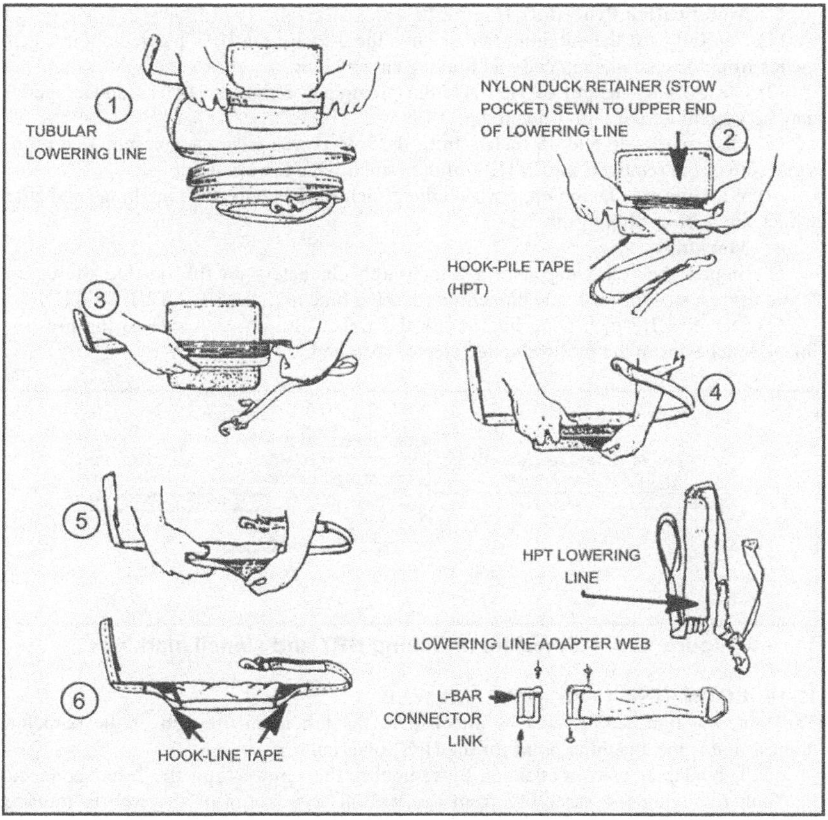

Figure 12-3. Hook-pile tape lowering line.

12-9. HOOK-PILE TAPE LOWERING LINE (MODIFIED)

The modified HPT lowering line (NSN 1670-01-067-6838) must be used when the DMJP and AT4JP are lowered as a tandem load. The modification is accomplished in the field.

a. **Materials Required**.

(1) Fastener tape, pile, color OD 106, 1 inch wide, Class 1, MIL-F-21840, NSN 8315-00-106-5974.

(2) Thread, nylon, color OD-S1, Type II, Class A, size E, V-T-295, 2,500-yard tube, NSN 8310-00-244-0609.

(3) Ink, marking, parachute, color strata blue, Type IV of MIL-I-6903 (or AA-59291), NSN 7510-00-286-5362.

b. **Stitching Requirements**.

(1) Stitching will conform to FED-STD-751, Type 301, 7 to 9 stitches per inch.

(2) Ends of stitching will be over-stitched not less than 1/2 inch.

c. **Modification Procedure** (Figure 12-4).

(1) Carefully cut the stitching that secures the 2-inch-long HPT located about 11 3/4 inches from the ejector snap end and remove cut stitching.

(2) Cut a 2-inch length of HPT. (If the previously removed HPT is undamaged, it may be used in lieu of replacement tape.)

(3) Place marks 46 and 48 inches from the folded web edge ejector snap end on the same side of the removed 2-inch HPT of the 1-inch-wide lowering line.

(4) With pile side facing up, position the 2-inch HPT between the markings and stitch with a single box stitch formation.

d. **Markings**.

(1) Stencil the following with 1/2-inch-high characters on the outside of retainer fabric using a stencil brush and parachute-marking blue ink: "DMJP/AT4JP MOD."

(2) Stencil a 1/8-inch-wide line across the web width on each side of the lowering line, 18 inches from the fold web edge ejector snap end.

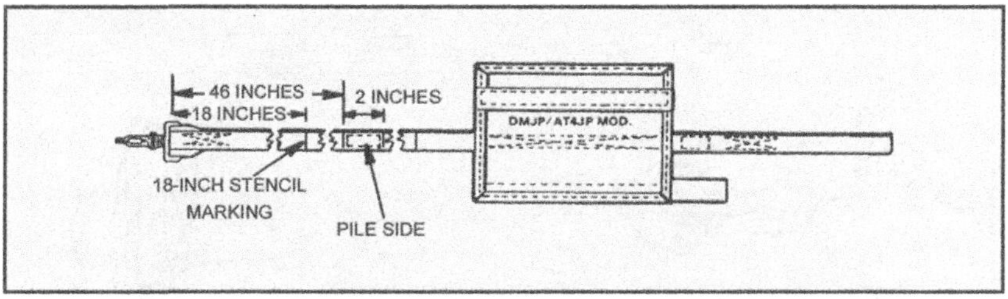

Figure 12-4. Method for attaching HPT and stencil markings.

12-10. LOWERING LINE ADAPTER WEB

The lowering line adapter web is attached to the left main lift web of the parachute harness and is the attaching point for the HPT lowering line.

a. The adapter web is attached by removing the screws from the L-bar connector link and removing the assembly from the web. The free end of the web is threaded through the left D-ring of the parachute harness and between the L-bar and main lift web. The link assembly is reattached through the loop on the free end of the adapter web and around the main lift web, and the screws are securely reset (Figure 12-5).

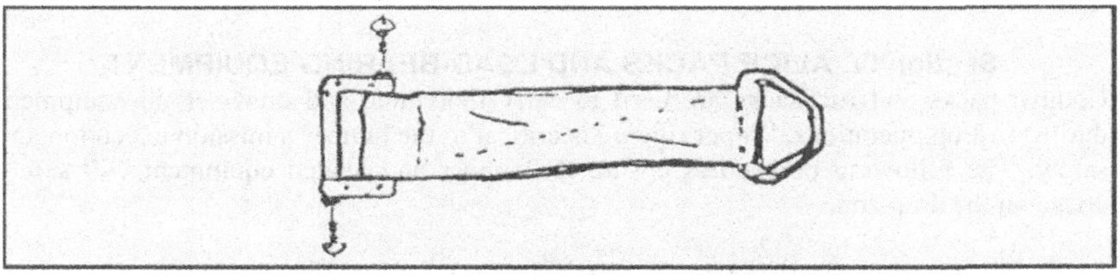

Figure 12-5. Lowering line adapter web.

b. When attaching equipment to be lowered, the web must be routed under the waistband. Type II or III nylon cord (gutted) is used to eliminate slack between the accessory attaching ring and the parachute harness main lift web. This is done by making one turn through the accessory attaching ring loop in the adapter web, and one turn around the main lift web of the parachute harness and the lowering line adapter web (Figure 12-6). The loose ends must be tied together using a suitable joining knot such as a square knot followed by two overhand knots or a surgeon's knot.

NOTE: The lowering line adapter web will become obsolete when the T-10-series modified parachute harness with triangle links is received by units.

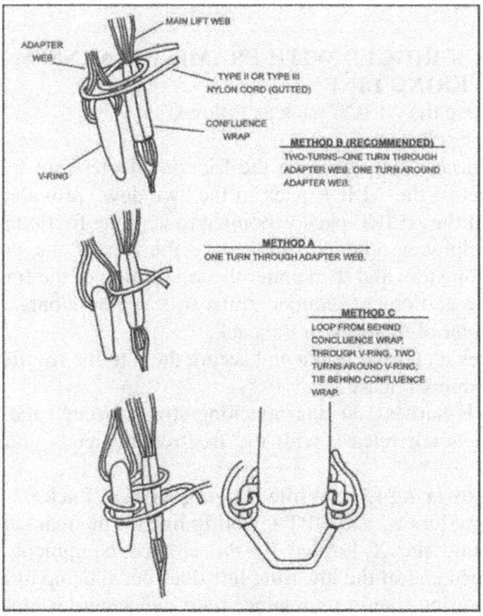

Figure 12-6. Securing lowering line adapter web with Type II or Type III nylon cord.

Section IV. ALICE PACKS AND LOAD-BEARING EQUIPMENT

Combat packs and rucksacks are used to carry individual and crew-served equipment during airdrop operations. Proper rigging is critical to the jumper's mission execution and safety. The following procedures ensure the jumper and loaded equipment will safely arrive on the drop zone.

12-11. ALICE PACKS (MEDIUM AND LARGE)

Medium and large ALICE packs are attached to the parachutist using the H-harness or the HSPR. They can be lowered during descent by attaching the HPT lowering line.

a. Items of equipment are inserted, and padding is placed between the load and the front portion of the pack. The outside pockets are filled with nonfragile items (full pockets help position the H-harness on the pack). The pack is closed by engaging the drawstrings and tie-down straps. The adjustable shoulder carrying straps are adjusted snugly. The excess webbing on the adjustable shoulder carrying straps is rolled and taped. The running ends of the waist straps are routed around the frame opposite the lower back pad, tightened, and secured in place by taping. This secures the adjustable shoulder carrying straps and reduces the possibility of entanglement onboard the aircraft.

b. If carried, the sleeping mat is rolled tightly to reduce its size. The mat is placed between the two top vertical tie-down straps and the top cover of the pack. It is secured by tightening the straps. The H-harness or HSPR encompasses the mat when it is routed around the ALICE pack.

12-12. ALICE PACK RIGGED WITH FRAME, H-HARNESS, AND HOOK-PILE TAPE LOWERING LINE

Rig, attach, and release the ALICE pack as follows:

a. **Rigging the Pack.**

(1) Lay the H-harness flat, ensuring the friction adapters are facing down. Place the center outside pocket of the ALICE pack in the "window" provided by the cross straps. Ensure the bottom of the ALICE pack is pointed toward the friction adapters.

(2) Run the equipment retainer straps over the top of the pack load outside the shoulder carrying loop strap and then under the top portion of the frame.

(3) Then, run the equipment retainer straps over the horizontal bar of the frame and cross them at the center of the back of the pack.

(4) Run the straps under the frame and secure them to the friction adapters, forming a two- to three-finger quick release.

(5) Thread the (H-harness) D-ring attaching straps through the intermediate friction adapters, forming a quick release with the free-running ends pointing away from the parachutist.

b. **Attaching Lowering Line While Rigging ALICE Pack.**

(1) Girth-hitch the looped end HPT lowering line (at the rear center of the pack from top to bottom) around the X formed by the crossed equipment retainer straps. This ensures that the looped end of the lowering line does not slide up or down the H-harness.

(2) Stow the lowering line in its retainer flaps and secure it to the left side (as worn) of the pack frame with two retainer bands.

NOTE: It is strongly discouraged that any additional items, such as canteens and entrenching tools, be attached to the outer portion of the ALICE pack. However, mission requirements may dictate that some items be rigged to the outside of the ALICE pack. Items such as canteens and entrenching tools may be attached to the outside of the ALICE pack in accordance with individual unit SOPs, providing they are secured by point of attachment (clips) and further secured with 1/4-inch cotton webbing (two turns) or Type II or III nylon cord.

c. **Attaching the ALICE Pack.**

(1) Attach the snap hooks of the D-ring attaching straps to the D-rings on the parachute main lift web outside the reserve parachute connector snaps. Secure the ALICE pack in one of two ways:

NOTE: This load must be snug under the reserve parachute.

(a) When jumping from the right door, route the lower tie-down tape on the weapons case around the left leg and frame of the ALICE pack. Tie it with a bowknot on the leading edge of the weapons case.

(b) When jumping from the left door, attach 1/4-inch cotton webbing to the ALICE frame on the right side (as worn) by means of a girth hitch, secure it around the right leg, and tie in a bowknot.

(2) Attach the HPT lowering line ejector snap to the accessory attaching ring on the lowering line adapter web (Figure 12-7).

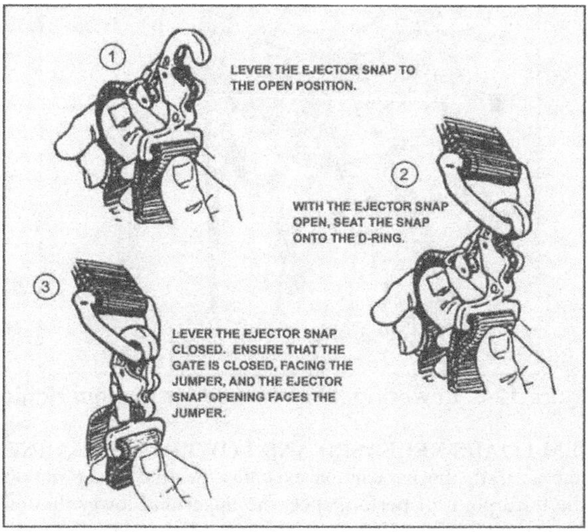

Figure 12-7. Ejector snap on HPT lowering line attached to accessory attaching ring on lowering line adapter web.

d. **Releasing the Pack.**

(1) Upon exiting the aircraft, execute the first two points of performance and, during the third point of performance, release all tie-downs.

(2) At 100 to 200 feet above the ground (during the fourth point of performance), check below for other parachutists, then pull both free-running ends of the D-ring attaching straps, allowing the ALICE pack to fall the length of the lowering line.

12-13. TANDEM LOAD AND LOWERING LINE

After the pack is rigged to the H-harness and frame, the lowering line is attached to the H-harness and stowed. The lowering line is secured to the left vertical bar of the frame (as worn) with two retainer bands. The ALICE pack and M1950 weapons case, when jumped together, are rigged as a tandem load and lowered on the same lowering line. After the ALICE pack and weapons case are attached to the parachute harness D-rings, the lowering line ejector snap is passed between the weapons case and the nylon chafe material of the case. The lowering line ejector snap is then attached to the accessory attaching ring on the lowering line adapter web (Figure 12-8).

Figure 12-8. Lowering line attached for tandem rigging.

12-14. TANDEM LOADS RELEASED AND LOWERED (H-HARNESS)

Upon exiting the aircraft, the parachutist executes the first two points of performance. Then, during the third point of performance, the upper and lower tie-down tapes on the weapons case are untied. The right leg tie-down on the ALICE pack is released (if jumping the left door). During the fourth point of performance at an altitude of about 100 to 200 feet above the ground, the ALICE pack is lowered by pulling at the same time on the free-running ends of the D-ring attaching straps, allowing the pack to fall to the end of the line. The activating arm of the quick-release assembly on the weapons case is activated, and the case slides down the lowering line to rest on top of the (lowered) pack.

NOTE: The D-ring attaching straps are removed from the parachute harness and secured to the H-harness before the parachute is returned to parachute maintenance after a jump exercise.

12-15. ALICE PACK RIGGED WITH FRAME USING HARNESS, SINGLE-POINT RELEASE AND HOOK-PILE TAPE LOWERING LINE

Before attaching the HSPR to the pack, the release handle and adjustable D-ring attaching straps are attached to the HSPR (Figure 12-9).

a. Route the two release handle cables between the two plies of the release handle cross strap. Attach the pile tape of the release handle to the hook tape attaching tab located between the plies of the release handle cross strap. Ensure that the release handle lanyard is not misrouted. Place the triangle links of the adjustable D-ring attaching straps on top of the white attaching loops. Route the white attaching loop up through the triangle link.

b. Route the green attaching loop up through the white attaching loop. Route the red attaching loop up through the green attaching loop. Route the red attaching loop through the grommet on the female portion leg strap release assembly. Ensure that the cable loop retainer on the female portion of the leg strap release assembly is facing up. Route the release handle cable through the red attaching loop and then through the cable loop retainer.

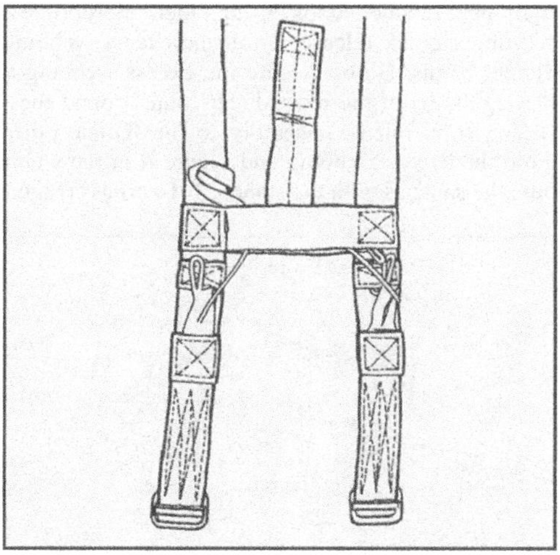

Figure 12-9. Release handle assembly.

c. Repeat the process for the other strap. Turn the harness over so that the adjustable D-ring attaching straps are on the bottom. Place the ALICE pack on top of the harness so that the middle outer cargo pocket is placed between the release handle cross strap and the adjustable cross strap. Ensure the top of the pack is facing the equipment retainer straps. Route the equipment retainer straps underneath the top of the frame, cross them on the back of the pack to form an X, then route them underneath the frame and the backrest of the pack (Figure 12-10).

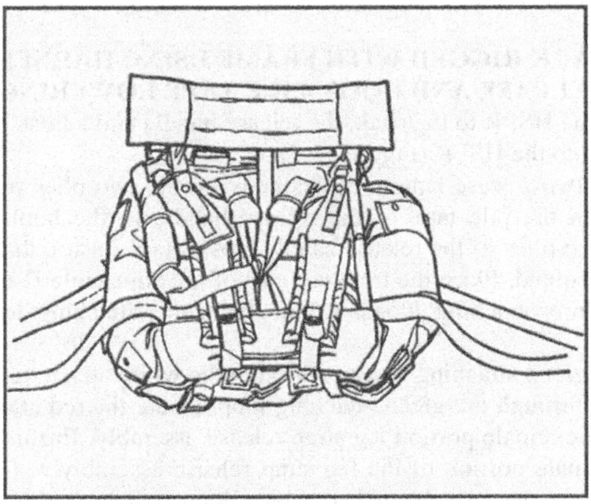

Figure 12-10. Harness routed.

d. Route the equipment retainer straps through their appropriate friction adapters and form a two- or three-finger quick release. S-roll the excess webbing and secure it with two turns of the retainer bands. Do not secure the excess webbing to the quick release. Route the adjustable leg straps in the most direct route around the pack and attach the male portion of the leg strap release assembly to the female portion of the leg strap release assembly. Fold the excess webbing and secure it in the webbing retainer. Attach the HPT lowering line the same as with the standard H-harness (Figure 12-11).

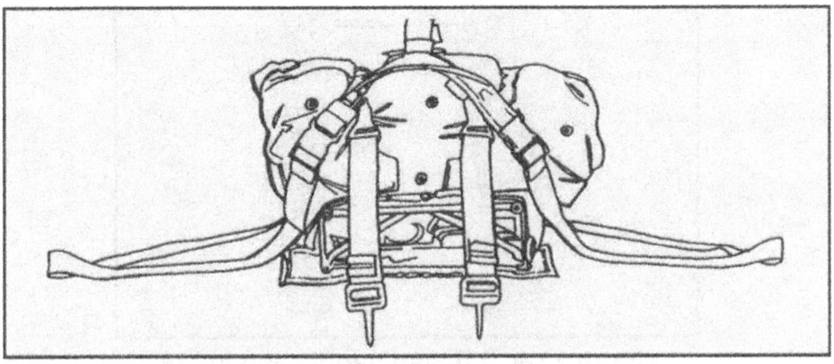

Figure 12-11. Leg straps routed.

12-16. ATTACHMENT OF HARNESS, SINGLE-POINT RELEASE AND ALICE PACK TO PARACHUTIST

When completely rigged, the HSPR is attached to the parachutist in the following sequence.

a. Using the buddy system, the parachutist stands in front of the HSPR-rigged ALICE pack. He grasps the harness by the two adjustable D-ring attaching straps and secures them to the D-rings on the main lift web outside of the connector snaps of the MIRPS/T-10 reserve (Figure 12-12).

b. The parachutist pulls on the free-running ends of the adjustable D-ring attaching straps and snugs the pack under the reserve. He then routes the ejector snap of the HPT lowering line through the nylon chafe material on the M1950 weapons case, from front to rear, then attaches the M1950 weapons case. He attaches the ejector snap on the HPT lowering line to the accessory attaching ring or triangle line as the outermost item on the D-ring. To attach the HPTL to the left D-ring, first open the ejector snap.

c. After the JM completes his JMPI, he routes the male portion leg strap release assemblies around the parachutist's legs and the M1950 weapons case, and attaches them to the female portion leg strap release. Adjustable leg straps may be tightly secured, but **must not** be so tight that they prevent the parachutist from executing a strong exit.

Figure 12-12. Parachutist adjusts leg straps.

12-17. TANDEM LOAD ATTACHED TO PARACHUTIST

The attachment of the rigged pack to the parachutist is identical to the procedure described in paragraph 12-16, with the following exceptions (Figure 12-13).

a. The lower tie-down tape is removed from the weapons case. If using the modified weapons case (with the HPT leg tie-down strap), the tie-down is routed around the case, the HPT is pressed together, and excess webbing is secured.

b. After the weapons case is attached to the parachutist, the leg strap of the HSPR is routed around the jumper's leg and the outside of the weapons case. Then the jumper tightens the adjustable leg strap after hooking the static line snap hook to the anchor line cable.

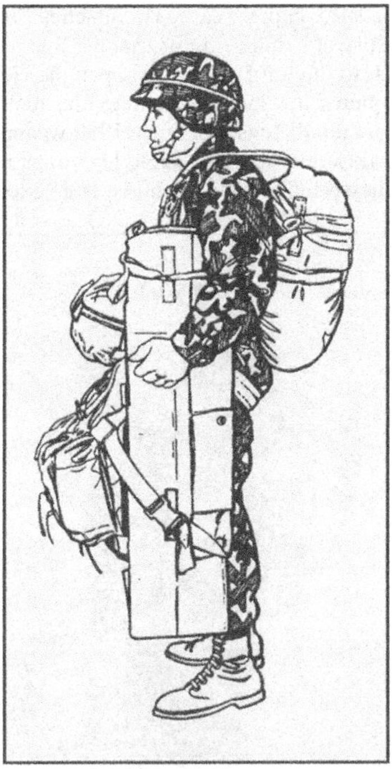

Figure 12-13. Tandem load attached.

c. The long end of the upper tie-down tape on the weapons case is routed around the case and main lift web directly above the chest strap of the parachute harness. It is tied with a bowknot on the front leading edge of the weapons case.

12-18. TANDEM LOADS RELEASED AND LOWERED (HARNESS, SINGLE-POINT RELEASE)

Upon exiting the aircraft, the parachutist executes the first two points of performance and then, during the third point of performance, unties the upper tie-down tape on the weapons case.

a. About 100 to 200 feet above the ground, the parachutist grasps the release handle and pulls up and out quickly, while at the same time releasing the load (and leg straps) and allowing it to drop the length of the lowering line. The activating arm of the quick-release snap on the weapons case is activated, and the case slides down the lowering line to rest on top of the (lowered) pack.

b. The release handle is released immediately following separation of the load from the jumper. The handle for releasing the load is secured to the HSPR with a release handle lanyard and stays with the HSPR to prevent its loss or separation.

c. To jettison the HSPR in an emergency, the parachutist lowers the pack, lowers the M1950 weapons case, then pulls out on the yellow safety lanyard (attached to the ejector snap on the HPT lowering line), which allows the pack to fall free.

12-19. JUMPING OF EXPOSED LOAD-BEARING EQUIPMENT

The procedures for rigging and wearing the protective mask carrier and the exposed load-bearing equipment during airdrop operations are as follows (Figure 12-14).

a. **Protective Mask Carrier**. The exposed protective mask carrier is worn in a different manner when airdrop operations are conducted.

(1) Wrap the narrow waist strap around the carrier, underneath the flap of the lower outside pocket, through the large O-ring, and hook it to the small D-ring to secure the quick opening flap securely closed.

CAUTION
Do not place masking tape on the mask carrier.

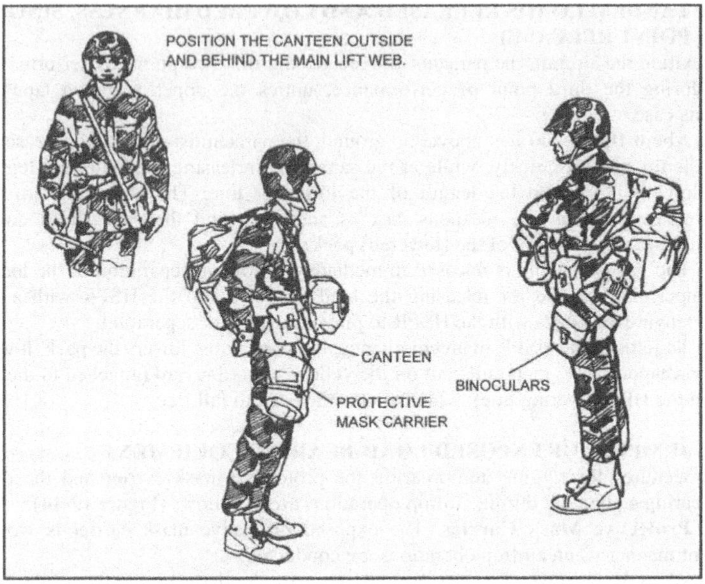

Figure 12-14. Binoculars and protective mask carrier rigged.

(2) Route the wide waist strap around the jumper's waist and hook it to the large D-ring, before donning the parachute and LBE. Don LBE and parachute, if LBE is to be jumped exposed.

(3) Place the aviator kit bag under the protective mask carrier. Route the parachute harness left leg strap through the carrying handle of the aviator kit bag and connect it to the left ejector snap. Route the parachute harness right leg strap directly to the right ejector snap. Do not attempt to secure the mask carrier with the leg straps.

(4) Adjust the length of the wide waist strap of the mask carrier so the entire mask carrier rests below the hip pad of the ALICE pack, but not so low that it interferes with the parachutist's ability to walk.

b. **Load-Bearing Equipment**. The LBE is worn in a standard manner when airdrop operations are conducted.

(1) Position the LBE just below the bottom of the parachute pack tray so that no portion of the parachute harness (for example, waistband, horizontal back strap, diagonal back straps, and so forth) rests on any major accessory item placed on the LBE (for example, canteen or ammunition pouch).

(2) Ensure the LBE is not covering any point of contact (for example, thighs, buttocks, or pull-up muscle) and is not buckled in front. JMs and jumpers must not wear knives on a point of contact.

(3) If a field pack is worn (for example, buttpack), make sure it does not contain any fragile items or equipment that is in a hard case.

(4) Tape any sharp edged piece of equipment, such as the wire cutter on the M9 bayonet sheath, with 100-mph tape to prevent snagging on other equipment.

(5) Ensure proper wear of the LBE by fitting it snugly against the jumper's body to prevent it from becoming entangled in risers, suspension lines, or equipment inside the aircraft.

(6) Route the LBE through the parachute harness before donning, or wear it before donning the parachute.

12-20. ENHANCED TACTICAL LOAD-BEARING VEST

The ETLBV must be in its full extended position. Both upper and lower chest buckles and the pistol belt remain unfastened. The jumper removes all slack from the chest buckle adjusting strap, S-folds or rolls the excess webbing, and secures the tape. After the jumper dons his parachute, the pistol belt and canteens are routed below the horizontal back strap and over the saddle. The ETLBV chest-mounted sewn-in magazine pouches are routed in the V formed by the horizontal back straps.

Section V. ADJUSTABLE INDIVIDUAL WEAPONS CASE (M1950)

The parachutist's M1950 individual weapons case is designed to carry individual or crew-served weapons. The M1950 may also be modified for other configurations. It is 10 inches wide and is adjustable in length from 33 1/2 inches to 50 1/2 inches. It is secured vertically by a quick-release snap attached to the left D-ring on the parachute harness.

12-21. M1950 SECURED TO PARACHUTIST

To prevent the M1950 from swaying during the opening shock of the parachute, two tie-downs fasten the case to the parachutist.

a. The upper tie-down tape is tied around the main lift web of the harness, and the lower tie-down tape goes around the parachutist's leg. If the weapons case has been modified, the lower tie-down strap (HPT) is secured around the leg.

b. Upon landing, the parachutist can secure his weapon quickly by opening the slide fastener, which is protected by a closing flap.

c. The slide fastener is designed as a quick release. To activate it (when the case is closed), a sharp tug on the slide fastener and tab thong (in the same direction as when zipping the container closed) causes the slide fastener to come apart.

12-22. M1950 ATTACHED TO PARACHUTIST

The M1950 is attached after the parachutist dons (and adjusts) the main and reserve parachutes.

a. The quick-release snap is attached to the quick-release link on the case. The opening gate of the quick-release snap faces the parachutist and is attached to the parachutist's left D-ring (to the outside of the reserve parachute connector snap).

b. The long end of the lower tie-down tape is passed around the outside of the case and in back of the left leg above the knee. Using a bowknot, the ends of this tape are tied together on the front leading edge of the weapons case. (The knot is untied before landing.) If the case has been modified with HPT leg straps, the piles are pressed together.

c. The long end of the upper tie-down tape is passed (from left to right) around the weapons case and main lift web of the parachute harness above the chest strap. Using a bowknot, the ends of this tape are tied together on the front leading edge of the weapons case.

Section VI. M16 RIFLE/M203 GRENADE LAUNCHER, EXPOSED AND PACKED

The M16 rifle or M203 grenade launcher can be jumped exposed or inside the M1950 weapons case. If the field-expedient method (weapon exposed) is used, the commander must consider that this may increase risk of injury to the parachutists and, therefore, hinder success of the mission. Inherent hazards of the exposed weapon include—

- The weapon becoming entangled with another jumper's parachute if a midair collision occurs.
- Possible injury to the parachutist during the PLF.
- Damage to the weapon during landing, which may cause a failure to fire.

12-23. M16 RIFLE/M203 GRENADE LAUNCHER EXPOSED

The sling is extended all the way, and the keeper is taped in place. The padding is secured over the side-mounted bolt forward assist and charging handle.

a. Place a plastic muzzle cap on the M16 muzzle, or pad and tape the muzzle and sight. This prevents the weapon from becoming entangled with the parachute suspension lines, or dirt from clogging the weapon during landing. Insert a magazine into the weapon and tape the magazine to the receiver, including the ejector port cover, to prevent loss of the magazine and debris from entering the bolt area. (Ensure that a round is **not** chambered and the weapon is placed on SAFE). Tape the hand guards to prevent loss upon impact when landing. (To aid in the removal of the padding and tape, fold and press the adhesive side on the running end of the tape together to form a quick-release pull tab.)

b. Further secure the weapon with two tie-downs of 1/4-inch cotton webbing (or a like item) with bowknots (Figure 12-15). Secure the sling to the diagonal back strap with the upper tie-down, which is a 12-inch tie strap. Secure the barrel of the rifle to the leg with a 24-inch strap to prevent possible entanglement with suspension lines. Remove the lower tie-down before landing to avoid personal injury.

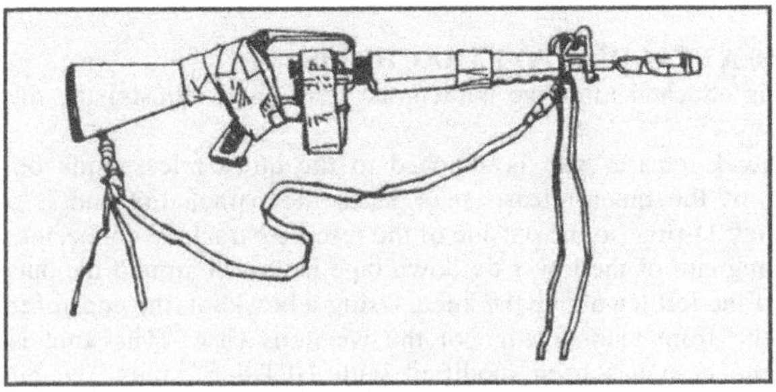

Figure 12-15. Tie-downs.

c. Sling the weapon over the left shoulder, with the muzzle down, and rotate it so that the pistol grip is facing the (parachutist's) rear (Figure 12-16). Place the sling from the lower keeper (stock) on the outside of the stock and over the left shoulder. Then run it under the chest strap of the main lift web. Thread the waistband through the carrying handle and into the metal adjuster on the waistband adjuster panel. Tighten the waistband securely so the weapon lies snugly against the parachutist's side.

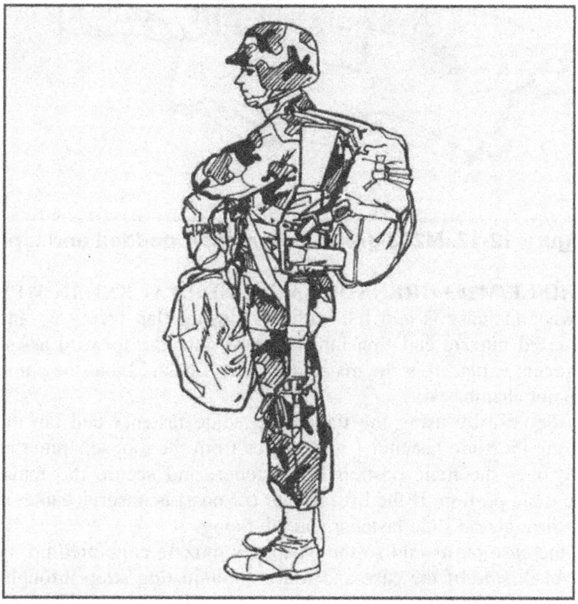

Figure 12-16. Rifle positioned with muzzle down.

d. Prepare the M203 grenade launcher as outlined above, but place additional padding and tape around it (Figure 12-17).

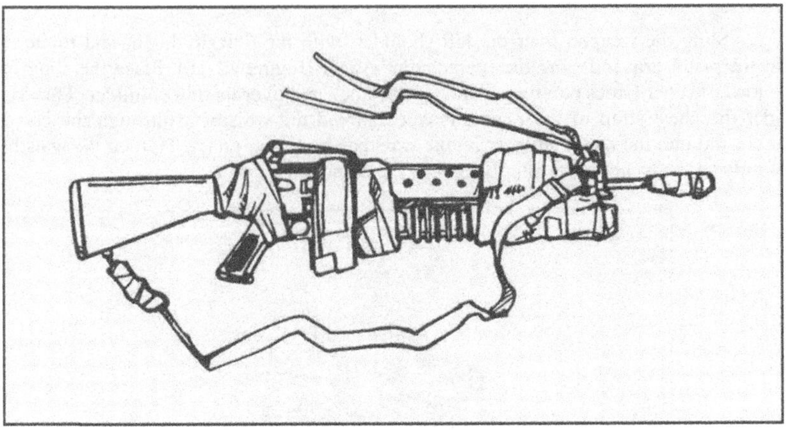

Figure 12-17. M203 grenade launcher, padded and taped.

12-24. M16 RIFLE/M203 GRENADE LAUNCHER PACKED IN M1950

The M1950 weapons case is laid flat with the closing flap facing up and opened. The weapon is inserted muzzle end first into the case with the forward assist facing up. A 20-round magazine is placed in the magazine well, if desired, but the parachutist ensures that a round is not chambered.

a. Close the case by using the flap thong, slide fastener and tab thong, and slide fastener. Leaving the slide fastener 1 to 2 inches from the top, separate the slide fastener and tab thong over the male portion lift fastener, and secure the female portion lift fastener to the male portion. If the lift fastener (or post) is unserviceable, route the upper tie-down tape through the slide fastener and tab thong.

b. Stand the case on its end so the weapon's muzzle is pointed up. Fold the excess case over the back side of the case and route the adjusting strap through the top set of adjusting strap connectors and secure it with a half hitch.

c. The M16/M203 may be jumped inside the case with a LAW or starlight scope. To do this, place the LAW or starlight scope on top of the M16/M203 and add padding (cellulose wadding) between the two items. It is usually rigged to the parachutist as a tandem load to be lowered on a single lowering line with the ALICE pack (paragraph 12-13).

Section VII. M60 MACHINE GUN

The M60 machine gun can be rigged on the individual parachutist or dropped as an accompanying load. If jumped on the individual, it is jumped as a team load.

12-25. M60 PACKED ASSEMBLED

The M1950 weapons case is laid down with the closing flap facing up. The muzzle of the M60 is placed into the lower-right corner of the weapons case. The weapon is pressed down until it is seated inside the case with the operating bolt up.

12-26. M60 PACKED DISASSEMBLED

The M1950 weapons case is shortened to 36 inches and laid down with the closing flap facing up (required when characteristics of a particular aircraft dictate length restrictions). It is packed as follows:

a. Disassemble the two groups by removing the barrel group.

b. Place the receiver group in the weapons case with the forearm assembly to the right and the cover facing down.

c. Place the barrel group in the case with the front sight to the left and pointing down.

d. Fit in the barrel group by sliding it to the right as far as possible so that the bipod-leg feet are not opposite the trigger housing group.

e. Add padding between the two groups.

NOTE: When the M60 is team loaded, the accessory bag, spare barrel, and tripod are placed in a separate weapons case and jumped by the assistant gunner. Both cases are usually tandem rigged and lowered (paragraph 12-13).

Section VIII. M249 SQUAD AUTOMATIC WEAPON

The SAW can be rigged for lowering as a tandem load or individual load. Tandem rigging is the same as for the M1950 weapons case and ALICE pack.

12-27. SAW MOD M1950 WEAPONS CASE

A modified M1950 weapons case is used for the SAW. The case is marked "SAW MOD" on the outside and has an extended 6 1/2-inch closing flap, which allows the weapon to fit in the case. A piece of cellulose wadding, about 20 inches long and 10 inches wide, is folded to form a pad about 10 inches long and 5 inches wide. This pad is placed in the muzzle end of the case. A 30-round magazine may be jumped with the SAW. The magazine is taped to the left side of the buttstock. The SAW is placed inside the case muzzle first. The pistol grip is toward the inside, and the carrying handle is facing up and away from the parachutist's body.

12-28. ATTACHMENT TO PARACHUTIST

When rigging the SAW as a single item to be lowered (Figure 12-18), the parachutist attaches the quick-release snap to the quick-release link. This ensures that the opening gate quick-release snap is facing away from the main body of the SAW MOD M1950 weapons case when the quick-release link is pointed toward the top of the case. The parachutist **does not** route the quick-release link through the metal V-ring. No safety tie is used.

a. Prepare the HPT lowering line by folding it in the normal manner. No HPT lowering line should protrude from the retainer flaps. Secure each end of the HPT retainer flap with the HPT tab.

b. Route the looped end HPT lowering line from top to bottom through the V-ring of the SAW MOD M1950 weapons case. Route the entire HPT lowering line through the looped end HPT lowering line, forming a tight girth hitch around the V-ring.

c. Place the HPT lowering line against the main body of the SAW MOD M1950 weapons case to the left of the nylon chafe material, which secures the V-ring and quick-release link. The lowering line ejector snap points toward the top of the SAW MOD M1950 weapons case. The 1-inch tubular nylon webbing, which forms the looped end HPT lowering line, should rest between the retainer flap of the HPT lowering line and the nylon chafe material.

d. Secure the HPT lowering line with four turns of 1-inch-wide masking tape--two turns are routed around the main body directly below the quick-release link, and two turns are routed around the main body directly above the upper set of the adjusting strap connectors.

e. To attach the SAW MOD M1950 weapons case to the jumper's parachute harness, attach the quick-release assembly to the left D-ring of the parachute harness as the outermost item of equipment. Route the lowering line ejector snap below the quick-release assembly and attach it to the accessory attaching ring on the lowering line adapter web, with the opening gate facing toward the parachutist. Secure the upper tie-down tape and lower tie-down strap.

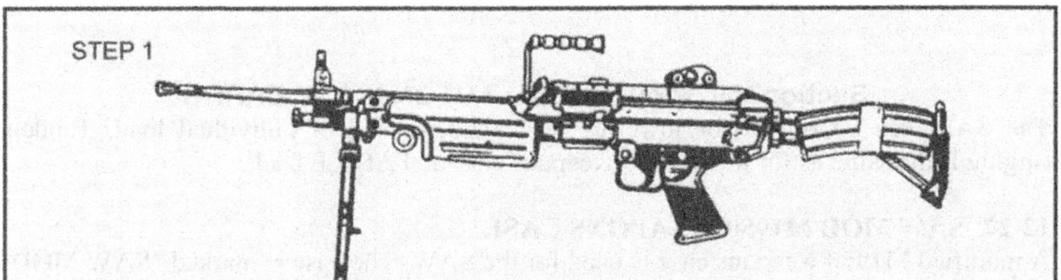

Figure 12-18. SAW rigged as a single item to be lowered.

Section IX. M224, 60-MM MORTAR

A modified M1950 weapons case is used for the M224 60-mm mortar. It is marked "60-mm MOD" on the outside and has an extended 11 3/4-inch closing flap.

12-29. MAJOR COMPONENTS

The components of the M224 are as follows (Figure 12-19).

- Aiming posts with case.
- M8 baseplate (small).
- M64 sight unit.
- M225 barrel.
- M170 bipod assembly.
- M7 baseplate (large).

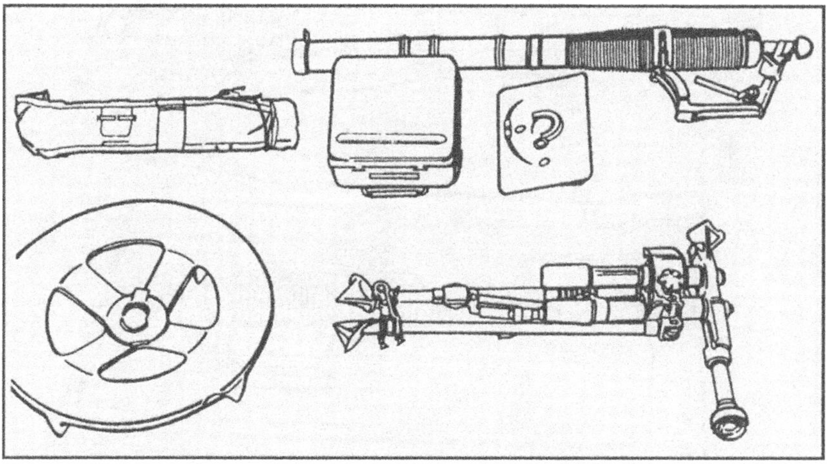

Figure 12-19. Components, M224 mortar.

12-30. LOAD DISTRIBUTION

The M224 crew members jump with the following components.

a. **Gunner**.
   - M224 barrel, in M1950 modified weapons case marked "60-mm MOD."
   - M8 baseplate, in ALICE pack.
   - M64 sight unit, centered in ALICE pack.

b. **Assistant Gunner**.
   - M170 bipod, in M1950 modified weapons gunner assembly case marked "60-mm MOD."

c. **Ammunition Bearer**.
   - Ammunition, in ALICE pack.

- M7 baseplate, in ALICE pack (outside of ALICE pack).
- Aiming posts, in ALICE pack (on top) with case.

12-31. INSTRUCTIONS FOR RIGGING

Following are instructions for rigging the M224 60-mm mortar.

a. **Gunner**. Place the barrel inside the case, muzzle down. Pack the small baseplate and sight unit inside the ALICE pack. Pad the sight unit with clothing or cellulose wadding to absorb the impact shock. Tandem-rig the pack and case for lowering (Figure 12-20).

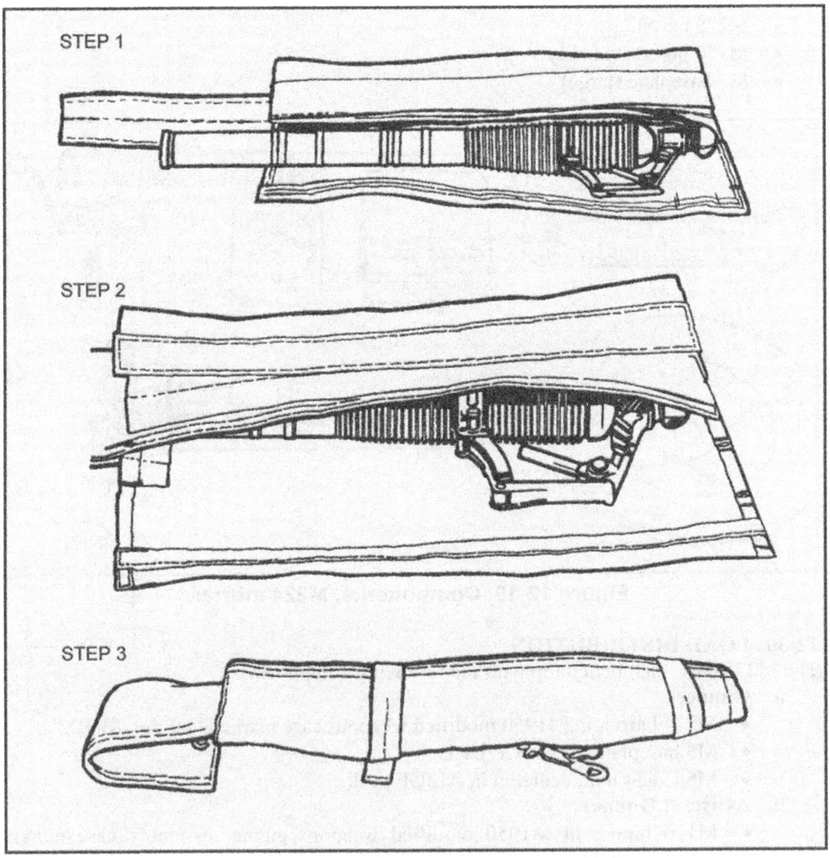

Figure 12-20. Barrel packed.

b. **Assistant Gunner**. Place the bipod assembly inside the case (Figures 12-21 and 12-22, page 12-26), ensuring that the traversing mechanism is in the middle of the case.

Secure the closing flap. Tandem-rig the case and pack for lowering (Figure 12-23 and paragraph 12-13).

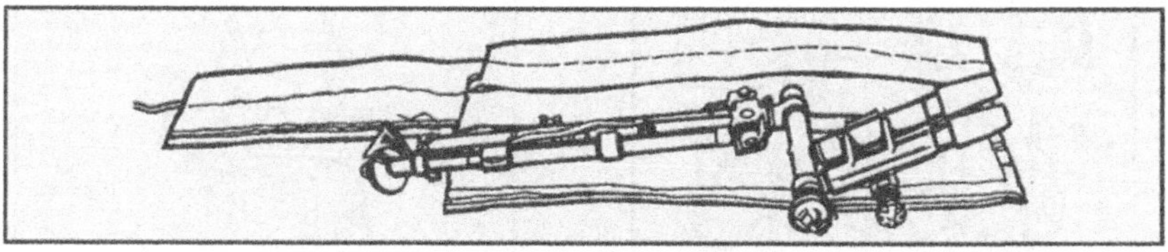

Figure 12-21. Bipod with case.

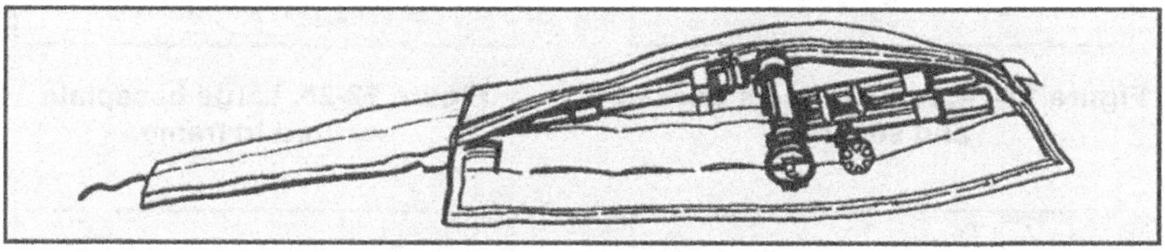

Figure 12-22. Bipod in case.

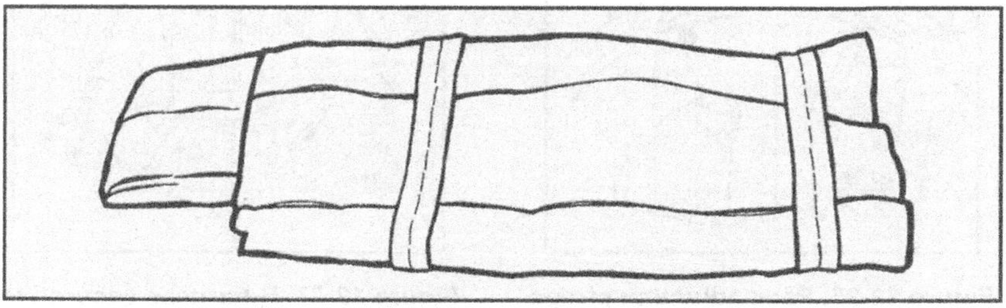

Figure 12-23. Assistant gunner's case completed for jumping.

c. **Ammunition Bearer**. Place the aiming posts under the top flap of the pack and secure with 1/4-inch cotton webbing to the top of the frame (Figure 12-24). Place the large baseplate over the outer accessory pouches and secure to the top of the frame with 1/2-inch tubular nylon webbing (Figure 12-25). Route the free-running ends of the pack adjusting straps through the baseplate (over, under, and over) and secure (Figure 12-26). Secure the H-harness to the pack and baseplate, ensuring that the equipment retainer straps are routed under the baseplate (top of the ALICE pack) (Figure 12-27) and over the baseplate (bottom of the ALICE pack).

NOTES:
1. Protect ammunition with cellulose wadding or clothing in ALICE pack.
2. Do not exceed weight limitations of the ALICE packs (large, 95 pounds; medium, 70 pounds).
3. Do not exceed weight limitation of parachute (360 pounds rigged weight for the T-10-series or the MC1-series).

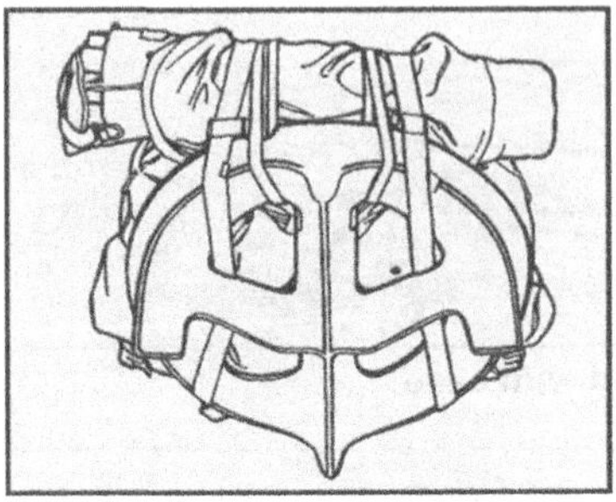

Figure 12-24. Aiming posts packed and secure.

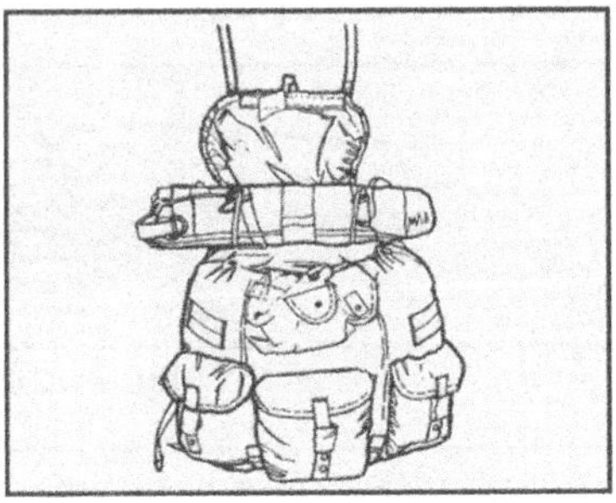

Figure 12-25. Large baseplate secured to frame.

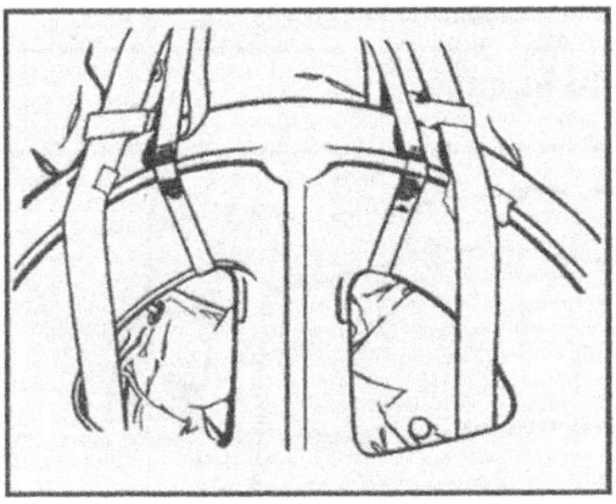

Figure 12-26. Pack adjusting straps routed through baseplate.

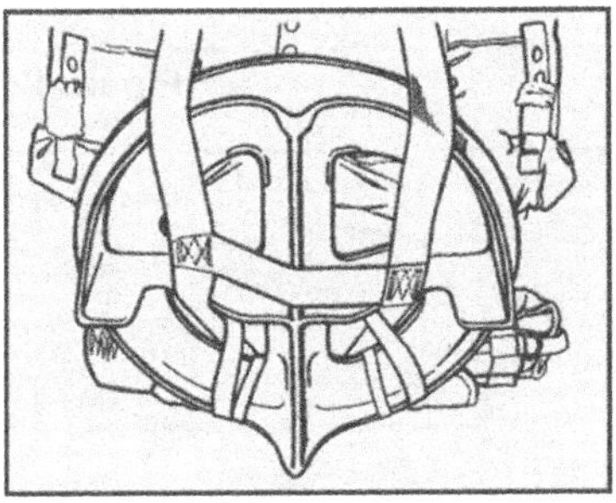

Figure 12-27. H-harness secured to pack and baseplate.

Section X. CONTAINER, WEAPON, INDIVIDUAL EQUIPMENT AND M202A1 ROCKET PACK

The CWIE is a general-purpose item used to carry designated combat equipment. It consists of a container and harness assembly, and when rigged for jumping, the container is attached to a harness assembly. The maximum dimensions for the CWIE are 12 inches wide, 12 inches deep, and 36 inches long. It can be adjusted to a minimum size of 12 inches wide, 6 inches deep, and 18 inches long. The CWIE weighs 16 1/2 pounds and, when packed, the container and contents must **not** exceed 95 pounds gross weight.

12-32. PREPARATION OF CONTAINER FOR PACKING

The container is laid down with the open portion facing up. A check is made to ensure that all securing straps and quick-fit adapters are present and serviceable.

a. **Packing Container**.

(1) Enough padding is used between items to prevent metal-to-metal contact. Special care is given to fragile items such as radios.

(2) Weapons that are too long to fit into the container are disassembled, and the parts are wrapped in padding.

(3) A combat pack or sleeping roll is placed in the bottom, which will hit the ground first to cushion the load.

(4) Related items are packed as one load.

(5) The heaviest items of equipment are loaded at the bottom of the container.

b. **Closing Container**.

(1) The side panels are folded over the contents.

(2) The bottom of the container is folded in an S-pattern to the desired length.

(3) The running ends of the three horizontal securing straps are passed over the container and fastened to the appropriate quick-fit adapters (using quick-release folds).

(4) The running ends of the two vertical securing straps are passed over the S-fold at the bottom of the container and fastened in a like manner.

12-33. HARNESS ASSEMBLY ATTACHED TO CONTAINER

The harness assembly is laid down with the inside portion facing up. All straps are straightened. The container is placed on the harness assembly. (The word TOP faces up. The arrow points toward the top of the harness assembly.)

a. The top and middle horizontal securing straps are placed around the container. The parachutist ensures that they are routed through (not over) the carrying straps. Each securing strap is fastened to the appropriate quick-fit adapters with a quick-release fold.

b. The bottom horizontal securing strap is placed around the container and secured to its quick-fit adapter. The two vertical securing straps are placed around the container. The parachutist ensures the securing straps are routed through (not over) the carrying straps and fastened to the appropriate quick-fit adapters with quick-release folds. The quick-fit adapters pass under the top carrying handle and the cable and conduit assembly.

c. The side securing strap is placed around the bottom of the container and over the two vertical securing straps. It is fastened to the appropriate quick-fit adapter with a quick-release fold.

d. The parachutist stands the entire assembly on end, top up. He tightens all securing straps and tapes any excess webbing. He then pulls out the release knob (red ball attached to the harness assembly).

e. The quick-release, quick-fit connecting links are inserted into the female part of the side fasteners on the harness assembly. The release knob, when pushed in, allows the cable to engage the claws in the hole of the quick-release, quick-fit connecting links.

f. The safety pin is inserted into the recess in the release knob from either side.

g. The looped end of the lowering line is threaded through the V-ring and under the carrying handle on top of the harness assembly. The entire length of the lowering line is pulled back through the loop and then pulled tight (Figure 12-28).

h. The HPT lowering line is folded, stowed in its container, and attached to the top carrying handle with two retainer bands.

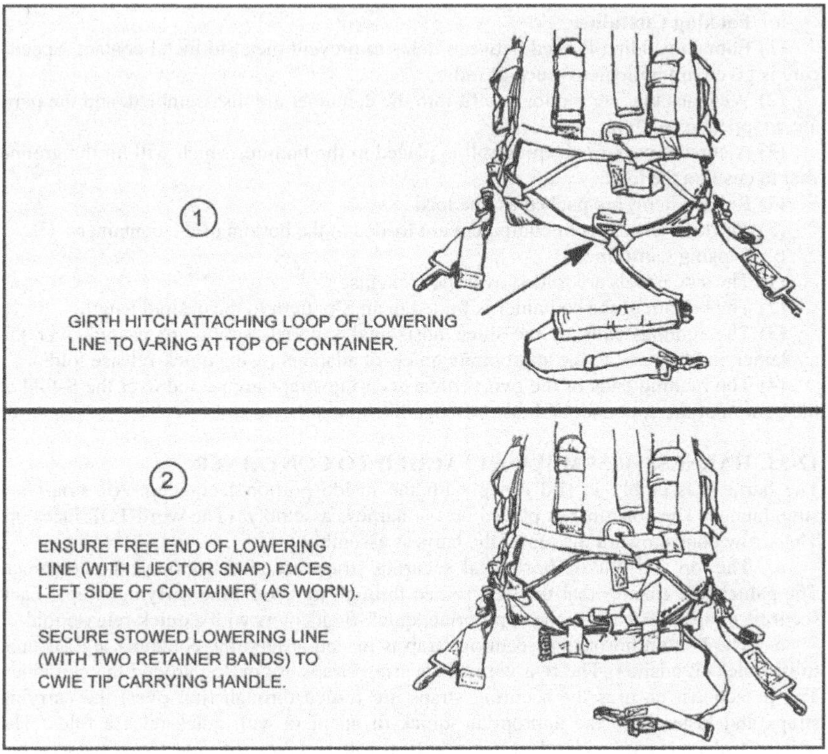

Figure 12-28. HPT lowering line attached to CWIE.

12-34. CONTAINER AND ASSEMBLY ATTACHED TO PARACHUTIST

The snap hooks of the quick-release straps are attached to the D-rings on the parachute harness. The parachutist ensures that the snap hooks are positioned to the outside of the connector snaps of the reserve parachute. The leg retaining strap is fastened around the leg (left leg, right door; right leg, left door), through the friction adapter, and a quick-release fold is made. The ejector snap on the HPT lowering line is attached to the lowering line adapter web on the left side of the parachute harness.

12-35. CONTAINER RELEASED

After exiting the aircraft, the parachutist executes the first three points of performance. After he executes the third point of performance, he releases the leg retaining strap and checks below to ensure that no other parachutists are in the way. The parachutist removes the safety pin from the red release knob when about 200 feet above the ground. He pulls the red release knob up and out, allowing the container to drop the full length of the lowering line.

Section XI. DRAGON MISSILE JUMP PACK

The Dragon (M47) missile jump pack (DMJP) is designed to carry one missile system and the M16 or M203 (Figure 12-29). The jump pack consists of a pack body constructed of nylon duck material with 1/4-inch felt material permanently sewn inside, nylon securing straps, lowering line, and quick-release assembly.

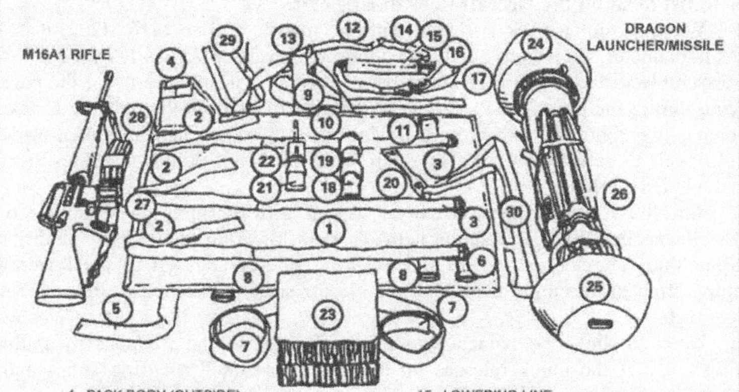

1. PACK BODY (OUTSIDE)
2. HORIZONTAL SECURING STRAPS
3. HORIZONTAL SECURING STRAPS QUICK-FIT ADAPTERS
4. RIFLE MUZZLE STOW POCKET
5. RIFLE BUTT SECURING STRAP
6. RIFLE BUTT SECURING STRAP QUICK-FIT ADAPTER
7. LAUNCHER FORWARD END SECURING STRAPS
8. LAUNCHER FORWARD END SECURING STRAP QUICK-FIT ADAPTERS
9. LAUNCHER AFT END BRIDLE
10. LAUNCHER AFT END BRIDLE SECURING STRAP QUICK-FIT ADAPTER
11. LAUNCHER AFT END BRIDLE SECURING STRAP QUICK-FIT ADAPTER
12. LOWERING LINE ATTACHING STRAP
13. LOWERING LINE ATTACHING STRAP LARGE LOOP END
14. LOWERING LINE ATTACHING STRAP SMALL LOOP END
15. LOWERING LINE
16. LOWERING LINE QUICK-EJECTOR SNAP
17. LOWERING LINE EJECTOR SNAP YELLOW SAFETY LANYARD
18. LOWERING LINE STOW POCKET
19. SECURING TABS
20. CARRYING HANDLE
21. QUICK-RELEASE LINK
22. QUICK-RELEASE SNAP
23. CARDBOARD HONEYCOMB ENERGY DISSIPATER
24. AFT END SHOCK ABSORBER (LARGE END CAP)
25. FORWARD END SHOCK ABSORBER (SMALL END CAP)
26. BIPOD
27. 20-ROUND MAGAZINE
28. 30-ROUND MAGAZINE
29. UPPER TIE-DOWN
30. LOWER TIE-DOWN

Figure 12-29. Nomenclature, DMJP.

12-36. MISSILE AND TRACKER

The missile and tracker can be jumped in two configurations. They can be rigged to be lowered as individual items, requiring two lowering lines, or tandem rigged on a single lowering line. The tandem rig requires the DMJP/AT4JP MOD HPT lowering line and modifications to the jump pack.

a. Due to the length of the missile and the difficulties in handling and moving the equipment around in the aircraft, the DMJP is restricted to parachutists who are at least 5 feet 6 inches tall. **The DMJP cannot be jumped from any aircraft that requires the parachutist to sit on the aircraft floor during exit.**

b. When the jump pack is rigged with the missile and an M16/M203, it is 11 1/2 inches in diameter and weighs about 40 pounds. It is secured vertically by a quick-release snap assembly to the left D-ring of the parachutist's harness. To prevent the pack from swaying during the parachutist's exit from the aircraft or from the opening shock of the parachute, two tie-downs are provided. The upper tie-down tape is routed around the main lift web, directly below the chest strap, while the lower tie-down secures the pack to the parachutist's left leg.

c. An HPT lowering line assembly is issued with the jump pack and can only be used to lower the DMJP as a single item. Tandem lowering requires the modified HPT lowering line. The looped end of the lowering line is attached to the lowering line attaching strap on the jump pack, and the ejector snap is attached to the lowering line adapter web.

d. Upon landing, the parachutist can rapidly get to the weapons by pulling the running ends of the quick releases on the three horizontal securing straps and three adjustable end straps.

12-37. DRAGON MISSILE JUMP PACK RIGGED

A detailed explanation of rigging the jump pack follows.

a. **Position the Dragon missile.**

(1) Place the pack with the felt side of the pack body facing up and extend all straps.

(2) Position the missile in the pack with the bipod facing up (Figure 12-30).

(3) Fit the aft shock absorber (large end) into the launcher aft end bridle of the jump pack.

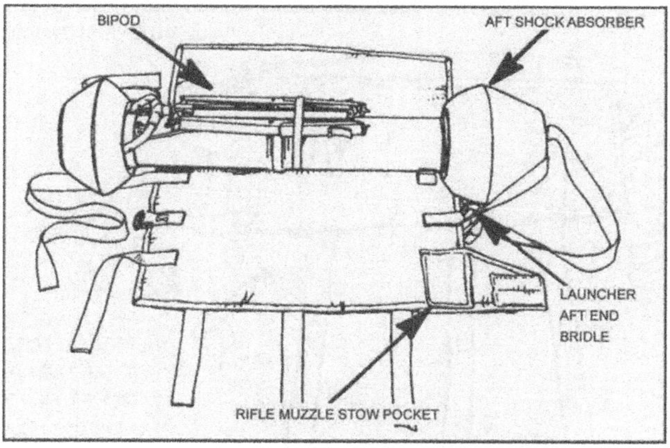

Figure 12-30. Missile being positioned.

b. **Position the weapon.**

(1) Insert a 20-round magazine into the magazine well (optional).

(2) Secure a 30-round magazine to the weapon sling, using adhesive tape (optional).

(3) Position the rifle on top of the pack, on top of the missile bipod, by inserting the rifle muzzle into the rifle muzzle stow pocket. The butt of the rifle pistol grip lies along the inner edge of the jump pack (Figures 12-31 and 12-32).

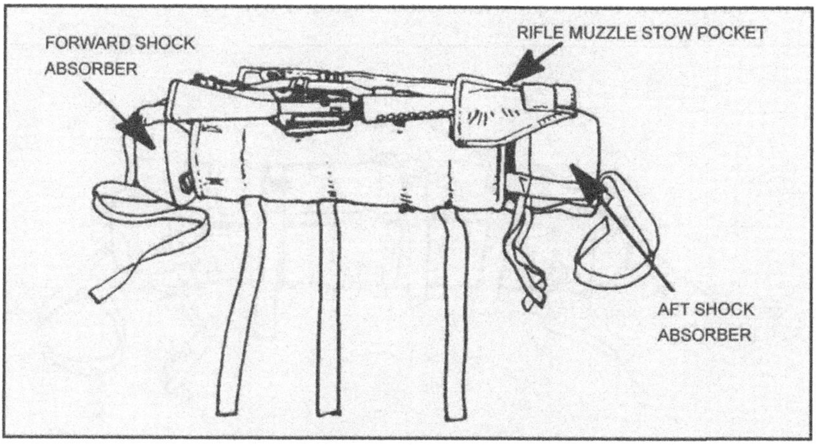

Figure 12-31. Rifle positioned.

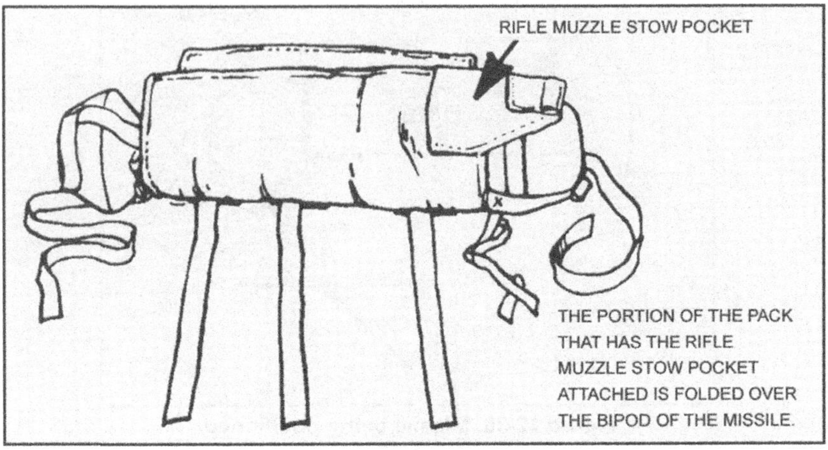

Figure 12-32. Pack folded.

c. **Close the pack.**

(1) Fold the outer flap of the jump pack over the weapon (if packed).

(2) Route the three horizontal securing straps through the quick-fit adapters using a quick-release fold, but do not tighten.

(3) Route the rifle butt securing strap through the quick-fit adapter. Tighten the strap as much as possible, using a quick-release fold (Figure 12-33).

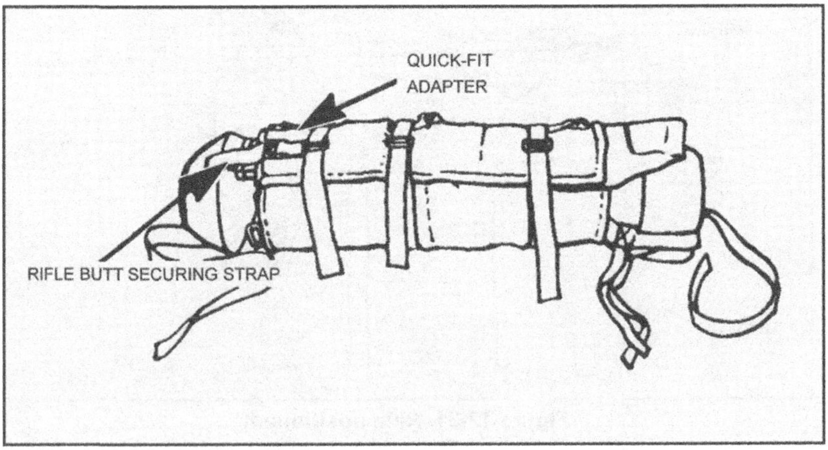

Figure 12-33. Pack closed.

d. **Secure honeycomb squares.**

(1) Route the missile launcher aft end bridle securing strap through its quick-fit adapter and tighten, using a quick-release fold.

(2) Secure two 9-inch-square blocks of cardboard honeycomb energy dissipater across the forward shock absorber of the missile, using the launcher forward end securing straps. Tighten the straps and form an X over the center of the honeycomb. Secure the straps using a quick-release fold (Figure 12-34). Tie one turn of 1/4-inch cotton webbing around the X formed by the launcher forward end, securing straps in a surgeon's knot and locking knot. Tie another one turn of 1/4-inch cotton webbing around the launcher forward end, securing straps below the cardboard honeycomb energy dissipator in a surgeon's knot and locking knot. If the missile forward end shock absorber is missing from the Dragon missile training round, it should be replaced with a 9- by 9-inch square of cardboard honeycomb.

WARNING

Never jump a live missile without this honeycomb in place and securely fastened. Place parachutists with DMJPs or CWIEs to the front of the stick. Staggered exits are difficult to control if individuals carrying either the DMJP or CWIE are not near the front of the sticks.

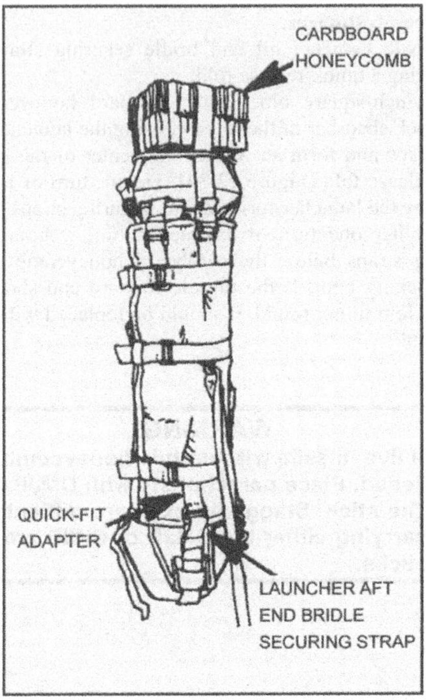

Figure 12-34. Honeycomb squares secured.

(3) If not already installed, attach the lowering line attaching strap to the launcher aft end bridle by means of a girth hitch with the large loop end (Figure 12-35, page 12-36). Run the small loop end through the large loop end and tighten (run the small loop end toward the lowering line stow pocket).

(4) Position the DMJP on the aft end shock absorber and retighten the launcher forward end securing straps using quick-release folds. All excess on the securing straps are folded back under the corresponding quick-fit adapter. Retighten one turn of 1/4-inch, 80-pound test cotton webbing in an X pattern around the launcher forward end securing straps, and tie securely with a surgeon's knot and locking knot (Figure 12-36, page 12-36).

(5) Tie one turn of 1/4-inch, 80-pound test cotton webbing around the launcher forward end, securing straps below the 9-inch square honeycomb. Tie in a surgeon's knot and locking knot to prevent the honeycomb from shifting.

(6) Position the DMJP on the honeycomb and tighten the launcher aft end bridle securing strap. Secure its running end by folding it back under the corresponding quick-fit adapter.

(7) Tighten the three horizontal securing straps as evenly as possible around the weapon(s). Incorporate a quick-release fold and secure with a retainer band or two turns of masking tape.

(8) Attach the quick-release assembly to the pack (Figure 12-37).

Figure 12-35.
Lowering line attaching strap installed.

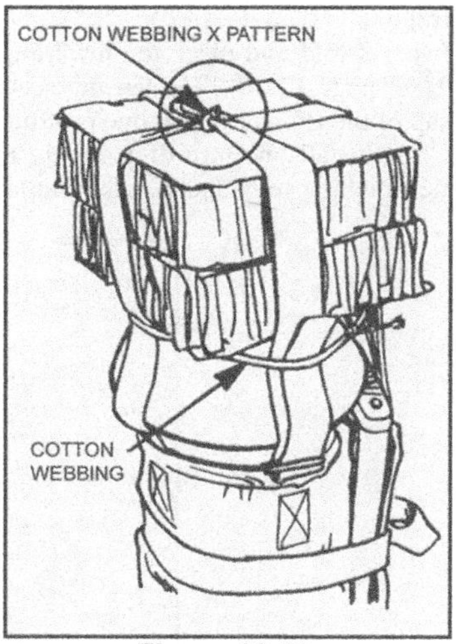

Figure 12-36.
X pattern.

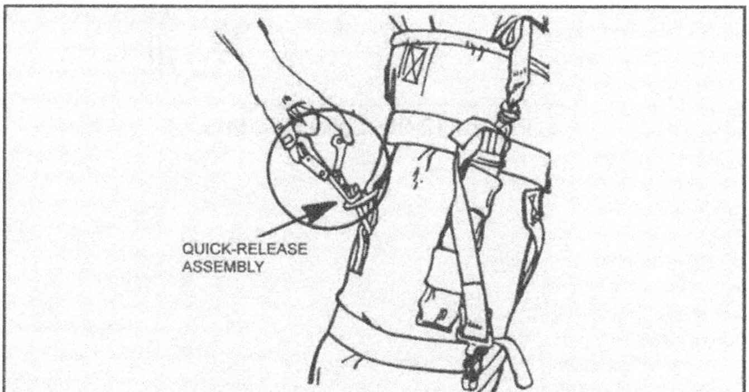

Figure 12-37. Quick-release assembly.

NOTE: Apply more tension to the end strap by tightening the three horizontal securing straps evenly. Uneven or insufficient tension can cause misalignment of the overlap portion of the pack and result in improper retention and protection of the rifle.

e. **Attach lowering line to DMJP.**

(1) Attach the lowering line to the lowering line attaching strap by routing the looped end of the HPT lowering line through the looped end of the lowering line attaching strap (Figure 12-38). Route the lowering line through its own loop and pull tight, forming a girth hitch.

(2) S-fold and place the lowering line inside the lowering line stow pocket (Figure 12-39, page 12-40). Tighten the securing tabs (Figure 12-40, page 12-41). The ejector snap of the HPT lowering line protrudes from the lowering line stow pocket.

(3) Install one turn of 1/4-inch, 80-pound test cotton webbing around the lowering line attaching strap and the adjacent launcher aft end bridle strap, and tie securely.

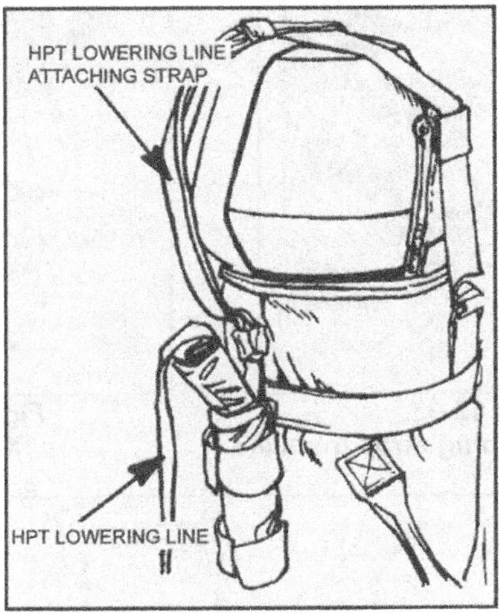

Figure 12-38. Lowering line.

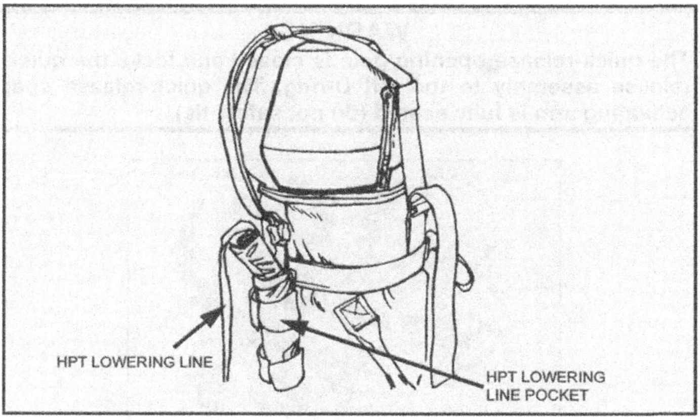

Figure 12-39. Lowering line inserted in pocket.

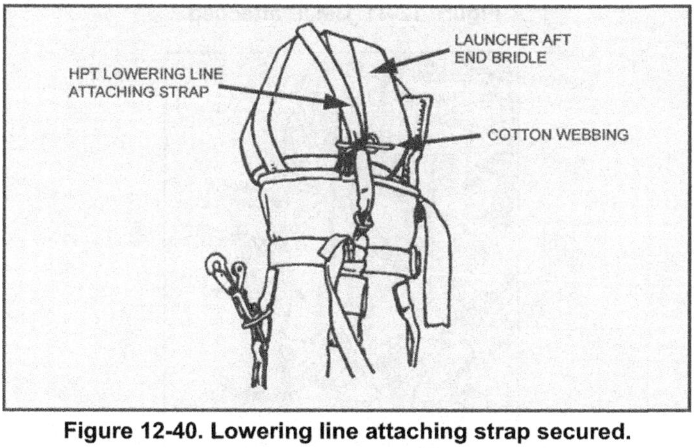

Figure 12-40. Lowering line attaching strap secured.

12-38. DRAGON MISSILE JUMP PACK ATTACHED TO PARACHUTIST

The DMJP is the last item attached to the parachutist.

a. The DMJP is secured to the parachutist by attaching the quick-release snap to the parachutist's left D-ring as the outermost item (Figure 12-41).

b. The lower tie-down tape is routed around the DMJP and the parachutist's left leg (Figure 12-42). A single-loop bowknot is tied on the front of the DMJP where it is easy for the parachutist to reach.

c. The upper tie-down tape is routed around the left main lift web directly below the chest strap. It is tied snugly with a single-loop bowknot (Figure 12-43, page 12-42).

WARNING

The quick-release opening gate is closed and locks the quick-release assembly to the left D-ring. The quick-release snap activating arm is fully seated (do not safety tie).

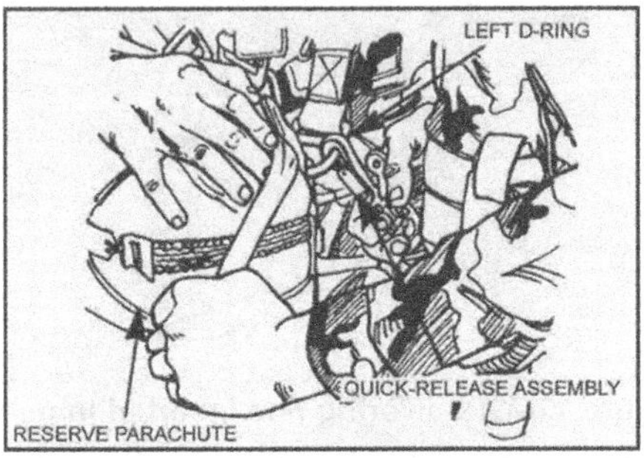

Figure 12-41. DMJP attached.

Figure 12-42. HPT lowering line attached.

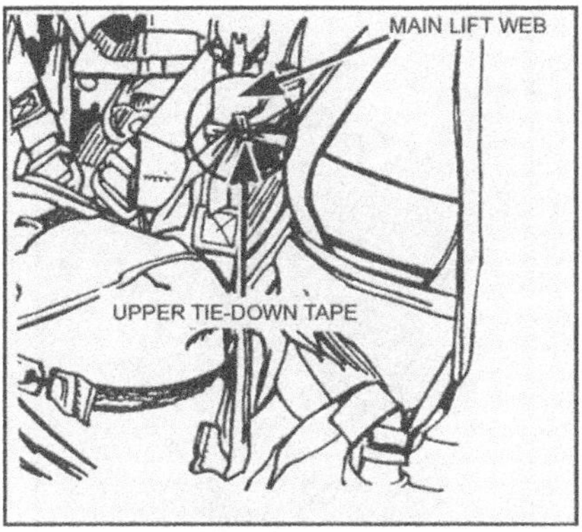

Figure 12-43. Upper tie-down.

12-39. INDIVIDUAL JUMP PROCEDURES

The DMJP is jumped with a front-mounted large ALICE pack (Figure 12-44).

a. At the 20-minute time warning, one of the safety personnel attaches the DMJP to the parachutist and inspects it.

b. After leaving the aircraft, the parachutist—

(1) Releases the lower tie-down tape, then the upper tie-down tape, in sequence. If either tie malfunctions and the parachutist must ride the missile down, it is important that the missile be vertical, not hanging directly below the feet.

(2) Looks to see that the area below is clear (about 100 to 200 feet above ground), and activates the DMJP quick-release assembly by using the left hand to push down and away on the activating arm, allowing it to fall the length of the lowering line.

NOTE: To jettison the DMJP in an emergency, the parachutist lowers the pack, then pulls out on the yellow safety lanyard attached to the ejector snap, allowing the missile pack to fall free.

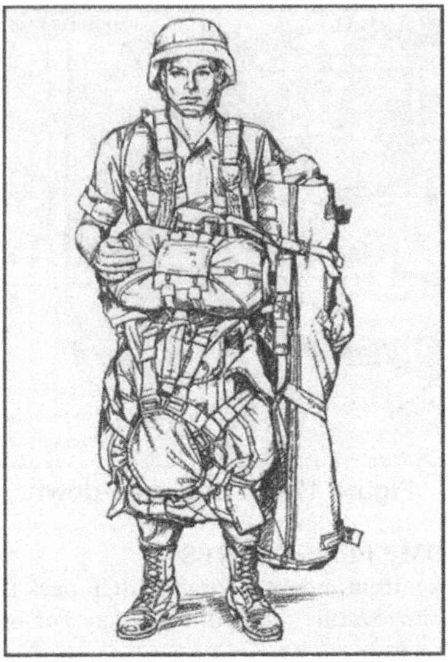

Figure 12-44. DMJP attached to parachutist.

12-40. DRAGON TRACKER

The tracker (sighting system) is placed inside the large ALICE pack and is attached to the parachutist. The system must remain in its protective case during the jump.

12-41. DRAGON MISSILE JUMP PACK AND ALICE PACK RIGGED AS A TANDEM LOAD

To rig the DMJP and ALICE pack for lowering, modifications are required along with the DMJP/AT4JP modified HPT lowering line.

a. **Materials**.

- Fastener tape, pile, color OD 106, 1-inch width, Class 1, MIL-F-21840, NSN 8315-00-106-5974.
- Fastener tape, hook, color OD 106, 1-inch width, Class 1, Type II, MIL-F-21840, NSN 8315-00-106-5973.
- Webbing, textile, nylon, Type X, Class 1, 1 23/32-inch width, color olive drab, MIL-W-4088, and Class R, MIL-W-27265, NSN 8305-00-261-8584.
- Thread, nylon, color OD-S1, conforming to Type II, Class A, Size 3, 800-yard tube of V-T-295, NSN 8310-00-559-5212.
- D-ring, parachute harness, drawing, No. 11-1-485.

b. **Stitching Requirements**.

(1) Stitching must conform to FED-STD-751, Type 301, 5 to 8 stitches per inch.

(2) Ends of stitching must be over-stitched not less than 1/2 inch.

c. **Modification Procedure**.

(1) ***Replacement of the V-ring with D-ring, 11-1-485.***

(a) Place the jump pack on a repair table, positioning it so the outside faces upward.

(b) Cut the box-X stitching on the chafe web end, which secures the V-ring to the pack. Remove the V-ring and cut the stitching.

(c) Pass the end of the chafe through replacement D-ring 11-1-485 and stitch with the box-X stitch formation to pack as in the original construction.

(2) ***Attachment of the D-ring 11-1-485 and web chafe to aft shock absorber harness cross adjustable strap.***

(a) If installed, remove the lowering line attaching web, which is girth-hitched to the aft shock absorber harness cross straps.

(b) Cut a 7 1/4-inch length of Type X nylon webbing and sear cut ends to prevent fraying.

(c) Place marks 3 inches from each end and in the center of the webbing.

(d) Place marks 3 5/8 inches from each side of the center of the aft harness cross adjustable strap.

(e) Pass the end of 7 1/4-inch Type X nylon webbing through the D-ring 11-1-485 loop and position it between the markings of the aft harness cross adjustable strap. Stitch with a 3-inch, 4-point WW stitch formation at each end.

(3) ***Attachment of hook-and-pile lowering line retainer.***

(a) Cut a 3 1/2-inch length of hook tape.

(b) Cut a 4-inch length of pile tape.

(c) With hook and pile facing each other, overlap tabs 1 3/4 inches and press together.

(d) Place marks 1 and 2 inches (on the inside) from the binding edge of the pack body on the aft cross adjustable strap.

(e) With pile side facing toward outside of pack and hook facing to inside of pack, position edges of tape overlap even with cross strap edges between markings and stitch to cross strap with a single box stitch formation.

12-42. ALICE PACK (LARGE) JUMPED WITH DRAGON MISSILE JUMP PACK

The large ALICE pack with frame must be jumped with the DMJP to stabilize the DMJP and to accommodate the Dragon tracker assembly.

a. Completely pad the tracker assembly with cellulose wadding to prevent damage.

b. Place the padded tracker assembly in the inside pouch with soft articles of clothing or equipment packed around it. Place additional items in the pouch, with hard and sharp items padded.

c. Fill the outside ALICE pockets with nonfragile items to capacity, since the filled pockets aid the positioning and prevent slippage of the H-harness during parachute opening and lowering line deployment.

d. Close the pack by use of the drawstring closure, engaging the securing straps and attachment of pocket snap fasteners.

e. Adjust the adjustable shoulder carrying straps snugly. S-fold and tape the excess webbing on the adjustable shoulder carrying straps. Route the running ends of the waist straps around the frame opposite the lower back pad, tighten, and secure in place by taping. This secures the adjustable should carrying straps and reduces the possibility of entanglement on board the aircraft.

f. To stow the Dragon modified lowering line, S-fold the lowering line neatly on top of the web inside the retainer, ensuring that ends are stacked evenly with the retainer outer edges. Secure the pile tab on web located 46 inches from the ejector snap end to hook the extension on the retainer. Fold the hook side of flap tightly over the S-folded lowering line and, holding it firmly, fold pile side of the flap over and secure hook and pile together. Secure the pile extension on the retainer flap to the hook tab at the looped end of the lowering line.

12-43. ALICE PACK RIGGED WITH FRAME

The pack is turned upside down.

a. Place the modified H-harness on the pack so that the cross straps are in front of the pack and the friction adapters are touching (or near) the bottom of the pack frame.

(1) Route the equipment retainer straps over the top of the pack and then under the top portion of the frame.

(2) Route the equipment retainer straps over the horizontal bar of the frame and cross at the center of the back of the pack.

(3) Route the straps under the frame and secure them to the friction adapters, forming a quick release.

b. To attach the lowering line after rigging pack, girth-hitch the lowering line loop at the intersection of the crossed equipment retainer straps (at the rear center of the pack).

(1) Secure the stowed lowering line to the left side (as worn on the parachutist) on the keepers of the pack with two turns of masking tape or two sets of retainer bands. The lowering line ejector snap faces up and to the right or left.

(2) Thread the (H-harness) D-ring attaching straps through the intermediate friction adapters, forming a quick release with the running ends that are pointing away from the parachutist.

12-44. DRAGON MISSILE JUMP PACK RIGGED

The DMJP is rigged as described above, omitting installation of the lowering line attaching strap.

12-45. ALICE PACK ATTACHED TO PARACHUTIST

The D-ring attaching strap snap hooks (on rigged ALICE) are attached to the harness D-rings outside of the reserve connector snaps.

12-46. DRAGON MISSILE JUMP PACK RIGGED FOR TANDEM LOAD

The rigged DMJP is placed next to the parachutist's left side.

a. To secure the 15-foot HPT lowering line, grasp the ejector snap of the HPT lowering line. Route it (from bottom to top) through the D-ring openings at the pack body and aft end locations. Position the 18-inch strata blue marking on the HPT lowering line even with the D-ring on top of the DMJP (Figure 12-45).

b. Secure the lowering line to the aft cross strap at the pack body end by folding the hook tab over the lowering line and fastening pile together.

c. Attach the quick-release snap to the quick-release link, which has been routed through the D-ring. The snap opening must face away from the pack.

12-47. DRAGON MISSILE JUMP PACK ATTACHED TO PARACHUTIST

The DMJP quick-release assembly is attached to the left harness D-ring outside of the reserve connector snap and H-harness or HSPR snap hook (Figure 12-46).

a. Attach the HPT lowering line ejector snap to the lowering line adapter web.

b. Route the lower tie-down tape around the DMJP and the parachutist's left leg.

c. Tie with a bowknot on the front of the DMJP so it is easy for the parachutist to reach.

d. Route the upper tie-down tape under and around the left main lift web directly below the chest strap and tie snugly with a bowknot.

e. Use the adjustable leg strap of the HSPR as a lower tie-down tape.

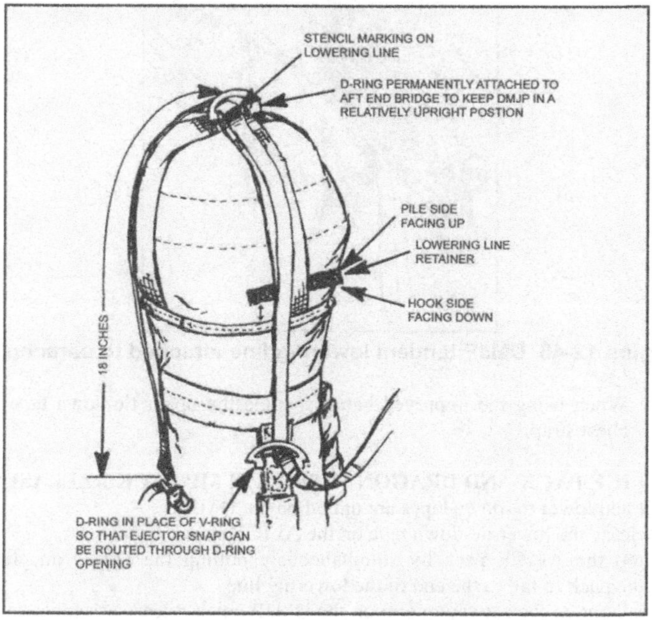

Figure 12-45. DMJP modified with D-ring.

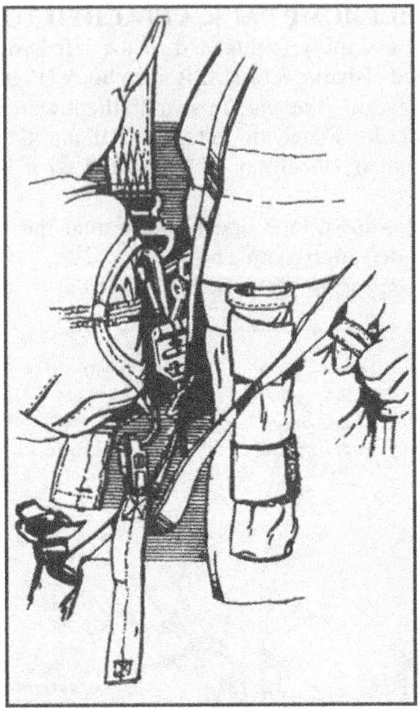

Figure 12-46. DMJP tandem lowering line attached to parachutist.

NOTE: When using the improved harness, route the upper tie-down tape below the chest strap.

12-48. ALICE PACK AND DRAGON MISSILE JUMP PACK RELEASED

The upper and lower tie-down tapes are untied on the DMJP.

a. Release the lower tie-down tape on the ALICE pack.

b. Drop the ALICE pack by simultaneously pulling the D-ring attaching straps, allowing the pack to fall to the end of the lowering line.

c. Push out on the activating arm on the DMJP quick-release assembly so the DMJP slides down the lowering line to the end.

NOTE: To jettison the ALICE pack and DMJP in an emergency (after performing the above), lift the reserve up and pull the yellow safety lanyard on the lowering line ejector snap, allowing the ALICE pack/DMJP to fall free.

WARNING

Under no circumstances will the parachutist release the DMJP before releasing the ALICE pack.

12-49. REMOVAL OF LOWERING LINE

Upon landing, the parachutist removes the lowering line by pulling the line through the two D-rings and releasing the girth hitch on the H-harness. The DMJP MOD lowering line is stored with the DMJP for reuse.

Section XII. AT4 JUMP PACK

The parachutist's AT4JP is designed to carry one AT4 weapon round (SM-136) and one M16 rifle. The jump pack consists of a pack body constructed of nylon and 1/4-inch thick felt material. When the jump pack is rigged with the AT4 weapon round, M16, and shock absorber, it is 47 inches long, 9 inches in diameter, and weighs about 29 pounds. It is secured vertically by a quick-release snap to the left D-ring of the parachutist's harness.

12-50. COMPONENTS

To prevent the jump pack from swaying during the parachutist's exit from the aircraft or the opening shock of the parachute, two tie-down tapes are provided. The upper tie-down secures the pack to the main lift web, while the lower tie-down secures the pack to the parachutist's left leg.

a. A lowering line stow pocket is attached to the exterior of the pack for retaining the HPT lowering line. A 24-inch lowering line attaching strap is issued with the jump pack and is required for assembling the pack and lowering line when rigged for delivery of the AT4JP only. (When the AT4JP and ALICE are rigged for tandem load, a modified [DMJP/AT4] lowering line is required.) The looped end of the lowering line or lowering line attaching strap (based upon the configuration to be lowered) is attached to the nonadjustable cross D-ring strap on the jump pack so it will be suspended below the parachutist and impact in a vertical position. The ejector snap on the opposite end is attached to the lowering line adapter web on the parachutist's harness.

b. The shock absorber (polyurethane/honeycomb) configuration at the aft end of the jump pack is intended to prevent damage to the AT4 weapon round during ground impact. Upon landing, the parachutist can rapidly gain access to the weapon round and M16 by pulling the quick releases rigged into the jump pack securing straps. The total assembly is shown in Figure 12-47.

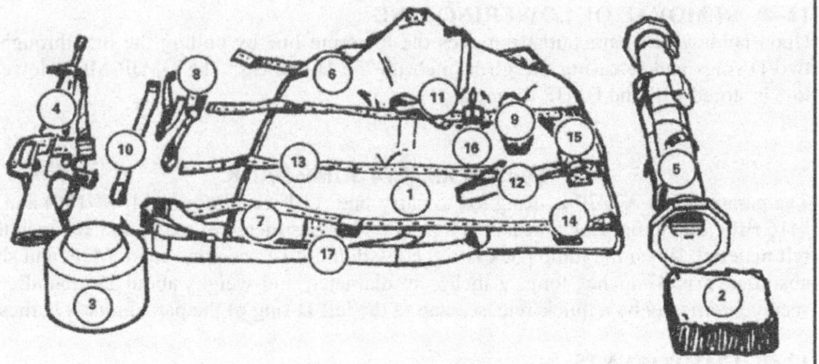

1. AT4 JUMP PACK
2. CARDBOARD HONEYCOMB
3. LAUNCHER AFT END SHOCK ABSORBER
4. M16 RIFLE
5. AT4 WEAPON
6. RIFLE MUZZLE STOW POCKET
7. RIFLE BUTT STOW POCKET
8. HPT LOWERING LINE
9. LOWERING LINE POCKET
10. LOWERING LINE ATTACHING STRAP
11. UPPER TIE-DOWN TAPE
12. LOWER TIE-DOWN TAPE
13. SIDE SECURING STRAPS
14. QUICK-FIT ADAPTER
15. CARRYING HANDLE
16. QUICK RELEASE ASSEMBLY
17. LAUNCHER AFT END SECURING STRAPS

Figure 12-47. AT4 nomenclature.

12-51. AT4 JUMP PACK RIGGED

To prepare the pack, the parachutist lays the pack down with the felt side of the pack facing up and extends all securing straps.

a. **Positioning the AT4 Weapon Round.**

(1) Position the weapon round on the pack with the carrying sling facing down, the launcher forward (small) end fitted into the forward end securing strap, and the launcher aft (large) end centered on the middle launcher aft end securing strap (Figure 12-48, page 12-48).

(2) Ensure that the launcher forward (small) end fits snugly into the nonadjustable cross D-ring strap.

(3) Fold the portion of the pack that has the rifle muzzle and butt stow pockets attached over the weapon round.

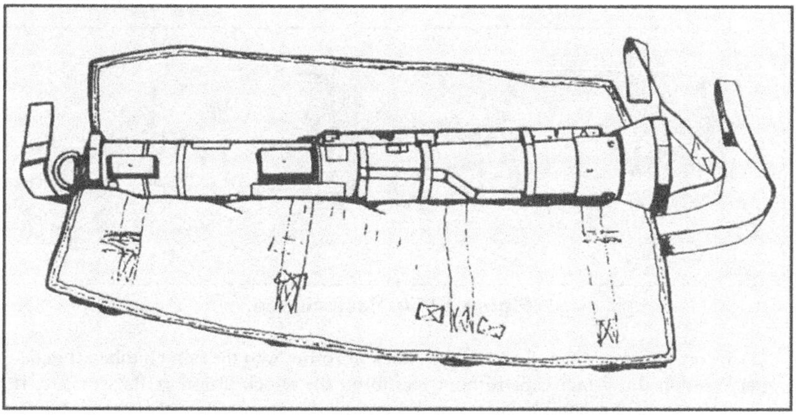

Figure 12-48. AT4 weapon round positioned.

b. **Positioning the M16.**

(1) Insert a 20-round magazine into the rifle (optional).

(2) Secure a 30-round magazine to the sling of the rifle, using adhesive tape (optional).

(3) Insert the rifle muzzle into the muzzle stow pocket and place the butt in the rifle butt stow pocket as shown in Figure 12-49. Ensure that the magazine well and pistol grip point toward the side securing straps, and the rifle muzzle is inserted into the muzzle pocket as far as possible.

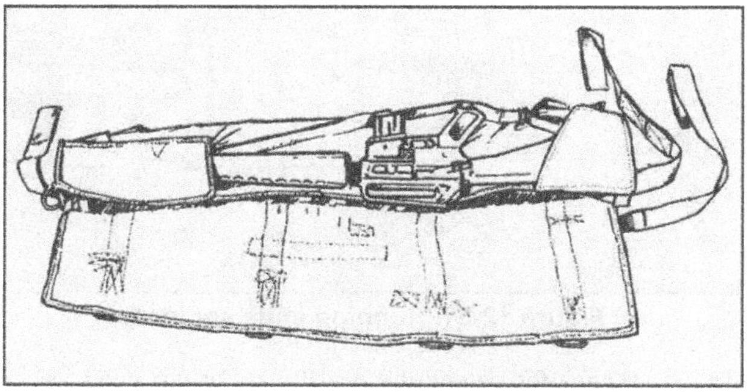

Figure 12-49. Positioning the M16.

c. **Closing the Pack.**

(1) Rotate the unfolded portion over the rifle and weapon round snugly.

(2) Route the four side securing straps through the quick-fit adapters and tighten, then thread the end through the adapter, forming a quick-release fold. Fold the strap excess back under its corresponding quick-fit adapter (Figure 12-50).

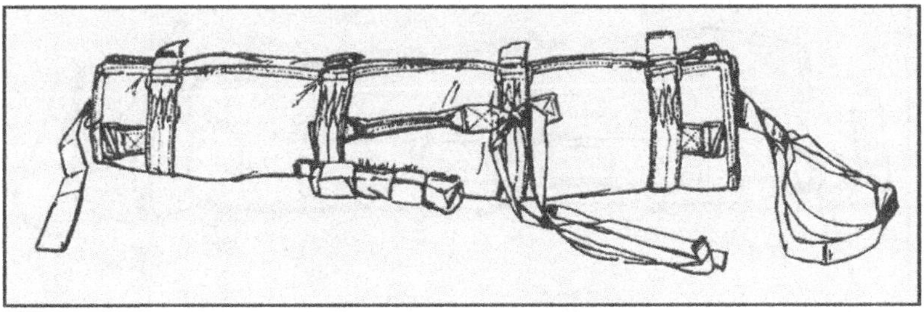

Figure 12-50. Pack closed.

(3) Insert the cone-shaped end of the shock absorber into the launch tube aft end.

(4) Position the 8-inch square honeycomb on the shock absorber flat surface. If the shock absorber is missing from the training round, replace it with two 8- by 8-inch cardboard honeycomb energy dissipaters. Thread the end of the securing strap through the launcher aft end securing strap keeper. Route the two aft end securing straps over the cardboard honeycomb through each quick-fit adapter and tighten, then thread the end through the adapter, forming a quick-release fold. Fold excess securing straps back under the corresponding quick-fit adapter (Figure 12-51).

(5) Route the launcher forward end securing strap under the nonadjustable cross D-ring strap through the quick-fit adapter and tighten. Then thread the end through the adapter, forming a quick-release fold. Fold excess securing straps back under the corresponding quick-fit adapter.

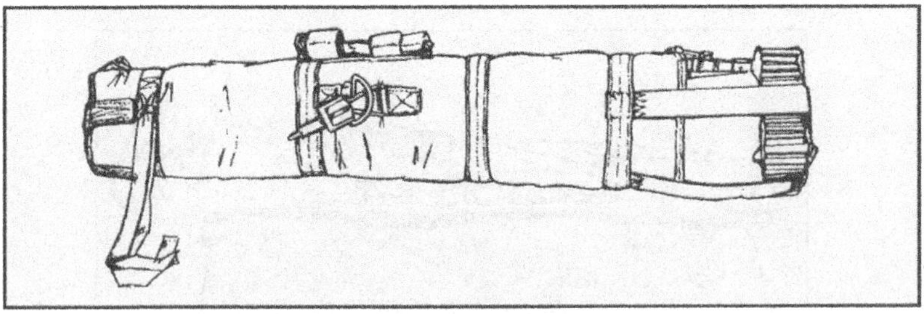

Figure 12-51. Running ends secured.

NOTE: The procedures for rigging the AT4JP vary at this point, depending on the method used to lower the parachutist's equipment. Follow paragraphs 12-51 and 12-52 if a single lowering line (tandem) is used.

12-52. AT4 AND ALICE PACK RIGGED

To attach the quick-release snap, the parachutist attaches it to the quick-release link that is routed through the D-ring. The opening gate should face away from the pack (Figure 12-52).

a. **Stowing the Standard Lowering Line.**

(1) S-fold the lowering line neatly on top of the web inside the retainer, ensuring that the ends are stacked evenly with the retainer flap outer edges. Secure the pile tab on the web located at the ejector snap end to hook extension on retainer.

(2) Fold the hook side of the flap tightly over the S-folded lowering line and, holding it firmly, fold the pile side of flap over and secure hook and pile together.

(3) Secure the pile extension on the flap retainer to the hook tab at the loop end of the lowering line (Figure 12-53).

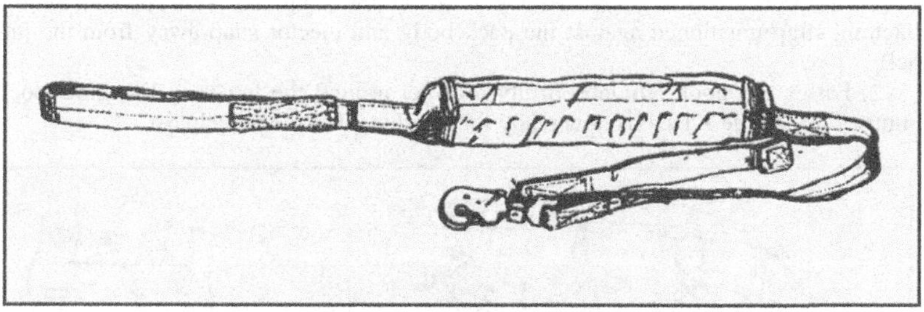

Figure 12-52. HPT lowering line stowed.

NOTE: The hook-pile tape located on the retainer flap and at each end of the stowed lowering line assembly must be secured firmly and without slack to prevent line spillage. To provide a secure closure of the hook and pile, press them firmly, evenly, and smoothly together without puckers.

b. **Securing the Attaching Web to the Standard Lowering Line.**

(1) Attach the lowering line to the lowering line attaching strap by routing the loop of the lowering line through the small loop end of the lowering line attaching strap.

(2) Route the lowering line through its own loop and pull tight (Figure 12-53).

c. **Fastening Attaching Strap to Nonadjustable Cross Strap D-Ring.**

(1) If not already installed, attach the lowering line attaching strap to the nonadjustable cross strap D-ring by routing the large loop around the D-ring.

(2) Route the lowering line through the attaching strap loop and pull tight (Figure 12-53).

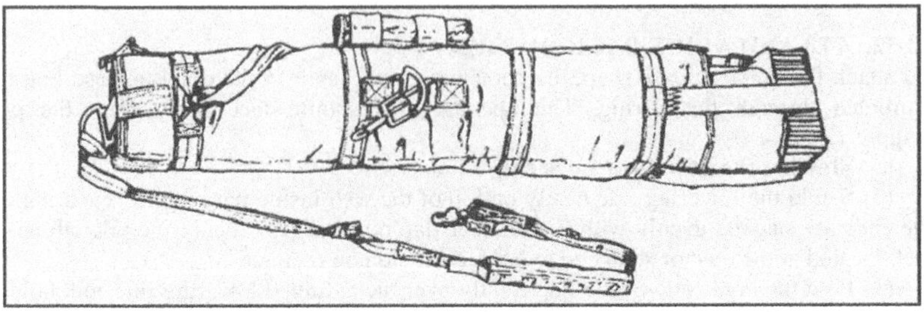

Figure 12-53. Attaching strap secured.

d. **Stowing the Standard Lowering Line in Stow Pocket.**

(1) Insert the lowering line assembly in the lowering line stow pocket with the attaching strap positioned against the pack body and ejector snap away from the jump pack.

(2) Fasten the hook-pile tabs firmly together around the lowering line stow pocket (Figure 12-54). The AT4JP is now ready for attachment to the parachutist.

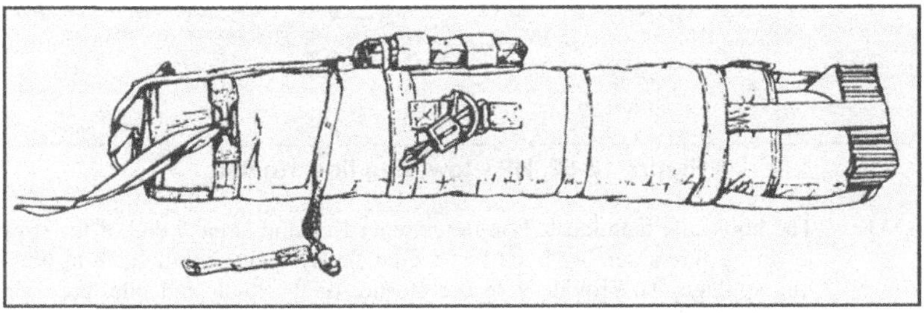

Figure 12-54. Standard lowering line stowed.

NOTE: Before attachment of the reserve parachute, individual equipment, and AT4JP, ensure that the canopy release assemblies of the T-10-series harness are in the hollows of the parachutist's shoulders for a proper fit. Adjust an improperly fitted harness by repositioning the diagonal back straps to the size channel that correctly fits the parachutist. A properly adjusted T-10-series harness on the parachutist ensures that the reserve parachute, individual equipment, and AT4JP are in the proper position. An improperly fitted harness with a full equipment and weapon load is uncomfortable and could result in problems during ground movement prior to enplaning, movement in the aircraft, aircraft exit, and main parachute deployment.

12-53. EQUIPMENT ATTACHED TO PARACHUTIST (STOWED LOWERING LINE)

To rig the ALICE pack with frame, the parachutist's ALICE pack with frame is prepared following procedures listed in paragraph 12-43, with the following exception: secure the stowed lowering line to the right side (as worn on the parachutist) on the vertical pack frame bar in two places with two turns of masking tape or two sets of retainer bands. The lowering line ejector snap faces up and to the right or left.

a. Attach the D-ring attaching strap snap hooks (on rigged ALICE pack) to the harness D-rings outside of the reserve connector snaps.

b. Secure ALICE pack to the parachutist's leg.

c. Attach the HPT lowering line ejector snap from the ALICE pack to the right D-ring on the harness.

d. Secure the AT4JP by attaching the quick-release snap to the left D-ring outside of the H-harness snap hook (Figure 12-55). Ensure that the quick-release safety gate is closed and the quick-release snap is locked to the D-ring, and that the quick-release assembly release attachment sequence is inboard to outboard as follows: reserve, H-harness, and AT4JP quick-release snap.

Figure 12-55. AT4JP secured.

e. Route the upper tie-down tape under and around the left main lift web directly below the chest strap. Tie snugly with a double-loop bowknot on the front of the AT4JP where it is easy for the parachutist to reach.

f. Route the lower tie-down tape around the AT4JP and the left leg. Tie with a double-loop bowknot on the front of the AT4JP where it is easy to reach. (If the HSPR is used, remove the lower tie-down on the AT4JP and route the left leg strap of the HSPR around the outside of the AT4JP, then retighten.)

12-54. ALICE PACK WITH FRAME AND AT4JP RIGGED

The quick-release snap is not attached at this point. This will be accomplished during the equipment attaching sequence. To stow the DMJP/AT4JP modified lowering line for tandem load, the following must be performed:

a. S-fold the lowering line neatly on top of the web inside the retainer, ensuring that ends are stacked evenly with the retainer outer edges.

b. Secure the pile tab on the web that is located 46 inches from the ejector snap end to hook extension on retainer.

c. Fold the hook side of the flap tightly over S-folded lowering line and, holding it firmly, fold the pile side of the flap over and secure hook and pile together.

d. Secure the pile extension on the retainer flap to the hook tab at the looped end of the lowering line (Figure 12-56, page 12-56).

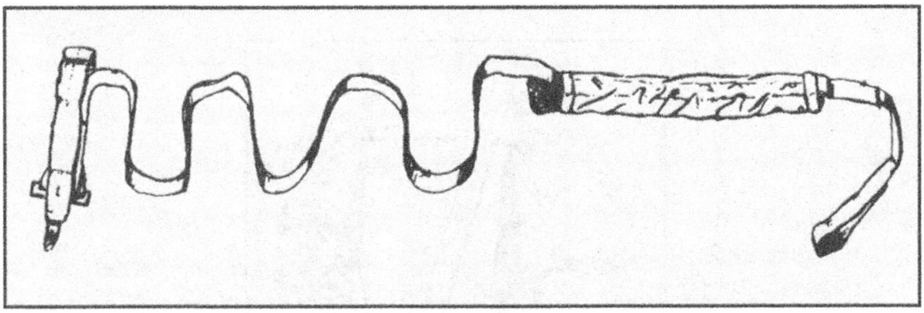

Figure 12-56. HPT lowering line stowed for tandem load.

12-55. EQUIPMENT ATTACHED TO PARACHUTIST (MODIFIED STOWED LOWERING LINE)

To rig the ALICE pack with frame, the parachutist's ALICE pack with frame is prepared following procedures listed in paragraph 12-43 with the following exception: secure the modified stowed lowering line to the left side (as worn on the parachutist) on the vertical pack frame bar in two places with two turns of masking tape or two sets of retainer bands. The lowering line ejector snap faces up and to the right or left.

Figure 12-57. Attaching equipment to parachutist.

a. Attach the D-ring attaching strap snap hooks (on rigged ALICE pack) to the harness D-rings outside of the reserve connector snaps.

b. Secure the ALICE pack to the parachutist's leg.

c. Stand the rigged AT4JP next to the parachutist's left side.

d. Grasp the ejector snap of the modified HPT lowering line already attached to the ALICE pack and route (from bottom to top) through the D-ring opening at the center of the pack, upper tie-down chafe loop, and top D-ring located at the aft end. Position the 18-inch strata blue marking on the modified HPT lowering line even with the D-ring on top of the AT4JP (Figure 12-57).

e. Attach the quick-release snap to the AT4JP. The snap opening faces away from the pack.

f. Attach the modified HPT lowering line ejector snap to the lowering line adapter web.

g. Secure the AT4JP by attaching the quick-release assembly to the left D-ring outboard of the H-harness snap. Ensure that the quick-release safety gate is closed, the quick-release snap is locked to the D-ring, and the quick-release assembly release lever is fully seated (do not safety tie). The harness left D-ring snap attachment sequence is inboard to outboard as follows: reserve, H-harness, and AT4JP quick-release snap.

h. Route the upper tie-down tape under and around the left main lift web directly below the chest strap. Tie snugly with a double-loop bowknot on the front of the AT4JP where it is easy to reach (Figure 12-58).

i. Route the lower tie-down tape around the AT4JP and the left leg. Tie with a double-loop bowknot on the front of the AT4JP where it is easy to reach. If the HSPR is used, remove the lower tie-down on the AT4JP and route the left leg strap of the HSPR around the outside of the AT4JP, then tighten (Figure 12-59).

Figure 12-58. Routing upper tie-down.

Figure 12-59. Lower tie-down tape routed around AT4JP.

12-56. ALICE PACK AND AT4JP RELEASED

The lower and upper tie-downs on the AT4JP are released.

a. Release the ALICE pack as follows:

(1) *H-harness (modified)*. Release the lower tie-down tape on the ALICE pack. About 100 to 200 feet above ground, drop the ALICE pack by simultaneously pulling the quick-release folds on the D-ring attaching straps, allowing the pack to fall to the end of the line.

(2) *HSPR*. About 200 feet above the ground, grasp the release handle and pull up and out quickly, simultaneously releasing the load from the harness and legs, allowing the load to drop the length of the lowering line. **Release the handle immediately following separation of the load from the parachutist.** Secure the handle for releasing the load to the HSPR with a length of tubular webbing and stay with the HSPR to prevent its loss and separation.

b. Push out on the activating arm of the quick-release assembly of the AT4JP to lower the AT4JP. The AT4JP will either fall to the end of its lowering line or slide down the tandem lowering line to the ALICE pack.

NOTE: To jettison the ALICE pack and AT4JP in an emergency, the parachutist (after performing the above) lifts the reserve up and pulls the yellow safety lanyard on the lowering line ejector snap, allowing the ALICE pack and AT4JP to fall free.

WARNING

Under no circumstances will the parachutist release the AT4JP before releasing the ALICE pack.

c. Upon landing, remove the HPT lowering line by pulling the line through the two D-rings and disassembling the girth hitch on the H-harness.

d. Store the AT4JP modified HPT lowering line and polyurethane shock absorber with the AT4JP for reuse.

Section XIII. ALL-PURPOSE WEAPONS AND EQUIPMENT CONTAINER SYSTEM (AIRPAC)

The AIRPAC is a lightweight nylon equipment pack system used to enclose a jump load. The AIRPAC front-mount container can accommodate all size packs, M47 tracker, radios, and 60-mm mortar with baseplate or bipod. The side-mount container can hold one M47 round of ammunition, the AT4, M4, M24, SAW, M60 machine gun, and other ancillary equipment.

12-57. COMPONENTS

The AIRPAC consists of the front-mount container, the side-mount container, and the parachutist's individual equipment rapid release (PIE/R2) release mechanism.

12-58. RIGGING LOADS IN THE FRONT-MOUNT CONTAINER

Loads in the front-mount container are rigged as follows:

a. **Rigging the ALICE Pack and the Field Pack, Large, Internal Frame (FPLIF).**

(1) Extend all retaining straps outward to their limits. (The retainer strap with the circular closing flap is the top of the container.)

(2) Place the container so that the retaining straps face the ground and the parachute recovery bag faces up.

(3) Fold the parachute recovery bag (PRB) so that no portions protrude outside the container limits.

(4) Place pack on the PRB. Ensure the circular release rings on the bottom of the container are not constrained by the pack. The ALICE pack is jumped upside down with the waist pad against the front of the thighs. The FPLIF is jumped right side up.

(5) Bring the right side of the container material up first, then the left side and the bottom flap, and then the top flap.

(6) Place the circular closing flap on top of the pack and open the circular closing flap by pulling the tab. Bring the left, right, and bottom retaining straps to the center of the circular closing flap and route the white grommet securing loop through each grommet of the retaining straps. No specific order is required. Bring the top flap material up last. Place the cotter pin through the white grommet securing loop.

(7) Tighten all retaining straps, ensuring that the circular closing flap remains centered on the ALICE pack. Adjust the protective flaps to stow excess fabric. Close the circular closing flap using three hook-pile tabs.

(8) Underhand fold the free-running ends of the retaining straps toward the friction adapters and secure them under the webbing retainers.

NOTE: The circular release rings should be exposed on the top flap of the front-mount container. If the rigged pack is above the rings and restricts the rings' movement, the load must be adjusted.

b. **Rigging the Release Mechanism of the Front-Mount Container.**

(1) Stand the rigged pack in an upright position with the bottom of the AIRPAC facing the jumper.

(2) Route release handle cables through the cross strap on top of the container and secure the HPT.

(3) With the attaching strap snap hook facing down and towards the jumper, route the circular ring of the attaching strap (from bottom to top) through the circular release ring on top of the container. Fold the circular ring of the attaching strap back over and route the red attaching loop through the attaching ring on the attaching strap. With the cable loop retainer on the female portion of the leg strap release assembly facing up, route the red attaching loop from bottom to top through the grommet on the female portion of the leg strap release assembly. Route the release handle cable through the red attaching loop and then through the cable loop retainer. Repeat this sequence for the other strap.

c. **Rigging the HPT Lowering Line.**

(1) Select the desired landing orientation of the front-mount container (horizontally or vertically) and attach the short bridle to the appropriate triangle link.

(2) Route the looped end of the HPT lowering line through the unsewn portion of the short bridle and form a girth hitch. Place a retainer band around the nylon tab located inside the closure flaps. Route the HPT lowering line through the retainer band and stow in the retention pocket. Secure the HPT.

(3) For transportation, attach the ejector snap to the primary short bridle triangle link.

12-59. RIGGING LOADS IN THE SIDE-MOUNT CONTAINER

Rigging procedures for the side-mount container are as follows:

a. Place the side-mount container on the ground with the friction adapters facing down and the free-running ends of the internal securing strap to the right. Then, position packing material flush with the bottom edge of the fabric flaps and place the weapon on the container in the following orientation (Table 12-3):

WEAPON	PACKED ORIENTATION	PACKING MATERIAL
M60 MG/SAW	Muzzle up	None required
AT4	Muzzle up	1 piece 6" by 6" honeycomb on bottom
60-mm Mortar	Muzzle down	1 piece 6" by 6" honeycomb on bottom
Dragon	Muzzle down	2 pieces 8" by 8" honeycomb on bottom

Table 12-3. Packed weapon orientation and packing material required for side-mount container.

b. Tape and pad exposed or protruding parts (that is, sight posts, grip stocks, or bipods) with a soft material such as cellulose wadding or an adequate substitute.

c. Center load on the internal securing strap and pull lower fabric flap over packing material. Route free-running end of internal securing strap through the friction adapter on the lower fabric cap. Place floating V-ring of internal securing strap on the top center of the load. Form a 5- to 6-inch quick release in the strap and tighten. From the top, fold excess weapons container material downward until flush with the top of the load.

d. Wrap unpadded fabric flap over weapon. Bring padded fabric flap over load and loosely secure container retaining straps to friction adapters; form a 5- to 6-inch quick release in the strap. Do not tighten strap.

NOTE: If jumping an individual weapon, place weapon muzzle down in container slide-fastener compartment. Secure the slide fastener.

e. Tighten the container retaining straps (with quick releases). S-fold excess webbing and secure to the quick-release loop with two turns of a retainer band, leaving a 3-inch tail. Secure drawstring on top of container with a quick-release knot, if necessary.

12-60. RIGGING AIRPAC AS TANDEM LOAD WITH HOOK-PILE TAPE LOWERING LINE

The AIRPAC is rigged as a tandem load with HPT lowering line as follows:

a. If the HPT lowering line is attached to the front-mount container, detach and place to the side. Using the extended bridle, place the looped end through unsewn portion of short bridle and secure with a girth hitch. Route looped end of HPT lowering line through floating V-ring of extended bridle and form a girth hitch.

b. Stow the HPT lowering line in the retention pocket. Stow extended bridle excess in the retention pocket, allowing for a 2-foot running end. Secure the retention pocket and attach the ejector snap of the HPT lowering line to the accessory attaching ring, triangle link, or left D-ring on the parachutist's harness.

c. After the side-mount container is attached to the jumper, the extended bridle snap hook is attached to the V-ring on the top of the side-mount container.

12-61. AIRPAC ATTACHED TO PARACHUTIST USING PIE/R2 RELEASE MECHANISM

AIRPAC is attached to parachutist using the PIE/R2 release mechanism as follows:

a. Place the front-mount container attaching straps under the reserve parachute and attach the front-mount container snap hooks to the enlarged D-rings on the outside of the reserve connector snaps (same as a single-point release system).

b. Attach the HPT lowering line ejector snap to the parachutist harness lowering line adapter web.

NOTE: When using the PIE/R2 release mechanism, ensure the harness has a lowering line adapter web or new harness triangle link.

c. Attach the PIE/R2 release mechanism to the side-mount container (located on jumper's left side) by placing the hook portion of the snap shackle with the hook through the quick-release link from top to bottom. The spring-loaded gate should face away from the container. Remember to hook the gate toward jumper's groin muscle.

d. Open the snap shackle with hook, attach to the D-ring on the outside of the attaching strap snap hook from back to front, and close the snap shackle.

e. Take the extended bridle snap hook and attach to floating V-ring on top of side container. Secure extended bridle under retention flap on container.

f. On the PIE/R2 release mechanism, route the upper attaching strap around the back of the side-mount container, through the main lift web, and above the chest strap. With orange release handle facing forward on the side-mount container, route the grommet tab through the looped end of the upper attachment strap, over the red attaching loop, and secure with cable. Tighten the upper strap, ensuring that the orange release handle stays on the front portion of the side-mount container.

g. Connect the white release lanyard snap hook to the short bridle attaching the V-ring on the top of the front-mount container. Listen for a metallic click and pull to ensure the snap hook is secure.

h. Attach the leg straps, ensuring the left leg strap is also routed around the side container. Tighten and secure. Stow excess webbing under webbing retainers.

Section XIV. STINGER MISSILE JUMP PACK

The Stinger missile jump pack provides a means for the air delivery of the Stinger missile system when it is attached to a jumper during airborne operations. The SMJP is an overpack body constructed of nylon duck material, shock-absorbing material, nylon securing straps, an HPT lowering line, and a quick-release snap shackle.

> **WARNING**
> **Parachutists carrying the SMJP will jump from the first two positions in the stick only. Parachutists who are less than 5 feet 8 inches in height are not permitted to jump the SMJP. The SMJP is authorized to be jumped from C-130 and C-141 aircraft only.**

12-62. COMPONENTS

Components of the Stinger missile jump pack are as follows:

a. The SMJP consists of a pack body constructed of nylon duck material, shock-absorbing material, nylon securing straps, an HPT lowering line, and a quick-release snap shackle.

b. When rigged with a Stinger missile round, the SMJP is 12 inches in diameter, 68 inches long, and weighs approximately 38 pounds. It is secured vertically by the quick-release snap shackle to the left D-ring on the parachutist's harness. To prevent the SMJP from swaying during exit and deployment of the parachute, the parachutist uses the nylon webbing tie-downs to secure the SMJP to the left leg and to the parachute harness at chest level.

c. The HPT lowering line with a 6-foot extension is part of the SMJP system. The web loop end of the lowering line extension is attached to the SMJP with the HPT lowering line ejector snap routed through the D-ring on the other end of the extension sling and to the lowering line adapter web on the parachute harness. The HPT lowering line is attached to a front-mount container (ALICE pack, large, or the FPLIF) which contains the detachable Stinger weapon round components (gripstock, IFF, and the three BCUs). The parachutist lowers the SMJP and ALICE pack by pulling the SMJP quick-release assembly. Upon landing, the parachutist can rapidly gain access to the Stinger.

12-63. RIGGING PROCEDURES

Rigging procedures for the Stinger missile jump pack are as follows:

NOTES:
1. Throughout the rigging procedure, duct tape and masking tape are used to tape various parts of the webbing. The jumper can use duct tape for masking tape, **except** where the procedures state "masking tape only."
2. Throughout the instructions, the phrases "forward end" and "aft end" are used to describe the SMJP, Stinger missile round, and FHT. The forward

end is nearest the gripstock and sight assembly; the aft end is furthest from the gripstock and sight assembly.

a. **Stinger Round**. Before rigging the SMJP, inspect the Stinger round or FHT IAW TM 9-1425-429-12. A Stinger round exhibiting damage must not be jumped under any circumstances. It must be returned to the ASP for repair. Field handling trainers that exhibit signs of general minor damage and no major ruptures or cracks may be jumped if the damage does not exceed safe or serviceable limits. Pad the Stinger round with a 6-inch-wide piece of cellulose wadding placed around the aft end of the round near, but not on, the blowout disk. Tape the wadding with **masking tape only**.

b. **Aft Foam End Cap**. Place the foam end cap on the aft end of the missile round, one half at a time, ensuring the foam posts slide into the holes in the opposite end cap half. The end cap should fit snugly. If not, check to see that both halves fit together without any gaps and there is enough cellulose wadding about the missile round. Tape the foam end cap halves together using duct tape.

c. **Front Cover**. Make sure the front cover is on firmly. If loose, use a pen (or similar object) to press in the three rubber spacers on the cover. Push each spacer in evenly to achieve a firm fit between the cover and the missile round. The cover should slide on snugly when properly adjusted. Make sure the tab on the cover aligns with the notch on the missile round. When the cover is properly fitted, the "lollipop" should be visible when looking through the sight.

d. **Forward Foam End Cap**. Place the foam end cap on the forward end of the missile round in the same fashion, ensuring that the "lollipop" on the front cover is positioned in the cutout of the foam end cap. Tape the foam end caps together using duct tape, ensuring that there are no gaps.

e. **Sight**. Flip up the sight on the missile round.

(1) Fold a piece of cellulose wadding, making a piece that is 6 inches by 24 inches. Place the wadding under the sight and lower the sight down onto the wadding. Make sure that the wadding is not placed over the clip. Firmly press down on the sight to make the wadding fill the gap between the sight and the missile round and lock the sight into position.

(2) Loosely wrap cellulose wadding around the entire sight area until the wadding measures about 11 to 12 inches in diameter, about the same as the diameter of the foam end cap. Tape the wadding in place using **masking tape only**.

f. **Stinger Round Positioning**. Place the SMJP on the ground, felt side up. Make sure all straps are extended. Position the missile in the SMJP with the aft foam end cap fitted in the SMJP nonadjustable end strap bridle and with the missile round carrying strap facing up.

g. **SMJP Closing**. Fold the pack around the missile, keeping the securing strap quick-fit adjusters on the outside of the pack.

(1) Route the side securing straps through the quick-fit adjusters using a quick-release fold. Loosely tighten the straps to form the SMJP around the missile. Do not tighten the straps.

(2) Place the SMJP on the aft end so the free-running end straps are on top. Route the end straps through the quick-fit adapter opposite of each strap. Sit on the ground, place

your foot on the forward end cap, and tighten the adjustable straps. Tighten the end straps. Do **not** use a quick-release.

(3) Place the pack on the side and tighten the side securing straps, using a quick release. Tighten the side securing straps so the pack overlaps the entire length of the round. S-fold excess webbing and tape, using **masking tape only**.

NOTE: If there is a gap near the sight assembly, remove cellulose wadding until the gap is closed.

(4) Stand the SMJP on the aft end cap. Center a 6-inch by 6-inch piece of honeycomb on the forward end cap. Route the free-running end straps over the honeycomb, through the opposite quick-fit adapters, and tighten to form an X over the honeycomb.

(5) Secure straps with a quick release.

(6) S-fold excess webbing and tape with duct tape.

h. **Attachment of the Quick-Release Assembly**. Secure the quick-release web and SMJP HPT. Route the coated wires into the fabric guides. Hook the quick-release shackle onto the lowest SMJP D-ring which, when the SMJP is attached to the jumper, keeps the SMJP from dragging the ground.

NOTE: The quick-release handle must face toward the jumper's rear.

WARNING

A parachutist harness with too much slack, or with the incorrect SMJP D-ring, will cause the SMJP to drag on the aircraft floor and can result in jumper injury when he exits the aircraft.

i. **Weapon Components Rigging**. The gripstock, IFF, and the three BCUs are packed in the ALICE pack or FPLIF where they can be quickly recovered on the DZ.

(1) To protect the weapon components, load hard items in the lower portion of the pack, place soft items on top of the hard items, and pack the Stinger weapon components on top of the soft items near the top of the pack.

(2) Wrap each component with one wrap of cellulose wadding and secure with **masking tape only**.

(a) Gripstock antenna. The gripstock antenna must be in the folded position. Secure the gripstock antenna attaching clip in the closed position, with **masking tape only**.

(b) IFF. The interconnecting cable must remain attached to the IFF and wrapped separately to avoid metal-to-metal contact between the cable and IFF. Secure the wrapped cable to the wrapped IFF. Ensure the electrical outlet protecting cap is in place.

12-64. STINGER MISSILE JUMP PACK ATTACHED TO PARACHUTIST

The SMJP is the last item attached to the parachutist harness.

a. Secure the SMJP to the parachutist harness by attaching the quick-release snap shackle to the left D-ring, ensuring it opens away from the jumper.

NOTE: Failure to place the snap shackle away from the jumper may cause the SMJP to hang when the release is activated.

b. Route the lowering line ejector snap hook through the SMJP lowering line extension sling D-ring and connect to the lowering line adapter web.

c. Secure the lowering line and SMJP HPT.

d. Secure the lowering line extension to the SMJP and tape with **masking tape only**. Make sure there is no slack in the lowering line extension.

e. Route the chest tie-down around the left main lift web directly below the chest strap and around the SMJP. Route the chest tie-down strap with grommet through the chest tie-down loop. Route the red fabric loop through the grommet and place the coated wire through the red loop.

f. Route the leg tie-down around the parachutist's left leg, through the frame on the ALICE pack or through the nearest equipment retaining ban on the FPLIF, and around the SMJP. Secure in the same manner as the chest tie-down.

g. Attach a 6- to 8-foot doubled piece of 1/4-inch, 80-pound cotton webbing to the SMJP carrying handle. Route the cotton webbing through a secure loop on or near the bottom of the front-mount ALICE pack and tie to the handle of the HSPR.

NOTE: The jumper must use the HSPR when jumping the SMJP.

12-65. INDIVIDUAL JUMP PROCEDURES

The SMJP must be jumped with a front-mount container.

a. At the 20-minute warning, a safety attaches the SMJP to the parachutist and inspects it.

b. At exit, the jumper must **not** assume a tight body position. A tight body position causes severe twists and may flip the jumper through his risers. It also causes entanglement in suspension lines.

c. About 200 to 100 feet above the ground, the jumper ensures the area below is clear and activates the release by pulling the orange release tab down and away from his body with the left hand. The SMJP falls the length of the lowering line.

NOTE: To jettison the SMJP, the parachutist lowers the SMJP and pulls out on the yellow lanyard attached to the quick-ejector snap.

Section XV. RANGER ANTIARMOR/ANTIPERSONNEL WEAPON SYSTEM PACKED IN AT4JP AND DRAGON MISSILE JUMP PACK

The AT4JP and Dragon missile jump pack can be used to secure the RAAWS during airborne operations.

12-66. COMPONENTS AND CONTAINER DESCRIPTION

The RAAWS consists of three main components plus the cleaning kit and gun-mount.

a. **Main Components**. The main components are the weapon system, the telescopic sighting system, and various types of loads.

(1) *Weapon and telescoping sighting system.* When carried by the parachutist, they are packed in the AT4JP.

(2) *Rounds of ammunition.* The missile round containers are constructed of molded polyethylene. The rounds come in two container sizes: 27.6 inches (TP and HEAT) and 21 inches (illumination, smoke, and HE). The TP and HEAT round containers must be rigged in a DMJP (modified). Illumination, smoke, or HE missile rounds are packed under the top pack flap of the large ALICE pack or FPLIF.

b. **Cleaning Kit and Gun-Mount**. The cleaning kit and gun-mount are packed separately near the top of the jumper's ALICE pack or FPLIF.

c. **Container Description**. When configured for RAAWS rigging, the AT4JP cannot be rigged with the M16/M203. The AT4JP is about 49 inches long and 9 inches in diameter, and it weighs about 26 pounds. It is secured vertically to the left D-ring on the parachutist harness. To prevent the AT4JP from swaying, two tie-down tapes are attached. The upper tie secures the AT4JP to the main lift web below the chest strap, and the lower tape secures the jump pack to the parachutist's left leg (provided the jumper is not using the HSPR). A 15-foot lowering line and 24-inch lowering line attachment web are used. Two 9-inch by 9-inch pieces of honeycomb are used at the venturi end of the weapon (on the bottom, when rigged) to cushion the weapon during impact. After landing, the parachutist rapidly gains weapon access by pulling seven quick releases.

12-67. RIGGING PROCEDURES

Rigging procedures for RAAWS components are as follows:

a. Rig the RAAWS weapon and sighting system.

(1) Lay the jump pack on the ground (felt side up) and fully extend all securing webs.

(2) Place the venturi and muzzle covers on the weapon.

(3) Position the telescopic sight package on the weapon (between the front grip and the shoulder pad) and alongside the firing mechanism. Secure the telescopic sight package to the weapon with two lengths of 1/4-inch cotton webbing.

(4) Place the weapon on the pack with the front grip pointed straight up, the muzzle (small end of weapon) fitted into the nonadjustable cross D-ring strap, and the venturi (large end of weapon) centered on the middle launcher aft end securing strap.

(5) Ensure the muzzle fits snugly into the nonadjustable cross D-ring strap.

(6) Fold the rifle and muzzle pockets portion of the pack over the weapon.

(7) Rotate the unfolded flap portion over the other flap and route the four side securing straps through the adapters and form quick releases. S-fold the running ends and secure by folding excess back under corresponding quick-fit adapter.

(8) Position two 9-inch by 9-inch pieces of honeycomb onto the venturi of the weapon. Route the two aft end securing straps over the honeycomb, through the quick-fit adapters, and tighten securely. Form a quick release with the running ends, S-fold the excess, and secure by folding excess back under the corresponding quick-fit adapter.

(9) Route one turn of 1/4-inch cotton webbing around the securing straps between the honeycomb and the pack; tie securely. (This tie will prevent the honeycomb from shifting.)

(10) Attach the HPT lowering line to the AT4JP using procedures described in paragraph 12-8. If the AT4JP lowering line is to be employed, a 24-inch lowering line adapter web must be attached to the parachutist harness.

b. Rig the RAAWS 84-mm TP and HEAT rounds in the DMJP MOD.

NOTE: Due to the length of the TP and HEAT twin-tube containers, the containers must be jumped inside the DMJP MOD. DMJP modification procedures are described in paragraph 12-70.

(1) Lay the DMJP MOD on the ground felt side up.

(2) Center the long twin tube containers on the DMJP MOD.

(3) Route the center vertical and horizontal securing straps over the container and secure with a quick release. Ensure the tubes are oriented so the container caps are at the upper portion of the pack.

(4) Fold the flap containing the M16/M203 rifle muzzle stow pocket over the tubes.

(5) Place the M16/M203 rifle muzzle into the stow pocket, ensuring the carrying handle is facing away from the tubes.

(6) Fold the unfolded flap over the M16/M203 and secure the three horizontal securing straps. Ensure the center securing strap is routed through the rifle carrying handle. Form quick releases in the running ends, S-fold, and fold the excess back under the corresponding quick-fit adapters.

(7) Route the rifle butt securing strap over the butt and through the quick-fit adapter. Pull as tight as possible and form a quick release.

(8) Cut two pieces of honeycomb 5 inches by 11 inches. Place the two pieces onto the end of the container with the exposed rifle butt. Route the two aft end securing straps over the honeycomb, through the quick-fit adapters, and secure as tight as possible. S-fold the excess back under the corresponding quick-fit adapters.

(9) Tie the two securing straps together at the point where they cross with one turn of 1/4-inch cotton webbing to prevent them from shifting.

(10) Cut two pieces of 5 1/2-inch by 9 1/2-inch honeycomb. Place the honeycomb on the other end (top) portion of the container.

(11) Route the center securing strap through a D-ring, over the honeycomb, and through the quick-fit adapter. Tighten securely, S-fold, and fold the excess back under the corresponding quick-fit adapters.

NOTE: The nonadjustable strap is not used. It is pushed into the top of the pack before placing the honeycomb on top.

(12) Secure the D-ring to the center of the strap by running a length of 1/4-inch cotton webbing through the lower portion of the D-ring, around the strap, and tying. This will prevent the D-ring from shifting on the strap.

(13) Attach the lowering line to the DMJP MOD.

NOTE: The DMJP MOD with the long tubes must be rigged with its own lowering line.

(a) Attach the lowering line to the D-ring on top of the DMJP MOD by routing the lowering line through the D-ring and then girth-hitch the lowering line through itself.

(b) Stow the lowering line into the stow pocket on the DMJP MOD by pushing it as far as possible into the pocket. It will be halfway into the pocket. Secure the upper HPT flaps around the stowed lowering line.

NOTE: Do not tighten the lower HPT. This may bind the lowering line and prevent the DMJP MOD from lowering.

c. Rig the RAAWS 84-mm (illumination, smoke, or HE) rounds (short twin tubes) in the large ALICE pack with frame.

(1) Place the short tubes on the top of the ALICE pack and under the top flap. Secure all straps.

(2) Route one turn single Type I or III nylon cord through the equipment attaching points on one side of the ALICE pack, over the tubes, and through the equipment attaching points on the other side of the ALICE pack, then secure the running ends together.

d. Attach the HSPR to the ALICE pack.

12-68. EQUIPMENT ATTACHED TO PARACHUTIST

The AT4JP is secured to the parachutist after the reserve parachute and ALICE pack or FPLIF have been attached to the parachute harness D-rings.

a. Secure the AT4JP to the parachutist by attaching the quick-release assembly to the left D-ring. Ensure that the safety gate is completely closed, secure, and the activating arm is fully seated.

b. Ensure the harness left D-ring snap attachment sequence is inboard to outboard as follows:

(1) Reserve parachute.

(2) H-harness or other equipment.

(3) AT4JP quick-release snap.

c. Secure lowering line to parachutist by attaching the ejector snap to the lowering line adapter web. Ensure activating lever is fully seated. Ensure the yellow safety lanyard is not misrouted or entangled and is immediately accessible to the parachutist.

d. Route the upper tie-down tape around the main lift web at a point below the chest strap. Secure the lower tie-down tape around the left leg, or use the HSPR leg strap.

12-69. MODIFICATION PROCEDURES FOR THE DRAGON MISSILE JUMP PACK

Equipment needed to modify a DMJP for RAAWS missile round rigging are as follows: DMJP, two each A-7A straps, markers, a cutting tool, suitable needle, and No. 6 nylon thread.

a. Lay the DMJP for modification flat, felt side up. Using a suitable marker, make a mark at the center of the pack and in line with the aft end bridle securing strap at the top of the DMJP (Figure 12-60).

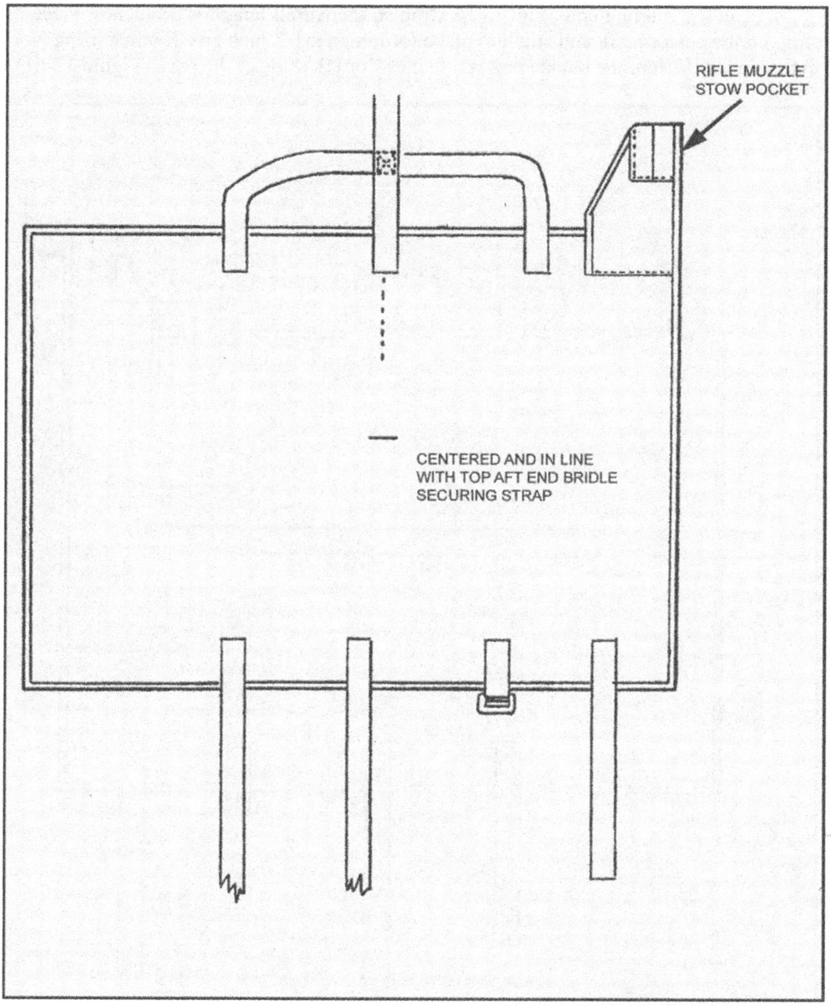

Figure 12-60. Marking the center of the DMJP.

b. Cut the running end of an A-7A strap so its overall length is 38 inches. Place the strap on the center mark and stitch in place forming a 2 1/2-inch box-X stitch using No. 6 nylon thread. Ensure the buckle end is 5 inches from the side of the pack (Figure 12-61).

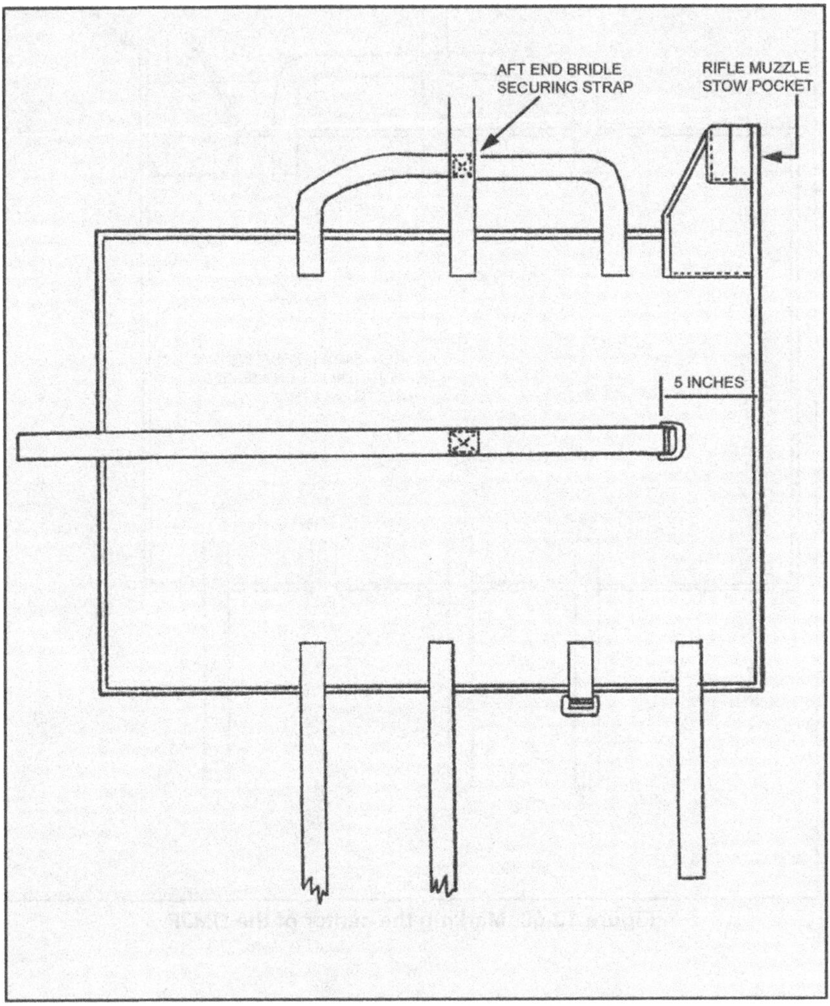

Figure 12-61. A-7A strap centered and stitched.

c. Locate the point where the aft end bridle securing strap crosses the U-shaped portion and is sewn. Cut and remove this stitching, being careful to not damage the straps (Figure 12-62).

d. From this side of the DMJP, measure and make two marks at 3 1/2 inches and 6 inches. From the other side of the DMJP, measure and mark at 6 inches and 8 1/2 inches (Figure 12-62).

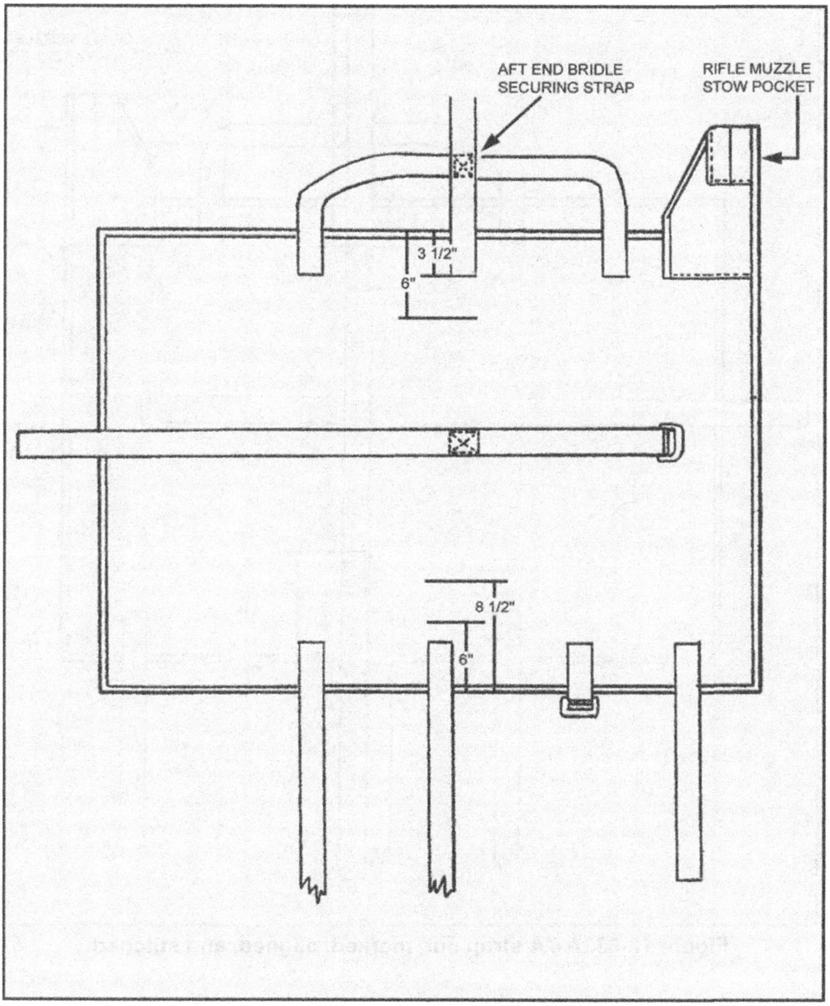

Figure 12-62. Stitching removed and DMJP markings.

e. Cut a second A-7A strap to an overall length of 68 inches. Mark the A-7A strap at a point 17 inches from the buckle. Align this mark with the 3 1/2-inch mark and stitch with two box-X stitch formations (Figure 12-63).

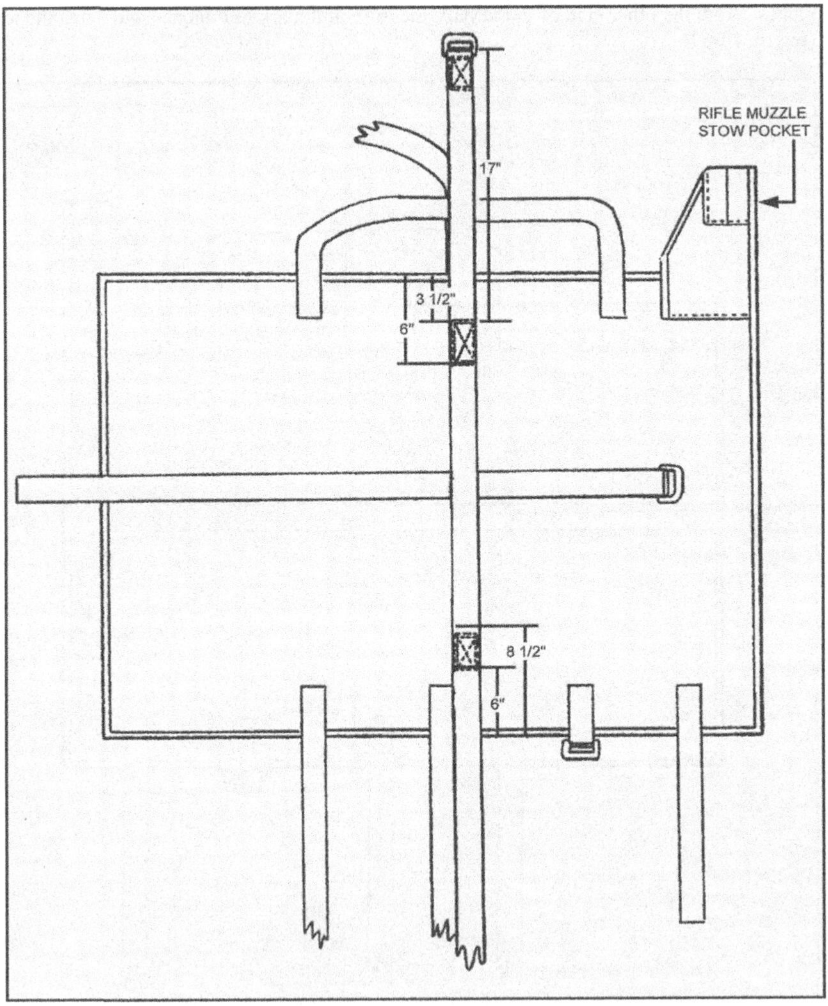

Figure 12-63. A-7A strap cut, marked, aligned, and stitched.

Section XVI. FIELD PACK, LARGE, INTERNAL FRAME

The FPLIF, with or without the patrol pack, can be rigged in the HSPR during airborne operations. However, the FPLIF is **not** recommended for jumping; substantial damage can occur to the pack and cause it to be unserviceable for further use.

12-70. RIGGING THE FIELD PACK, LARGE, INTERNAL FRAME WITHOUT PATROL PACK

Rigging instructions for the FPLIF without patrol pack, using the HSPR, are as follows:

a. Before attaching the harness to the FPLIF, attach the release handle assembly, adjustable D-ring attaching straps, and leg strap release assembly to the harness. S-fold, or accordion fold, all excess on the equipment retainer straps and secure with two turns of retainer band. No additional equipment will be attached to the outside of the FPLIF.

WARNING

The packed FPLIF can be no longer than 30 inches when rigged with the standard HSPR. The maximum recommended total rigged weight is 75 pounds, including the patrol pack. because of its internal frame, the FPLIF should be jumped in its fully extended position. THE H-harness *cannot* be used with the FPLIF.

b. Shorten the shoulder straps to their minimum length, roll and tape the free-running ends, route the cross-chest strap around the shoulder straps, and then fasten the cross-chest strap buckle. Stow upper patrol pack attaching straps inside the FPLIF by tucking them through the antenna closure flaps.

c. Extend the waistband straps, turn the FPLIF over with the shoulder straps down, route the waistband straps through the bottom loops formed by the lower patrol pack attaching straps, and buckle the waistband fastener.

d. Lay the HSPR on top of the FPLIF with the release handle assembly toward the top of the FPLIF, the adjustable cross-strap facing up, and the leg straps above the equipment retainer straps. Route the running ends of the HSPR equipment retainer straps from the top of the FPLIF down through the FPLIF's second and third equipment attaching loops. Route the adjustable leg straps through the second equipment attaching loops only. Route the HSPR equipment retainer straps under the waistband. Extend the HSPR adjustable cross-strap halfway.

e. Turn the FPLIF over with the shoulder straps facing up. Route the HSPR equipment retainer straps over the bottom of the FPLIF, under the waistband (through the outer waistband frame loops), and up over the comfort pad; form an X with the running ends. Ensuring that the release handle assembly is positioned on the top of the FPLIF, continue to route the HSPR equipment retainer straps under the shoulder straps, through the loops sewn on the back of the FPLIF near the top of the frame staves, and connect the free-running ends to the friction adapters, forming quick releases. S-fold the free-running ends and secure with retainer bands.

f. To attach the HPT lowering line to the HSPR, route the looped end of the HPT lowering line under the X formed by the HSPR equipment retainer straps, from top to bottom. Route the ejector snap through the loop, forming a girth hitch. Route the HPT lowering line over the top of the right FPLIF shoulder strap. Secure the HPT lowering line on the right side of the FPLIF to the horizontal FPLIF compression straps with two retainer bands, routing the ejector snap from bottom to top so that the HPT lowering line ejector snap is at the top.

g. Temporarily attach the HPT lowering line ejector snap to the triangle link of the left attaching harness strap. Attach the adjustable leg straps to the leg strap release assemblies, take up the slack in the adjustable leg straps, and fold and stow the excess using the webbing retainers. Attach the adjustable D-ring attaching strap snap hooks to the top carrying handle on the FPLIF.

12-71. RIGGING THE FIELD PACK, LARGE, INTERNAL FRAME WITH PATROL PACK

Rigging instructions for the FPLIF with patrol pack, using the HSPR, are as follows: (See Figure 12-64, page 12-75 and Figure 12-65, page 12-76.)

a. Before attaching the harness to the FPLIF, attach the release handle assembly, adjustable D-ring attaching straps, and leg strap release assembly to the harness. Fold and tape the running ends of all straps. No additional equipment will be attached to the outside of the FPLIF.

b. Shorten the shoulder straps to their minimum length, roll and tape the free-running ends, route the cross-chest strap around the shoulder straps, and then fasten the cross-chest strap buckle.

c. Extend the waistband straps, turn the FPLIF over with the shoulder straps down, route the waistband straps through the bottom loops formed by the lower patrol pack attaching straps, and buckle the waistband fastener.

d. Place the patrol pack on the outside front of the FPLIF and fasten the upper patrol pack attaching straps. Lay the HSPR on top of the patrol pack with the release handle assembly toward the top of the FPLIF, with the adjustable cross-strap facing up, and with the adjustable leg straps above the equipment retainer straps. Route the running ends of the HSPR equipment retainer straps and adjustable leg straps over the top of the patrol pack and down through the bottom equipment attaching loops of the patrol pack. Then route the equipment retainer straps through the second and third equipment attaching loops on the FPLIF. Route the adjustable leg straps through the second equipment attaching loops, only. Route the HSPR equipment retainer straps under the waistband. Route the lower patrol pack attaching straps under the third FPLIF equipment retainer loops and fasten them to the bottom of the patrol pack.

e. Turn the FPLIF over with the shoulder straps facing up. Route the HSPR equipment retainer straps over the bottom of the FPLIF, under the waistband (through the outer waistband frame loops), and up over the comfort pad; form an X with the running ends. Ensuring that the release handle assembly is positioned on the top of the FPLIF, continue to route the HSPR equipment retainer straps under the shoulder straps, through the loops sewn on the back of the FPLIF near the top of the frame staves, and connect the free-running ends to the friction adapters, forming quick releases. S-fold the free-running ends and secure with retainer bands.

f. To attach the HPT lowering line to the HSPR, route the looped end of the HPT lowering line under the X formed by the HSPR equipment retainer straps, from top to bottom. Route the ejector snap through the loop, forming a girth hitch. Route the HPT lowering line over the top of the right FPLIF shoulder strap. Secure the HPT lowering line on the right side of the FPLIF to the horizontal FPLIF compression straps with two retainer bands, routing the ejector snap from bottom to top so that the HPT lowering line quick-ejector snap is at the top.

g. Temporarily attach the HPT lowering line ejector snap to the triangle link of the left attaching harness strap. Attach the adjustable leg straps to the leg strap release assemblies, take up the slack in the adjustable leg straps, and fold and stow the excess using the webbing retainers. Attach the adjustable D-ring attaching strap snap hooks to the top carrying handle on the FPLIF.

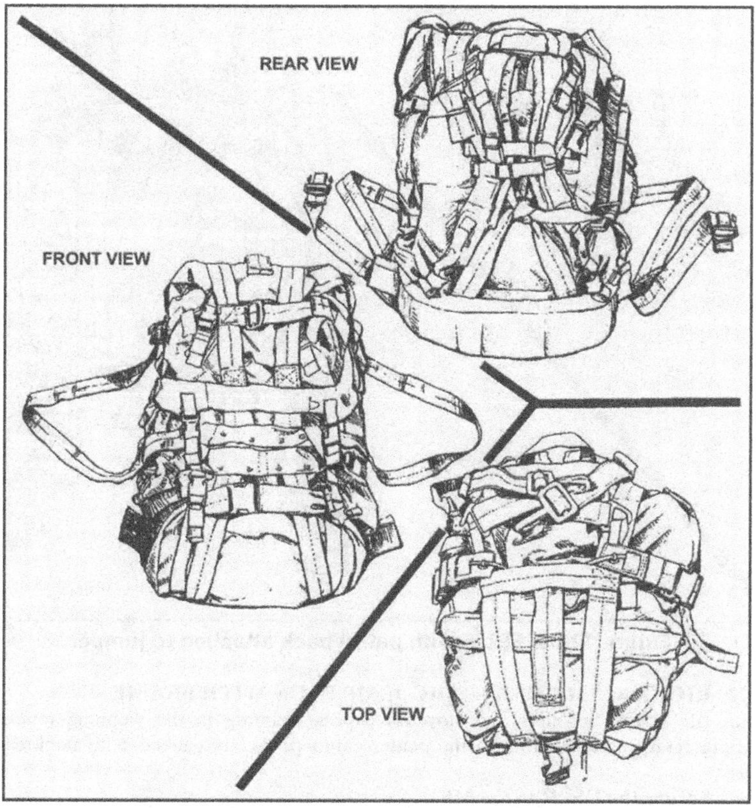

Figure 12-64. FPLIF with patrol pack on front.

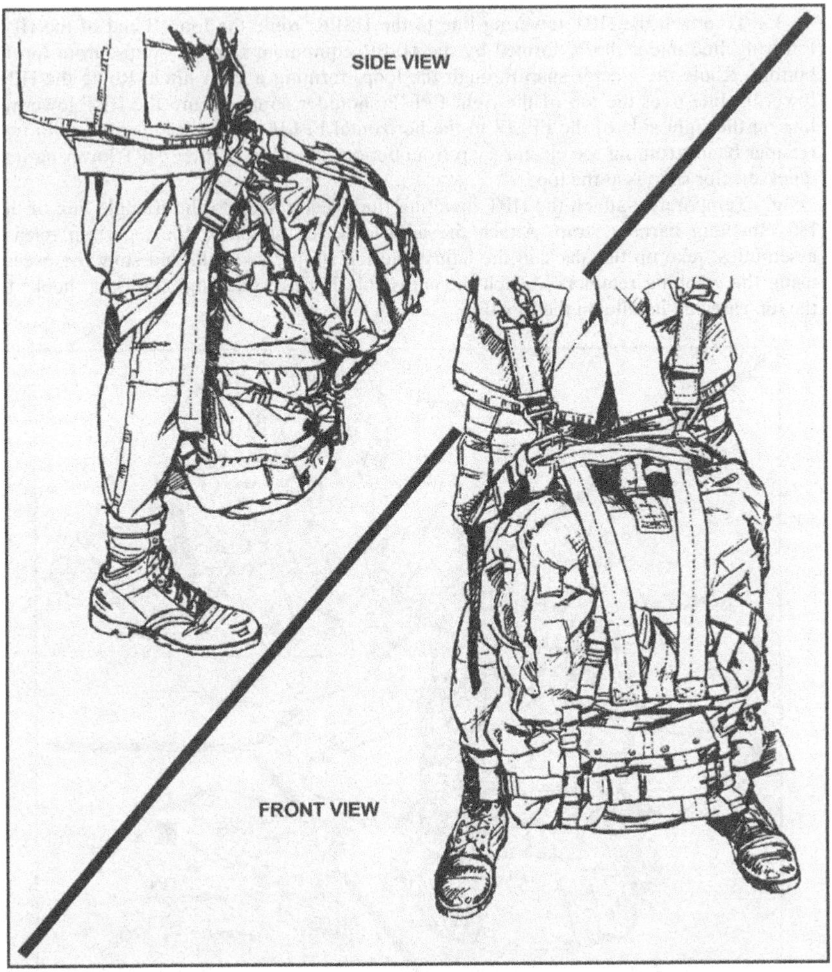

Figure 12-65. FPLIF with patrol pack attached to jumper.

12-72. RIGGING THE M82, MEDIC JUMP PACK WITH FRAME

Secure all Fastex fasteners and stow all excess webbing in the webbing retainer or webbing retainer bands. Ensure the center cargo pocket is packed with nonbreakable items.

a. **Secure the HSPR to the M82.**

(1) Lay out the HSPR with the attaching loops facing up.

(2) Route the two release handle cables between the two plies of the release handle cross strap. Attach the pile tape of the release handle assembly to the hook tab located between the plies of the release handle cross strap. Ensure the release handle lanyard is not misrouted.

(3) Place the triangle links of the adjustable D-ring attaching straps on top of the white attaching loops, with the opening gates of the snap hooks facing down. Route the white attaching loop through the triangle link; the green loop through the white attaching loop; the red attaching loop through the green attaching loop and then through the grommet on the female portion of the leg strap assembly. Ensure the cable loop retainer on the female portion of the leg strap assembly is facing up. Route the release handle cable through the red attaching loop and then through the cable loop retainer. Repeat this procedure for the other adjustable D-ring attaching strap.

(4) Turn the HSPR over so that the opening gates on the snap hooks of the adjustable D-ring attaching straps are facing up.

(5) Place the M82 upside down on the HSPR so that the center cargo pocket is in the window formed by the adjustable cross strap and the release handle cross strap.

(6) Route the equipment retainer straps under the frame and cross them in the center forming an "X", then route them to the appropriate friction adapter. Form a two- to three-finger quick release and tighten. S-roll the free-running ends and secure with two turns of retainer band (do not secure the excess to the quick releases).

(7) Route the male portion of the leg strap release assembly around the M82. Ensure the leg straps are fully extended with no twists and then attach them to the female portion of the leg strap release assembly.

NOTE: When stowing the HPT lowering line within the retainer flap, ensure stows are even with the retainer flap. No excess should exceed either end of the retainer flap.

b. **Attach the HPT Lowering Line to the HSPR.**

(1) Route the looped end of the HPT lowering line under the "X" formed by the equipment retainer straps from top to bottom (bottom to top), north to south (south to north).

(2) Route the remainder of the HPT lowering line through the loop end of the HPT lowering line, then pull it tight forming a girth hitch around the "X" formed by the equipment retainer straps.

(3) Secure the HPT lowering line to the left side of the frame by routing the ejector snap and HPT lowering line within the closed retainer flap through two retainer bands from bottom to top as the M82 is worn by the jumper.

CHAPTER 13

ARCTIC RIGGING

When using arctic rigging, the number of personnel who can be parachuted from a single aircraft is reduced by the bulk of equipment and cold weather clothing. When computing weight factors, the cold weather-equipped parachutist is estimated to weigh 310 pounds.

Section I. ARCTIC EQUIPMENT SPACE CONSIDERATIONS

Exiting interval between each parachutist is increased to two seconds when using arctic rigging. Aircraft compartment space required for a parachutist is 1 1/2 times more in cold regions than in temperate climates. Commanders must be familiar with the airborne operations portion of FM 31-71 for successful arctic rigging operations. However, plane-side parachute issue and rigging are impossible during winter months due to harsh temperatures.

13-1. WEIGHT FACTORS

Aircraft must be within 200 meters of the parachute rigging facility to keep rigged jumpers from walking through deep snow or over ice during winter months when temperatures are low and the individual parachutist's equipment is the heaviest. The serviceability of the activating lever on the ejector snap of the HPT lowering line should be checked, since there is an increased risk of the lever malfunctioning due to the heavy loads.

13-2. MODIFICATIONS

Modifications of standard equipment must be made for airborne operations under cold weather conditions.

a. **Waistband**. A modified waistband strap is used in lieu of the standard waistband when parachuting with snowshoes or skis. The strap consists of two pieces: a 6-foot A-7A strap and a 16-inch strap with a buckle at each end. The skis or snowshoes are attached to the side of the jumper opposite the static line (to prevent fouling). The modified waistband allows the buckle for the quick-release fold to be located on the same (either) side with the snowshoes or skis (Figure 13-1, page 13-2). Tandem loads are dropped on a single lowering line. Rigging or lowering procedures are contained in this chapter. Under arctic conditions, most individual equipment is lowered during descent due to the weight.

b. **Reserve Parachute**. Deployment of the reserve parachute under arctic or extreme cold weather conditions may be hindered by the bulkiness of the gloves worn. To overcome this obstacle, the rip cord is inverted as an optional requirement. Commanders requesting the MIRPS or T-10 reserve parachute for an arctic airborne operation must allow enough time for the reserves to be modified.

c. **Mittens**. Arctic mittens are not worn during the parachute jump; their bulkiness interferes with deployment of the reserve parachute and the lowering of equipment. The mittens are tucked inside the front of the jacket or under the parachute harness. They are not attached to, or packed in, a container. Trigger-finger mittens are stowed inside the jacket for wear as soon as the jumper is on the ground.

d. **Arctic Canteen**. The arctic canteen poses a hazard due to its long neck and metal body, which can injure a jumper if the PLF is executed on top of the canteen. Commanders should consider packing it in the ALICE pack to prevent personal injury or damage to the canteen.

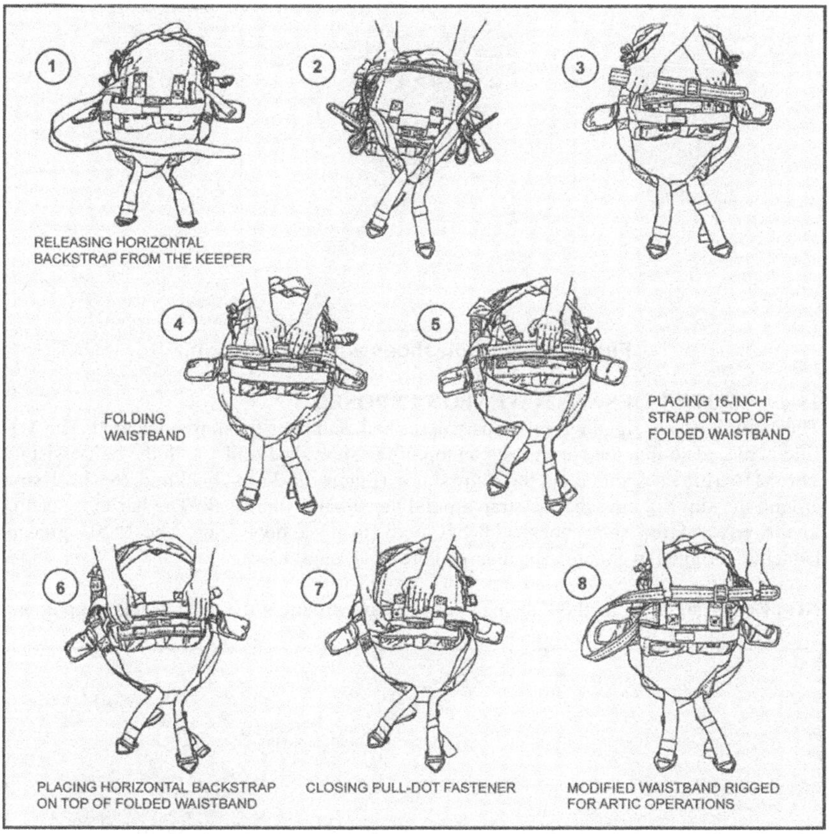

Figure 13-1. Modified waistband routed on parachute harness.

Section II. SNOWSHOES AND INDIVIDUAL WEAPON

Snowshoes are usually rigged on the parachutist to allow for immediate access.

13-3. SNOWSHOES WITHOUT WEAPON

The prefitted snowshoes are placed one on top of the other. The heel strap of the lower snowshoe is run underneath the lower shoe and up between the frame and webbing of both snowshoes. The heel strap buckle is brought up similarly on the other side of the snowshoes. A sling may be fabricated using 550-cord (or other suitable material). The snowshoes are secured with an additional tie-down, using 550-cord at the toe (Figure 13-2).

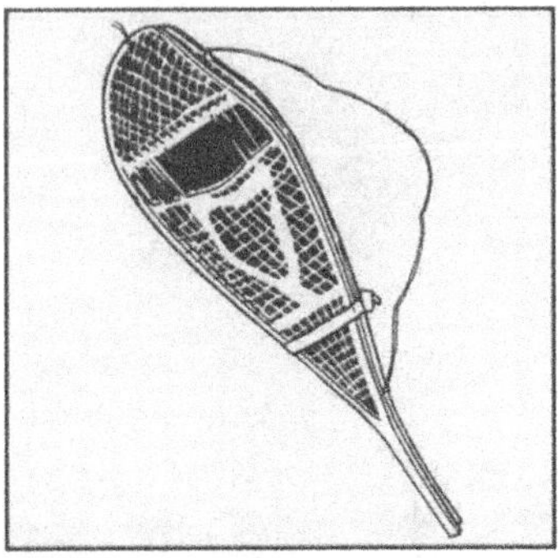

Figure 13-2. Snowshoes without weapon.

13-4. SNOWSHOES WITH WEAPON EXPOSED

The snowshoes are rigged as previously described, and the rifle sling is secured. The M16 rifle is placed so that the barrel rests on top of the snowshoe trails, with the bolt-assist up. The M16 rifle is secured to the snowshoes (Figure 13-3) by buckling the heel strap around the slip ring and the toe strap around the small of the stock. The barrel is secured to the yoke of the snowshoes with 550-cord, using a bow knot. The M203 grenade launcher is rigged in a similar manner (Figure 13-4, page 13-4).

NOTE: Using this method, the parachutist may exit either the right or left jump door.

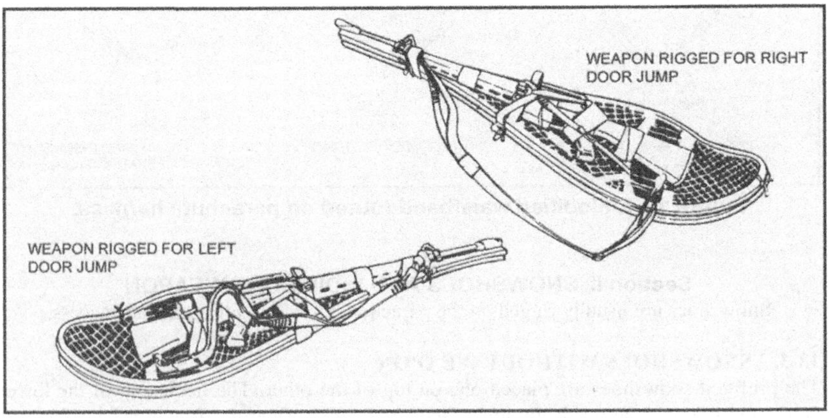

Figure 13-3. Snowshoes with weapon.

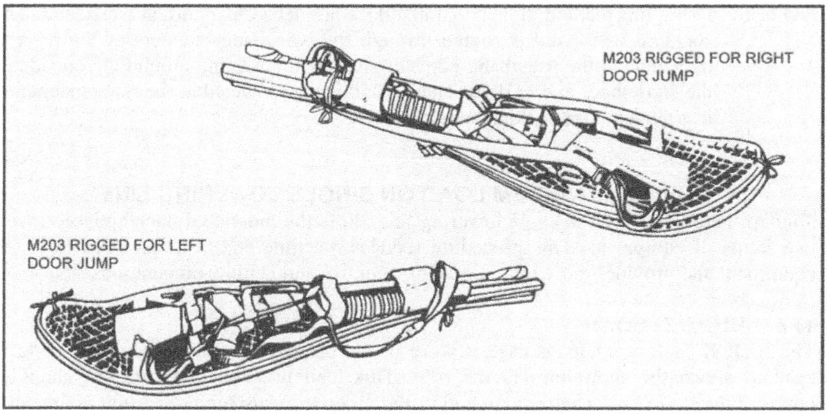

Figure 13-4. Snowshoes with M203.

13-5. JUMPING SNOWSHOES WITH M1950 WEAPONS CASE

Snowshoes are attached to the outside of the M1950 with the tails down and the tips of the snowshoes facing the jumper. The 550-cord is used to secure the snowshoes to the M1950 through both the upper and lower tie-down tape retaining bars. The running ends of the upper tie-down tape are routed through the toe window on the snowshoes and then around the M1950 to the left side of the main lift web. The M1950 is then attached as the outermost item to the left D-ring (Figure 13-5).

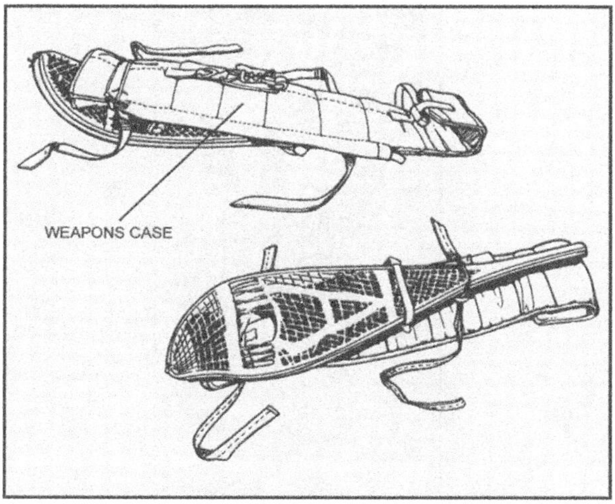

Figure 13-5. Snowshoes with weapons case.

NOTE: Using this method, the parachutist does not need the modified waistband. The standard waistband is routed through the waistband retainers of the reserve parachute to the waistband adjuster panel without going around the outside of the snowshoes and M1950. The M1950 is then lowered in the same manner as in a tandem load without snowshoes.

Section III. TANDEM LOAD ON SINGLE LOWERING LINE

Tandem loads rigged on a single lowering line allow the individual parachutist to lower two items of equipment. This procedure reduces the time and tasks required to lower equipment and provides more time for canopy control and landing preparations.

13-6. RIGGED LOAD

The ALICE pack or weapons case is worn in the prescribed manner. The rifle sling is used to attach the snowshoes to the rifle. This load is suspended over the shoulder opposite the static line. The running end of the modified waistband extension is threaded through the waistband retainers of the reserve parachute and around the snowshoes, and the end is made into a quick-release fold (Figure 13-6).

NOTE: The rifle sling is adjusted to fit the parachutist snugly. The sling adjustment is also small enough so that it does not come off the load when released to slide down the lowering line.

Figure 13-6. Parachutist rigged to jump left door.

13-7. HOOK-PILE TAPE LOWERING LINE

The lowering line is threaded through the rifle sling and attached to the main lift web of the parachutist (on the side the snowshoes are attached).

a. A length of 80-pound test tape is attached around the sling and the main lift web with a bow knot (just below the canopy release assembly). The lowering line adapter web is attached (Figure 13-7) to the left (right) side, corresponding to the side to which the snowshoes and rifle are to be attached.

b. When jumping with the ALICE pack or weapons case, the parachutist routes the lower tie-down on the case through the metal frame and around the leg. A separate tie-down tape for each is not necessary (Figure 13-8, page 13-8).

c. The stowed lowering line is secured with two retainer bands to the left (right) side of the vertical bar of the combat pack. After the pack, snowshoes, and rifle are attached to the parachutist, the lowering line ejector snap is passed through the rifle sling and attached to the accessory attaching ring of the adapter web. If the adapter web is not used, the ejector snap is attached directly to the D-ring on the harness. If the weapons case is attached to the snowshoes (Figure 13-9, page 13-8), the lowering line ejector snap is passed between the case and the cotton chafe material, which is attached to the case V-ring.

d. The upper tie-down tape (above the chest strap and below the canopy release assembly) and the lower tie-down tape around the parachutist's leg are untied, and the load is dropped by pulling the white release handle located on the cross strap of the HSPR. This ensures that the pack load falls the length of the line. The modified waistband quick-release fold is pulled and releases the snowshoes, which slide down the lowering line on top of the pack (Figure 13-10, page 13-9).

Figure 13-7. Lowering line assembly attached for a tandem load.

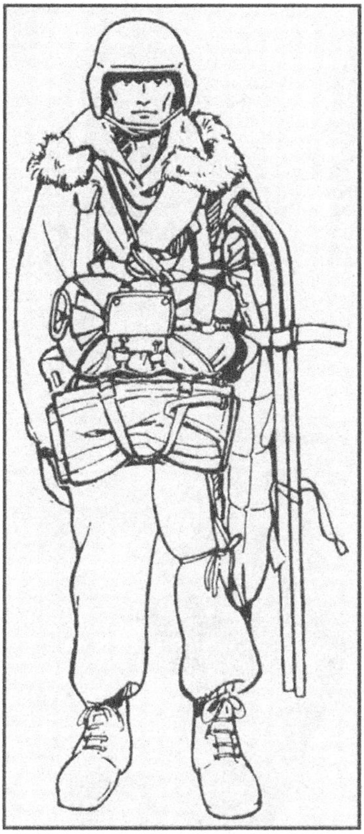

Figure 13-8. Upper and lower tie-downs.

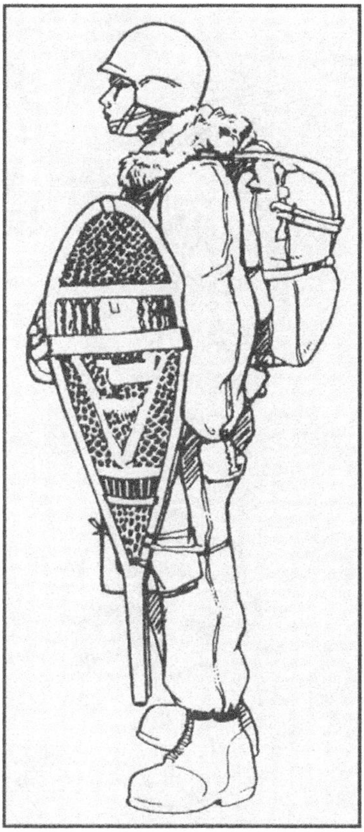

Figure 13-9. Left side view with weapons case.

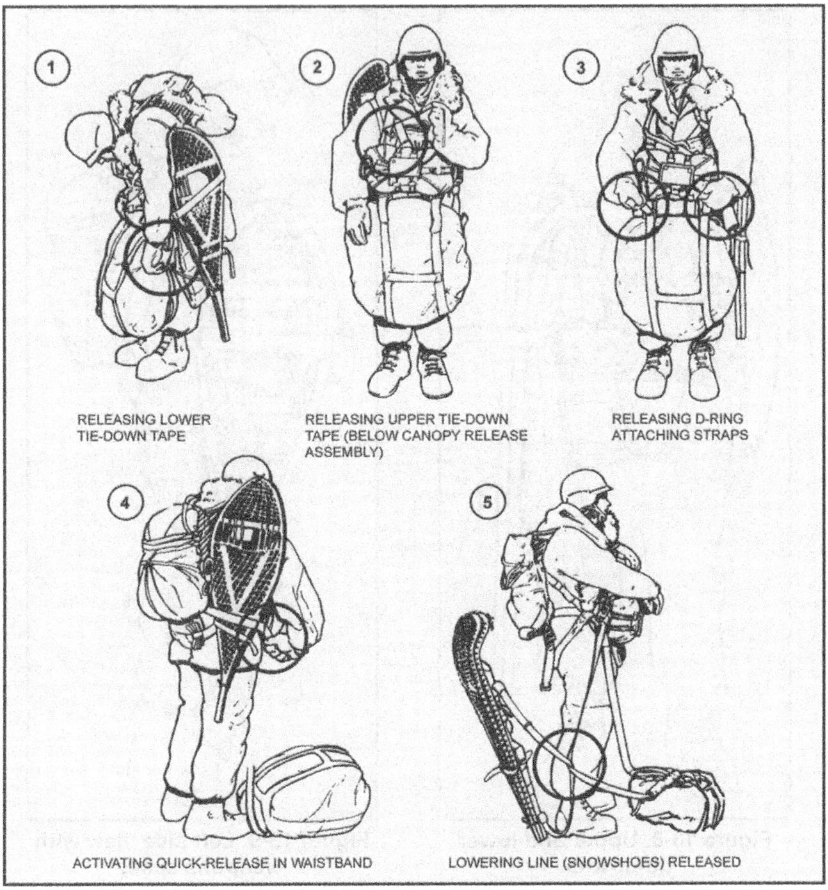

Figure 13-10. Arctic equipment released and lowered.

Section IV. SKIS JUMPED WITH RIFLE OR ALICE PACK

Skis, with the rifle attached, can be jumped using the procedures outlined.

13-8. SKIS AND RIFLE

When the parachutist jumps with skis, or with skis as part of the individual jump load, the jump must be from a rear platform or ramp (Figure 13-11).

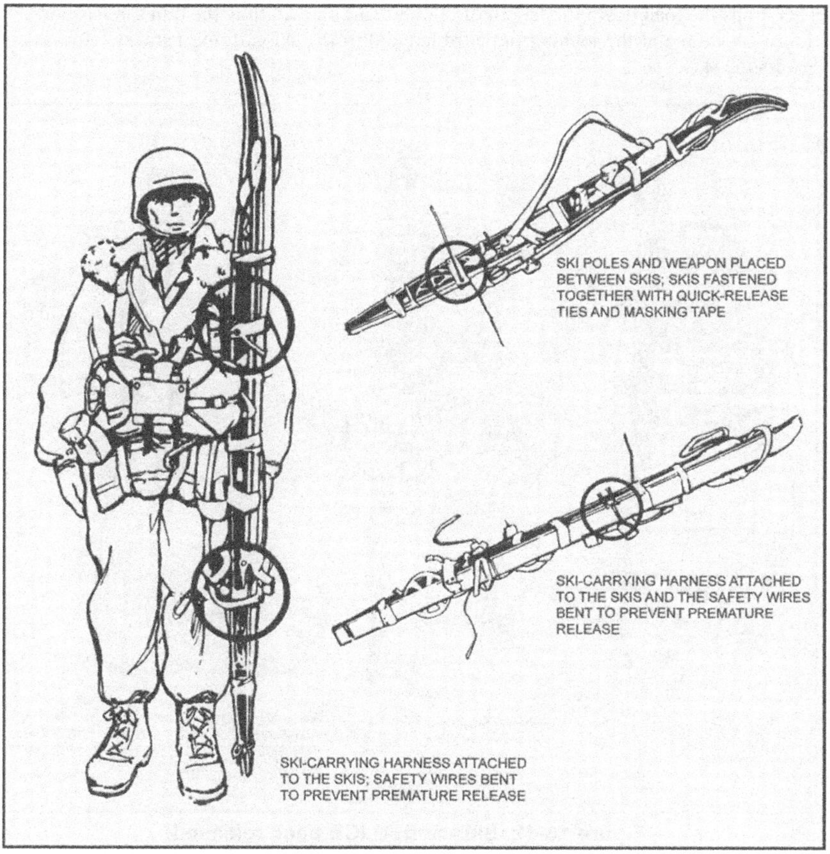

Figure 13-11. Skis and rifle rigged.

13-9. SKIS AND ALICE PACK OR WEAPONS CASE

The ALICE pack (or weapons case) lowering line is threaded between the skis and attached to the main lift web on the side to which the skis are attached. The lowering line is routed the same way, and the ejector snap is fastened to the lowering line adapter web or to the D-ring on the parachute harness (Figure 13-12). To secure and stabilize the skis, a length of 80-pound test is attached to the top and bottom of the skis with a bow knot. The skis and ALICE pack (or weapons case) are lowered by pulling the release handle. This drops the load down the length of the lowering line. Pulling the quick-release fold of the A-7A strap and the locking-pin cord lanyard of the ski-carrying harness releases and lowers the skis.

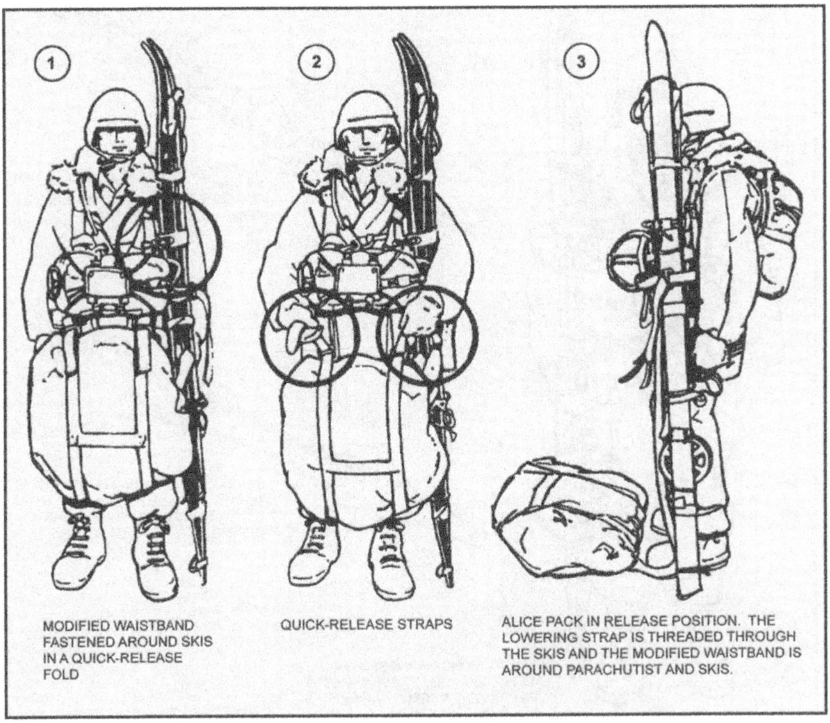

Figure 13-12. Skis and ALICE pack released.

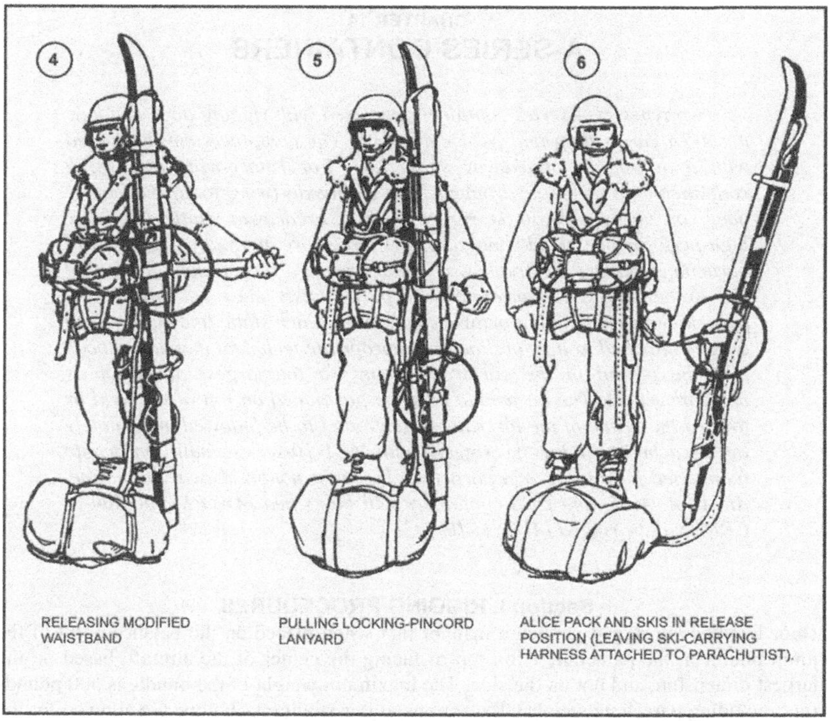

Figure 13-12. Skis and ALICE pack released (continued).

CHAPTER 14
A-SERIES CONTAINERS

Two types of A-series containers are used with rigging door bundles: the A-7A cargo sling and A-21 cargo bag. The containers can be rigged with a drogue or breakaway static line. For rotary-wing aircraft, a container load to be air-dropped from a shackle (wing load), helicopter door, or utility aircraft is rigged with a breakaway static line. For high-performance fixed-wing aircraft, loads are normally rigged with parachutes having nonbreakaway static lines. Paratroop door loads that are to be followed immediately by parachutists must be rigged with parachutes having nonbreakaway static lines. Each static line must have a drogue attached to it as outlined in appropriate technical manuals. Loads must be placed in the paratroop doors so the largest dimension is upright/vertical. The parachute must be positioned on top of the load or toward the inside of the aircraft. A ramp load to be followed immediately by parachutists must be rigged with a T-10-series main parachute (converted for cargo) or a parachute having a nonbreakaway static line. All A-7A, A-21, and CDS containers will be rigged IAW FM 100-500-3; CRRC will be rigged IAW FM 10-542.

Section I. RIGGING PROCEDURES

Door bundles are rigged in such a manner that when placed on the balance point of the jump platform, the parachute is on top or facing the center of the aircraft, based on the largest dimension, and not on the side. The maximum weight of the bundle is 500 pounds (not including parachute weight). Exceptions to this rigging technique are allowed for the 90-mm recoilless rifle and the Stinger missile. In both cases, the bundle is placed upright with the parachute facing the center of the aircraft. Both are rigged using the A-21 container. The skid board on the Stinger is placed inside the canvas cover.

14-1. ASSEMBLIES

When rigging an item, all components needed for its assembly must be packed in the same airdrop bundle. (For example, a radio and battery are packed in the same bundle.) When items such as radio equipment are rigged, each item is individually wrapped. Padding or honeycomb is placed under the item being prepared and inserted between the items comprising the load to prevent contact. Cellulose wadding, felt, or other suitable material must be used to avoid metal-to-metal or metal-to-wood contact.

14-2. WEBBING

All excess lengths of webbing are rolled and tied with 1/4-inch cotton webbing in a surgeon's knot and locking knot. This reduces the danger of bundles becoming snagged when ejected or released from the aircraft.

14-3. HAZARDOUS MATERIALS

If hazardous materials are placed inside bundles, they must have a shipper's certificate completed IAW AFJMAN 24-204/TM 38-250.

NOTE: The shipper's certificate is attached to the manifest, not the bundle.

Section II. A-7A CARGO SLING

The A-7A consists of the following components:

- Four straps, 188 inches long, constructed of Type X cotton or Type VII nylon.
- A strap fastener at one end of each strap.
- Four D-rings.

14-4. CHARACTERISTICS

Container components weigh 8 pounds with a maximum weight of 500 pounds (not including the parachute). The minimum weight depends on the parachute used. The dimensions are a maximum 30 inches wide by 48 inches deep by 66 inches high (to include cargo parachute).

14-5. TWO-STRAP BUNDLE

The jumpmaster lays out one strap perpendicular (lengthwise) to the bundle with the thick lip portion of the friction bar on the strap fastener facing down. He lays out one strap parallel (width) to the bundle with the thick lip portion of the friction bar on the strap fastener facing down and over the top of the perpendicular strap. When the straps are in place, they are ready to receive the bundle.

a. Center the bundle on the perpendicular strap. Route the perpendicular strap over the top of the bundle and through the single D-ring (through the rectangular portion of the D-ring), fold, and secure. When using the G-14 or T-10-series, two D-rings are used on the perpendicular strap. They are tied in place with Type II or Type III nylon cord gutted.

b. Route the parallel strap through the D-ring (through the rectangular portion of the D-ring), roll, and secure. Tie all excess webbing onto itself using one turn of 1/4-inch cotton webbing tied in a surgeon's knot and locking knot.

c. Tighten all straps. Tie off the excess above the strap fastener; ensure that the excess webbing is not above the top of the bundle. The bundle has one smooth side for ease in ejecting from the aircraft.

14-6. THREE-STRAP BUNDLE

The jumpmaster lays out one strap parallel (lengthwise) to the load. He then lays out two more straps parallel to each other and perpendicular to the load. He ensures that the strap fasteners for these straps are on the same side of the load and that straps are at least 16 inches from each other, and are centered on the load.

a. Center the load to be rigged on the straps. Route the strap that is running parallel (lengthwise) to the load over the top of the load and through the two D-rings. Center the two D-rings on top of the load. Route the strap through the strap fastener and tighten it down. Ensure that the strap fastener does not rest on the top of the load. Roll the excess hand over hand towards the load and secure it with a non-slip knot above the strap fastener, but not on top of the load. Route the straps that are running perpendicular to load through the D-rings on the top of the load, from the inside toward the outside so that the D-rings are pointing towards each other. Route the straps through the strap fasteners, and tighten them down. Roll the excess hand over hand towards the load and secure it with a non-slip knot above the strap fastener, but not on top of the load.

b. Once all straps are tightened and secured, the free running ends should not rest on top of the load. The bundle should have one smooth side for ease in ejecting it from the aircraft.

14-7. FOUR-STRAP BUNDLE

The jumpmaster lays out two straps parallel (lengthwise) to the load and centered, ensuring that both strap fasteners are on the same side of the load. He then lays out two straps parallel to each other and perpendicular to the load, ensuring that both strap fasteners are on the same side of the load and that all straps are centered on the load.

a. Route the two straps that are running lengthwise on the load over the top of the load and through the two D-rings (one D-ring per strap). Center the D-rings on top of the load. Route the straps through the appropriate strap fasteners and tighten them down. Roll the excess webbing hand over hand towards the load and secure it with a non-slip knot above the strap fastener. Route the straps that are running perpendicular to the load from inside to outside through the appropriate D-rings (ensuring that both D-rings point in the same direction). Then, route the straps through the appropriate strap fasteners and tighten them down. Roll the excess webbing hand over hand towards the load and secure it with a non-slip knot above the strap fastener.

b. When tying off the free running ends of the straps, tie them in a non-slip knot above the strap fastener; ensure that the excess webbing does not rest on top of the load. The load should have one smooth side for ease in ejecting it from the aircraft.

Section III. A-21 CARGO BAG

The A-21 cargo bag consists of the following components.

- Canvas cover: Cotton duck material, 97 inches by 115 inches, with eight strap keepers.
- Sling assembly with scuff pad: One 188-inch main strap, two 144-inch side straps, scuff pad 30 inches by 48 inches, and four lifting handles.
- Quick-release assembly: Quick-release device with safety clip, three quick-release straps, and one fixed quick-release strap.
- Two-ring straps: The ring strap has one 9-inch strap that has a friction adapter, and one 7-inch strap with a D-ring.

14-8. CHARACTERISTICS

Container components weigh 18 pounds with a maximum weight of 500 pounds (not including the parachute). (The minimum weight depends on the parachute.) Dimensions are a maximum 30 inches wide by 48 inches deep by 66 inches high, or 69 inches high for the Stinger missile (see FM 10-550). Dimensions include the cargo parachute.

NOTE: See FM 10-500-3 for further information on rigging containers.

14-9. METHOD OF RIGGING

The jumpmaster spreads the canvas cover on a level surface with all strap keepers facing up. He positions the sling assembly webbing straps down on the canvas cover and threads the straps through the keepers. The sling and canvas cover are turned over as a unit so the sling is beneath the cover. The parachutist centers the load on the canvas cover, using cushioning material, as needed. He wraps the load in the canvas cover, side flap first, and folds all excess material under.

a. Attach the two-ring straps to the 188-inch main strap, keeping the D-ring-to-D-ring contact and ensuring they are centered. Attach the four quick-release straps to the 144-inch side straps. Ensure that the rotating disk is facing up when the quick-release assembly is placed on top of the load (thick-lip portion of the friction bar facing out).

b. Thread the fixed, quick-release strap with the quick-release assembly attached through the nearest steel rod ring. Thread the remaining quick-release straps through the nearest steel rod rings. Insert the lugs into the quick-release assembly.

c. Tighten the quick-release straps and the two-ring straps; roll all excess webbing. Ensure that it is tied off below the friction adapter with a surgeon's knot and locking knot and that the quick-release device is centered on the bundle.

Section IV. CARGO PARACHUTE RIGGING ON A-SERIES CONTAINERS

After the A-series containers are rigged, the jumpmaster inspects the cargo parachutes and attaches them to the load.

14-10. INSPECTION

The cargo parachute is placed on the center of the bundle and is inspected for—

- Four tie-down straps.
- Two risers complete (clevis, clevis pin, safety wire).
- Static line complete with drogue device (clevis, clevis pin, safety wire). The drogue device must be attached to the breakcord attaching loop, unless a breakaway static line is used.

14-11. ATTACHMENT

The jumpmaster ensures that the risers go directly to the attaching point (D-ring), tie-downs are attached (tied to side straps), static line is free to deploy, and risers are not routed around or under any part of the bundle.

NOTE: The cargo parachute should be attached with the side of the pack where the risers come out, collocated to the rough side of the bundle.

PART FOUR
Aircraft Used in Airborne Operations

CHAPTER 15
AIRCRAFT AND JUMP ALTITUDES

This chapter contains general aircraft descriptions, jumpmaster procedures, and aircraft preparations. Critical elements of airborne operations are the aircraft and drop altitudes. These aircraft are service tested and approved for troop jumping. Minimum jump altitudes and considerations that apply to basic airborne jumps, tactical jumps, and combat jumps are discussed.

15-1. TYPES OF AIRCRAFT

Commonly used types of aircraft for airborne operations are:

a. **High-Performance Aircraft**. The C-5, MC/C-130, KC-130, C-17, and C-141 are high-performance fixed-wing aircraft used for paratroop drops. (Nonstandard jump aircraft are found in Chapter 19.)

b. **Utility Rotary-Wing Aircraft**. The UH-1 series, UH-60, CH-47, CH-53 (USMC), and CH-46 (USMC) are the most commonly used rotary-wing aircraft for troop airdrop.

15-2. JUMP ALTITUDES

The minimum criteria discussed below include a 125-foot aircraft altimeter error and a 100-foot canopy control requirement for the MC1-series.

a. **Minimum Jump Altitudes**. The minimum jump altitudes for **all** aircraft are as follows.

(1) ***Peacetime Tactical Training***.

- Basic airborne training—1,250 feet AGL.
- Tactical training—1,000 feet AGL; however, 800 feet AGL drop altitude may be employed after a mature Command Decision Risk Assessment has been completed.
- Aircraft with a drop speed of 90 knots or less—1,500 feet AGL.

(2) ***Combat (Wartime).*** (Restricted to aircraft with a drop speed of 90 knots or higher.)

(a) *Rotary-Wing Aircraft with a Drop Speed of 90 Knots or Greater.*

- T-10-series parachute—600 feet AGL.
- MC1-series parachute—600 feet AGL.
- Minimum airspeed—90 knots.

(b) *Fixed-Wing Aircraft with a Drop Speed of 125 Knots or Higher.*

- T-10-series parachute—435 feet AGL.
- MC1-series parachute—475 feet AGL.
- Minimum airspeed—125 knots.
- Reserve parachute—optional.

b. **Prejump Training**. Sustained prejump training is modified to ensure that the individual parachutist accomplishes the following tasks:

(1) Upon receiving the opening shock of the main, immediately lowers individual equipment.

(2) Assumes the landing attitude (T-10-series) or canopy control (MC1-series).

c. **Data Summarizing Parachute Reliability**. Table 15-1 summarizes parachute reliability data at varying altitudes with personnel parachutes (mains only—not reserves).

	T-10-Series PARACHUTE		MC1-Series PARACHUTE	
Exit altitude in feet AGL	**Percentage of canopies that would be open**	**Percentage of parachutists who would have 100 feet or more to prepare to land**	**Percentage of canopies that would be open and would provide canopy control (no twists).**	**Percentage of parachutists who would have canopy control plus 100 feet or more to land**
200	76.11	4.00	13.00	NONE
300	99.92	76.11	90.66	13.00
400	99.98	99.92	99.98	90.66
500	99.98	99.98	99.98	99.98
600	99.98	99.98	99.98	99.98
700	99.98	99.98	99.98	99.98
800	99.98	99.98	99.98	99.98

Table 15-1. Parachute reliability data for varying altitudes.

d. **Jump Altitude Risk Assessment Decision-Making**. When making a training parachute jump altitude risk assessment decision, the airborne commander should consider soldier experience levels, soldier fatigue at jump time, whether the jump altitude is critical to mission success, and the data from Table 15-2 below. Risk reduction measures include alternate methods of entry for personal equipment (CDS, door bundle, and so on) and action-in-the-aircraft rehearsals, with particular attention devoted to jumper exit procedures.

Planned Altitude (Feet AGL)	**Possible Aircraft Altimeter Error** (Feet)	**Actual Jump Altitude Feet** (AGL)	**Total Time Available to Activate Reserve** (Seconds)	**Time Available Minus 4000-Count** (Seconds)
1,250	+125	1,375	9.4	5.4
	0	1,250	8.8	4.8
	-125	1,125	8.2	4.2
1,000	+125	1,125	8.2	4.2
	0	1.000	7.6	3.6
	-125	875	6.9	2.9
800	+125	925	7.1	3.1
	0	800	6.4	2.4
	-125	675	5.6	1.6

Table 15-2. Time available to activate reserve parachute.

15-3. HIGH-ELEVATION JUMPING

The term high-elevation jumping refers to airdrop operations that begin at normal altitude above ground level (that is, 800 feet AGL) but where the DZ is 5,000 to 10,000 feet above mean sea level, such as in mountainous terrain. Commanders must consider that lower air density or higher altitude will increase the canopies' rate of descent.

a. **Parachutes**. Standard troop-type parachutes are suitable for the airdrop of personnel onto DZs with ground elevations up to 10,000 feet. Current jump procedures are valid.

b. **Injuries**. Combat or training exercises onto DZ elevations of 5,000 to 10,000 feet that place safety secondary to tactical considerations can produce injury rates of up to four times those expected for similar DZ operations near sea level. Injury rates can be reduced by intensive instruction, training, and practice to include the following.

(1) Exit body position and PLF upon ground impact.

(2) Jumper awareness of increased opening shocks and faster rates of descent.

(3) The ability to control the parachute during sudden wind shifts and changes in wind velocity.

(4) Wearing equipment correctly and ensuring serviceability.

CHAPTER 16

HIGH-PERFORMANCE TRANSPORT AIRCRAFT

The C-5, MC/C-130, C-17, and C-141 are US Air Force high-performance aircraft for paratroop drops. These high-performance transport aircraft, provided by the USAF, are configured to meet the needs of the unit mission request. These configurations are termed tactical airdrop personnel (TAP) for the C-130 and the C-141 in USAF publications. The initial JA/ATT mission request from the airborne unit determines aircraft configuration. The USMC version of this aircraft is the KC-130; characteristics and parachute procedures are similar.

- *Tactical mass airdrop (both jump doors)—full seating configuration.*
- *In-flight rigging—tactical mass airdrop (both doors)—full seating configuration; comfort pallet with/without litters.*
- *Other load considerations—combination of air-land and airdrop mission; single door—reduced seating configurations.*
- *Over-the-ramp—combination equipment and personnel.*

Section I. C-130 HERCULES

The C-130 (Figure 16-1) is a medium-range, high-wing transport aircraft powered by four turboprop engines. Parachutists may be dropped using either the two jump doors or the ramp. (See notes regarding over-the-ramp operations at paragraph 16-3.)

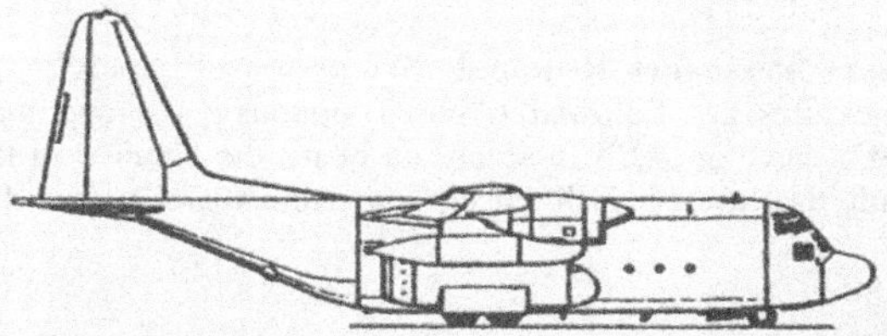

CONFIGURATION	PARACHUTIST LOAD
TAP 1 - MASS DROP	64
TAP 2 - IN-FLIGHT RIGGING	44
TAP 3 - HALO RAMP OPERATIONS AND WEDGE	42
TAP 4 - ARCTIC OPERATIONS	40

Figure 16-1. C-130 Hercules.

16-1. SEATING CONFIGURATION

The seating configuration for the C-130 Hercules is as follows:

a. **Peacetime Training (TAP 1)**. A total of 64 parachutists can be seated in two sticks of 32. Numbers 1, 2, and 3 are seated outboard aft of the wheel well; 4 through 23 are seated on the inboard seats; and numbers 24 through 32 are seated outboard forward on the wheel well (Figure 16-2). Jumpers are normally loaded over the aft end loading ramp.

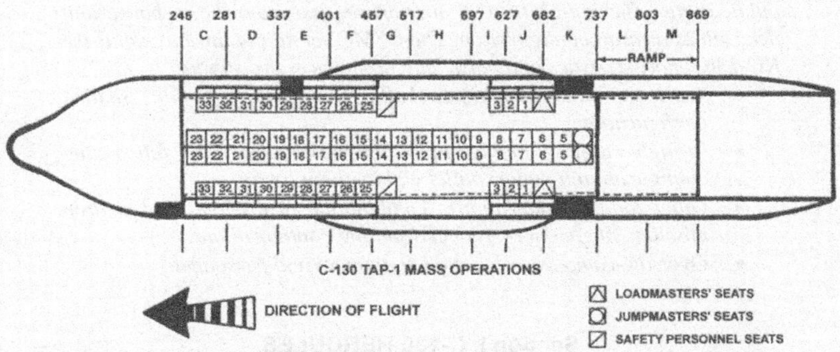

1. 68 TROOP SEATS - SEAT BELTS ON 24-INCH CENTERS - (66 TROOP SEATS - 2 LOADMASTER SEATS).
2. NORMALLY SEATS #10, LEFT AND RIGHT, ARE NOT RIGGED. SEATS #11, 12, AND 13, LEFT AND RIGHT, ARE RIGGED ON 20-INCH SPACING VERSUS 24-INCH. THIS PROVIDES 4 SEATS AFT WHEEL WELL BOTH SIDES.

Figure 16-2. C-130 seating configuration.

b. **Supervisory Personnel Required**. Six personnel supervise parachutists and ensure safety measures are followed (ADEPT options 1, 2, and mass exit. (USAF operations may not have an AJM or safety on board the aircraft. In these cases, prior coordination with the aircrew will ensure procedures are known by all personnel involved.)

- One JM.
- One AJM.
- Two safety personnel.
- Two airdrop certified loadmasters.

NOTE: USMC and USN utilize one door, the other door remains closed and unmanned, therefore only 1 JM and 1 safety are used. These units do not fall under AR 350-2.

c. **Jump Commands**. Jump commands are given in the following sequence:

GET READY.
OUTBOARD PERSONNEL, STAND UP.
INBOARD PERSONNEL, STAND UP.

HOOK UP.
CHECK STATIC LINES.
CHECK EQUIPMENT.
SOUND OFF FOR EQUIPMENT CHECK.
STAND BY.
GO.

16-2. IN-FLIGHT RIGGING PROCEDURES

In-flight rigging procedures include the following:

a. **Personnel**. These procedures provide in-flight rigging for the airdrop of 44 personnel (with equipment) including one JM and four AJMs. Three AJMs are designated from the parachutists onboard to assist in rigging. Two nonjumping safety personnel are also required. For USAF operations, a safety is not used and either the JM or loadmaster fulfills the duty after coordination.

b. **Briefing**. All parachutists must be briefed and rehearsed on their actions before executing this type mission. The preferred method for in-flight rigging is **buddy rigging**. This allows faster rigging and reduces parachutists' movement in the aircraft. The other method, **station rigging**, is seldom used. (See paragraphs 16-2e and f.)

c. **Aircraft Configuration**. The aircraft is TAP 2 configured to seat a total of 48 personnel to include 2 loadmasters, 2 nonjumping safeties, and 44 parachutists (Figure 16-3).

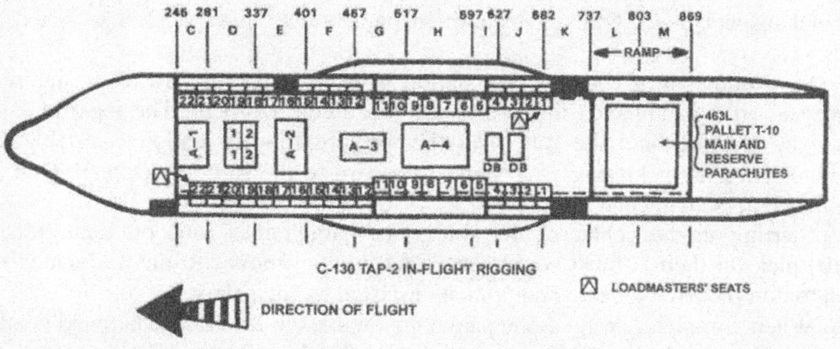

1. 48 TROOP SEATS-SEAT BELTS ON 20-INCH CENTERS-46 TROOP SEATS OFFERED. THIS CONFIGURATION IS FOR IN-FLIGHT RIGGING OF PARACHUTES ONLY (LONG-RANGE MISSIONS).

2. A-1: STORAGE AREA FOR ROLLER CONVEYORS, TOOL BOXES, AND CREW BAGGAGE.

   A-2: 22 EACH MAIN AND RESERVE PARACHUTES IN KIT BAGS TO BE RELOCATED FROM CARGO RAMP TO AREA A-2 PRIOR TO RIGGING.

   A-3: M1950 WEAPONS CASE STORAGE FOR WEAPONS BELONGING TO TROOPS THAT OCCUPY WHEEL WELL SEATS.

   A-4: WEAPONS IN EQUIPMENT CONTAINERS.

   DB: DOOR BUNDLES STORAGE AREA.

Figure 16-3. C-130 configuration for in-flight rigging (48 seats).

d. **Storage of Equipment**. Storing equipment involves the following:

(1) Forty-four parachutes and reserves, in kit bags, are palletized (covered with a cargo net or tie-down devices) on the ramp.

(2) Door bundles are placed in the center aisle just forward of the ramp.

(3) Special items of equipment are placed in the center aisle just forward of the door bundles.

(4) M1950 weapons cases should be placed behind individual seats. Weapons cases for personnel occupying wheelwell seats are placed on the floor at Station 477.

(5) Individual equipment should be placed under seats or, if too large, may be placed in the center aisle forward of the special items of equipment.

(6) All equipment placed in the center aisle must be secured.

e. **Buddy Rigging**. Buddy rigging begins 2 hours and 20 minutes before drop time. Rigging must be completed by the 20-minute warning.

(1) The main and reserve parachutes, with kit bags, are passed forward until each man has a parachute. Once everyone has his parachute, buddy rigging begins under the supervision of the JMs, designated JM qualified parachutists, and safety personnel. Each parachutist must know who his rigging partner is.

(2) Once a parachutist is rigged, he sits down and waits to be inspected by one of the JMs. The safety personnel serve as roving correction inspectors. The attachment of the static line snap hook to the top carrying handle of the reserve signifies an inspected parachutist.

(3) The JM supervises the entire rigging operation and assists, as needed.

(4) Once all the parachutists have been inspected, the JMs don equipment, and safety personnel inspect them. Other JM qualified parachutists may be used to speed up the rigging process.

f. **Station Rigging**. One rigging station is established forward of the jump doors; another station is established forward in the cargo compartment. The forward station is manned by a safety and the AJM. As the forward rigging station is established, 22 parachutes (in aviator kit bags) are passed forward to the safety personnel. One JM kit bag must be present at each station.

(1) Starting in the center of the stick, two parachutists (one on each side of the aircraft) pick up their combat equipment. Each jumper moves to the designated rigging station to don the parachute and equipment, assisted by the safety.

(2) When completely rigged, the parachutist moves to the AJM (stationed nearby) for inspection. The AJM attaches the static line snap hook to the top carrying handle of the reserve parachute when he reaches that portion of the inspection sequence. Only the JM or AJM removes the static line snap hook from the reserve before the command HOOK UP, since this indicates that the jumper has received JMPI.

(3) After JMPI, the jumper returns to his proper seat (or stick position). To ensure minimum time loss, the next parachutist is waiting to be rigged by safety personnel.

16-3. OVER-THE-RAMP OPERATIONS

These procedures provide for over-the-ramp airdrop of 40 personnel including one JM, one AJM, and two nonjumping safety personnel (Figure 16-4, page 16-6). For USAF operations, a safety is not used and either the JM or loadmaster fulfills the duty after coordination.

NOTES:

1. IAW message Headquarters, USAF AMC, DTG 021431Z MAY 1990, subject: C-130 Tailgating; over-the-ramp C-130 airdrop operations are restricted "to combination airdrops, tests requiring over-the-ramp operations to satisfy a specific objective, jumpers required to jump with snow skis, support for special operations forces training, MC-130, pathfinder and SOLL II operations." SOF were defined as "USA Special Forces (Green Berets), Ranger units, and Ranger Training Brigade, Navy Seals, USAF Combat Control Teams, Pararescue Teams, and those assigned to special tactics units." Message further stated, "These forces will continue to train for over-the-ramp operations which is their normal method for deployment."
2. IAW message Headquarters, USAF AMC, DTG 111345Z JUL 1991, subject: Authorization of Arctic Equipped Parachutists to Tailgate; authorization was granted "to allow arctic equipped parachutists with large field pack and/or skis to tailgate from C-130 aircraft." (Large field pack mentioned in message DTG 111345Z JUL 1991 is the FPLIF.)
3. IAW letter CG, Marine Corps Combat Development Command, C42 over 5600, dated 22 AUG 95, USMC Force Reconnaissance and Air Naval Gunfire Liaison Company (ANGLICO) personnel are authorized to conduct C-130 over-the-ramp parachute operations if required by their mission. These units train for over-the-ramp parachute operations as a normal method of employment.
4. Based on Airborne materiel and training proponency issues, the 1st Battalion (Abn), 507th Infantry, 11th Infantry Regiment, United States Army Infantry School is authorized to plan and conduct over-the-ramp operations from the C-130 aircraft.

a. **Equipment Drop**. Over-the-ramp operations can include the combination of an equipment drop followed by parachutists. Equipment is defined as heavy equipment rigged for airdrop, or equipment packaged and contained in the A-7A or A-21 containers. Parachutists may be dropped over the ramp without an equipment drop (maximum of 20 parachutists for each pass).

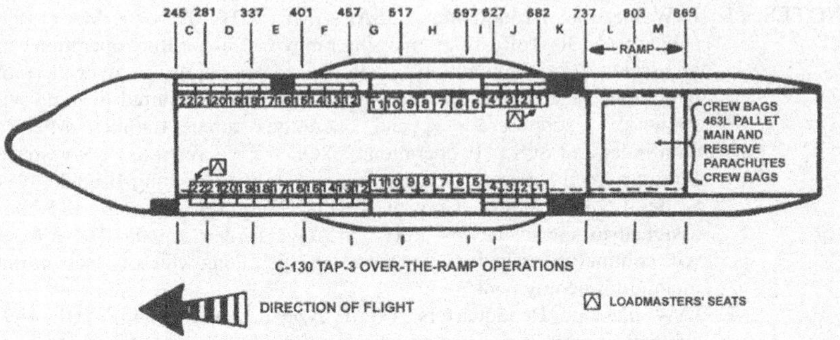

1. 44 TROOP-SEATS - SEAT BELTS ON 20-INCH CENTERS-42 SEATS OFFERED. THIS CONFIGURATION IS FOR TROOP DOOR HALO, TAILGATING, AND WEDGE.
2. EVERYTHING DOWN THE CENTER WILL BE REMOVED. PALLET ON THE RAMP WILL BE THE WEDGE (AS APPLICABLE).
3. 27 EACH MAIN AND RESERVE PARACHUTISTS IN KIT BAGS TO BE RELOCATED FROM AFT PALLET PRIOR TO RIGGING IN-FLIGHT.

Figure 16-4. C-130 configuration for over-the-ramp operations.

b. **Aircraft Operation**. The anchor line cables (only two are used—one on each side) are rigged from the forward outboard anchor line cable attachments to the aft inboard anchor line cable attachments. The anchor line cable stop (a small clevis, padded and taped) must be installed on the anchor line cable 20 inches aft of the center anchor line cable support bracket. The center anchor line cable support brackets at the door are disconnected and secured at the top of the fuselage. The static line retriever cables are tied or taped to the sides of the fuselage aft of the doors to ensure that they remain secured.

c. **Jump Commands**. Jump commands are given in the following sequence:

GET READY.
STAND UP.
HOOK UP.
CHECK STATIC LINES.
CHECK EQUIPMENT.
SOUND OFF FOR EQUIPMENT CHECK.
STAND BY. (Parachutists maintain a reverse bight in the static line.)
GO. (On the command GO, movement onto the ramp is a normal walking or shuffle pace. Exits are made at an angle of about 30 degrees toward the side of the aircraft and are not vigorous. Parachutists place their hands on the ends of their reserve parachute prior to exit.)

d. **Jump Procedures**. If the JM and AJM elect to jump, they will be the number 1 parachutist of each stick. Therefore, the safeties control the flow of the parachutists.

(1) An oral 10-minute warning is given, and the JM begins the jump commands. Prior to the 10-minute warning, the JMs hook up to the anchor line cable, hand the static line to the safety, and announce, SAFETY, CONTROL MY STATIC LINE. The JM then issues the jump commands.

(2) Jumpers hook up to the appropriate anchor line cable with the open gate of the static line snap hook facing toward the skin of the aircraft. They maintain a reverse bight (Figure 16-5).

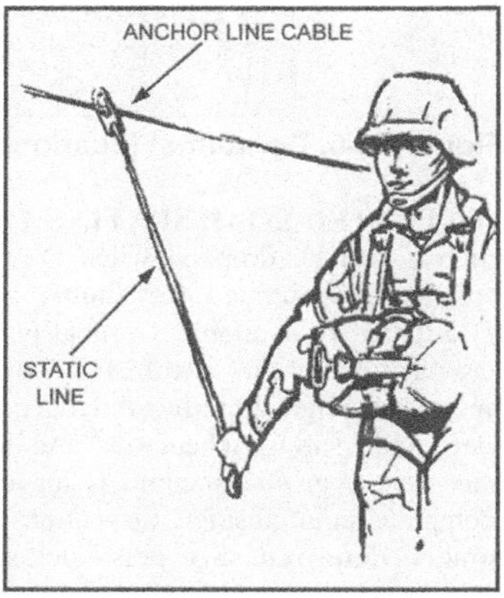

Figure 16-5. Static line grasped with reverse bight.

(3) Door check is not required. If using GMRS, the JM must spot the ground marking from the left side of the ramp.

(4) After giving the command STAND BY, the JM moves to the center edge of the ramp and exits on green light. The safety positions himself immediately behind the hinged portion of the ramp and controls the flow of parachutists.

(5) Each parachutist walks off toward the center of the ramp at an angle away from the anchor line cable (Figure 16-6, page 16-8).

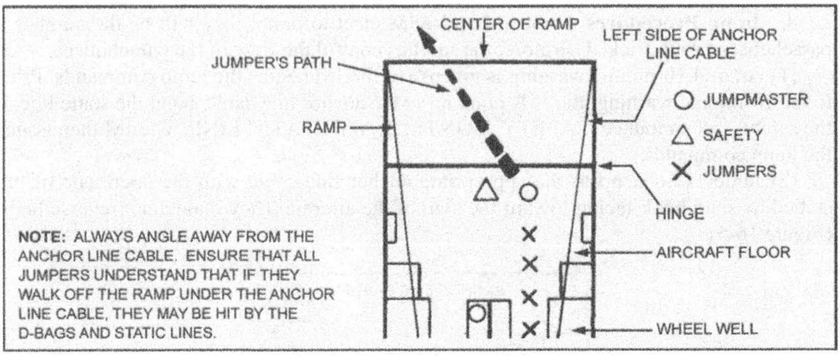

Figure 16-6. Personnel locations.

16-4. COMBAT CONCENTRATED LOAD SEATING CONFIGURATION

A maximum of 80 parachutists can be dropped when these procedures are used for wartime emergency operations. Concentrated parachutist loading is used when not enough C-130 and C-141B aircraft are available. Combat procedures reduce individual space inside the aircraft; crowded conditions restrict freedom of movement and present potential hazards if not supervised. Flight time beyond 1.5 hours may intensify adverse effects on parachutists. Data and seating schematics are included under appropriate aircraft titles. When the sticks are over 40 parachutists for each pass, additional safety personnel may be used to complete required safety inspections within the time warnings.

a. **Supervisory Personnel Required**. Six personnel supervise parachutists and ensure safety measures are followed.

- One JM.
- One AJM.
- Two (jumping) safety personnel.
- Two loadmasters.

b. **Loading Procedures and Seating Arrangement**. The aircraft is configured to provide 19 outboard seats and 21 inboard seats with seat belts installed on 20-inch centers (Figure 16-7).

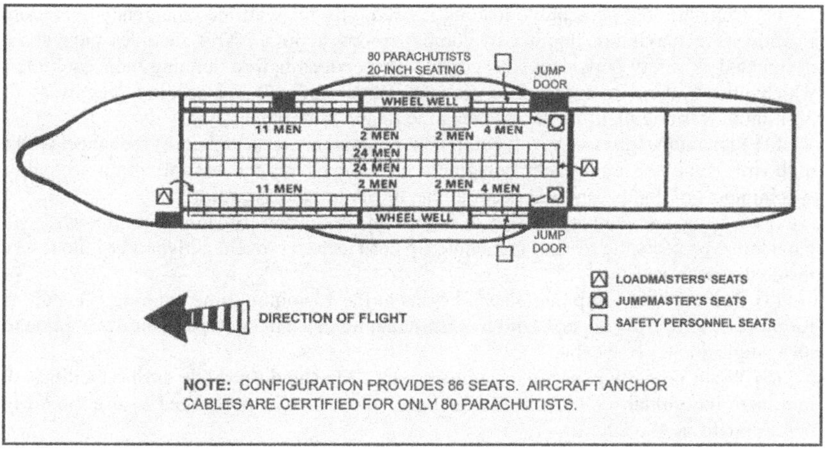

Figure 16-7. C-130 combat concentrated load.

(1) Jumpers are loaded through the aft end loading ramp and seated in two sticks of 40 jumpers each. Before entering the aircraft, all jumpers release their adjustable leg straps so they can lift their equipment over the seats. These safety straps are retied when the jumper is seated with seat belt in place.

(2) Each jumper must be assisted in seating by the AJM or safety personnel. They ensure that jumpers select the correct seat belt and that it is properly fastened.

NOTE: Parachutists sitting in outboard wheel well seats must place their knees forward or aft in the space of the raised seat. The area is not spacious enough to allow inboard or outboard parachutists to face each other.

c. **Jump Procedures**. The jump procedures for the C-130 remain the same with the following exceptions:

(1) The 30-minute time warning replaces the 20-minute time warning, and the 15-minute time warning replaces the 10-minute time warning.

(2) On the command STAND UP, inboard parachutists stand up and then stand on their seats, supporting themselves by grasping the center stanchion seat support rail. They remain in this position until outboard parachutists stand up and raise and secure their seats in the up position. On the command GET DOWN, inboard parachutists step off their seats, then raise and secure them in the up position. Standard jump commands are then resumed.

(3) On the command HOOK UP, no more than 20 parachutists hook up and exit on any single anchor line cable.

(4) When the command GO is given, outboard personnel exit in numerical sequence, followed by inboard personnel.

d. **Safety Procedures**. The standard safety procedures for the C-130 apply, with the following changes:

(1) Concentrated parachutist loading is used only for wartime emergency operations. It allows the maximum number of combat troops to jump. This includes parachutists designated as safety personnel who conduct inspections before donning their equipment, which allows them to move freely about the aircraft. This inspection begins at the 30-minute warning to allow the safeties time to don their parachutes.

(2) Parachutists jumping with equipment containers are seated aft of the wheel well in outboard seats; the equipment containers are stowed on the aircraft ramp. Attaching special items of equipment will occur at the 30-minute time warning.

(3) During movement in the aircraft and execution of jump commands, each parachutist protects the reserve parachute rip cord grip to prevent activation of the reserve inside the aircraft.

(4) Execution of jump commands begins at the 15-minute time warning. This allows for more time to stow all seats and to ensure that no more than 20 parachutists are hooked on a single anchor line cable.

(5) When preparing for airborne operations, JMs should provide latrine facilities. To minimize individual movement during flight, jumpers are encouraged to use the latrine before boarding the aircraft.

16-5. C-130 JUMPMASTER CHECKLIST

The JM inspects the aircraft for the following:

a. **Seats.**

- Enough seats are available for troop load.
- All seats have safety belts.
- Seats are not torn.
- No seat projections are present.
- Seat legs are locked into floor.
- Diagonal leg brace is attached in each set section (if required).
- Seats along the wheel well are removed (if required).

b. **Floor.**

- Nonskid covering is in good condition.
- Floor is clean and safe to walk on.
- All cargo compartment roller conveyors and dual rail system floor sections are removed when jumping the door.
- Loose equipment is lashed and does not interfere with movement or comfort.

c. **Jump Platform.**

- Nonskid surface is present.
- No cracks or bends are present.
- Hinge plate slots are engaged by tie-down studs and screw (with washer) so that the hinge fittings do not slide off the tie-down studs. When the screw and washer are missing, the hinge plate must be safety wired with four turns of stainless steel wire to the tie-down ring bracket.
- Two spring-down lock catches engage the edge of the door.
- Clutch mechanism is present on the spring-down lock catches.

d. **Jump Doors**.

- No excessive grease is on the door tracks.
- No sharp edges or protrusions are on door frames.
- Pip pin in top, forward edge of door frame is present.
- Auxiliary hydraulic ramp pump handle is secured (after takeoff).
- Doors open and close easily. (Doors are operated in flight by aircrew members.)

e. **Air Deflectors**.

- No sharp edges are on trailing edge.
- Deflectors operate electrically.
- No items or trash are stored in wells.

f. **Jump Caution Lights**. (Check all seven sets for operation.)

- Set 1—crew entrance door.
- Sets 2 and 3—top leading edges of right and left doors.
- Sets 4 and 5—trailing edges of right and left doors, waist high.
- Sets 6 and 7—right and left anchor cable aft supports.

g. **Anchor Cable System**.

(1) ***Forward Support Beam.***

- Four U-bolts, with self-locking nuts or nuts with cotter pins, are attached.
- Anchor cables are attached to first and second U-bolts right and left of center line for personnel jump.
- Forward latch assembly is in the locked position and secured with locking pins.

(2) ***Anchor Cable.***

- No breaks (within acceptable tolerances of TO 1C-130A-9), frays, or kinks exist.
- Cable is clean and free of rust.

(3) ***Anchor Cable Intermediate Center Support.***

- Cables run through slots after ramp is closed.
- Quick-release retaining pins are present

(4) ***Anchor Cable Aft Support.***

- Aft latch assembly is closed.
- U-bolts, nuts, and safety pins are present.
- Support anchor bolts, nuts, and safety pins are present.

(5) ***Static Line Retrievers.***

- Motor is operational.
- Retriever cables are not broken (within acceptable tolerances of TO 1C-130A-9), frayed, kinked, dirty, or rusty.
- Spool clamp and shackle are attached forward of intermediate cable support and are tied to support with two turns of 1/4-inch cotton webbing.
- Retriever cables are secured with two turns of 1/4-inch cotton webbing to litter brackets at station 627. If cable clips are installed on the wheel well, the tie at station 627 is not used. (Retriever cable must be at least 4 inches above the anchor line cable.)

h. **Emergency Equipment.**
- First aid kits (4) are present.
- Fire extinguishers (3) are present.
- CGU1-B cargo tie-down straps (2) (for retrieval of towed parachutists) are present.
- Alarm system is operational.
- Emergency exits are operational and accessible.
- Sufficient emergency parachutes are available.

i. **Miscellaneous.**
- Lighting system is operational.
- Airsickness bags are available.
- Comfort facilities are available.
- JM kit bag (extra equipment) is onboard.

Section II. C-141B STARLIFTER

The C-141B (Figure 16-8) is a swept-wing, long-range transport powered by four turbofan jet engines and equipped for in-air refueling. The aircraft can be configured in different peacetime troop-carrying modes for airborne operations. Parachutists jump from the two aft doors of the aircraft.

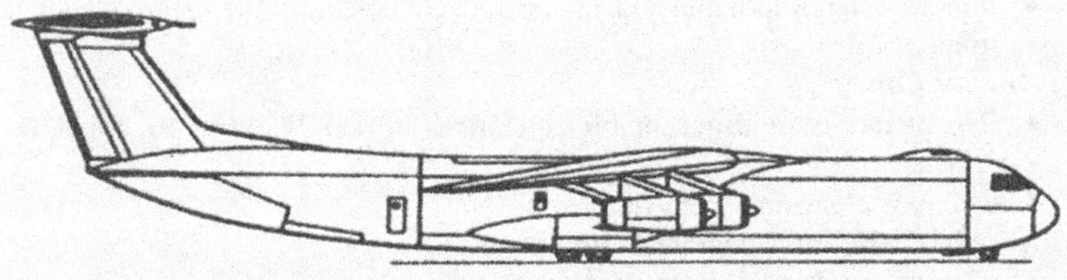

TACTICAL TRAINING	
CONFIGURATION	**PARACHUTIST LOAD**
TAP 1 - WITHOUT COMFORT PALLET	152
TAP 2 - IN-FLIGHT RIGGING WITH COMFORT PALLET	130
COMBAT OPERATIONS	
CONFIGURATION	**PARACHUTIST LOAD**
CONCENTRATED LOADING	180
NOTE: THE C-141B IS NOT USED FOR PERSONNEL JUMPING (STATIC LINE) OVER THE RAMP.	

Figure 16-8. C-141B Starlifter.

16-6. SEATING CONFIGURATION WITHOUT COMFORT PALLET

The seating configuration for the C-141B is as follows:

a. **Peacetime Training (TAP 1)**. In this configuration, the C-141B seats 156 personnel: 152 parachutists, 2 static safety personnel, and 2 airdrop qualified loadmasters (Figure 16-9). The maximum number of personnel for each anchor line cable is 45.

b. **Supervisory Personnel Required**. Six personnel supervise parachutists and ensure safety measures are followed.

- One JM.
- One AJM.
- Two static safety personnel.
- Two airdrop qualified loadmasters.

NOTE: USMC and USN utilize one door, the other door remains closed and unmanned. Therefore, only one JM and one safety are used. These units do not fall under AR 350-2.

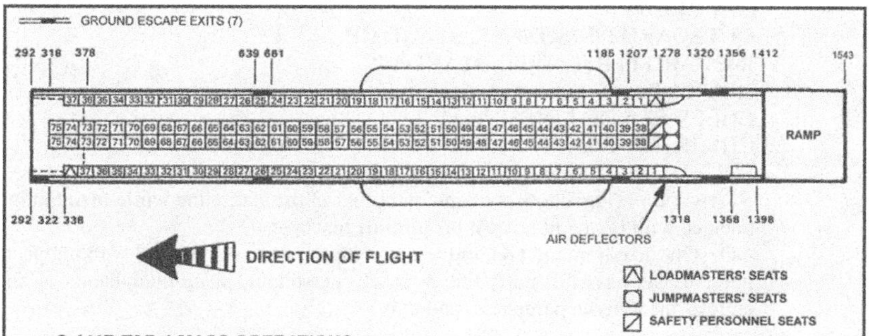

C-141B TAP-1 MASS OPERATIONS

1. 156 TROOP SEATS--SEAT BELTS ON 24-INCH CENTERS--(154 TROOP SEATS, 2 LOADMASTER SEATS), 160 MAXIMUM ALLOWED ON BOARD OVER WATER (INCLUDING CREW). ADDITIONAL LIFE RAFTS WILL BE REQUIRED.
2. PORTABLE URINALS--STATIONS 1358-1398.
3. SEATS ARE NUMBERED FOR IDENTIFICATION AND ARE ONLY REFERRED TO AS SIDE-WALL SEAT 1 LEFT/SEAT 1 RIGHT OR CENTER LINE SEAT 1 LEFT/SEAT 1 RIGHT, AND SO FORTH.
4. THE DOTTED LINES DENOTE THE CREW REST FACILITY.

Figure 16-9. C-141B seating configuration without comfort pallet.

c. **Time Warnings**. Time warnings include the following:

(1) The 20-minute time warning may be increased to 30 minutes to provide enough time for static safety personnel to complete safety checks and to rig equipment containers.

(2) The 10-minute time warning may be increased to 15 minutes to allow time for parachutists to release and stow troop seats and for static safety personnel to complete safety checks before the 1-minute time warning. Prior to the 10-minute time warning, the JMs hook up to the inboard anchor line cable (or anchor line cable if only one is available), hand the static line to the safety, and announce, SAFETY, CONTROL MY STATIC LINE. The JM then issues the jump commands. If the aircraft is configured with only one anchor line cable, the JMs hook up to the one cable.

d. **Jump Commands**. The loadmaster gives the JM an oral 10-minute time warning. Emphasis is on using a public address system to give jump commands, since parachutists in the forward end of the cargo compartment may not be able to see JM hand-and-arm signals.

> GET READY.
> OUTBOARD PERSONNEL, STAND UP.
> INBOARD PERSONNEL, STAND UP.
> HOOK UP.
> CHECK STATIC LINES.
> CHECK EQUIPMENT.
> SOUND OFF FOR EQUIPMENT CHECK.
> STAND BY. (Parachutists maintain a bight in the static line while maintaining balance with the hand nearest the aircraft fuselage.)
> GO. (On the command GO, movement into the door is a normal walking pace. Parachutists pass the static line to safety personnel, place their hands on the ends of the reserve parachute, and exit.)

e. **Safety Considerations**. Safety personnel are seated aft to aid the JMs, at the 20-minute time warning, in positioning door bundles and performing other duties.

(1) Static safety personnel must complete their 20-minute checks and arrive at the forward end of the cargo compartment before the 10-minute time warning.

(2) No more than 45 parachutists are hooked to any one anchor cable.

(3) Deployment bags may be trailed between passes on single drop zones, or when dropping on multiple drop zones with less than 10 minutes flight time between drop zones.

16-7. IN-FLIGHT RIGGING SEATING CONFIGURATION WITH COMFORT PALLET

In this TAP 2 configuration, the C-141B seats 134 personnel: 130 parachutists, 2 static safety personnel, and 2 loadmasters. Floor space is provided forward, midway, and aft for stowage of parachute assemblies. Seats are on a 24-inch center (Figure 16-10).

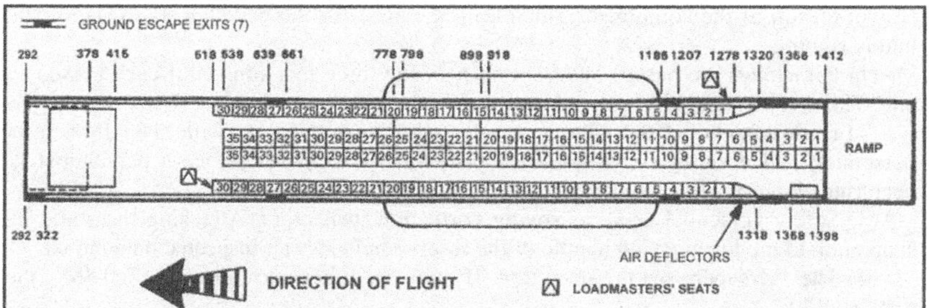

C-141B TAP-2 IN-FLIGHT RIG WITH COMFORT PALLET

1. 130 TROOP SEATS--SEAT BELTS ON 24-INCH CENTER (128 TROOP SEATS, 2 LOADMASTER SEATS, 160 MAXIMUM ALLOWED ON BOARD OVER WATER [INCLUDING CREW]). ADDITIONAL LIFE RAFTS WILL BE REQUIRED.

2. SEATS ARE NUMBERED FOR IDENTIFICATION ONLY AND ARE REFERRED TO AS SIDE-WALL SEAT 1 LEFT/SEAT 1 RIGHT OR CENTER LINE SEAT 1 LEFT/SEAT 1 RIGHT.

3. THE DOTTED LINES DENOTE THE CREW REST FACILITY.

4. FOR IN-FLIGHT RIGGING, DELETE CENTER LINE SEATS FORWARD OF FS 538, BETWEEN FS 778-898 AND AFT OF FS 1318. LEFT AND RIGHT OUTBOARD SEATS WILL BE RAISED FORWARD OF FS 518 AND BETWEEN FS 798-918. 102 TROOP SEATS--SEAT BELTS ON 24-INCH CENTERS--100 TROOP SEATS OFFERED (39 SEATS IN FORWARD SECTION AND 61 SEATS IN AFT SECTION).

5. IF REQUIRED, SEAT BELTS MAY BE INSTALLED ON 20-INCH CENTERS FOR IN-FLIGHT RIGGING ONLY. DELETE CENTER LINE SEATS FORWARD OF FS 538, BETWEEN FS 778-898 AND AFT OF FS 1318. LEFT AND RIGHT OUTBOARD SEATS WILL BE RAISED FORWARD OF FS 518 AND BETWEEN 798-918. 130 TROOP SEATS--SEAT BELTS ON 20-INCH CENTERS--128 SEATS OFFERED (51 SEATS IN FORWARD SECTION AND 77 SEATS IN AFT SECTION).

Figure 16-10. C-141B seating configuration with comfort pallet.

a. **Supervisory Personnel Required**. Fifteen personnel supervise parachutists and ensure safety measures are followed.

- One JM.
- Six AJM.
- Six safety personnel (two static).
- Two loadmasters.

b. **Storage of Equipment**. Parachute assemblies (133) are placed in aviator kit bags and stowed and secured in three locations on the cargo floor. There are 130 parachutists seated in two sticks of 65 each on the left and right.

(1) Door bundles are stowed on the cargo ramp.

(2) Individual equipment, web gear, ballistic helmets, and ALICE packs are stowed under the troop seats; weapon containers are stowed behind the seats; and individual equipment containers are stowed in the aisles or with the parachutes.

(3) One JM kit bag is stowed at each of the six rigging stations (two forward, two midway, and two aft).

c. **Buddy Rigging**. The JM initiates in-flight buddy rigging 2 hours before the 20- or 30-minute time warning. Parachutists unzip the seats, place their equipment on the seats, and sit on top of the equipment. This clears the aisle for the parachutists to stand while buddy rigging.

(1) Parachutes are passed out from each of the three locations until each person has one. Then buddy rigging begins.

(2) Each parachutist must know exactly who he will rig up with. The JMs, safety personnel, and designated JM qualified parachutists supervise. Once a parachutist has been rigged, he sits down and waits to be inspected by one of the JMs.

(3) Safety personnel serve as roving correction inspectors. Attaching the static line snap hook to the top carrying handle of the reserve indicates an inspected parachutist.

(4) The JM supervises the operation. If there are other currently qualified JMs, they may be used.

(5) Once all the parachutists have been inspected, the JMs rig up, and safety personnel inspect them.

d. **Station Rigging**. Station rigging involves the following:

(1) Six AJMs and six safety personnel (two static and four jumping safety personnel) are required to man the six rigging stations.

(2) Parachutists are divided into three segments and assigned to rigging stations.

(3) The JM initiates in-flight rigging 2 hours before the 20- or 30-minute time warning.

(4) Before initiation of in-flight rigging, jumpers are instructed to unzip the troop seats, place their equipment on the seats, and sit on it. Placing equipment on the seats clears the aisles for jumpers moving to and from the rigging stations.

(5) AJMs and safety personnel at each rigging station stow their seats to provide more rigging space.

(6) Starting with two parachutists from the end of the divided sticks, each parachutist moves to a designated rigging station and is rigged by safety personnel. When completely rigged, the parachutist moves to the AJM (stationed nearby) for inspection. The AJM attaches the static line snap hook to the top carrying handle of the reserve parachute when he reaches that portion of the inspection sequence.

(7) Only the JM or AJM removes the static line snap hook from the reserve (before the command HOOK UP) since this indicates that the parachutist has received the JM's inspection.

16-8. COMBAT CONCENTRATED LOAD SEATING CONFIGURATION

In this configuration, the C-141B seats 182 personnel: 180 parachutists and 2 loadmasters (Figure 16-11). Personnel are seated on the side-facing seats on 20-inch centers. Parachutists (180) are seated in two sticks of 90 each on the left and right sides. If 45 parachutists cannot be seated on the outboard seats, the overflow may be seated inboard aft on available center seats. The JM must ensure that only 45 parachutists hook up to the outboard/inboard anchor cables.

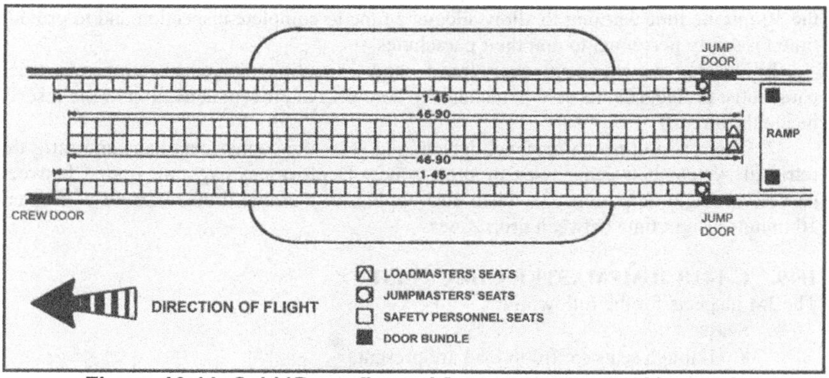

Figure 16-11. C-141B configured for combat load (182 seats).

NOTE: Concentrated parachutist loading is used in wartime emergency operations when not enough C-130 and C-141B aircraft are available. These procedures reduce individual space inside the aircraft. Crowded conditions restrict freedom of movement and present a potential safety hazard if jumpers are not supervised. Flight time beyond 1.5 hours may intensify adverse effects on parachutists. Data and seating schematics are included under appropriate aircraft titles. Additional safety personnel may be used (if the sticks have over 40 personnel for each pass) to complete the required safety inspection within the time warnings.

a. **Supervisory Personnel Required**. Six personnel supervise parachutists and ensure safety measures are followed.

- One JM.
- One AJM.
- Two (jumping) safety personnel.
- Two airdrop qualified loadmasters.

b. **Jump Procedures**. Jump procedures involve the following:

(1) Jump procedures for the C-141B remain the same with the following exceptions: the 30-minute warning replaces the 20-minute time warning, and the 15-minute time warning replaces the 10-minute time warning.

(2) Jump commands are echoed by jumpers since parachutists in the forward end of the cargo compartment cannot see the JM's arm-and-hand signals.

c. **Safety Considerations**. At the 30-minute time warning, safety personnel are seated aft to aid the JMs in positioning door bundles and in performing other duties. The JMs ensure that no more than 45 parachutists are hooked to any one anchor line cable.

(1) The standard safety procedures for the C-141B apply with the following changes:

(a) Concentrated parachutist loading is used only for wartime emergency operations and allows the maximum number of combat troops to jump. This includes parachutists designated as safety personnel who conduct safety inspections before donning their equipment, which allows them to move freely about the aircraft. This inspection begins at the 30-minute time warning to allow adequate time to complete inspection and to provide time for safety personnel to don their parachutes.

(b) During movement in the aircraft and execution of jump commands, each parachutist protects the reserve parachute rip cord grip to prevent activation of the reserve inside the aircraft.

(2) There is no time to retrieve, detach, and stow the deployment bags and rerig the retrieval system between passes or drop zones. Deployment bags are trailed between passes on single drop zones, or when airdropping on multiple drop zones with less than 10 minutes flight time between drop zones.

16-9. C-141B JUMPMASTER CHECKLIST

The JM inspects for the following:

a. **Seats**.
- Enough seats for troop load are present.
- All seats have safety belts.
- Seat backs are secured.
- Seats are not torn.
- No projections exist through seats.
- Legs are locked into floor.

b. **Floor**.
- Nonskid covering is in good condition.
- Floor is clean and safe to walk on.
- Roller conveyors are stowed.
- Loose equipment is secured in the cargo ramp area and does not interfere with troops.

c. **Jump Platforms**.
- Nonskid surface is present.
- No cracks or bends exist.
- Studs are locked in seat track receptacles.
- Tie-down fitting is locked.
- All bolts and nuts are present.
- Platforms swing in and out easily.
- The two spring-down lock catches engage the edge of the door.

d. **Jump Doors**.
- No excessive grease on the door tracks exists.
- No sharp or protruding edges exist on door frames or on loading strut door (aft and down from jump door).
- Doors open and close easily.
- Door lever catches are operational.

e. **Air Deflectors**.
- No sharp edges present.
- Deflectors operate electrically.

f. **Jump Caution Lights**. (Check seven sets for operation.)
- Set 1—crew entrance door.
- Sets 2 and 3—top leading edges of right and left doors.

- Sets 4 and 5—trailing edges of right and left doors, waist high.
- Sets 6 and 7—right and left anchor cable aft supports.

g. **Anchor Cable System.**

(1) ***Forward Support Beam.***

- Bolts and nuts are present and tight.
- Anchor cables are attached to first and third anchor points right and left of enter line for personnel jumps.
- Cable, bolts, nuts, and safety wire are present.
- Turnbuckle is secured with safety wire.

(2) ***Anchor Cable.***

- No breaks (within acceptable tolerances of TO 1C-141B-9), frays, or kinks exist.
- Cable is clean and free of rust.
- Cable swage is present.

(3) ***Anchor Cable Intermediate Support.***

- Cables run through slots after ramp is closed.
- Quick-release retaining pip pins are present.

(4) ***Static Line Retrievers.***

- Motor is operational.
- Retriever spools are secured forward of intermediate cable support and tied to supports with one turn of double 1/4-inch cotton webbing.
- Retriever cables are not broken (within acceptable tolerances of TO 1C-141B-1), frayed, or kinked.
- Retriever cables are secured in spring clips.
- Retriever bars are available (one for each door).

NOTE: Retriever cable must be at least 4 inches above the anchor line cable.

h. **Emergency Equipment**.

- Public address system is operational.
- First aid kits (4) are present.
- Fire extinguishers (3) are present.
- Alarm system is operational.
- Emergency exits are operational and accessible.
- Sufficient emergency parachutes are available.

i. **Miscellaneous**.

- Lighting system is operational.
- Airsickness bags are available.
- Comfort facilities are present.
- JM kit bag (extra equipment) is onboard.

Section III. C-5 A/B/C GALAXY

The C-5 A/B/C Galaxy is a swept-wing, long-range transport powered by four turbofan jet engines and equipped for in-flight refueling. The aircraft can be configured in different peacetime troop-carrying modes for airborne operations. Parachutists jump the two aft doors of the aircraft.

16-10. SEATING CONFIGURATION WITHOUT COMFORT PALLET

The configuration for the C-5 A/B/C for peacetime training is TAP-1. This configuration offers seats for 73 parachutists in the troop compartment and provisions for personnel airdrop as follows: 71 jumpers and 2 static safety personnel. The maximum number of personnel for each anchor line cable is 40.

16-11. IN-FLIGHT RIGGING SEATING CONFIGURATION WITH COMFORT PALLET

This configuration offers a partial palletized seat kit (eight seat pallets and one comfort pallet) for 78 jumpers in the cargo compartment and provisions for personnel airdrop. It does not provide ditching exits or sufficient ground emergency exits. Therefore, this configuration should be carefully evaluated when considered for use. This configuration should be used only during over-land missions.

16-12. JOINT PREFLIGHT INSPECTION

The purpose of the joint preflight inspection (JPI) is to verify the readiness of the aircraft for the conduct of aircraft missions and to take actions necessary to achieve this readiness.

a. **Supervisory Personnel Required**. To ensure command and control, a JM, AJM, and two safety personnel are used for C-5 A/B/C airdrops.

b. **Jumpmaster/Aircraft Commander/Loadmaster Coordination**. It is the responsibility of the JM to inform the aircraft commander and loadmaster of the exact time sequence of prejump procedures. Following preliminary orientation and before loading personnel onboard the aircraft, the JM and loadmaster make a joint inspection of the aircraft.

c. **Exterior Inspection**. An external inspection of the aircraft is made to detect hazards to the airdrop of personnel. Particular attention is directed to the rear of the aft paratroop doors. Any protruding objects and sharp edges must be removed or padded and taped.

d. **Interior Inspection**. This inspection is conducted to detect and correct any interior safety hazards.

(1) Any sharp edge or protrusion is securely taped and padded, as required.

(2) All equipment in the cargo compartment is securely stowed and lashed.

(3) The floor is clean and free of lubricants; no obstructions are on the walkway or along the paratroop exit route (outboard area between the safety fence and fuselage).

(4) Anchor line cables are installed and under correct tension. Maximum deflection of the cable at midpoint (FS 1465) must produce a minimum cable height of 73 inches above the cargo compartment floor.

(5) A seat and seat belt are available in the troop compartment for each parachutist.

(6) The retrieval system is installed in the aircraft with the winch cable retained in clips and free of the anchor line cable. The retriever spool is secured to the aft support frame with one turn of double 80-pound, 1/4-inch cotton webbing. A phenolic (plastic) block is installed on the leading edge of the paratroop doors at the retriever bar level.

(7) Troop compartment lavatory units are installed and operational.

(8) Air deflector systems are installed.

(9) Jump platforms are installed.

(10) Jump signal lights are operational.

(11) Cargo compartment and troop compartment lights are operational.

(12) The public address intercomm system is operational.

NOTE: To reduce confusion in loading, the safety fence can be numbered at this time.

e. **Jumpmaster Preload Inspection of Parachutists**. The JM/AJM inspects each parachutist, parachute, and parachutist's equipment prior to loading the aircraft. The JM/AJM inspects—

(1) Parachutist's helmet.

(2) Parachutist's ID tags and ID card.

(3) Rigging of parachutist's equipment:

- ALICE pack with H-harness, AIRPAC, and PIE/R2, or CWIE containers.
- Dragon missile jump pack, Stinger missile jump pack, or AT4 jump pack.
- M1950 weapons case.
- Lowering lines.

f. **Parachute Stowage**. Personnel stow parachutes in aviator kit bags, secure all equipment, and move to the aircraft.

g. **Aircraft Loading Sequence**. Loading is determined by the USAF loadmaster.

h. **Aircraft Loading Through Forward Ramp**. When loading through the forward ramp, parachutists enter in normal sequence (1 through 36 right side, 1 through 37 left side). When loading through the aft right door, parachutists enter in reverse sequence (36 through 1 right side, 37 through 1 left side).

16-13. PERSONNEL AND EQUIPMENT CONFIGURATION

Personnel seating configuration and equipment stowage for the C-5B Galaxy are as follows:

a. **Seating Configuration**. Parachutists are assembled into two sticks: 37 parachutists for the left door and 36 parachutists for the right door. Each parachutist is assigned a number in the stick. This number is the parachutist's rigging station and seat number. Each parachutist is issued a main and reserve parachute. Each parachutist must inspect his own parachute for safety wires and for fitting of the parachute harness.

b. **Equipment Configuration**. Personnel face inboard, cover on the nylon net assembly equipment stowage sections, and place their equipment on the floor of the cargo compartment. Equipment is stowed on the safety fence under the parachutist's assigned stick number. To facilitate equipment stowage on the safety fence, the two top retaining straps of the nylon net assembly must be fully extended. The individual parachutist's equipment (load-bearing equipment, weapons, and main and reserve parachutes) is stowed under the retaining straps.

16-14. MOVEMENT TO THE TROOP COMPARTMENT

On jumpmaster order, parachutists begin movement to the troop compartment as follows:

a. **Left Stick**. The left stick faces aft, moves in stick order, and ascends the aft stair/ladder to the troop compartment. Upon reaching the troop compartment, the stick leader faces the forward end of the aircraft and moves to the forward row of seats on the left side. The left stick is seated row by row. Seats are occupied in the following forward to aft order: outboard, center, and aisle.

b. **Right Stick**. As the left stick begins to ascend the aft stair/ladder, the right stick faces aft, moves past the end section of the safety fence, crosses the airdrop system guide rails/rollers to the left side of the cargo compartment, and moves to the base of the stair/ladder. Parachutists must exercise caution to avoid tripping over the rails and rollers of the airdrop system. The right stick ascends the stair/ladder into the troop compartment and is seated on the right side, row by row, forward to aft, outboard to inboard in the same manner as the left stick.

NOTE: The troop compartment contains 75 seats: 73 seats are for parachutists and the remaining 2 are reserved for USAF loadmasters.

16-15. LOADMASTER BRIEFING

As soon as all parachutists are seated, the loadmaster uses the public address/intercomm system to brief the parachutists on aircraft safety, emergency procedures, and comfort facilities.

16-16. MOVEMENT TO THE CARGO COMPARTMENT FOR IN-FLIGHT RIGGING PROCEDURES

Movement to the cargo compartment for in-flight rigging proceeds as follows:

a. **Movement of Left and Right Sticks**.

(1) The loadmaster alerts the jumpmaster 1 hour 20 minutes prior to drop time. At that time, the JM moves the two sticks to the cargo compartment, right stick followed by left stick, both sticks in reverse stick order.

(2) Each stick descends the stair/ladder to the floor of the cargo compartment and moves to the equipment stowed on the safety fence. Routes used are the same as those used in moving to the troop compartment. The right stick must exercise care in crossing the airdrop system rails and rollers.

NOTE: The 1-hour 20-minute time warning can be modified to a shorter time depending on the situation and number of parachutists.

CAUTION

Each parachutist must descend to the cargo compartment facing the troop ladder with both hands on the handrail, using caution at each step.

b. **In-Flight Rigging Procedures.**

(1) Individuals face inboard and cover on their stowed equipment.

(2) Using the buddy system, half of the parachutists in each stick don their parachutes and equipment. After the first group have completed donning their equipment, the second half don their parachutes and equipment.

(3) JMPI begins as soon as rigging is complete. The JM and AJM inspect each parachutist and correct any errors. As the JM/AJM completes inspection of each parachutist, the rigged parachutists are seated on the walkway. Safeties assist the JM/AJM during JMPI.

> **CAUTION**
> When on, lights located under the walkway can get hot enough to melt nylon.

16-17. JUMP COMMANDS

The following eight jump commands are used:

GET READY.
STAND UP.
HOOK UP.
CHECK STATIC LINES.
CHECK EQUIPMENT.
SOUND OFF FOR EQUIPMENT CHECK.
STAND BY.
GO.

16-18. JUMP PROCEDURES

The jump procedures for the C-5 A/B/C are as follows:

a. **Jumpmaster and Assistant Jumpmasters**. The JM and AJM hook up to the anchor line cable and stand adjacent to the rear of their respective aft personnel door. The JM and AJM face their sticks to give jump commands. The JM initiates the jump commands, which are then relayed by the AJM.

b. **Parachutists**. All parachutists are hooked up to the paratroop anchor line cable. Personnel must close up tightly, reserve to backpack, to the aft end of the aircraft to permit sufficient room for all parachutists to hook up to the anchor line cable. The number 1 parachutist in each stick does not move aft of the air deflector while it is positioned inside the aircraft. When parachutists are rigged with combat equipment, it may become necessary to stagger the parachutists to have enough room for all parachutists to hook up.

c. **Commands**. The following commands are used:

(1) ***SOUND OFF FOR EQUIPMENT CHECK.*** At this command, the two safety personnel move from forward to aft, checking each parachutist's static line. The safety personnel report to the JM or AJM at the completion of the check, hook up, and then position themselves by the aft personnel door to control the static lines of exiting

parachutists. The two safety personnel must be JM qualified and may jump at the end of the stick. The JM may jump from either aft personnel door. He should jump from the door from which he can best observe checkpoints and the drop zone.

(2) ***STAND BY.*** At this command, the number 1 parachutist moves toward the door and assumes a normal number 2 man position.

(3) ***GO.*** At this command, parachutists' movement into the door will be at a normal walking pace, without shuffling. Exits from the troop doors are not vigorous and are made at an angle of about 70 to 80 degrees to the tail of the aircraft. Each parachutist passes his static line to the safety, places both hands on the ends of his reserve parachute, and exits the aircraft.

16-19. TIME WARNINGS

The following actions occur at each time warning:

a. **1-Hour 20-Minute Time Warning**. Parachutists are moved from troop compartment to cargo compartment under the supervision of the JM, AJM, and safeties. In-flight rigging and JMPI begins.

b. **20-Minute Time Warning**. The JM makes necessary checks of personnel and equipment. Special items of equipment are attached to the parachutists, and door bundles are moved to the vicinity of the personnel doors.

c. **6-Minute Time Warning**. JMPI of all parachutists is complete. The loadmaster gives the JM a verbal and visual 6-minute warning, and the JM begins his jump commands.

d. **Slow-Down Warning**. Approximately 3 minutes from drop time, the jump commands are completed. Personnel doors are opened, air deflectors deployed, and jump platforms extended and locked.

e. **1-Minute Time Warning**. Troops are alerted. The JM and AJM make safety checks from personnel doors. The AJM informs the JM that the AJM's side is clear and that it is safe to jump.

f. **10-Second Time Advisory**. The loadmaster gives the JM a visual 10-second time advisory. At this time, the JM gives the command STAND BY, and the first parachutist assumes a normal number 2 position. The remainder of the stick moves aft to close up the stick.

g. **Green Light, GO**. At the green light signal, the JM taps out the first man. If T-10-series type parachutes are issued, the AJM taps out the first man in the opposite door.

16-20. SAFETY PRECAUTIONS

Safety precautions for the C-5 A/B/C are as follows:

a. **Platforms, Air Deflectors, Aft Cargo Door**. The JM ensures that jump platforms are extended and locked and that the air deflectors are fully deployed. The aircraft will **not** be jumped unless these conditions are met. These conditions are mandatory for each aft personnel door that is to be used.

b. **Door Bundles**. When personnel are to follow door bundles, the door bundle static line is outfitted with three drogue devices, a breakaway static line, or a T-10-series parachute.

c. **Movement into the Door**. Parachutists must exercise caution as they move to the door to avoid becoming entangled with the static lines of preceding parachutists. This precautionary action may slow movement into and out of the door.

d. **MC1-Series Type Parachutes**. Only single door exits will be executed when MC1-series type parachutes are used. However, both doors must be open to conduct outside air safety checks.

e. **Ramp Exits**. Static line parachutists are not authorized to make aft ramp exits of the C-5 A/B/C aircraft.

16-21. C-5 A/B/C JUMPMASTER CHECKLIST

The jumpmaster inspects the following at the departure airfield:

a. **Seats**.
- Enough seats are available for troop load.
- All seats have safety belts.

b. **Safety Fence**.
- Safety fence is installed.
- Sufficient equipment restraint nets are available for troop load.

c. **Floor**.
- Nonskid surface covering is in good condition.
- Floor is clean and safe to walk on.
- Roller conveyors are stored.
- Loose equipment is secured in the cargo ramp area and does not interfere with troops.

d. **Jump Platforms**.
- Nonskid surface covering is present and in good condition.
- No cracks or bends are present.
- Studs are locked in seat track receptacles.
- Tie-down fitting is locked.
- All bolts and nuts are present.
- Platforms swing in and out easily.

e. **Jump Doors**.
- No excessive grease is on door tracks.
- No sharp or protruding edges are on door frames or on loading strut door (aft and down from jump door).
- Doors open and close easily.
- Door level catches operate.
- Phenolic (plastic) block is installed on leading edge at retriever bar height.

f. **Air Deflectors**.
- No sharp edges are present.
- Deflectors operate electrically.

g. **Jump Lights**. (Check five sets for operation.)
- Set 1—crew entrance door.
- Sets 2 and 3—leading edges, top of right and left doors.
- Sets 4 and 5—trailing edges, right and left doors, waist high.

h. **Anchor Line Cable System**.

(1) ***Forward Support Beam.***
- Bolts and nuts are present and tight.
- Cable, bolts, nuts, and safety wire are present.
- Turnbuckle is secured with safety wire or tied with 1/4-inch cotton webbing.

(2) ***Anchor Line Cable.***
- Cable has no breaks.
- Cable has no frays.
- Cable has no kinks.
- Cable is clean and free of rust.
- Cable swage is present.
- Cable is at least 73 inches above floor.

(3) ***Static Line Retrievers.***
- Motor is operational.
- Retriever spools are secured and tied to supports with one turn of double 1/4-inch cotton webbing.
- Retriever cables are not broken, frayed, or kinked.
- Retriever cables are secured in spring clips.
- Retriever bar is available.

NOTE: Retriever cable must be in position, not less than 4 inches above the anchor line cable.

i. **Emergency Equipment**.
- Public address system is operational (upper deck only).
- First aid kits (4) are present.
- Fire extinguishers (3) are present.
- Alarm system is operational.
- Emergency exits are operational and accessible.
- Sufficient emergency parachutes are available.

j. **Miscellaneous**.
- Lighting system is operational.
- Airsickness bags are available.
- Comfort facilities are available.
- JM kit (extra equipment) is onboard.

k. **Aircraft Slowdown Warning** (about 3 minutes).
- Doors are opened and locked in place.
- Air deflectors are extended.
- Jump platforms are locked in place.

l. **Loadmaster/Jumpmaster Safety Checks**.

(1) ***Equipment.***
- Door bundles employ 15-foot-long static lines with three drogue devices, breakaway static lines, or T-10-series cargo parachute.

(2) ***Aircraft.***
- Jump platform is secure and will sustain parachutist's weight.

NOTE: The C-5 A/B/C aircraft cannot be used for aft ramp static line jumping. HALO personnel may conduct aft ramp jumps from this aircraft. Due to the number of jumpers that can be hooked up to one anchor line cable, the C-5 A/B/C aircraft have not been certified for combat concentrated loads. Any deviation to the listed configurations requires authorization from HQ, AMC/DOT/DOV.

Section IV. C-17A GLOBEMASTER III

The C-17A (Globemaster III) is a swept-wing, long-range transport aircraft powered by four turbofan engines capable of airlifting large payloads over intercontinental ranges without refueling. An in-flight refueling capability increases the deployment range. The cargo compartment is designed to permit safe and efficient operation with one loadmaster for any mission. The aircraft accommodates outsize or oversize cargo, tactical vehicles, container loads (which can be configured to conduct a combination of airland, personnel and equipment airdrop), and MEDEVAC operations. The C-17A can be configured with or without a comfort pallet. The jump platforms and air deflectors are an integral part of the aircraft fuselage. A dedicated antenna system enables use of organic Army tactical satellite (TACSAT) communication systems en route.

16-22. SEATING CONFIGURATION

The seating configuration for the C-17A (Globemaster III) is as follows:

a. 102 parachutists, includes seating for JM, AJM, and static safety personnel.

b. Two static safety personnel seated in crew compartment.

c. One loadmaster seated in forward loadmaster station.

d. Two door bundles (one per door) placed on the aircraft floor aft of the inboard seats.

16-23. SUPERVISORY PERSONNEL REQUIRED

The following personnel are required for all C-17A personnel airdrop operations:

- One JM.
- One AJM.
- Two JM qualified and current static safety personnel.
- One loadmaster.

16-24. TIME WARNINGS

Time warnings include the following:

a. **2-Hour 20-Minute Time Warning** (used only on in-flight rigging missions).

b. **20-Minute Time Warning**. At the 20-minute time warning, all in-flight rigging is complete. All jumpers are alert with ballistic helmets fastened. Door bundles are positioned in the vicinity of the jump doors, hooked up to the outboard anchor line cable and inspected. Special items of equipment are attached to their respective jumpers and inspected.

c. **10-Minute Time Warning**. The aircrew turns on the red jump caution lights, then notifies the JM and all that the inspections are complete. The JMs hook up to the inboard anchor line cables and begin issuing jump commands. If both doors are used on one pass, then both JMs issue the commands at the same time.

d. **6-Minute Time Warning and Aircraft Slowdown**. The aircrew completes their 6-minute slowdown check: aircraft deck is set between 6 and 7 degrees, aircraft slows to a drop speed of 130 knots (plus or minus 3 knots indicated airspeed), air deflectors are deployed, and troop doors opened. The loadmaster takes a position between both jump doors and gives control of the doors to the JMs. The JMs perform troop door safety checks, outside air safety checks, and spot for checkpoints en route to the drop zone.

e. **1-Minute Time Warning**. The JM issues the time warning to the parachutists with his lead hand and then spots for checkpoints.

f. **30-Second Time Warning**. At the 30-second reference point, the JMs make a final outside safety check, return inside the aircraft, gain eye-to-eye contact with each other, and give each other a "thumbs up" indicating that all conditions outside the aircraft are safe for the jump. The amber jump caution light is turned on. The JMs face the jumpers and issue the eighth jump command, STAND BY.

g. **Green Light**. The green light is the final time warning on an Air Force aircraft. When it comes on, the JM issues the ninth jump command, GO. The JM's first jumper exits the aircraft. The AJM taps his first jumper out a half-second after the JM's number 1 jumper exits.

16-25. JUMP COMMANDS

Jump commands are given in the following sequence:

a. **GET READY**. Jumpers undo their seatbelts.

b. **OUTBOARD PERSONNEL, STAND UP**. All outboard jumpers stand, place their seatbelts in their seats, and push in the seats to retract them to the raised position. (This allows the jumpers to use the handhold straps on the cargo wall and assists in keeping the seatbelts off the floor, reducing tripping hazards.)

c. **INBOARD PERSONNEL, STAND UP**. Jumpers stand up. (The inboard seats do not fold.) Safety moves forward in the aircraft and inspects each jumper's quick release in the waistband, quick-release snap on the M1950 weapons case, ejector snap of the hook-pile tape lowering line, and the adjustable leg straps of the HSPR.

d. **HOOK UP**. During mass tactical operations, a maximum of 27 jumpers on the outboard anchor line cables and 24 jumpers on the inboard anchor line cables are allowed.

e. **CHECK STATIC LINES**. The last two jumpers turn toward the skin of the aircraft. The second to the last jumper checks the last jumper's static line. Each jumper checks the static line of the jumper to his front.

f. **CHECK EQUIPMENT**. The JMs immediately check their own equipment.

g. **SOUND OFF FOR EQUIPMENT CHECK**. At this command, both safeties move from the forward portion of the aircraft to the aft end. They check the jumpers' static lines and tell the jumpers to gain eye-to-eye contact when handing their static lines to the safety.

h. **STAND BY**. The amber light activates 10 seconds before the green light activates, and both number 1 jumpers assume a position in the door but not on the jump platform. Each JM takes a position facing his door to control the flow of the jumpers to the platform. The safeties stand directly under the aft end of the anchor line cable support bracket facing the jumpers. They control the static lines by vigorously pushing them back and to the upper trailing corner of the door and not allowing slack to fall down into the door.

i. **GO**. During mass exits, on the command GO, the JM exits his number 1 jumper. The AJM taps out his number 1 jumper a half second after the JM's number 1 jumper exits the aircraft. Movement to the door is a normal walking pace without shuffling. The jumper keeps his arms straight and elbows locked while passing his static line to the safety; he places both hands on the ends of the reserve parachute and exits the aircraft. Exits are made at a 90-degree angle off the jump platform. The AJM follows his last jumper out the aircraft. The JM waits until the AJM clears the platform, then exits the aircraft. The safeties immediately conduct a towed jumper inspection and turn the door over to the loadmaster with a "thumbs up." (If using ADEPT options 1 or 2, use the appropriate exit procedures.)

16-26. DOOR CHECK PROCEDURES

After the JM receives the announcement "All okay, jumpmaster," from the jumpers, he regains control of his static line from the safety and takes the number 1 jumper position. Once the aircrew has completed their slowdown checks, the loadmaster transfers control of the door by giving the JM a "thumbs up" signal and moves to the center of the cargo floor area where he can observe both doors. It is important for the safeties to control the JM's static line with two hands and also to keep an eye on the loadmaster and the jumpers. The JM is now ready to perform his door safety check. He transfers control of his static line back to the safety. The JM then grasps the troop door guide tracks with his trail hand and conducts the door safety check procedures.

a. **Troop Door Up-Lock**. With the lead hand, the JM grasps the troop door lifting bar, raises the door up, and then pulls it back down to the door up-lock. A visual inspection of the door up-lock verifies that it is in the locked position.

b. **Trail Edge of the Troop Door**. With the lead hand, the JM starts at the top of the trail edge of the troop door and traces down to the inside edge, then back up to the door clearance faring guard cover, then back down the trail edge of the door clearance faring. At this point, the JM ensures the troop door clearance faring is fully retracted and locked in the stowed position. He traces back up to the top of the door. Then he secures a handhold with his lead hand by grasping the troop door guide or the handhold provided on the lead edge of the door

CAUTION
The JM must be aware that the reserve parachute ripcord grip can contact the trail edge of the door and could cause inadvertent reserve parachute deployment.

c. **Air Deflector**. To check the air deflector, the JM leans toward the trail edge of the door and performs a visual inspection of the air deflector, ensuring it is fully deployed.

d. **Outside Air Safety Check**. The JM places either foot on the center of the platform with his trail foot a comfortable distance behind the lead foot. He leans out by bending forward at the waist and locking both elbows. He maintains a firm grasp with both hands at all times and performs his initial 360-degree outside air safety check:

- Direction of flight.
- Overhead.
- Rear.
- Straight down.
- Straight to the front.
- Back toward the direction of flight.

The JM continues his duties at subsequent time warnings.

WARNING

At no time during airborne operations should the cargo ramp door be open when the troop doors are open. If this situation exists, all jumpers, including JMs and safety personnel must be seated and strapped in, guarding the ripcord grip of their reserve parachutes. If Army personnel must work around the open troop doors, caution must be taken to prevent personnel from being swept out of the aircraft. Reserve activation with troop doors and cargo door open can cause serious injury or death.

16-27. DOOR BUNDLE PROCEDURES AND EJECTION

Bundles may be pushed out the door. Use the same procedures as for the C-130 or C-141 aircrafts.

16-28. SAFETY PRECAUTIONS

In accordance with a memorandum from Headquarters, US Army Test and Evaluation Command (Subject: Safety of Personnel Formation Airdrop Operations from the C-17, Date: 8 March 1997), personnel parachutes MUST be equipped with a 20-foot static line.

a. **Ramp Jumps**. Static line personnel airdrops from the cargo ramp are prohibited. The cargo ramp must be closed during static line jumps from the paratroop doors.

b. **Aircraft Configuration**.

(1) The aircraft deck angle is set between 6 and 7 degrees during static line personnel airdrops with an air speed of 130 knots, plus or minus 3 knots indicated airspeed.

(2) The total aircraft gross weight should not exceed 360,000 pounds at the time static line personnel airdrops are initiated.

(3) Troop air deflector is deployed.

c. **Bundle Static Line**. All door bundles are equipped with a 20-foot static line, with one static line drogue device attached.

d. **Static Safety Personnel**.

(1) Static safety personnel are seated in the crew compartment, but they move to the troop compartment after takeoff.

(2) Static safety personnel must complete their 20-minute checks and arrive at the forward end of the cargo compartment before the 10-minute time warning.

(3) The safeties stand directly under the aft end of the anchor line cable support bracket facing the jumpers. They control the static lines by vigorously pushing them back and to the upper trailing corner of the door and not allowing slack to fall down into the door.

e. **Jumper Exit Technique**. All jumpers must be trained to exit the aircraft with a strong effort in leaving the platform at a 90-degree angle straight out the door as close to the lead edge of the door as possible.

f. **Aircraft Inspection.** Before the flight, the JM or safety inspects the exterior aft of the troop doors for any protrusions, sharp edges, or snag hazards. He closes the cargo ramp during this inspection to allow for examination of the actuator door bulb seal and jack pads.

g. **Paratroop Retrieval System**. Static line personnel airdrops are restricted to C-17A aircraft with installed paratroop retrieval system (Canadian retrieval systems).

h. **Anchor Line Cable**. The JMs ensure that no more than 27 jumpers are hooked up to the outboard anchor line cables and no more than 24 jumpers are hooked up to the inboard anchor line cables during mass tactical operations.

i. **Formation**. For personnel airdrop operations from separate aircraft, the formation separation time between aircrafts is 2.5 to 3 minutes or 26,000 feet to avoid any aircraft wake vortex interference with jumpers.

16-29. IN-FLIGHT RIGGING PROCEDURES

The C-17A seats 102 jumpers, 2 static safeties, and 1 loadmaster. Ramp space is provided for storage of parachute assemblies on pallets and door bundles.

a. **Supervisory Personnel Required**. The following are required for in-flight rigging:

- One JM (supervises in-flight rigging).
- Eight AJMs.
- Two static safety personnel.
- One loadmaster.

b. **Equipment Stowage**. Parachute assemblies (mains and reserves, 106 each) are placed in kit bags and stowed and secured in two pallets on the ramp. Jumpers are seated in two sticks each on the left and right sides of the aircraft. All ALICE packs and weapons containers are placed in the center aisles and are strapped down by the loadmaster. All DMJPs, SMJPs, AT-4s, and door bundles are strapped down to the cargo floor section aft of the inboard seats or on the ramp. The jumper keeps his ballistic helmet, protective mask, and LBE under the seat.

c. **Parachute Issue**. All ALICE packs and weapons containers are retrieved from the center aisles and placed on the jumper's lap. Starting with the left outboard stick, each jumper stands up, places his ALICE pack and weapon on his seat, walks to the pallets on the ramp, draws an aviator kit bag, continues in a counterclockwise motion between the two sticks of jumpers seated on the right side of the aircraft, and returns to his original seat. The left inboard stick repeats the same procedure following the outboard jumpers. After the left side is complete, the right outboard stick jumpers stand up, place their ALICE packs and weapons on their seats, and draw aviator kit bags in the same manner except that they return in a clockwise motion moving through the left door center aisles. Right inboard jumpers follow the outboard jumpers.

d. **Buddy Rigging**. The JM initiates in-flight buddy rigging two hours before the 20-minute time warning.

(1) Once the parachutes are issued, buddy rigging begins. The JM, safety personnel, and designated AJMs supervise. Once a parachutist has been rigged, he sits down and waits to be inspected by one of the JMs.

(2) The JM personnel serve as roving inspectors. Attaching the static line snap hook to the top carrying handle of the reserve designates inspected jumpers.

(3) The JM supervises the operation and ensures all personnel are inspected. If there are other current and qualified JMs onboard the aircraft, they may be used.

(4) Once all the jumpers have been inspected, the JMs don their equipment, and the safety personnel inspect them.

16-30. JUMPMASTER AIRCRAFT INSPECTION

The JM inspects the aircraft as follows:

a. **Exterior of Aircraft Fuselage**.
- Inspect aft of doors for any protrusions, sharp edges, or snag hazards.
- Inspect ramp area with the ramp in the closed position. Special attention must be made to the actuator door bulb seal on both sides of the exterior of the cargo ramp. Ensure that the rubber seal is in place and that there are no sharp edges or snag hazards.

b. **Seats**.
- There should be enough seats for troop load (102, normal load; 108, maximum), and all seats should be serviceable.
- All seats should have safety belts.
- Seats are secured to the floor or sidewall of the aircraft.

c. **Floor**.
- Nonskid covering is in good condition, clean, and safe to walk on.
- Roller conveyors are stowed.
- Loose equipment is secured in the cargo area.
- Deck tie-down ring covers (4) in place and secured just before each door.

d. **Jump Platforms**.
- Nonskid surface present.
- Platform lights operate for night jumps.

e. **Jump Doors**.
- No excessive grease on the door tracks.
- No sharp or protruding edges on door frames.
- Doors open and close without excessive force.
- Door lever catches operate in the up-lock position.

f. **Troop Door Clearance Faring**.
- No sharp edges, retracts properly, and can be locked into place.

g. **Air Deflector**.
- No sharp edges.
- Deploys properly to 40 degrees, as indicated by the interior angle gauge.

h. **Jump Caution Lights**.
   - Check for proper operation: red, amber (10 seconds), to green.

i. **Anchor Line Cable System**.
   - Properly installed with no breaks, kinks, or frays. Ensure the cables are clean and free of rust.
   - Cables run through slots.
   - Quick release retaining pip pins are present and installed.
   - Paratroop retrieval system (Canadian retrieval system) attachment point on aft anchor line cable stanchion present.

j. **Static Line Retrievers**.
   - Paratroop retrieval system assembly present and complete.
   - Motor is operational.
   - Retriever cables are not broken, frayed, or kinked.
   - Retriever cables are secured in spring clips.

k. **Emergency Equipment**.
   - Public address system is operational.
   - First aid kits and fire extinguishers present.
   - Alarm system is operational.
   - Emergency exits are operational and accessible.
   - Sufficient emergency parachutes are available.

l. **Miscellaneous**.
   - Lighting system is operational, especially red lights for night operations.
   - Air sickness bags and hearing protection are available.
   - Comfort facilities, water, urinals, and so forth, are available for in-flight rigging.
   - Complete JM kit bag is on the aircraft.

16-31. TOWED JUMPER PROCEDURES

When a jumper becomes towed, follow these procedures:

a. **Supervisory Personnel Actions**. When a jumper becomes towed, the JM stops the stick and notifies the loadmaster. The loadmaster notifies the pilot, and requests that the drop altitude and speed be maintained. The JM identifies how the jumper is being towed. If the jumper is being towed by anything other than his static line, the JM attempts to free the jumper. If the jumper is being towed by the static line, recovery procedures are initiated. The JM observes the jumper and recommends whether to retrieve or cut the jumper free. The recommendation is relayed by the loadmaster to the pilot. The pilot makes the decision to retrieve or cut the jumper free from the aircraft. If directed by the pilot, the USAF loadmaster cuts the jumper free from the aircraft.

b. **Jumper Actions**. If a towed jumper is conscious, he maintains a good tight body position with both hands on the ends of the reserve, right hand protecting the rip cord grip, and he is prepared to activate the reserve if cut free from the aircraft.

c. **Priority for Action**.

(1) ***First Priority (Jumper is Towed by the Static Line):*** Retrieve the jumper using the paratroop retrieval system.

(2) ***Second Priority:*** Cut the jumper free.

(3) ***Third Priority (Least Preferred):*** Land with the jumper outside the aircraft. When using this method, if the jumper cannot be completely retrieved into the aircraft, he should be retrieved as close as possible to the door and secured. The aircraft lands at the closest available runway.

d. **AJM and Static Safety Actions**. If the jumper is to be retrieved, the AJM and safeties move the remaining jumpers toward the front of the aircraft. All personnel stay clear of the door and line of travel of the static line retriever cable. When the jumper has been retrieved to the door, the JM and the safety gain physical control of the jumper. At no time should the JM or the safety grasp the jumper's reserve parachute. The retrieved jumper is moved all the way forward in the aircraft, and he does not exit the aircraft during the airdrop operation. Instruct the jumper not to touch any of his equipment or de-rig in any way. If emergency medical treatment is needed, cancel the airdrop operation during training. During combat operations, the mission may be continued.

e. **Emergency Jettisoning**. To jettison the HSPR in an emergency, the parachutist lowers the ALICE pack, lowers the M1950 weapons case, and then pulls out on the yellow safety lanyard (attached to the ejector snap on the HPT lowering line), which allows the ALICE pack to fall free.

CHAPTER 17

ROTARY-WING AIRCRAFT

Rotary-wing aircraft can be used for airdrop operations when special missions are conducted to deploy small-unit forces. The aviation unit supporting the airdrop is responsible for preparing the aircraft for equipment and personnel drops to include seat and door removal (if required) and installation or rearrangement of seat belts. The installation of the field-expedient anchor line system is the JM's responsibility. Aircraft preparation is usually accomplished jointly by the crew chief and JM.

Section I. SAFETY CONSIDERATIONS

Although safety considerations for each aircraft are discussed, the requirements below apply to all Army aircraft (unless otherwise indicated).

17-1. GROUND TRAINING

Unit commanders require all personnel to participate in ground training immediately before the jump. The parachutists are shown the correct movement procedures inside the aircraft and the exit procedures. Parachutists are required to practice and demonstrate these procedures to JM satisfaction before the jump. Different techniques are involved in jumping from rotary-wing aircraft; failure to conduct ground training may result in a serious jump accident.

17-2. MOVEMENT IN AIRCRAFT

The pilot is briefed to expect rapid shifts in the aircraft's center of gravity during stand up, hook up, and exit of parachutists.

17-3. RESERVE PARACHUTE

Crowded conditions inside the cargo compartment could cause accidental activation of a reserve parachute, creating an extremely hazardous situation. During movement, the rip cord grip of the reserve parachute is protected by placing the right hand and forearm over the front of the reserve. This method allows the parachutist to control the pilot chute and canopy in case of accidental activation.

17-4. SPACE LIMITATIONS

The total number of parachutists and air delivery containers must conform to the weight and space limitations of the specific aircraft involved.

17-5. 6-SECOND COUNT

Due to the slow forward speed of helicopters and the downward rotor wash, the time interval between exit and full deployment of T-10-series or MC1-series parachutes requires about 100 feet more altitude. Due to the longer opening time, the parachutist extends the normal 4000-count to a 6000-count.

17-6. STATIC LINES AND DEPLOYMENT BAGS

Static lines and deployment bags are retrieved by the JM or crew chief immediately after the last parachutist is clear. The static lines and deployment bags are secured as soon as they are retrieved inside the aircraft. If the door on the aircraft can be closed, the static lines can be removed from the anchor cable or attaching point; otherwise, the static lines are not detached until the aircraft is on the ground.

17-7. CROWDED CONDITIONS

Crowded conditions inside these aircraft dictate that caution be used to prevent entanglement or misrouting of static lines during the parachutist's exit. Each parachutist is cautioned to watch the static line of the preceding parachutist and to observe all the static lines trailing from the lower aft corner of the cargo or personnel door. This precaution ensures that succeeding parachutists do not jump until the parachute of the preceding parachutist has deployed, and that the deployment bag has trailed to the rear of the aircraft.

17-8. CONTAINER LOADS

If container loads are to be airdropped from bomb shackles (wing load), helicopter door, bomb bay, or the doors of utility airplanes, they must be rigged using parachutes equipped with breakaway static lines. A-7A, A-21, and CDS container loads using breakaway static lines may be airdropped from the ramp or rear end (tailgate) of cargo and transport-type aircraft without jumpers following. CRRC loads may be dropped followed by jumpers with the 15-foot extraction parachute packed in a main parachute deployment bag IAW FM 10-542. Container loads with breakaway static lines are not rigged for airdrop from the troop door (side door) of cargo and transport-type aircraft. IAW message DTG 311313ZOCT00, for paratroop door drops, loads that will be followed immediately by parachutists will be rigged with cargo parachutes having nonbreakaway static lines. Each static line must have a drogue attached to it as outlined in the parachute's technical manual. When using the T-10-series cargo parachute or the 68-inch pilot parachute for the paratroop door load, the deployment bag serves as the drogue. Parachutists are **not** dropped at the same time as bundles that are rigged for release from bomb shackles.

17-9. HOOKUP PROCEDURES

When using rotary-wing and small fixed-wing aircraft for airborne operations, parachutists might use different hookup procedures from the standard hookup procedures used in USAF large fixed-wing troop carrier aircraft. This difference is due to the location of the anchor cables. Also, the JM may hook up the individual jumper. Unless otherwise specified in the hookup procedures for specific aircraft, the rule is to hook the open portion of the snap hook to the front of the aircraft with all static line snap hooks facing the same direction. This permits rapid, visual inspection before the jump and easy removal of the static lines after the jump.

17-10. TOWED PARACHUTIST PROCEDURES

In the event of a towed jumper on a rotary-wing aircraft, the jumpmaster will prevent any other jumpers from exiting and will notify the pilot. The parachutist stays in a tight body position and protects the rip cord grip. The jumpmaster will ensure the jumper is securely attached to the aircraft and will not break free during descent. If the jumper is not securely attached, the jumpmaster will attempt to shake or cut him free. If the jumper is attached, the aircraft will slowly descend to the DZ and come to a hover, and the jumper will be freed from the aircraft.

Section II. UH-1H IROQUOIS/UH-1N HUEY

The Army's UH-1H is powered by a single gas turbine engine; the USMC UH-1N has two gas turbine engines. Up to 8 combat-equipped parachutists can jump from the UH-1H/UH1-N, consistent with weight limitations (Figure 17-1). The JM is a static JM.

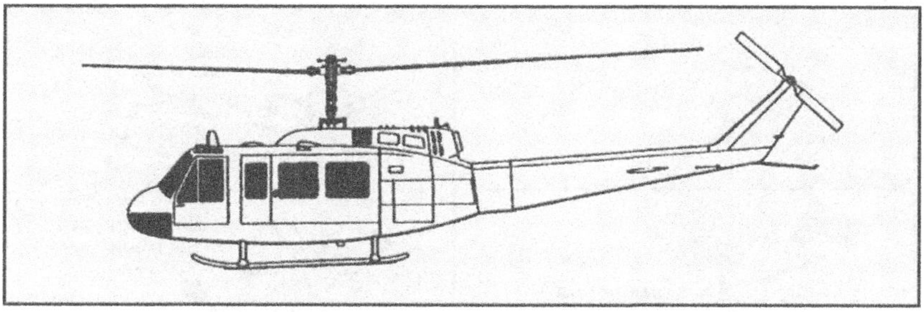

Figure 17-1. UH-1H Iroquois/UH-1N Huey.

17-11. PREPARATION AND INSPECTION

Prepare and inspect the UH-1H as follows:

a. **Preparation**. The following steps prepare the UH-1H for jumping:

(1) Both cargo compartment doors are locked in the open position. If the doors cannot be locked, they are removed.

(2) All troop seats are removed except one seat on each side (located to the rear of the pilot and copilot seats). These two seats are installed so they are facing to the rear of the aircraft. If the parachutists are equipped with combat equipment and eight parachutists are to jump, all seats in the cargo compartment are removed.

(3) The door and frame are inspected to ensure there are no sharp edges that could cut or fray static lines. If these are determined to be hazardous, corrective action is taken before the helicopter is jumped.

(4) Under field conditions, the door and frame can be padded and taped to preclude a mission abort. Otherwise, the aircraft is returned to maintenance for correction of the deficiency.

(5) Safety belts are attached to the tie-down rings on each side of the compartment for floor-seated parachutists.

(6) The door gunner/crew chief foot-operated radio switch may be unscrewed (by hand) before jumping. The exposed radio switch wires are taped to prevent an electrical short. If the switch is not removed, it is padded and taped. The ground-handling wheel-mount brackets on both landing skids are padded with cellulose wadding and taped

(Figure 17-2, page 17-4). Some aviation units have fabricated special covers that may be used to cover the wheel-mount brackets.

b. **Anchor Line Systems**. Two anchor line systems are available with the UH-1H aircraft for airdrop of personnel. They are the **standard overhead system** and the **expedient system** (Figure 17-3, page 17-4). The expedient system (modified STABO strap or Type XXVI nylon webbing anchor line cable assembly) consists of a nylon A-7A strap, four D-rings, and four connector snaps (TM 10-1670-298-20&P authorized the fabrication of the modified STABO strap or Type XXVI nylon webbing strap from Type XIII nylon webbing). Nylon modified STABO straps or Type XXVI nylon webbing must be used with cotton buffers on the D-rings and connector snaps.

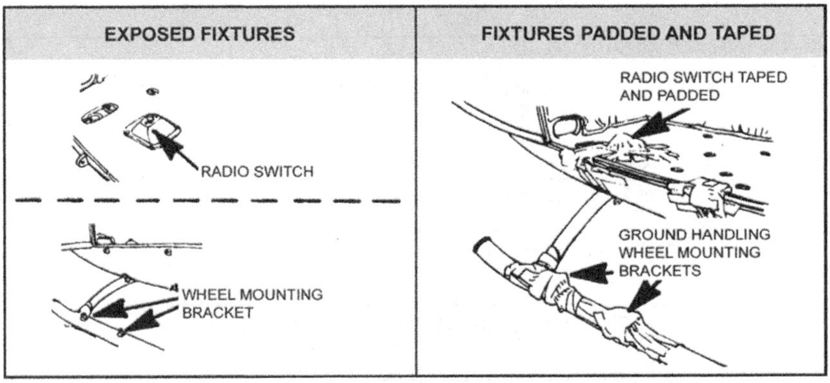

Figure 17-2. UH-1 exposed fixtures padded.

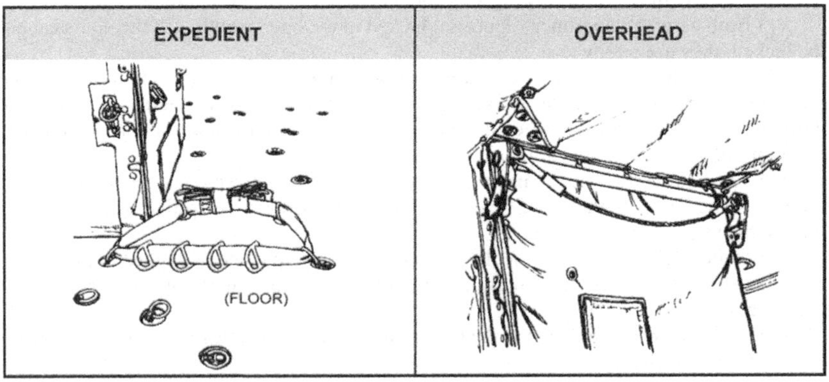

Figure 17-3. UH-1 anchor line system.

c. **Anchor Line Assembly Installation**. An anchor line assembly is installed on each side of the aircraft. It can be installed quickly by means of four tie-down rings located on the floor on the right and left sides of the aircraft compartment. The modified STABO strap or Type XXVI nylon webbing strap is threaded through the D-rings, which are used for attachment of the static line snap hooks (Figure 17-4).

(1) ***Left Door.*** For the left door, one connector snap on the modified STABO strap or Type XXVI nylon webbing strap is attached to the tie-down ring number G2. The strap is connected to the tie-down ring number F4. Four D-rings are on the strap with the round part of the rings facing outboard (of aircraft). The strap is then connected to tie-down ring number K3 and tie-down ring number J4. The free end of the strap is secured to the strap fastener, and any excess between tie-down rings number J4 and number G2 is taped. All connector snaps must be safety wired.

(2) ***Right Door.*** The same procedures apply to the right door as the left except that the A-7A strap is attached to tie-down ring number G1, then to F2. Four D-rings are on the strap with the round part of the rings facing outboard (of aircraft). The free end of the strap is secured to tie-down rings number K2 and number J3, and the strap fastener is secured. Excess strap between tie-down rings number J3 and number G1 is taped.

d. **Inspection**. Before enplaning, the JM and pilot, or pilot's representative, jointly inspect the aircraft to determine the following:

(1) All protruding objects near the cargo compartment doors are removed or taped.

(2) The lower right and left aft edges of both the cargo compartment doors are padded and taped.

(3) The anchor line cable or field-expedient anchor line system is secure, serviceable, and properly installed.

(4) A safety belt is available for each parachutist.

(5) A headset is available for the jumpmaster to effect coordination among the jumpmaster, the pilot, and the ground.

17-12. LOADING TECHNIQUES AND SEATING CONFIGURATION

During loading, jumpers should not approach directly from the front or sides but at a 45-degree angle to the nose of the aircraft. Jumpers 1 through 4 enter the cargo compartment through the right door, are hooked up by the JM in numerical order, and seat themselves (Figures 17-4 and 17-5). Jumpers 5 through 8 enter the cargo compartment through the left door, are hooked up by the JM in numerical order, and seat themselves.

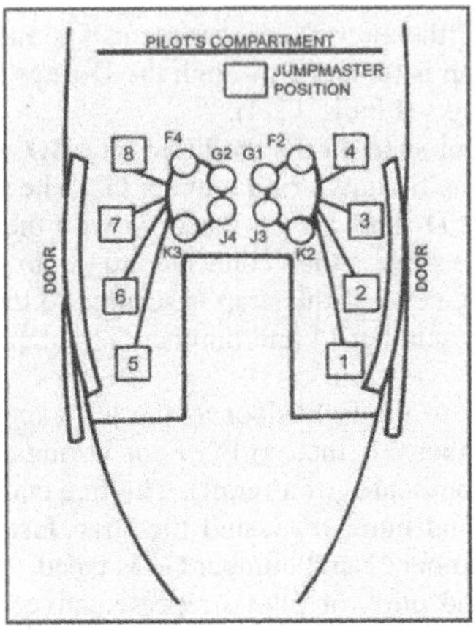

Figure 17-4. UH-1 seating configuration, expedient anchor line system.

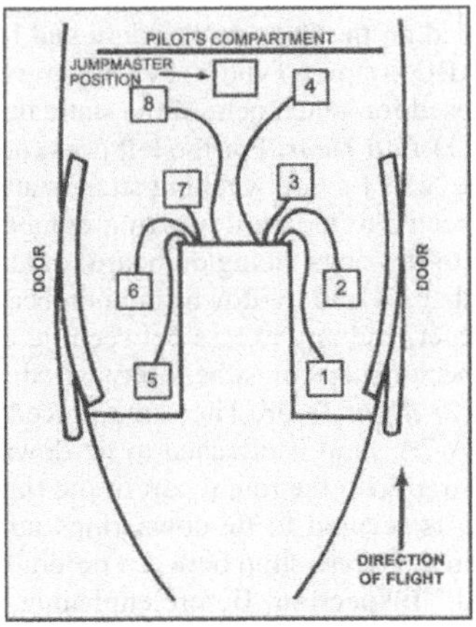

Figure 17-5. UH-1 seating configuration, overhead anchor line system.

a. The JM ensures excess static line is stowed as he hooks up each parachutist.

b. The open portion of the static line snap hook faces the front of the aircraft.

c. For flights less than 25 minutes long, jumpers may sit in the door with their feet outside the cargo compartment.

17-13. JUMP COMMANDS

The JM issues the following commands:

a. **GET READY**. This command is given 4 minutes or less from drop time, and the aircraft is level for final approach. All seat belts are unlatched and moved to the rear of the parachutists. The JM visually inspects each safety belt to ensure that it is clear of the parachutist and the equipment.

b. **CHECK STATIC LINES**. The JM rises and checks the routing of static lines from the point of attachment to the pack tray to ensure they are properly routed and hooked up.

c. **CHECK EQUIPMENT**. All parachutists check their equipment.

d. **SOUND OFF FOR EQUIPMENT CHECK**. On this command, number 1 parachutist orally indicates "Okay" to the JM. The remaining parachutists follow in order.

e. **SIT IN THE DOOR**. This command is given by the JM 30 seconds from drop time. Numbers 1 and 2 swing their legs to the right and take sitting positions in the door with feet together outside the cargo compartment. (Numbers 3 and 4 extend their legs outside and move to sitting positions.) They place both hands, palms down, on the floor alongside their thighs, turn their heads toward the JM, and wait. Numbers 5 and 6 swing their legs to the left, take sitting positions in the left door, and follow the same procedure as numbers 1 and 2. (Numbers 7 and 8 extend their legs outside and move to sitting positions.) This command is omitted if the parachutists are already sitting in the door on a short flight.

f. **STAND BY**. This command is given 8 to 10 seconds before the command GO.

g. **GO**. At this command, the following occurs:

(1) ***Personnel.*** The JM controls the jumper's exit and ensures a 1-second interval between jumpers by giving each jumper the oral command GO after the preceding jumper has exited and cleared the aircraft. The jumpers exit in numerical sequence.

(2) ***Air Delivery Containers.*** When an air delivery container is being released from the cargo hook, the pilot releases the container and informs the JM when the load has cleared the aircraft. The jumpers exit as explained above.

17-14. ARCTIC OPERATIONS

If the helicopter has skis, the ski attaching bolts and the sharp edges of the skis are padded and taped on the outboard side of the landing skids aft of the leading edge of the cargo door. Due to the bulk and weight of arctic clothing, individual equipment is not worn. The equipment is dropped either as an internal or external load.

17-15. SAFETY PRECAUTIONS

Safety precautions on the UH-1H are as follows:

a. **Parachutists**. During movement inside the aircraft, the parachutist protects the rip cord grip. Crowded conditions inside the cargo compartment and the open doors on both sides of the fuselage pose a hazardous situation regarding accidental activation of the reserve parachute.

b. **Jumpmaster**. The JM ensures all parachutists remain secured by their safety belts until the command GET READY is given. The JM prevents (or corrects) excessive static line from flopping about the aircraft. (The JM does not jump from this aircraft.) The JM wears a safety harness.

c. **Equipment**. Equipment prescribed in Chapter 12 can be worn by parachutists when jumping this aircraft.

(1) Standard air delivery containers rigged with G-14 cargo parachutes can be delivered from the cargo hook, using the breakaway static line (FM 10-500-2). The snap hooks of the static lines are hooked to the anchor line system before lift-off. Door bundles reduce the number of parachutists that can be carried, depending on the size and number of bundles.

(2) When CWIEs are jumped (two per load maximum), they are attached to number 3 and number 7 parachutists (one for each door). The DMJP may **not** be jumped from aircraft that require parachutists to sit on the floor.

d. **Aircraft**. The indicated airspeed of the aircraft during jumps is not less than 50 knots or more than 70 knots. The minimum drop altitude is 1,500 feet AGL. After the last parachutist has cleared the aircraft, the static lines are retrieved inside the aircraft and secured in an aviator kit bag or secured by a safety belt to the aircraft floor. The static line snap hooks **are not removed** from the anchor line cable until the aircraft lands.

Section III. UH-60A BLACK HAWK

The UH-60A is a twin-turbine, medium-speed, single-main-rotor helicopter (Figure 17-6). Eight combat-equipped parachutists can jump from this aircraft.

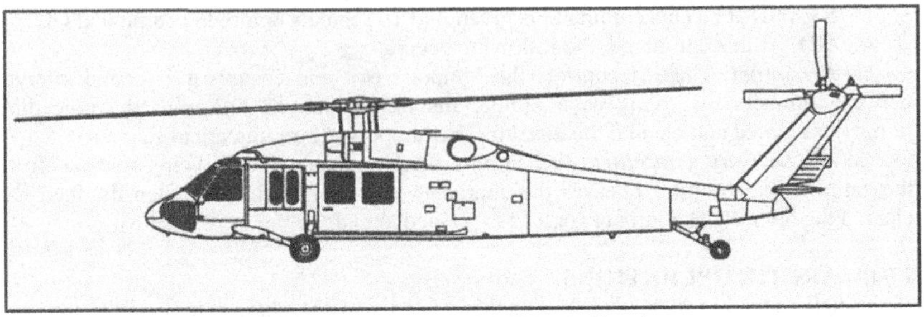

Figure 17-6. UH-60A Black Hawk.

NOTE: On missions requiring a window gunner, the maximum number of parachutists is reduced to six. Static line parachute operations require a static JM.

17-16. PREPARATION AND INSPECTION

The JM prepares and inspects the UH-60A as follows:

a. **Preparation.** To prepare the UH-60A for jumping, adhere to the following procedure (Figure 17-7):

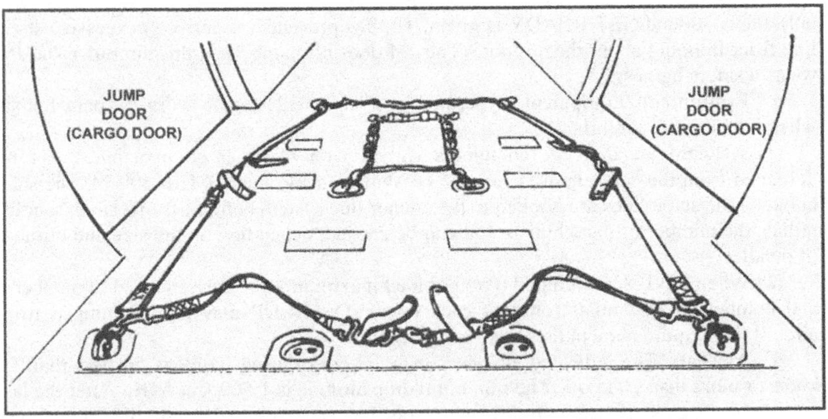

Figure 17-7. UH-60A compartment prepared for jumping.

(1) Lock both cargo doors in the open position.

NOTE: For arctic or other cold-weather operations, or during flights of long duration, the aircraft doors may be closed and locked. Doors cannot be opened during flight. The aircraft must either land or hover near the ground to open the doors. This procedure requires coordination between the supporting aviation and airborne units for the jump.

(2) Remove seat belts in the cargo compartment (except as required by aircraft crew).

(3) Tape cargo floor troop seat and tie-down fitting wells in front of the cargo doors.

(4) Tape sharp edges and tie-down fitting wells on the cargo floor and door jambs that could cut or fray static lines or snag parachutists' equipment.

(5) Tape the weather stripping on cargo doors below the door catch (Figure 17-8).

(6) Tape the entire trail edge of the door.

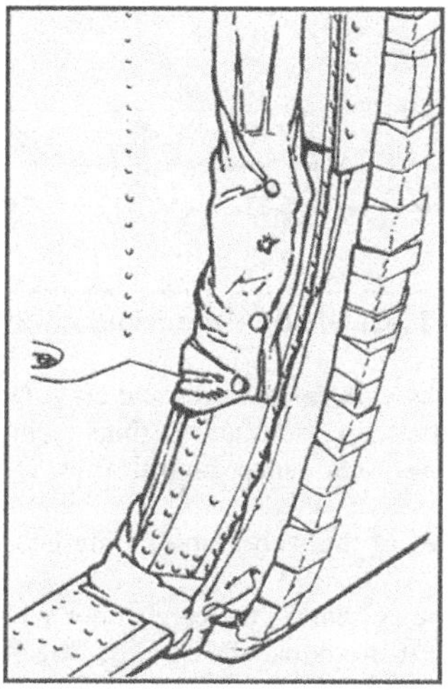

Figure 17-8. UH-60A door edge padded and taped.

NOTE: Tape must not interfere with closing, locking, unlocking, or opening cargo doors in flight. If the weather stripping below the cargo door catch is missing, pad the door edge with felt and tape in place. Padding must not preclude closing the cargo doors.

b. **Modified Anchor Line System**. Install a floor-mounted anchor line system (Figure 17-9, page 17-10) using a modified STABO extraction system anchoring strap assembly made from type XXVI nylon webbing (NSN 1670-00-999-3544, TM 10-1670-262-12&P and TM 10-1670-251-12&P). To modify the STABO, remove two of the connector snaps (leaving four) and add two D-rings (NSN 1670-00-360-0466). The cotton buffers may be locally manufactured.

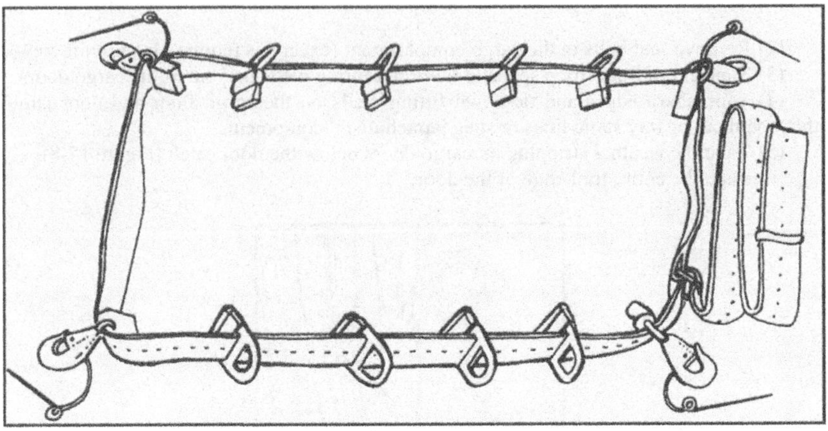

Figure 17-9. UH-60A modified anchor line.

(1) Install four snap hooks with safety wires and eight D-rings with cotton buffers on the anchor web loop, with the snap hooks and D-rings facing out in the following order: one snap hook, four D-rings; two snap hooks, four D-rings; and one snap hook (Figure 17-10).

(2) Insert about 30 inches of the web loop running end into the quick-fit adapter to secure the loop.

(3) Center the anchor line system on the cargo floor with the quick-fit adapter to the rear. Attach the snap hooks to tie-down fittings 3B, 3C, 4B, and 4C. Insert the safety wires and tape the snap hooks.

(4) Center the quick-fit adapter between tie-down fittings 4B and 4C, and tighten the web loop by pulling on the loop running end. Secure the web loop running end with an overhand knot. Fold and tape excess webbing to the web loop.

c. **Safety Belt Installation**. Install three floor-mounted safety belts.

(1) Attach a standard safety belt to tie-down fittings 5A and 5C for the JM. (This is necessary only if a seat has not been left for the JM.)

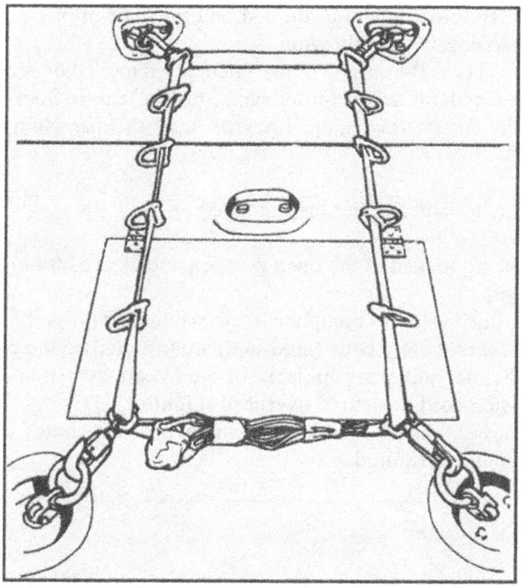

Figure 17-10. UH-60A modified anchor line secured to floor.

NOTE: The UH-60A cargo compartment configuration and floor tie-down fitting pattern preclude use of standard (individual) safety belts. Therefore, parachutists are restrained in groups of two and three, using modified safety belts.

(2) Attach an 86-inch-long (extended) safety belt to forward tie-down fittings 1A and 1D.

(3) Attach a 112-inch-long (extended) safety belt to tie-down fittings 1A and 5A, left door.

(4) Attach a 112-inch-long (extended) safety belt to tie-down fittings 1D and 5C, right door.

(5) Ensure that a serviceable safety harness is available for the JM (and the crew chief, when required). The JM's safety line is attached to tie-down fitting 5B. The crew chief's safety line is attached to tie-down fitting 1A or 1D, as required. If safety harnesses are not available, a backpack-type parachute may be used.

WARNING

Movement in the cargo compartment must be minimized to preclude inadvertent parachute activation.

d. **Inspection**. Before enplaning, the JM and pilot, or pilot's representative, jointly inspect the aircraft to ensure the following.

(1) All loose objects in the cargo compartment are removed or secured forward.

(2) Sharp edges and tie-down fitting wells on the cargo floor and doorjambs (or anything that could cut or fray static lines or snag the parachutists' equipment) are padded and taped.

NOTE: Door catches and handles are not taped.

(3) Cargo doors are locked in the open position and cleared for closing, depending on mission requirements.

(4) The anchor line system is complete, serviceable, and properly installed.

(5) Three serviceable safety belts (modified) are installed on the cargo floor.

(6) A headset/helmet and intercom jack for the JM are available and operational, and the intercom extension cord is secured overhead (Figure 17-11).

(7) Safety harnesses and backpack-type emergency parachutes are available for the JM and the crew chief, as required.

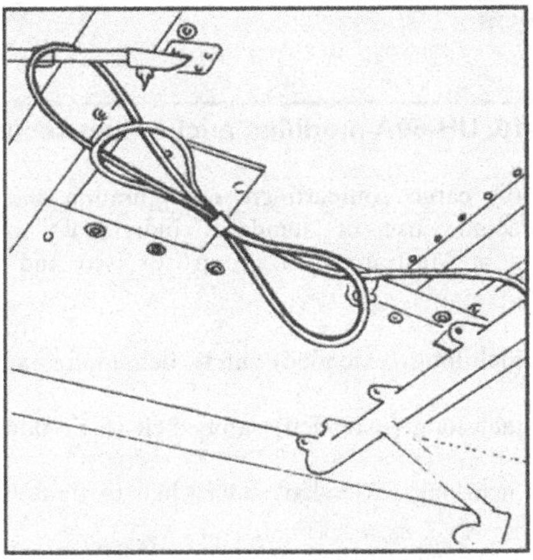

Figure 17-11. UH-60A JM's intercomm stowed overhead.

17-17. LOADING TECHNIQUES AND SEATING CONFIGURATION

Personnel are organized into a stick of eight parachutists. They approach the aircraft from the left or right side at a 90-degree angle in reverse order: numbers 8, 7, 6, 5, 4, 3, 2, 1 (Figure 17-12).

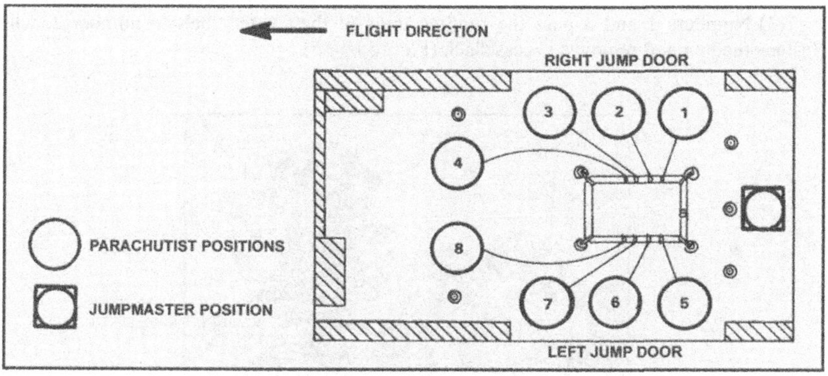

Figure 17-12. UH-60A seating and static line routing.

WARNING

Do not approach the aircraft directly from the front because the lowest arc of the turning rotor blades occurs at that point.

a. Number 8, followed by numbers 7, 6, and 5, enter the left door on command from the static JM. Numbers 4, 3, 2, and 1 enter the right door on command from the static JM. They are seated and hooked up by the JM in reverse numerical sequence, beginning with parachutist number 8, as they enter the aircraft. The open portion of static line snap hooks face the front of the aircraft.

b. Numbers 4 and 8 hold their static lines with a reverse bight: number 4 with the right hand and number 8 with the left hand. The static lines of the remaining parachutists, seated in the left and right doors, are routed directly behind them and down to the anchor line.

c. The JM ensures that any excess static line is stowed in the pack tray retainer band and that numbers 4 and 8 have correctly routed their static lines with the proper reverse bight.

NOTE: To preclude binding during exit, excess static lines of numbers 1, 2, 3, 5, 6, and 7 are stowed through the static line slack retainer on the parachutist's backpack.

d. When the JM commands FASTEN SAFETY BELTS, parachutists do the following:

(1) Numbers 4 and 8 pass the running ends of their safety belt to the center, fasten the belt, and remove excess slack (Figure 17-13, page 17-14).

(2) Numbers 5 and 7 pass the running ends of their safety belt to number 6, who fastens the belt and removes excess slack (Figure 17-14, page 17-15).

(3) Numbers 1 and 3 pass the running ends of their safety belt to number 2, who fastens the belt and removes excess slack (Figure 17-15).

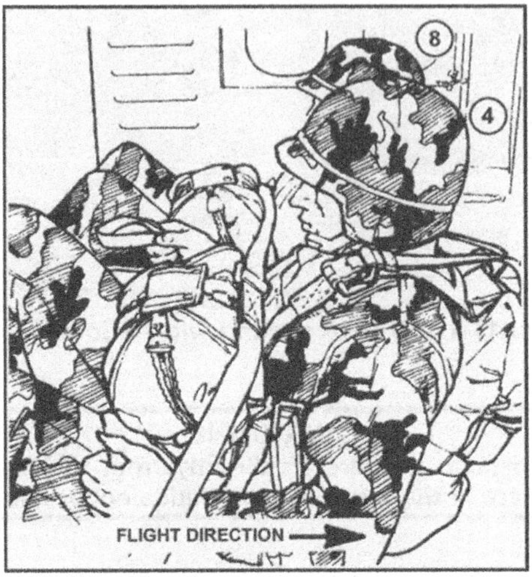

Figure 17-13. Numbers 4 and 8 with static line bight and safety belt secured.

e. The JM inspects all safety belts to ensure that they are securely fastened and properly fitted. He is seated aft with his safety belt fastened for lift-off and landing. (One seat should have been left in place for the JM.)

f. For airdrop operations requiring the crew chief and window gunner (seat installed), the number of combat-equipped parachutists is reduced to six. The seating configuration is modified—positions 4 and 8 are deleted, and positions 5, 6, and 7 are renumbered 4, 5, and 6.

Figure 17-14. Parachutists (left door) with safety belt secured.

Figure 17-15. Parachutists (right door) with safety belt secured.

17-18. JUMP PROCEDURES

If the cargo doors are to be closed en route to the drop zone, the JM briefs numbers 3 and 7 on door opening procedures before loading. At 6 minutes before the drop, the pilot either lands or brings the aircraft to a hover (near the ground) and notifies the JM to open the cargo doors. The JM directs numbers 3 and 7 to open them. He ensures that the cargo doors are opened and locked. A 4-minute, a 30-second, and an 8- to 10-second warning are relayed to the JM by the pilot through the intercom system.

17-19. JUMP COMMANDS

The JM issues the following commands:

a. **GET READY**. This command is given at the 4-minute warning to alert the parachutists. All safety belts are removed at 1,000 feet AGL.

NOTE: Safety belts are removed when directed by the JM. Numbers 2, 6, and 8 release them. The running ends are stowed forward and aft to clear the static lines and the exit path.

b. **CHECK STATIC LINES**. The JM checks the routing of all static lines (from pack trays to anchor cable) to ensure they are correctly routed and hooked up. He ensures excess static line is stowed through the slack retainer on the backpacks of numbers 1, 2, 3, 5, 6, and 7, and that numbers 4 and 8 have the prescribed reverse bight in their static lines.

c. **CHECK EQUIPMENT**. All parachutists check their equipment.

d. **SOUND OFF FOR EQUIPMENT CHECK**. Number 1 indicates orally (and with a hand signal) to the JM the status of his equipment, followed by the remaining parachutists in numerical order.

e. **SIT IN THE DOOR**. This command is given by the JM at the 30-second warning. Numbers 1, 2, 3, 5, 6, and 7 assume door positions (Figure 17-16) with feet together outside the cargo compartment. Numbers 4 and 8 remain in place, ensuring that their feet are clear of their static lines. (This command is omitted if the parachutists are already sitting in the door.)

f. **STAND BY**. This command is given at the 8- to 10-second warning. The JM ensures that all parachutists hear and understand this command, particularly number 1, who places both hands, palms down, on the cargo floor alongside his thighs and awaits the next command. Numbers 2, 3, 5, 6, and 7 place both hands, palms down, on the cargo floor and await the next command; numbers 4 and 8 remain in place.

g. **GO**. The JM gives this command by an oral GO and a sharp tap on the rear of the parachutist's helmet. Each parachutist is tapped out. The jump sequence is in numerical order, 1 through 8. As soon as number 3 clears the door, number 4 moves into the door and assumes the door position before being tapped out. The static JM assumes control of number 4 jumper's static line as the parachutist begins moving to the door. Numbers 5, 6, 7, and 8 repeat the sequence in the left door.

Figure 17-16. Numbers 1, 2, 3, 5, 6, and 7 exit positions.

17-20. SAFETY PRECAUTIONS

Safety precautions on the UH-60A are as follows:

a. **Parachutists**. If CWIEs will be jumped, they may be attached to numbers 1 or 5 parachutists, or both. No more than two CWIEs, one for each door, can be jumped. DMJPs may not be jumped from aircraft that require parachutists to sit on the floor. (Procedures for towed parachutists are in paragraph 17-10.) Crowded conditions inside the cargo compartment make accidental activation of the reserve parachute more likely. During movement inside the aircraft, parachutists must protect the rip cord grip.

b. **JM**. The static JM wears a safety harness that is attached to the aft cargo floor tie-down fitting (5B). Backpack-type emergency parachutes may be used if a safety harness is not available. The JM is equipped with a headset or flight helmet that allows direct communications with the aircraft crew. The static JM immediately notifies the pilot of a towed parachutist.

c. **Equipment**. Parachutists can wear combat equipment when jumping this aircraft. Without detaching the static lines, the JM retrieves static lines and D-bags, places them inside an aviator kit bag, and secures the kit bag until the aircraft has landed. The static line snap hooks are then removed from the anchor line attaching points. The UH-60A is **not** used for static line parachute operations with the cargo doors removed. The static line anchor line cable is **never** rigged to the cargo door or overhead tie-down rappelling rings, since trailing D-bags might foul the main rotor system (due to the high position in which the bags would trail).

d. **Aircraft**. The indicated airdrop speed of the aircraft should not be less than 65 knots or more than 75 knots. The minimum jump altitude is 1,500 feet AGL.

NOTE: The pilot must maintain level flight and airdrop speed during D-bag retrieval to preclude D-bag entanglement with the cargo doors.

17-21. SAFETY BELT MODIFICATION

The UH-60A cargo compartment floor configuration does not provide a specific design of tie-down fittings for restraining personnel seated on the cargo floor. The safety belts used for restraining personnel are part of the troop seat assembly and are removed when conducting parachute operations.

a. The three modified C-3A troop-type safety belts, using the cargo floor tie-down fittings, restrain parachutists in groups of two and three by a single safety belt (Figure 17-17).

b. Two safety belts, 112 inches long and adjustable to 86 inches, restrain parachutists numbers 1 through 3 and 5 through 7, who are seated in the left and right cargo doors.

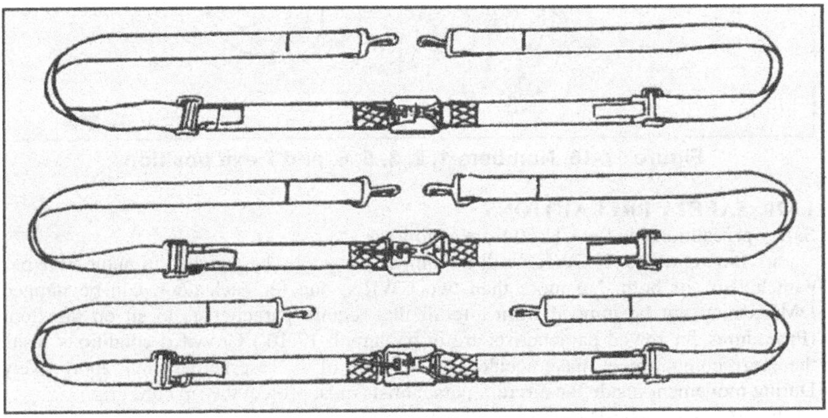

Figure 17-17. Modified C-3A troop safety belts.

c. One safety belt, 86 inches long and adjustable to 60 inches, restrains numbers 4 and 8 seated in the cargo compartment. Belt modifications are as follows:

(1) Place three standard C-3A troop-type safety belts (NSN 1670-00-447-9504) on a flat surface with hardware facing up.

(2) Remove the 8-inch lengths of webbing located between the end snap hooks and the quick-fit adapters of each belt.

(3) Cut two 32-inch and four 46-inch lengths of number 3 nylon webbing and heat-sear the ends.

(4) Reassemble one belt using the two 32-inch lengths of webbing.

(5) Thread the running ends of the webbing up through the bar of the snap hooks and quick-fit adapters. Make a 5-inch fold-back and tack in place.

(6) Sew a 4-inch, 4-point, WW stitch formation on each fold-back using number 3 nylon thread and a medium-duty machine (TM 10-1670-298-20&P).

(7) Reassemble the other two belts as indicated, using the 4-inch lengths of webbing.

Section IV. CH-47 CHINOOK

The CH-47 is a tandem-rotor, medium-transport helicopter. Twenty-eight combat-equipped parachutists can jump from this aircraft (Figure 17-18). The jumpmaster may be a jumping JM or a static JM.

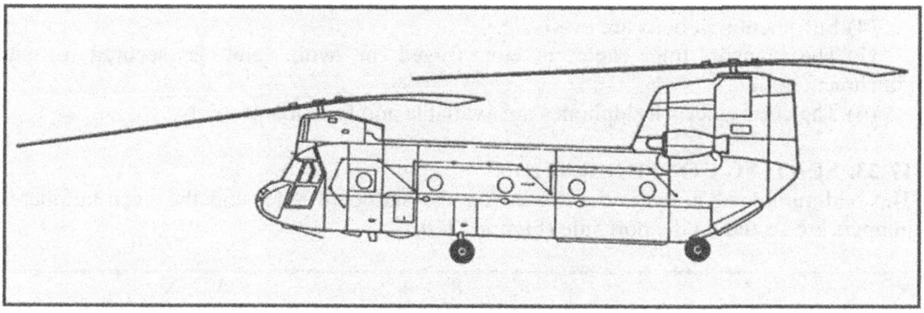

Figure 17-18. CH-47 Chinook.

17-22. PREPARATION AND INSPECTION

The JM prepares and inspects the CH-47 as follows:

a. **Preparation**. The following steps prepare the CH-47 for jumping.

(1) Install safety belts for each parachutist and extend all the way out to ensure positive hookup while seated.

(2) Secure the permanently installed anchor line cable to the attachment points on the starboard side of the aircraft (Figure 17-19).

(3) Incline the ramp for personnel parachute drops during flight.

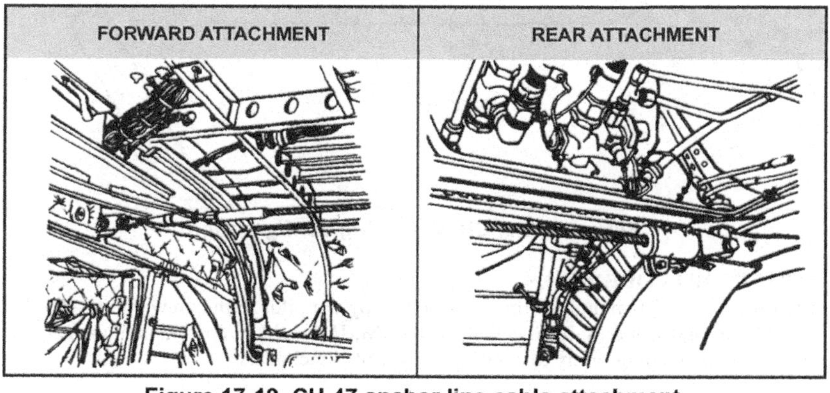

Figure 17-19. CH-47 anchor line cable attachment.

NOTE: The best incline is 3 degrees below the horizontal. Scribe marks may be placed on the ramp to show this degree of incline.

b. **Inspection**. Before enplaning, the JM and the pilot, or pilot's representative, jointly inspect the aircraft to determine the following:

(1) Troop seats can be easily lifted and secured before jumping.

(2) The ramp is clean and free of oil and water.

(3) Seats are securely fastened in the down position.

(4) Sufficient seat belts are available.

(5) The anchor line cable is not frayed or worn and is secured to the attachment points.

(6) The crew chief's headphones are available and function properly.

17-23. SEATING CONFIGURATION

The odd-numbered jumpers are seated on the starboard side, and the even-numbered jumpers are seated on the port side (Figure 17-20).

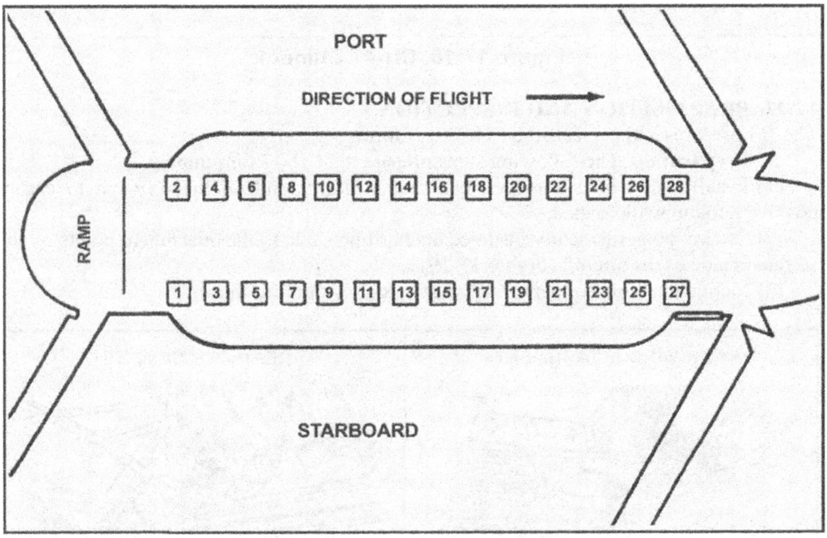

Figure 17-20. CH-47 seating configuration.

17-24. JUMP PROCEDURES

The 6-minute and 1-minute warnings are given by the pilot to the crew chief, who then relays them orally and by hand signals to the JM. If the JM jumps, he is number 1; this requires a nonjumping safety to control the flow of parachutists.

17-25. JUMP COMMANDS

The JM issues the following commands:

a. **GET READY**. This command is given at the 6-minute warning to alert parachutists.

b. **PORT SIDE PERSONNEL, STAND UP**. Parachutists seated on the port side of the aircraft stand up and secure their seats in the up position.

c. **STARBOARD SIDE PERSONNEL, STAND UP**. Parachutists seated on the starboard side of the aircraft stand up and secure their seats in the up position.

d. **HOOK UP**. On this command, odd-numbered personnel hook up, followed by the even-numbered personnel, who hook up (the open portion of the snap hook facing starboard) between the odd-numbered personnel to form one continuous stick of 28 parachutists.

NOTE: After hooking up, each parachutist controls the static line in a reverse bight at waist-level (left hand).

e. **CHECK STATIC LINES**. The JM or safety checks the routing of all static lines.

f. **CHECK EQUIPMENT**. All parachutists check their equipment.

g. **SOUND OFF FOR EQUIPMENT CHECK**. Beginning with number 28, the jumpers pass the status of their equipment toward the aft end of the aircraft. Number 1 orally (and with a hand signal) indicates to the JM the status of his and all other jumpers' equipment by stating, "All OK, jumpmaster."

h. **STAND BY**. The command is given 8 to 10 seconds before the command GO. Number 1 assumes a standing position at the ramp hinge (near center) of the aircraft. The remaining personnel close up interval behind the first parachutist.

i. **GO**. Number 1 walks off the port side rear corner of the ramp. The remaining parachutists follow at 1-second intervals.

NOTE: The JM or safety controls the flow from his position on the port side near the ramp hinge. Less than a 1-second interval between parachutists may result in entanglement of parachutists and static lines.

17-26. SAFETY PRECAUTIONS

Safety precautions on the CH-47 are as follows:

a. **Parachutists**. Parachutists ensure that seats are secured in the up position with seat legs rotated inside the seats. When following internal drop loads, parachutists exit between the ramp roller conveyor sections, staying as close to the port side section as possible. The parachutists jumping after external load drops, who are forward of the open floor hatch (used to check a load drop), remain clear of the opening until the load leaves the aircraft and the hatch is closed by the crew chief.

b. **Jumpmaster**. The JM or safety personnel ensure that parachutists are hooked up consecutively, 1 through 28 (Figure 17-21, page 17-22). If the JM does not jump, he wears a safety harness or back emergency parachute. He checks each parachutist after they hook up and controls the flow of parachutists. When an external load is delivered, the JM ensures the external load is clear and the aircraft has accelerated to a safe airdrop speed before dropping cargo bundles from inside the aircraft or before permitting parachutists to exit.

c. **Safety Personnel**. If the JM jumps, one nonjumping safety is required; the safety wears an emergency parachute.

d. **Equipment.** When cargo bundles are delivered, JMs use 15-foot breakaway static lines with cargo parachutes. The ramp roller conveyor section is installed on the starboard side of the ramp and is used to help eject the bundles from the cargo ramp; numbers 1 and 2 push the bundles out.

e. **Aircraft.** Aircraft safety requires that the speed during jumps is not less than 80 knots or more than 110 knots, with 90 knots being optimum speed. No special preparation is required if the aircraft has skis. Minimum jump altitude is 1,500 feet AGL when jumping below 90 knots. Minimum jump altitude 90 knots and above is 1,250 feet AGL. After the last parachutist has cleared the aircraft, the static lines are retrieved (using the static line retriever) inside the aircraft and secured in an aviator kit bag. The ramp must not be lowered until all jumpers have hooked up to the anchor line cable.

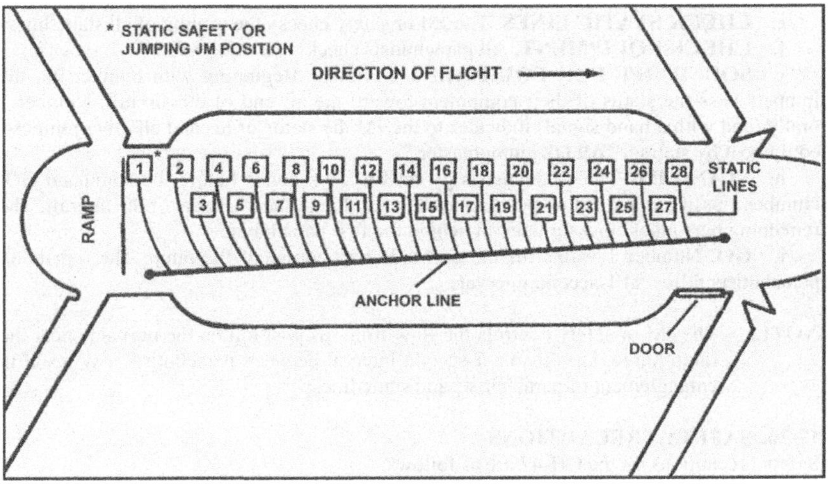

Figure 17-21. CH-47 static line routing.

CHAPTER 18

OTHER SERVICE AIRCRAFT

Other service rotary-wing aircraft can be used for parachuting operations. In addition to the procedures described here, Chapter 17, Section I, also applies.

Section I. CH-53 SEA STALLION (USMC)

The CH-53 is a twin-engine, single-rotor, medium-transport helicopter. Twenty combat-equipped parachutists, using the ramp, can jump from this aircraft (Figure 18-1).

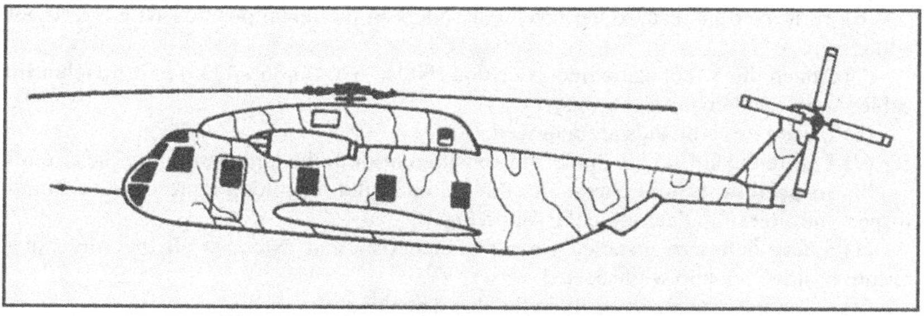

Figure 18-1. CH-53 Sea Stallion (USMC).

18-1. PREPARATION AND INSPECTION

The jumpmaster prepares and inspects the CH-53 as follows:

a. **Preparation**. The following procedures prepare the CH-53 for jumping:

(1) Install the anchor line cable on the port side of the floor (Figure 18-2, page 18-2) using the appropriate 5,000-pound tie-down fittings for the length of cable used (forward attachment point) and station number 522 (rear attachment point).

(2) Use one 1/4-inch or 3/8-inch steel cable (3/8-inch preferred), of an acceptable length, with four 1/4-inch or 3/8-inch cable clamps, lock washers, nuts, and bolts. The minimum installed length of the anchor line cable will be 8 feet between forward and rear attaching points (four to five jumpers). Ensure the installed anchor line cable is long enough so that a jumper's static line extends from the jumper to the anchor line cable without touching other jumpers in front of him in the stick.

(3) Route the anchor line cable directly through the forward and rear attaching point deck rings or use a steel stubai 85 snap link (with locking gate) between the anchor line cable and the aircraft attaching points. Ensure the locking gates on the snap links are facing up, locking to the aft of the aircraft, and taped. Ensure all tape is removed from the anchor line cable components and that the clamps are tight prior to every operation.

CAUTION
Do not use a clevis assembly (G-13) to attach the anchor line cable to the aircraft. If the cable is loose, the clevis may rotate sideways and drastically reduce the strength of the anchor line cable.

(4) Place a 4- by 4- by 6-inch wooden block between the anchor line cable and floor and attach a clevis assembly (G-13) to the tie-down ring at station number 502 for use as a static line snap hook stop.

(5) Remove excess slack from the anchor line.

(6) Turn over and secure the conveyor rollers in the down position with the smooth surface up.

(7) Attach the 5-foot static line extension (NSN 1670-00-368-4225) to the anchor line cable; insert and bend safety wires.

(8) Ensure drive-on aids are removed.

(9) Ensure the tail skid is in the "up" position prior to the jumpers exiting the aircraft.

b. **Inspection**. Before enplaning, the JM and pilot, or pilot's representative, jointly inspect the aircraft to determine the following:

(1) Safety belts are installed for all parachutists and extended all the way out to ensure positive hookup while seated.

(2) Seats are fastened securely in the down position.

(3) Seats are lifted and secured before jumping.

(4) The anchor line cable is not worn or frayed and is secured to the attachment points in the prescribed manner.

(5) The ramp and deck are clean and free of oil and water.

(6) All protruding objects near the ramp are removed or taped.

(7) The crew chief's headphones are available and function properly.

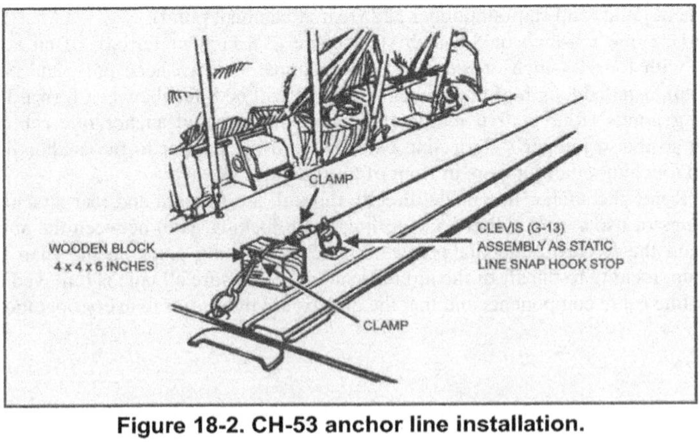

Figure 18-2. CH-53 anchor line installation.

18-2. LOADING TECHNIQUES AND SEATING CONFIGURATION

Parachutists enter the aircraft over the ramp with the static line over the right shoulder. Odd-numbered parachutists are seated on the port side, and even-numbered parachutists are seated on the starboard side (Figure 18-3).

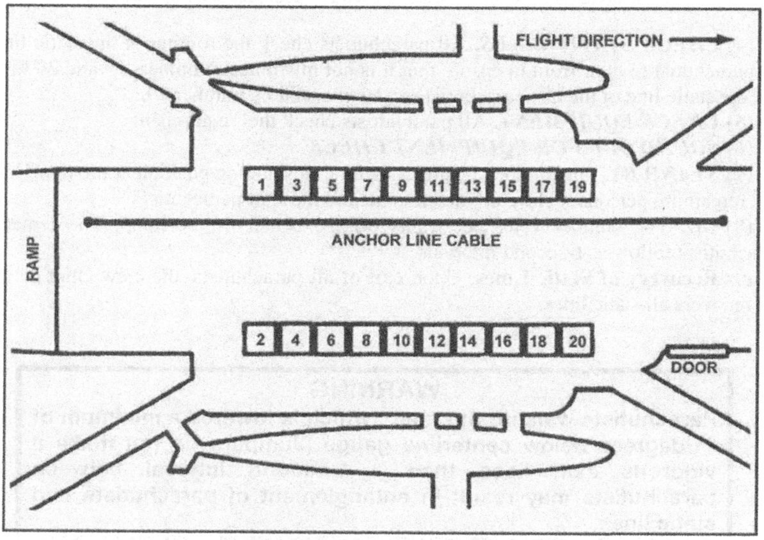

Figure 18-3. CH-53 seating configuration.

18-3. JUMP COMMANDS AND PROCEDURES

Jump commands and procedures for the CH-53 are as follows:

a. **Time Warnings**. The 6-minute and 1-minute warnings are given by the pilot to the crew chief, who in turn relays them orally and by hand signals to the JM.

b. **Jump Commands**.

(1) ***GET READY.*** All parachutists remove seat belts.

(2) ***STAND UP.*** All parachutists stand up and secure their seats in the up position.

(3) ***HOOK UP.*** The 5-foot static line extensions are attached to the anchor line cable by the snap hooks; insert and bend safety wires. The JM or AJM passes the static line extension to each jumper starting with the last jumper and working forward. When the command HOOK UP is given, the jumpers ensure the cotton sleeve is on the static line extension, attach their static line snap hooks, insert and bend safety wires, and hold the static line snap hook exposed for inspection by the JM. Once inspected, place the sleeve over the snap hook and take a reverse bite at waist level.

WARNING
The JM must ensure the jumper has a proper reverse bite. If an improper reverse bite is taken, the static line could become misrouted under the arm and cause a towed jumper or severe injury.

(4) ***CHECK STATIC LINES.*** All parachutists check the routing of the static line of the parachutist to their front to ensure that it is not misrouted. Numbers 19 and 20 turn so that the static line of the last parachutist can be checked by number 19.

(5) ***CHECK EQUIPMENT.*** All parachutists check their equipment.

(6) ***SOUND OFF FOR EQUIPMENT CHECK.***

(7) ***STAND BY.*** The number 1 jumper assumes a standing position at the ramp hinge. The remaining personnel close up intervals behind the first parachutist.

(8) ***GO.*** The number 1 jumper walks off the center of the ramp. The remaining parachutists follow at 1-second intervals.

c. **Recovery of Static Lines**. Upon exit of all parachutists, the crew chief or static JM recovers all static lines.

WARNING
Parachutists walk off the ramp, which is lowered a minimum of 11 degrees below centerline gauge. Jumpers do *not* make a vigorous exit. Less than a 1-second interval between parachutists may result in entanglement of parachutists and static lines.

18-4. SAFETY PRECAUTIONS

No more than two parachutists should jump with CWIE in one pass over the DZ. These parachutists should be numbers 1 and 2 in the stick. The static JM or safety personnel ensure that parachutists are hooked up consecutively (1 through 20), and that all seats are secured out of the way of the jumpers. The speed of the aircraft during jumps is 90 to 110 knots. The minimum drop altitude for aircraft with a 90- to 110-knot drop speed is 1,250 feet AGL. Either the JM or AJM must remain with the aircraft to handle the D-bags and towed jumpers. The JM spots the aircraft using WSVC, VIRS, or GMRS. The JM may spot from the crew chief door or the ramp as long as a safety is controlling the jumpers. If spotting from the ramp and jumping, the JM leads the stick and the AJM stays with the aircraft.

Section II. CH-46 SEA KNIGHT (USMC)

The CH-46 is a tandem-rotor, medium-transport helicopter. Twelve combat-equipped parachutists can jump from either the personnel door or the ramp (Figure 18-4).

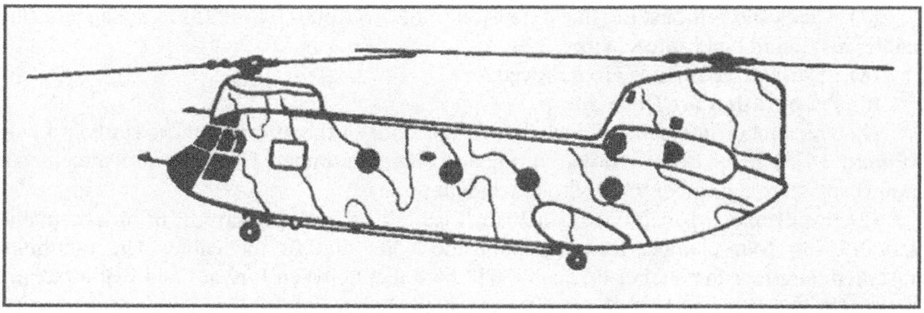

Figure 18-4. CH-46 Sea Knight.

18-5. PREPARATION AND INSPECTION

The jumpmaster prepares and inspects the CH-46 as follows:

a. **Preparation for a Ramp Jump**. The following procedure prepares the CH-46 for jumping:

(1) Install the anchor line cable on the port side of the floor using the appropriate 3,000-pound tie-down fittings for the length of the cable used (forward attachment point) and station number 410 (rear attachment point) (Figure 18-5, page 18-7).

(2) Use one 1/4-inch or 3/8-inch steel cable (3/8-inch preferred) of an acceptable length with 1/4-inch or 3/8-inch cable clamps and nuts. The minimum installed length of the anchor line cable will be 8 feet between forward and rear attaching points (four to five jumpers). Ensure the installed anchor line cable is long enough so that the jumper's static line goes from the jumper to the anchor line cable without touching other jumpers in front of him in the stick.

(3) Route the anchor line cable directly through the forward and rear attaching point deck rings or use a steel stubai 85 snap link (with locking gate) between the anchor line cable and the aircraft attaching points. Ensure the locking gates on the snap links are facing up, locking to the aft of the aircraft, and taped. Ensure all tape is removed from the anchor line cable components and that the clamps are tight prior to every operation.

CAUTION

Do not use a clevis assembly (G-13) to attach the anchor line cable to the aircraft. If the cable is loose, the G-13 clevis may rotate sideways and drastically reduce the strength of the anchor line cable.

(4) Place two 4-s by 6-inch wooden blocks between the anchor line cable and floor and attach one clevis assembly (G-13) to the tie-down ring at station number 390 for use as a static line snap hook stop.

(5) Remove excess slack from the anchor line.

(6) Turn the conveyor rollers over and secure them in the down position with the smooth surface up.

(7) Attach the 5-foot static line extension (NSN 1670-00-368-4225) to the anchor line cable; insert and bend safety wires.

(8) Ensure drive-on aids are removed.

b. **Preparation for Door Jump.**

(1) The anchor line cable is installed on the floor of the aircraft on the starboard side (Figure 18-5) using the tie-down fittings at station number 170 (forward attachment point) and station number 410 (rear attachment point).

(2) Install one 1/4-inch or 3/8-inch steel cable (3/8-inch preferred), of an acceptable length, with four clamps, lock washers, and bolts that fit the cable. The minimum installed length of the anchor line cable will be 8 feet between forward and rear attaching points (four to five jumpers). Ensure the installed anchor line cable is long enough so that the jumper's static line goes from the jumper to the anchor line cable without touching other jumpers in front of him in the stick.

(3) Route the anchor line cable directly through the forward and rear attaching point deck rings or use a steel stubai 85 snap link (with locking gate) between the anchor line cable and the aircraft attaching points. Ensure the locking gates on the snap links are facing up, locking to the aft of the aircraft, and taped. Ensure all tape is removed from the anchor line cable components and that the clamps are tight prior to every operation.

CAUTION

Do not use a clevis assembly (G-13) to attach the anchor line cable to the aircraft. If the cable is loose, the G-13 clevis may rotate sideways and drastically reduce the strength of the anchor line cable.

(4) Place two 4- by 6-inch wooden blocks between the anchor line cable and floor and attach a clevis assembly (G-13) to the tie-down ring at station number 190 for use as a static line snap hook stop.

(5) Remove excess slack from the anchor line.

(6) Turn the conveyor rollers over and secure them in the down position with the smooth surface up.

(7) Attach the 5-foot static line extension (NSN 1670-00-368-4225) to the anchor line cable; insert and bend safety wires.

(8) Ensure drive-on aids are removed.

(9) Remove the personnel door and install a plywood cover over the aft side of the door frame and secure it with a metal plate. Pad and tape the bottom and aft edges of the door frame (Figure 18-6, page 18-8).

c. **Inspection**. Before enplaning, the JM and pilot, or pilot's representative, jointly inspect the aircraft to determine the following:

(1) Seat belts are installed for all parachutists and extended all the way out to ensure positive hookup while seated.

(2) Seats are fastened securely in the down position.

(3) Seats can be lifted and secured before jumping.

(4) The anchor line cable is not worn or frayed and is secured to the attachment points in the prescribed manner.

(5) The wooden cover is installed and secured to the aft edge of the door frame when jumping the personnel door.

(6) The bottom edges of the door frame are properly padded and taped.

(7) The ramp and deck are clean and free of oil or water.

(8) All protruding objects near the ramp and personnel door are removed or taped.

(9) The crew chief's headphones are available and function properly.

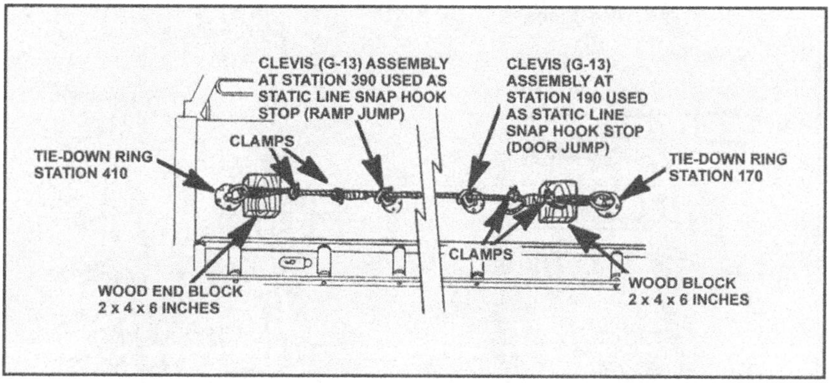

Figure 18-5. CH-46 anchor line cable installation.

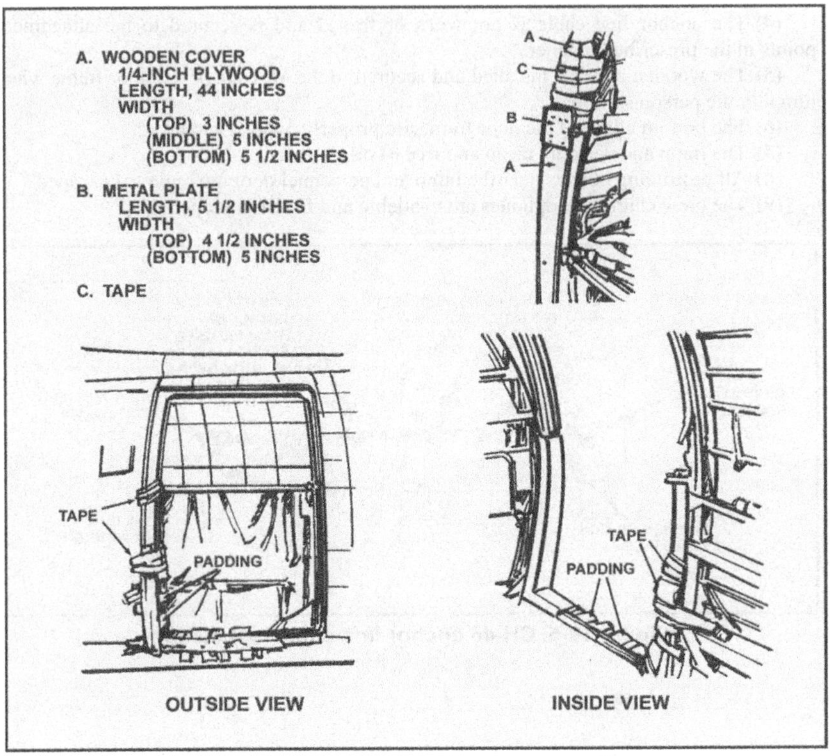

Figure 18-6. CH-46 door frame preparation.

18-6. LOADING TECHNIQUES AND SEATING CONFIGURATION

There are two ways a parachutist can jump from the aircraft: the ramp or the door. The parachutists enter the aircraft with the static line over the right shoulder; they are seated on the port and starboard side.

18-7. JUMP COMMANDS AND PROCEDURES

The 6-minute and 1-minute warnings, whether jumping the door or ramp, are given by the pilot to the crew chief, who in turn relays them orally and by hand signals to the static JM or AJM.

a. **Jump Commands for Door Jump.**

(1) ***GET READY.*** All parachutists in the first stick unfasten their seat belts.

(2) ***STAND UP.*** Parachutists stand up and move to the starboard side (Figure 18-7, page 18-10).

(3) ***HOOK UP.*** The 5-foot static line extensions are attached to the anchor line cable by the snap hooks; insert and bend safety wires. The JM or AJM passes the static line extension to each jumper starting with the last jumper and working forward. When the

command HOOK UP is given, the jumpers ensure the cotton sleeve is on the static line extension, attach their static line snap hooks, insert and bend safety wires, and hold the static line snap hook exposed for inspection by the JM. Once inspected, place the sleeve over the snap hook and take a reverse bite at waist level.

WARNING

The JM must ensure the jumper has a proper reverse bite. If an improper reverse bite is taken, the static line could become misrouted under the arm and cause a towed jumper or severe injury.

(4) ***CHECK STATIC LINES.*** The parachutists take a reverse bight (at waist level) of about 8 inches in the static line with the right hand, keep the arm close to the side, and check the static line of the parachutist to the front.

(5) ***CHECK EQUIPMENT.*** All parachutists check their equipment.

(6) ***SOUND OFF FOR EQUIPMENT CHECK.***

(7) ***STAND IN THE DOOR.*** Number 1 parachutist moves to the door and assumes the door position. He must crouch low to allow at least a 2-inch clearance between his helmet and the top of the door. The other parachutists close up behind number 1 at normal intervals.

(8) ***GO.*** Number 1 exits by jumping straight out the door and assuming the proper body position. The succeeding parachutists move up, make a 90-degree turn at the door, take up a correct door position, and exit in the same manner as number 1, maintaining a 1-second interval. Upon exit of all parachutists, the crew chief or static JM recovers all static lines.

WARNING

Parachutists must not spring upward when jumping this helicopter.

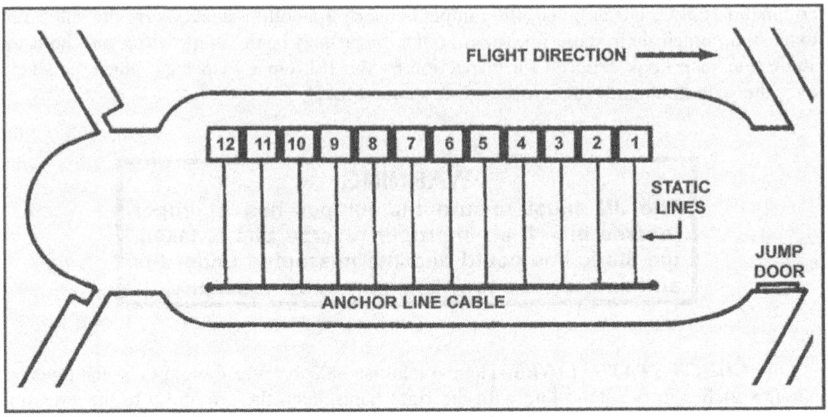

Figure 18-7. CH-46 door jump.

b. **Jump Commands for Ramp Jump**.

(1) ***GET READY.*** All parachutists in the first stick unfasten their seat belts.

(2) ***STAND UP.*** Parachutists stand up and move to the port side (Figure 18-8).

(3) ***HOOK UP.*** When the command HOOK UP is given, the jumpers ensure the cotton sleeve is on the static line extension, attach their static line snap hooks, insert and bend safety wires, and hold the static line snap hook exposed for inspection by the JM. Once inspected, place the sleeve over the snap hook and take a reverse bite at waist level.

(4) ***CHECK STATIC LINES.*** The parachutists take a reverse bight at waist level of about 8 inches in the static line with the right hand, keep the arm close to the side, and check the static line of the parachutist to the front.

(5) ***CHECK EQUIPMENT.*** All parachutists check their equipment.

(6) ***SOUND OFF FOR EQUIPMENT CHECK.***

(7) ***STAND BY.*** Number 1 assumes a standing position at the ramp hinge. The remaining parachutists close up the interval.

(8) ***GO.*** Number 1 walks off the starboard rear of the ramp and assumes a normal body position. The remaining parachutists follow and exit in the same manner as number 1, maintaining a 1-second interval between parachutists. Upon exit of all parachutists, the crew chief or static JM recovers all static lines.

WARNING

Parachutists walk off the ramp (which is lowered to 11 degrees below centerline gage) in a crouched position to avoid hitting their heads on the upper ramp door. They do *not* make a vigorous exit. Less than a 1-second interval between parachutists may result in entanglement of parachutists and static lines. Upon exit of all parachutists, the crew chief or static JM recovers all static lines.

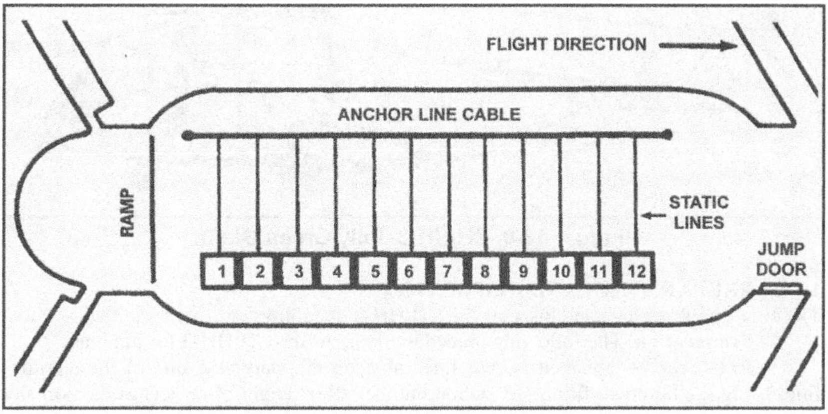

Figure 18-8. CH-46 ramp jump.

18-8. SAFETY PRECAUTIONS

On a single pass over the DZ, only the number 1 parachutist jumps with a CWIE from either the ramp or the door. The static JM or AJM ensures that parachutists are hooked up consecutively (1 through 12). The JM ensures that seats along the starboard side are secured in the up position when parachutists are jumping from the personnel door, or that seats along the port side are secured in the up position when they are jumping from the ramp. The speed of the aircraft is 80 to 90 knots when jumping. The minimum drop altitude is 1,250 feet AGL when jumping at 90 knots and 1,500 feet AGL when jumping at 80 knots. Either the JM or AJM must remain with the aircraft to handle the D-bags and towed jumpers. The JM spots the aircraft using WSVC, CIRS, or GMRS. The JM may spot from the crew chief door or the ramp as long as a safety is controlling the jumpers. If spotting from the ramp and jumping, the JM leads the stick and the AJM stays with the aircraft.

Section III. CH/HH-3 JOLLY GREEN GIANT (USAF)

The CH/HH-3 is a twin-engine, single-rotor, medium-transport helicopter. Fifteen combat-equipped parachutists can jump from this aircraft (Figure 18-9, page 18-12).

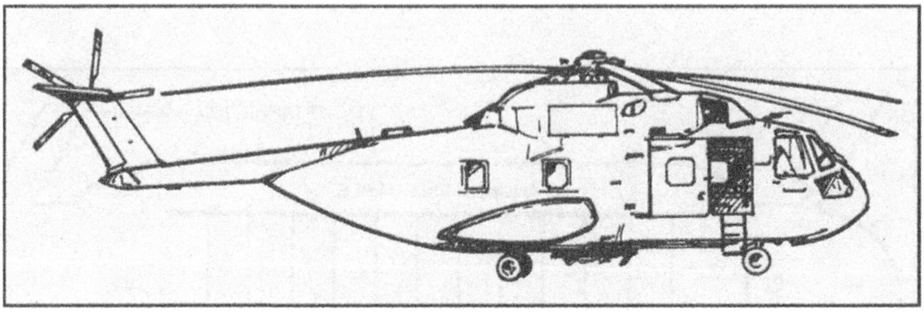

Figure 18-9. CH/HH-3 Jolly Green Giant.

18-9. PREPARATION AND INSPECTION

The crew chief prepares and inspects the CH/HH-3 as follows:

a. **Preparation**. The following procedure prepares the CH/HH-3 for jumping:

(1) Install the oval-shaped anchor line cable on the starboard side of the aircraft's floor using the tie-down fittings at station number 193.5 (right of center) as the port side forward attachment point, station number 212.5 as the starboard side forward attachment, station number 256.5 as an intermediate starboard side attachment point, and station number 276.5 as the starboard side rear and port side rear attachment points (Figure 18-10).

(2) The anchor line cable is constructed of 1/4-inch diameter, 6,400-pound test steel cable. Thread the anchor line cable through four static line snap hooks, where the static line is normally attached. These static line snap hooks connect the anchor line cable to the tie-down fittings. Complete the oval by overlapping both ends of the steel cable, then by securing the overlap with four cable clamps spaced intermittently between the swaged cable ends.

(3) Manufactured cables have the date of initial manufacture and weight testing capacity (2,500 pounds) permanently marked on the starboard side forward static line snap hook. Inspect cables each time the anchor line cable is installed for jumping. Remove from service cables showing excessive wear, corrosion, or more than three broken strands per inch. Weight test cables to a 2,500-pound capacity within each 12-month calendar period. (For example, cables weight tested on 1 January will be due weight testing by 31 January the following year.) Document annual weight testing on a DD Form 1574, which is attached to the cable.

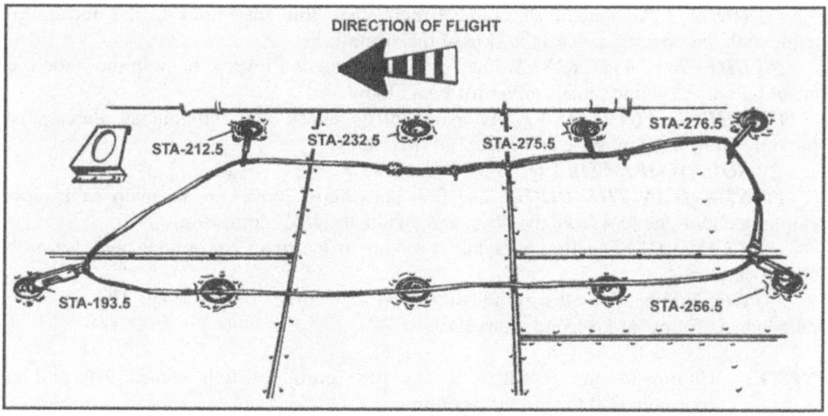

Figure 18-10. CH/HH-3 anchor line cable.

b. **Inspection**. Before enplaning, the JM and pilot, or pilot's representative, jointly inspect the aircraft to determine the following:

(1) The main cabin door is secured to the rear and taped, including the door handle and latch.

(2) Any external cargo slings are removed before conducting jump operations.

(3) All protruding objects near the doors are removed or taped. The penetrator is removed before conducting jump operations.

(4) Safety belts are installed for all parachutists and are extended completely to ensure positive hookup while seated.

(5) Seats are fastened securely in the down position with backs loose (except the four seats raised for anchor line cable installation).

(6) The anchor line cable is tight and free of frays.

(7) The deck is clean and free of oil and water.

(8) The flight engineer's headphones and the JM intercom cord are available and function properly.

18-10. LOADING TECHNIQUES AND SEATING CONFIGURATION

Parachutists enter the aircraft through the starboard side cabin door with their static line over their right shoulder. They enter the aircraft in reverse stick order with numbers 1 through 8 seated on the port side and numbers 9 through 15 seated on the starboard side.

18-11. JUMP COMMANDS AND PROCEDURES

Jump commands and procedures on the CH/HH-3 are as follows:

a. **Time Warnings**. The 6-minute and 1-minute warnings are given by the pilot to the flight engineer, who in turn relays them orally and by hand signals to the JM. Due to the limited space available to hook up, only four parachutists are airdropped each pass.

b. **Jump Commands**. The commands on the CH/HH-3 are—

(1) ***GET READY.*** The first four parachutists remove seat belts.

(2) ***STAND UP.*** Parachutists stand up and move to the anchor line cable.

(3) ***HOOK UP.*** Parachutists connect their static line snap hook to the anchor line cable with the opening toward the skin of the aircraft.

(4) ***CHECK STATIC LINES.*** Each parachutist takes a reverse bight in the static line, ensuring the static line remains over his bent elbow.

(5) ***CHECK EQUIPMENT.*** All parachutists check their equipment. Parachutists jumping equipment must be at the front of their stick.

(6) ***SOUND OFF FOR EQUIPMENT CHECK.***

(7) ***STAND IN THE DOOR.*** The first parachutist moves to the main cabin door, stopping about one foot from the door, and awaits the JM's commands.

(8) ***STAND BY.*** The first parachutist moves to the main cabin door and awaits the JM's commands.

(9) ***GO.*** Number 1 walks off the starboard rear corner of the ramp. The remaining parachutists follow at 1-second intervals.

NOTE: Commands are repeated for the next group of four parachutists and are repeated until the aircraft is empty.

WARNING

Parachutists walk out the door 90 degrees to the aircraft in a crouched position to avoid hitting their heads on the upper door frame. They do not make a vigorous exit.

c. **Exits**. Parachutists step out the same as exiting the tailgate of a fixed-wing aircraft, maintaining about a 1-second interval between parachutists. Less than a 1-second interval may result in entanglement of parachutists and static lines. Upon exit of all parachutists of each pass, the JM or flight engineer recovers all deployment bags and static lines.

18-12. SAFETY PRECAUTIONS

Approaching or loading the aircraft is performed only after visual clearance by the pilot or flight engineer. Before clearing any parachutists to jump, the JM confirms that the main gear is in the up position. The speed of the aircraft during all jump operations is between 70 knots and 90 knots indicated air speed.

CHAPTER 19

NONSTANDARD AIRCRAFT USED DURING AIRBORNE OPERATIONS

This chapter contains aircraft descriptions, JM procedures, and aircraft preparation techniques for nonstandard rotary-wing and fixed-wing aircraft. The aviation supporting unit prepares the aircraft for equipment and personnel drops to include seat and door removal and installation or rearrangement of seat belts. The installation of a field-expedient anchor line cable is the jumpmaster's responsibility. Aircraft preparation is usually accomplished jointly by the loadmaster/crew chief and JM. These aircraft are service tested and approved for personnel airdrop operations.

Section I. MODIFICATIONS TO JUMP COMMANDS AND JUMPERS' MOVEMENT IN NONSTANDARD AIRCRAFT

On some nonstandard aircraft, jumpers are required to shuffle in the aircraft and assume a stand-in-the-door position. The standard jump commands are modified by substituting the command STAND IN THE DOOR for STAND BY. The parachutists execute the shuffle and the stand-in-the-door position in the following manner.

19-1. SHUFFLE

The shuffle is a method of moving to the jump door without losing balance or tripping. To perform the shuffle—

a. The jumper's outboard arm is extended down and out to assist in maintaining balance and to assume the door position. The other hand grasps the static line in the correct bight for the aircraft.

b. Both feet are slightly spread, directly beneath the body, and staggered 6 to 8 inches. The jumper faces the rear of the aircraft and places his foot nearest the side of the aircraft forward; this foot is the shuffle foot. The foot nearest the center of the aircraft is the trail foot.

c. The jumper moves by stepping forward with his shuffle foot 6 to 8 inches and then with his trail foot. He keeps both feet staggered in the same relative heel-and-toe position throughout the shuffle.

19-2. STAND IN THE DOOR

At the command STAND IN THE DOOR—

a. The jumpers shuffle toward the jump door, and the number 1 jumper hands his static line to the safety.

b. The number 1 jumper ensures his arm is not entangled with the static line and shuffles into the door so that the toe of his outboard foot is extended about 2 inches over the edge of the doorsill or jump platform and his trail foot is about 6 inches to the rear.

c. The number 1 jumper keeps his feet shoulder-width apart. His weight is equally distributed on both feet, knees are bent, upper body is straight, head and eyes are to the front, and hands are outside the aircraft, with fingers extended and joined. He is in a position of "coiled alertness" and is ready to exit the aircraft without further weight adjustment. His hands are not grasping the aircraft but are used to maintain balance.

d. The number 2 jumper is in the shuffle position, roughly even with the leading edge of the jump door and facing the rear of the aircraft.

e. Follow-on jumpers close up behind the preceding jumper, using the shuffle position to maintain balance.

19-3. GO

At the command GO, the number 1 jumper makes a vigorous up-and-out exit, 6 inches up and 36 inches out, depending on aircraft requirements. The exit action is gained from the legs alone, and the hands are used only to guide the jumper. On some aircraft, the vigorous exit is essential to avoid hitting the aircraft or coming in contact with another jumper beneath or behind the aircraft. The number 2 jumper and all following jumpers—

a. Shuffle toward the jump door.

b. Hand the static line to the safety and ensure the arm is not entangled with the static line.

c. Shuffle into the door, take up a proper door position, and exit the aircraft without command, maintaining a 1-second interval between jumpers.

Section II. C-7A CARIBOU

The C-7A is a high-wing transport powered by two piston engines (Figure 19-1). A total of 24 parachutists may be dropped using the ramp or doors. The ramp is normally used for dropping parachutists.

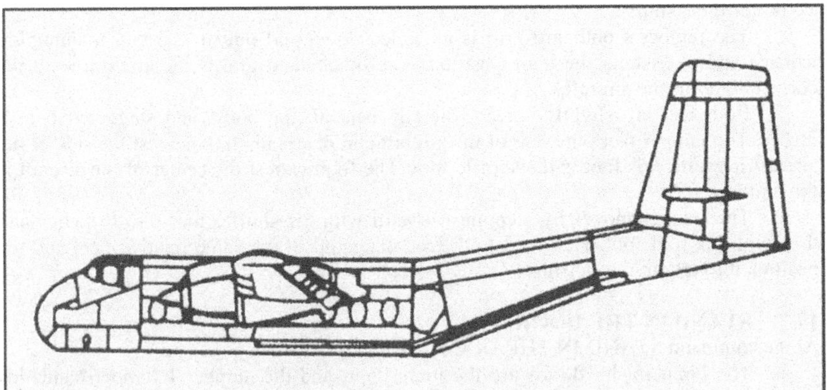

Figure 19-1. C-7A Caribou.

19-4. SEATING CONFIGURATION

Twenty-four parachutists are seated in two 12-parachutist sticks. Parachutists are loaded over the loading ramp or through the doors. The odd-numbered personnel are seated on the starboard side, and even-numbered personnel are seated on the port side.

19-5. SUPERVISORY PERSONNEL REQUIRED

Three personnel supervise safety procedures: one JM who performs standard aircraft check procedures, one safety NCO, and one loadmaster.

19-6. ANCHOR LINE CABLE ASSEMBLIES

There are two anchor line cable assemblies in the C-7A.

a. The anchor line cable for ramp jumps runs from the reinforced anchor line attachment plate on the forward bulkhead to the anchor line connector near the right side of the aft starboard door.

b. The anchor line cable for door jumps runs from the reinforced anchor line attachment plate down the center of the cargo compartment. It is permanently installed.

19-7. JUMP COMMANDS

Jump commands for the C-7A are as follows:

a. **GET READY**. Jumpers respond in the same manner as for other fixed-wing aircraft.

b. **PORT SIDE PERSONNEL, STAND UP**. Jumpers on the left side of the aircraft stand up, raise and secure their seats, and face the ramp of the aircraft.

c. **STARBOARD SIDE PERSONNEL, STAND UP**. Jumpers on the right side of the aircraft stand up, raise and secure their seats, and face the ramp of the aircraft.

d. **HOOK UP**. Even-numbered jumpers hook up between the odd-numbered jumpers to form a continuous stick of jumpers. The jumpers detach the static line snap hook from the top carrying handle of the reserve parachute and hook up to the anchor line cable with the open portion of the snap hook facing outboard, ensuring that the snap hook locks properly. The safety wire is inserted in the hole, pointing toward the rear of the aircraft, and folded down. The static line is controlled by each parachutist in a reverse bight at waist level in the left hand.

e. **CHECK STATIC LINES, CHECK EQUIPMENT, and SOUND OFF FOR EQUIPMENT CHECK**. These are executed in the same manner as with other fixed-wing aircraft.

f. **STAND BY/STAND IN THE DOOR**.

(1) ***STAND BY.*** The number 1 jumper, upon receiving the command STAND BY, assumes a standing position near the starboard side of the ramp hinge. The number 2 jumper stands on the port side of the aircraft slightly to the right of jumper number 1. The remaining personnel close up the interval behind the number 1 jumper.

(2) ***STAND IN THE DOOR.*** The proper door position is taken by the parachutists in both doors, with the appropriate foot resting on the elevated doorsill. There is no jump platform.

g. **GO**. Personnel exit the aircraft at 1-second intervals.

19-8. RAMP JUMPING

The number 1 jumper, upon receiving the command GO, walks off the port side rear corner of the ramp. The remaining jumpers follow at a 1-second interval. After the command GO, each jumper visually checks his body position for correctness and begins the 4000-count. (See Figure 19-2, page 19-4, for C-7A configuration for ramp jumping.)

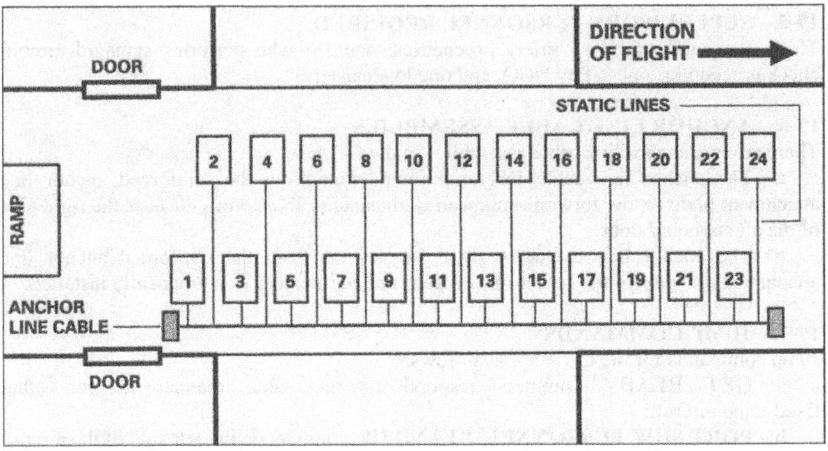

Figure 19-2. C-7A configuration for jumping from the ramp.

19-9. DOOR JUMPING

When the troop doors are used, simultaneous exits must not be made. Number 1 jumper exits the starboard door and number 2 exits the port door 1 second after number 1. The remaining parachutists alternate in numerical order at 1-second intervals. (See Figure 19-3 for C-7A configuration for door jumping.)

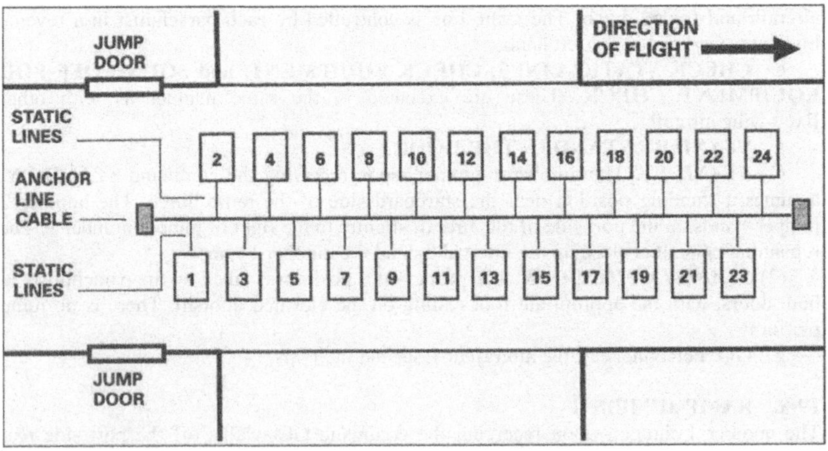

Figure 19-3. C-7A configuration for jumping from the doors.

19-10. SAFETY PRECAUTIONS

Safety precautions for the C-7A are as follows:

a. **Parachutists.**

(1) Parachutists ensure that all seats are secured in the up position when they stand to hook up. During extreme air turbulence, parachutists take a short bight on the static line and use the center anchor line to steady themselves.

(2) All parachutists remain off the ramp while it is being lowered to the 15-degree incline for aft end jumping. Parachutists walk down the ramp with feet spread wide to prevent striking the side of the aircraft. Upon exit from the aircraft, the parachutist brings the feet and knees together to form a tight body position. When following heavy equipment loads, parachutists exit between the roller conveyers of the aerial unloading kit.

b. **Jumpmaster.**

(1) The JM or safety ensures personnel are hooked up in an alternating manner to the same anchor line cable and form one continuous stick of jumpers.

(2) For door jumping, the JM or safety taps out the jumpers alternately to preclude a simultaneous exit from both sides of the aircraft.

(3) Normally, a safety is required on the aircraft, but if no safety personnel are in the aircraft, the JM jumps last. He must hook up to the center anchor line cable and exercise caution to control his own static line and ensure it does not become fouled.

NOTE: The left troop door may be removed before the operation to allow the JM to look for the DZ. If worn, the restraint harness is attached to the centerline anchor cable as a safety measure.

c. **Equipment.**

(1) When adjustable individual weapon cases are jumped from the doors, they must be reduced to 36 inches in length.

(2) When accompanying supplies and equipment are dropped from the doors, the bundles must be standard air delivery containers no larger than 40 by 24 by 36 inches.

(3) When ramp bundles are dropped, either the 15-foot static line with drogue or the breakaway static line may be used. When door bundles are dropped, the 15-foot static lines with drogues are used with cargo parachutes. The ramp roller conveyor section for the air unloading kit is installed on the port side of the ramp and is used to assist in ejecting the bundles from the ramp. Parachutists number 1 and number 2 push the bundles.

d. **Aircraft.**

(1) The speed of the aircraft during the jump will not be less than 90 nor more than 120 knots.

(2) When jumping from the doors, the crew chief must remove the doors and tape the rear portion of the door frames prior to takeoff.

(3) The rear tie-down ring, located beneath the tail section, should be removed prior to jumping. The ring can be unscrewed with a breaker bar or similar device.

Section III. C-23B/B+ SHERPA

The C-23B/B+ Sherpa is a twin-engine, nonpressurized, turboprop aircraft (Figure 19-4). It has a cruise speed of 180 knots with a range of 800 nautical miles. The aircraft can drop 12 combat-equipped parachutists or 16 Hollywood parachutists in the airdrop configuration. This is a base planning figure; actual troop capacity may vary due to aircraft limitations based on weight, density, altitude, and fuel loads. Troops may be loaded over the ramp or through the port side door.

This section outlines the procedures regarding the conduct of airborne operations by USASOC units from the C-23B/B+ Sherpa. The primary jump door for personnel and cargo is the rear ramp door. Only over-the-ramp procedures are authorized for static line personnel parachute operations. Only military free fall operations are authorized from the port side door. **Use of the port side door is *no longer authorized* for static line personnel jump operations.** Static line personnel drops using the port side jump door are authorized ***only for emergencies***.

Figure 19-4. C-23B/B+ Sherpa.

19-11. DROP PROCEDURES

The primary method for determining the exit point for jumpers for static line operations is using wind drift indicators (WDI). GMRS and VIRS can also be used based on the situation and the mission. The aircraft is also capable of a GPS release if the pilot is given the release point coordinates. A thorough briefing between the aircrew and all key personnel is mandatory before any operations involving the C-23B/B+. Standard drop altitude and speed is 1,500 feet AGL at 105 knots. Military free fall operations can be conducted up to 17,500 feet MSL.

19-12. SEATING CONFIGURATION

Parachutists are seated in two sticks along the port and starboard side of the aircraft (Figure 19-5). Numbers 1 through 8 are seated on the port side and numbers 9 through 16 are seated on the starboard side.

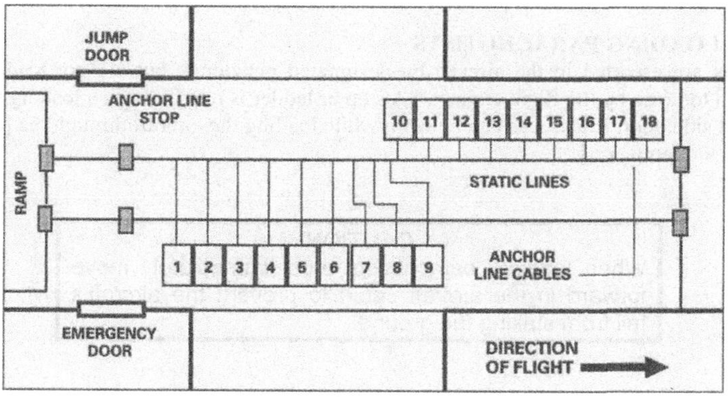

Figure 19-5. Seating configuration.

19-13. ANCHOR LINE CABLE ASSEMBLIES

There are two anchor line assemblies located overhead and running the length of the cabin down the center. The cables run from the reinforced anchor line attachment plate on the forward bulkhead to the anchor line connector at the center of the ramp hinge. Only the starboard side anchor line cable is used for personnel para-drop operations. Either cable can be used for cargo para-drop operations.

19-14. STATIC LINE RETRIEVAL SYSTEM

The static line retrieval system is a 5,000-pound winch located forward in the cabin on the floor and against the bulkhead. The retrieval cable runs up from the winch and along the port side anchor line cable and is attached to the starboard side anchor line cable forward of the anchor line stop. The retrieval system is operated from the rear of the cabin by the flight engineer. It is only operated in case of a towed jumper.

19-15. SUPERVISORY PERSONNEL REQUIRED

The supervisory personnel required for the C-23B/B+ Sherpa are as follows:

a. **Jumpmaster.** The jumpmaster will lead the stick out when jumping. The aircraft may be jumped using a nonjumping/static jumpmaster.

b. **Safety.** The safety will be nonjumping. Either one or two safeties may be used when jumping this aircraft.

c. **Flight Engineer.** The flight engineer is responsible for all operations in the cabin.

19-16. PREPARATION AND INSPECTION

The jumpmaster and the flight engineer jointly inspect the aircraft. The JM should follow basic aircraft inspection criteria as outlined in the JM handbook. At a minimum, inspection should include:

- Seat configuration and seatbelts (correct number and location).
- Static line retrieval cable and winch (attached correctly and secured).
- Jump lights (may have to wait until the aircraft is powered up).

19-17. LOADING PARACHUTISTS

Jumpers are escorted to the aircraft by designated personnel, but are not loaded until directed to do so by the flight engineer. A step or ladder is required when loading through the port side door. Rucksacks can be worn while loading the aircraft through the port side door or the ramp door.

> **CAUTION**
> When loading, parachutists must immediately move forward in the aircraft cabin to prevent the aircraft's tail from striking the ground.

a. **Cold Loading.** Cold loading is when the aircraft is shut down and the engines are not turning. Jumpers may be loaded over the rear ramp or the port side door. The C-23B/B+ has a double-hinged ramp that operates differently from conventional ramps on other aircraft. During ground operations, the ramp door can be lowered to its lowest position (resting on the ground) and equipment and jumpers can easily be loaded. The ramp can also be opened to the half-lowered position. The port side door with steps may also be used during cold loading.

b. **Hot Loading.** Hot loading is when the engines are turning. During multilift operation, the aircraft may be hot loaded in order to expedite the airborne operation. Jumpers may be loaded by either the ramp door or through the port side door. The ramp cannot be lowered to the ground during hot-load operations. After loading the aircraft, the jumpers must immediately take their seats and fasten their seatbelts. The safety ensures the jumpers are secured and signals the JM when completed. The JM then signals the flight engineer that he is ready for takeoff.

> **WARNING**
> **Hot loading aircraft is dangerous. Special control measures should be implemented to ensure jumpers and ground personnel remain clear of the propellers.**

19-18. JUMP COMMANDS AND TIME WARNINGS

Jump commands and times warnings are as follows:

a. **Jump Commands.**

- GET READY!
- PORT SIDE PERSONNEL, STAND UP!
- STARBOARD SIDE PERSONNEL, STAND UP! (if required).
- HOOK UP! (gates toward starboard skin).
- CHECK STATIC LINES!

- CHECK EQUIPMENT!
- SOUND OFF FOR EQUIPMENT CHECK!
- STAND BY! (30 seconds).
- FOLLOW ME! (jumping JM).
- GO! (static JM).

b. **Time Warnings.**

(1) ***20-Minute Warning.***

(a) *Flight Engineer.* Gives JM verbal and visual time warning.

(b) *Jumpmaster.* Gives time warning to jumpers.

(c) *Safety.* If jumping rucksacks, releases the cargo tie-down straps and facilitates the attachment of the jumpers' rucksacks.

(d) *Jumpers.* If jumping rucksacks, stand up; secure seatbelts and fold seats to the upright position and secure; attach rucksacks for jumping.

(2) ***10-Minute Warning.***

(a) *Flight Engineer.* Gives JM verbal and visual time warning.

(b) *Jumpmaster.* Hooks up his static line to the **starboard side** anchor cable. Hands static line to the safety. Begins jump commands.

(c) *Safety.* Takes static line from JM.

(d) *Jumpers.* All rigging completed.

(3) ***6-Minute Warning.***

(a) *Flight Engineer.* Gives JM verbal and visual time warning. After aircraft slow-down, the port side door and the rear ramp door are opened. When the rear ramp door and port side door are open and secure, the JM is signaled that the door and ramp are ready. Red light comes on.

(b) *Jumpmaster.* Continues giving jump commands. Upon completion of jump commands, moves to the port side door to begin spotting procedures, if required.

(c) *Safety.*

- Procedure with one safety: After the command of CHECK EQUIPMENT, hands the JM's static line to the JM and begins inspecting the jumpers from forward to the rear. After completing the inspection, he again secures the JM's static line. After the JM completes the jump commands, moves to the port side door with the JM and controls the JM's static line.
- Procedure with two safeties: Rear safety continues to maintain control of JM's static line while forward safety conducts standard jumper safety checks.

(d) *Jumpers.* Stand up and secure the seats. Upon the command of HOOK UP, jumpers hook up to the **starboard side** anchor line cable, gates facing starboard fuselage. Follow jump commands from JM. Take **standard bight** on static line.

(4) ***1-Minute Warning*** (all procedures).

(a) *Flight Engineer.* Gives verbal and visual time warning to the JM.

(b) *Jumpmaster.* Continues to monitor flight path from the port side door and identifies the DZ. Announces, "ONE MINUTE" to the jumpers.

(c) *Safety.* Controls JM's static line.

(d) *Jumpers.* Keep eyes on JM.

(5) ***30 Seconds*** (JM release procedures using GMRS or WDI).

(a) *Flight Engineer.* Ensures green light comes on.

(b) *Jumpmaster.* Checks green light. Tracks panels/exit point. Gives command of STAND BY.

(c) *Safety.* Controls JM's static line.

(d) *Jumpers.* Keep eyes on JM.

(6) ***30 Seconds*** (pilot release procedures using GPS/VIRS).

(a) *Flight Engineer.* Moves to starboard side of aircraft.

(b) *Jumpmaster.* Gives the command "STAND BY," turns toward the open ramp, takes control of his static line from the safety, and keeps eye on jump lights.

(c) *Safety.* Gives JM his static line and moves to starboard side of aircraft next to flight engineer.

(7) ***10 Seconds*** (JM release procedures).

(a) *Flight Engineer.* Ensures he is clear of the ramp.

(b) *Jumpmaster.* Keeps the panels in sight. Stands up and turns to face the ramp. Takes the static line from the safety. Waits until panels are 90 degrees from aircraft.

(c) *Safety.* Gives static line to the JM and moves to the starboard side of the aircraft.

(d) *Jumpers.* Keep eyes on JM.

(8) ***Exit*** (JM release procedures using GMRS/WDI).

(a) *Flight Engineer.* Ensures he is clear of the ramp.

(b) *Jumpmaster.* Identifies exit point, gives the command "FOLLOW ME," and walks straight out the ramp.

(c) *Safety.* Controls the jumper interval (one second). Retrieves static lines.

(d) *Jumpers.* Exit the ramp straight at one-second intervals.

(9) ***Exit*** (pilot release procedures using GPS/VIRS).

(a) *Flight Engineer.* Ensures he is clear of the ramp.

(b) *Jumpmaster.* When green light is illuminated, gives the command "FOLLOW ME" and walks out the ramp along the port side.

(c) *Safety.* Controls the jumper interval (one second). Retrieves static lines.

(d) *Jumpers.* Exit the ramp straight along the port side at one-second intervals.

CAUTIONS

Low overhead clearance may require jumpers to duck their heads while exiting the ramp door.

Jumpers must walk STRAIGHT out the ramp and along the port side fuselage and NOT at a 45-degree angle towards the center of the ramp.

Safety must remain clear of the ramp door and against the starboard side fuselage while jumpers exit.

19-19. CARGO OPERATIONS

The C-23B/B+ Sherpa is capable of both low level and high altitude cargo delivery operations. Bundle weight on the ramp should not exceed 500 pounds.

a. **Cargo Airdrops Without Personnel.** The following are procedures for cargo airdrops without personnel.

(1) The pilot sets up the approach, airspeed, and altitude. He commands the flight engineer to "STAND BY."

(2) The flight engineer ensures the bundle static line is hooked up and moves the cargo to the edge of the ramp.

(3) The pilot gives countdown to the flight engineer—"5, 4, 3, 2, 1, Now!"

(4) The flight engineer releases the cargo over the ramp and retrieves the static line/clevis.

b. **Cargo Airdrops with Personnel.** The following are procedures for cargo airdrops with personnel.

(1) The JM and safety coordinate and rehearse cargo release procedures with the flight engineer before the mission.

(2) Cargo may be released before or after jumpers exit.

(3) The safety retrieves static line/clevis after the drop.

NOTE: Based on the mission, the JM selects either the breakaway or nonbreakaway 15-foot static line.

CAUTIONS

No more than four jumpers are authorized aft of the port side jump door prior to exit.

Excessive weight load and cargo shift in the ramp door area prior to exit should be avoided.

19-20. MILITARY FREE-FALL OPERATIONS

All MFF operations are conducted IAW FM 31-19.

Section IV. C-27A (AERITALIA G-222)

The C-27A is a pressurized, medium transport aircraft developed from the Aeritalia G-222. It is a twin-engine, high-wing-mount, tailgate-equipped aircraft that is similar to a downsized C-130. The C-27A can carry 34 fully equipped combat troops, 28 static line parachutists, 34 military free fall (MFF) parachutists, or 16 MFF parachutists on oxygen. It can airdrop up to six CDS bundles. Typical internal loads are two HMMWVs or three full-sized 463L pallets that are turned sideways. Static line parachutists may be dropped using either of the two jump doors but may not use the ramp. MFF personnel may use both jump doors or the ramp.

19-21. SEATING CONFIGURATION

The seating configuration for the C-27A is as follows:

a. **Stick Configuration**. Parachutists are assembled into two sticks of jumpers. Jumper number 28 (the JM) is seated on the port side of the aircraft forward of the jump door. Forward of him is number 13, then numbers 1 through 12. Jumper number 27 (the AJM) is seated on the starboard side of the aircraft just forward of the jump door. Forward of him is jumper number 26; then jumpers 14 through 25. The safeties sit on each side to the rear of the jump doors.

b. **Anchor Line Cables**. There are two anchor line cable assemblies in the C-27A. The anchor line cables are from the attachment point on the forward bulkhead, through the anchor line support bracket just behind both doors, then to the side of the aircraft over the tailgate. Each parachutist is issued main and reserve parachutes. Each parachutist is responsible for inspecting his parachute for safety wires and for fitting of the parachute harness.

19-22. SUPERVISORY PERSONNEL REQUIRED

To ensure command and control when jumping one jump door, one jumpmaster, one nonjumping safety, and an airdrop-certified USAF loadmaster are required. These personnel requirements double when using both troop doors.

19-23. JUMP COMMANDS

Jump commands for the C-27A are as follows:

a. **Jump Commands**. The following nine jump commands are used whether the doors or the ramp are jumped.

(1) ***GET READY.*** Jumpers respond in the same manner as for other fixed-wing aircraft.

(2) ***PORT SIDE PERSONNEL, STAND UP.*** Jumpers on the left side of the aircraft stand up, raise and secure their seats, and face the ramp of the aircraft.

(3) ***STARBOARD SIDE PERSONNEL, STAND UP.*** Jumpers on the right side of the aircraft stand up, raise and secure their seats, and face the ramp of the aircraft.

(4) ***HOOK UP.*** The jumpers detach the static line snap hook from the top carrying handle of the reserve parachute and hook up to the anchor line cable with the open portion of the snap hook facing inboard, ensuring that the snap hook locks properly. The safety wire is inserted in the hole, pointing toward the rear of the aircraft, and folded down.

(5) ***CHECK STATIC LINES, CHECK EQUIPMENT, and SOUND OFF FOR EQUIPMENT CHECK.*** These commands are executed in the same manner as with other fixed-wing aircraft.

(6) ***STAND IN THE DOOR.*** A proper door position is taken by the parachutist.

(7) ***GO.*** Personnel exit the aircraft at 1-second intervals.

NOTE: Port side personnel exit first. After all port side personnel except the JM have cleared the aircraft, the starboard side personnel (except the AJM) exit the aircraft. JMs "clear to the rear" of the aircraft, the AJM exits, and the JM follows. All jumpers exit using the stand-in-the-door type door exit.

b. **Modification of Jump Commands**. At the 10-minute warning, the JM and AJM send jumpers number 13 and number 26 to the forward end of the aircraft to take their correct place in the stick. The two seats forward of both jump doors are folded upright and secured. The two safeties fold their seats upright and secure them.

19-24. SAFETY PRECAUTIONS

Safety precautions for the C-27A are as follows:

a. **Jumpmasters**. The JMs inspect the door platforms after the doors are opened. The JMs hook up to the cables on their side of the aircraft. They control and observe the personnel as they exit. JMs exit last.

b. **Safeties**. The safeties assist the loadmaster in installation of the door platforms if they are to be installed in-flight. They ensure personnel hook up **consecutively** and that jumpers number 13 and 26 are in the correct position. The safeties control the static lines as the jumpers approach the door to exit. They assist the loadmasters when retrieving the deployment bags.

c. **Equipment**. Standard combat equipment can be jumped out of the doors, which are 36 by 75 inches. Standard door bundles (that is, A-7A/A-21) can be dropped out of the doors. The 15-foot static line with drogue is used. Troops may follow.

d. **Aircraft**. The drop speed of the aircraft is 125 knots. Both doors **cannot** be jumped at the same time. The ramp of this aircraft cannot be used for static line ramp exits.

19-25. OVER-THE-RAMP OPERATIONS

Considerations for over-the-ramp operations follow.

a. **Static Line Operations**. The erratic behavior of the deployment bags poses a serious safety hazard; the C-27A **cannot** be used for static line over-the-ramp operations.

b. **Equipment Drop**. Door bundles can be pushed off the ramp. The rollers can be installed on the ramp to aid in handling larger bundles.

c. **Military Free Fall**. MFF exits can be made over the ramp when both doors are closed. The C-27A will hold 34 MFF jumpers; however, it is recommended the number of MFF jumpers be limited to 16 when oxygen consoles are installed, due to overcrowding in the aircraft. The using unit must provide two console positions for the loadmasters to use during MFF jumps above 10,000 feet. It is very difficult for the JM to spot the release point from the aircraft during ramp exits. Therefore, the JM should not wear an ALICE pack for this type operation, and the unit should use a nonjumping JM.

19-26. JOINT PREFLIGHT INSPECTION

The C-27A is inspected as follows:

a. **JM/Aircraft Commander/Loadmaster Coordination**. The JM is responsible for informing the aircraft commander and loadmaster of the exact time sequence of prejump procedures. Following preliminary orientation, and before loading personnel onboard the aircraft, the JM and loadmaster make a joint inspection of the aircraft. The purpose of the joint inspection is to verify the readiness of the aircraft for the conduct of the mission and to take actions necessary to achieve this readiness.

b. **Exterior Inspection**. An external inspection of the aircraft is made to detect hazards to the airdrop of personnel. Particular attention is directed to those areas to the rear of the aft paratroop doors. Any protruding objects and sharp edges are removed, or padded and taped.

c. **Interior Inspection**. An interior inspection checks for the following:

(1) Any sharp edge or protrusion is securely taped and padded, as required.

(2) All equipment in the cargo compartment is securely stowed and lashed.

(3) The floor is clean and free of lubricants; no obstructions are on the walkway or along the paratroop exit route (outboard area between the safety fence and fuselage).

(4) Anchor line cables are installed.

(5) A seat and seat belt are available in the troop compartment for each parachutist.

(6) The retrieval system is installed in the aircraft with the winch cable retained in clips and free of the anchor line cable.

(7) Windscreen systems are available/installed.

(8) Jump platforms are available/installed.

(9) Jump caution lights are operational.

(10) Troop compartment lights are operational.

d. **Jumpmaster Preload Inspection of Parachutists**. The JM/AJM inspects each parachutist, parachute, and parachutist's equipment prior to loading the aircraft.

19-27. LOADMASTER BRIEFING

As soon as all parachutists are seated, the loadmaster briefs them on aircraft safety, emergency procedures, and comfort facilities.

19-28. TIME WARNINGS

Time warnings are as follows:

a. **20-Minute Warning**. JMs check personnel and equipment. Missile jump packs are attached to the parachutists, their HSPR leg straps are secured, and door bundles are moved near the personnel doors.

b. **10-Minute Warning**. Final onboard JMPI of all parachutists is complete. A verbal and visual 10-minute warning is given to the JMs by the loadmaster, and the JMs begin jump commands.

c. **Slow-Down Warning**. About 3 minutes from drop time, the jump commands are completed. Personnel doors are opened, and jump platforms are extended and locked.

d. **1-Minute Warning**. JMs alert troops and make safety checks from personnel doors. The AJM informs the JM that his side is clear and that it is safe to jump.

e. **10-Second Warning**. The loadmaster gives the JMs a visual 10-second warning. At this time the JM gives the command STAND IN THE DOOR, and the number 1 jumper assumes a proper door position. The remainder of the stick shuffles aft to close up the stick.

f. **Green Light, GO**. At the green light, the JM taps out the first man. Port side personnel exit first. After all port side personnel have cleared the aircraft, the starboard side personnel exit the aircraft. JMs "clear to the rear" of the aircraft, the AJM signals to the JM that all jumpers are clear of the aircraft, the AJM exits, and then the JM exits. All jumpers exit using the stand-in-the-door exit.

19-29. ADDITIONAL SAFETY PRECAUTIONS

Further safety precautions for the C-27A follow.

a. **Platforms, Air Deflectors, Aft Cargo Door**. The JMs must ensure that jump platforms and windscreens are available. This equipment is mandatory for each aft personnel door that is to be used.

b. **Door Bundles**. When personnel follow door bundles, the door bundle static line will be outfitted with a drogue.

c. **Movement into the Door**. Parachutists exercise caution as they move to the door to avoid becoming entangled with the static lines of preceding parachutists. This precautionary action may slow movement into and out of the door.

19-30. C-27A JUMPMASTER CHECKLIST

The jumpmaster follows this checklist.

a. **Seats**.

- Adequate seats for troop load are onboard.
- All seats have safety belts.
- Seat backs are secure.
- Seats are serviceable.
- There are no projections through seats; pairs of seats forward of each troop door have a strap attached to secure them in the upright position.

b. **Floor**.

- Nonskid surface covering is in good condition.
- Floor is clean and safe to walk on.
- Roller conveyors are stored.
- Loose equipment is secured in the cargo ramp area and does not interfere with troops.
- Equipment tie-down rings are depressed into their recesses.

c. **Jump Platforms**.

- Nonskid surface covering is present and in good condition.
- There are no cracks or bends.
- Studs are locked in seat track receptacles.
- Tie-down fitting is locked.
- All bolts and nuts are present.
- Platforms swing in and out easily.

d. **Jump Doors**.

(1) ***Ground Check.***

- There are no sharp or protruding edges on door frames.
- Doors open and close easily.

(2) ***Prior to Exit.***

- The platforms are locked into the two "keyholes" on the floor and slid to the rear of the aircraft. The large portion of the keyhole slot should be visible.
- The platform locking lever on the leading edge of the door should be in its locked position. The lug this lever controls should be engaged to the door frame.
- The platform locking lever should be taped in place to help prevent any jumpers from inadvertently unlocking it.
- The flange on the trailing side of the platform must overlap the inside of the door frame approximately 1/2 inch.

e. **Jump Lights** (five total).

- Rear at the forward left door.

- Rear of both troop doors.
- High above and to the rear of the ramp on both sides.

f. **Static Line Anchor Cable System.**
- Forward end of cable is firmly secured to bracket on bulkhead with three threads showing on turnbuckle.
- Rear of cable has a pin in it with tape.

g. **Anchor Cable.**
- Cable has no breaks.
- Cable has no frays.
- Cable has no kinks.
- Cable is clean and free of rust.
- Static line stop is present.
- Support bracket at the trailing edge of the door is locked in place to support the cables.

h. **Static Line Retrievers.**
- Motor is operational.
- Cable is secured to ceiling with one turn of double 1/4-inch cotton webbing.
- The Y attachment is in place so one retrieval cable can pull in two groups of deployment bags. (Y cable is used if both jump doors are used.)
- Retriever cables are not broken, frayed, or kinked. They are clean and free of rust.
- Phenolic block/anchor cable spool is installed on both anchor cables with the pull ring on the forward end.
- Retriever equipment is available.

i. **Emergency Equipment.**
- First aid kit is onboard (one).
- Fire extinguishers are onboard (two).
- Alarm system is operational.
- Emergency exits are operational and accessible.
- Sufficient emergency parachutes are available.

j. **Miscellaneous.**
- Day lighting system is operational.
- Night lighting system is operational.
- Airsickness bags are available.
- JM kit (extra equipment) is onboard.
- Earplugs are available.
- Heavy tape is available to secure the platform and windscreen locking lever.
- If jump platforms and windscreen are not installed in the doors, they must be secured to the upper ramp.
- All equipment and crew baggage is secured to the floor.
- During the jump briefing, the jumpers are warned to avoid striking or grabbing the door platform or windscreen locking lever on the leading edge of the door.

> **WARNING**
> **It is a serious hazard to the exiting parachutist if the windscreen locking lever swings into the open door.**

k. **Tailgate Drops (MFF and Bundles Only).**
- Ensure that the loadmaster installs the stops on both sides of the tailgate so that it will be level with the aircraft floor when open.
- Disengage the support bracket near the door for bundle drops that use the retrieval system to pull in the static lines.
- Secure the retrieval cable against the anchor line in several places with breakaway ties starting at the rear of the cable and ending at the tailgate hinge. This will prevent the tailgate from cutting the retrieval cable during operation.

l. **Aircraft Slow-Down Warning at 3 Minutes.**
- Doors are opened and locked in place.
- Air deflectors are extended.
- Jump platforms are locked in place.

m. **Loadmaster/JM Safety Checks.**
- Door bundles employ approximately 15-foot-long static lines with three drogue parachutes.
- Jump platform is secure and will sustain parachutist's weight.

Section V. C-46 COMMANDO/C-47 SKYTRAIN

The C-46 and the C-47 are twin-engine, short-range transport aircraft. Because of the similarity in characteristics and procedures, both aircraft are discussed in this section.

19-31. SEATING CONFIGURATIONS

Seating configurations for the C-46 and C-47 are as follows:

a. **C-46.** A total of 27 parachutists can be jumped from the C-46 using both aft troop doors. This aircraft can accommodate two sticks: a 14-man stick sits on the starboard side, and a 13-man stick sits on the port side.

b. **C-47.** A total of 24 parachutists can be jumped from the C-47 using the aft troop door. This aircraft can accommodate two sticks: a 12-man stick sits on the starboard side and a 12-man stick sits on the port side.

19-32. JUMP PROCEDURES

Jump procedures for the C-46 and C-47 are as follows:

a. **C-46 Jump Commands.**

(1) ***GET READY.*** Jumpers respond in the same manner as for other fixed-wing aircraft.

(2) ***STAND UP.*** The parachutists near the doors stand and steady themselves by firmly grasping a structural member of the cargo compartment wall.

(3) ***HOOK UP.*** The open portion of the static line snap hook is away from the parachutist and toward the floor of the aircraft when he hooks up. The elbow of the arm holding the static line is kept close to the body. The static line is controlled by each parachutist in a reverse bight at waist level in the left hand. When the anchor line cable is installed along the top of the aircraft, the static line is controlled in a standard bight held at shoulder level.

(4) ***CHECK STATIC LINES.*** Each jumper checks the man in front to ensure the static line snap hook cover on the 5-foot static line extension covers the snap hook.

(5) ***CHECK EQUIPMENT.*** Each jumper checks his equipment and checks the man in front to ensure that the man's elbow is close to his body and the static line extension hangs below and behind the arm.

(6) ***SOUND OFF FOR EQUIPMENT CHECK.*** This command is executed in the same manner as with other fixed-wing aircraft.

(7) ***STAND IN THE DOOR.*** A proper door position is taken by the parachutists in both doors.

(8) ***GO.*** Personnel exit the aircraft at 1-second intervals.

b. **C-47 Jump Commands**.

(1) ***GET READY.*** Jumpers respond in the same manner as for other fixed-wing aircraft.

(2) ***PORT SIDE PERSONNEL, STAND UP.*** The odd-numbered jumpers seated on the left side of the aircraft stand up, secure their seats in the down position, and face the rear of the aircraft.

(3) ***STARBOARD SIDE PERSONNEL, STAND UP.*** The even-numbered jumpers seated on the right side of the aircraft stand up, secure their seats in the down position, and face the rear of the aircraft.

(4) ***HOOK UP.*** The even-numbered jumpers hook up between the odd-numbered jumpers to form a staggered stick of jumpers. The static line is controlled by each parachutist in a reverse bight at waist level in the left hand.

(5) ***CHECK STATIC LINES.*** Each jumper checks the man in front to ensure the static line snap hook cover on the 5-foot static line extension covers the snap hook.

(6) ***CHECK EQUIPMENT.*** Each jumper checks the man in front to ensure that the man's elbow is close to his body and the static line extension hangs below and behind the arm.

(7) ***SOUND OFF FOR EQUIPMENT CHECK.*** Jumpers respond to this command in the same manner as with other fixed-wing aircraft.

(8) ***STAND IN THE DOOR.*** A proper door position is taken by the parachutists in both doors.

(9) ***GO.*** Personnel exit the aircraft at 1-second intervals.

19-33. SAFETY PRECAUTIONS

Safety precautions for both aircraft are as follows:

a. Parachutists ensure all seats are in the **down** position when they stand to hook up.

b. Upon exit from the aircraft, parachutists bring their feet and knees together to form a tight body position.

19-34. SAFETY PERSONNEL AND JUMPMASTER RESPONSIBILITIES

Responsibilities for the safety and jumpmaster follow.

a. **C-46.**

(1) The JM or safety ensures personnel hook up in an alternating and consecutive fashion.

(2) One safety is required when jumping the C-46. After the safety has checked each parachutist, the safety then moves aft of the door and physically controls the JM's static line.

(3) Due to the movement of the safeties and length of the aircraft, it is advisable to use a USAF BA-18 emergency parachute when using one safety for each door. When using the safety harness, two safeties are required for each door. Safety number 1 is positioned forward to inspect jumpers. Safety number 2 is positioned aft to assist the JM.

NOTE: Although the C-46 has two aft troop jump doors, normally only the **left** aft troop door is used for jumping. When both doors are used, one safety for each door is required. The JM (or safety) taps out parachutists alternately at one-second intervals to preclude simultaneous exits from both sides of the aircraft.

(4) Normally, the JM exits first when jumping this aircraft. When the JM gives the command, STAND IN THE DOOR, he moves to the door and assumes a proper door position. On the command GO, the JM looks at the number 2 man, gives the command GO, and then exits the aircraft. Once the JM exits, the safety controls the static lines and the parachutists' flow out of the aircraft.

b. **C-47.**

(1) The JM or safety ensures personnel hook up consecutively.

(2) One safety is required when jumping the C-47. After the safety has checked each parachutist, the safety then moves aft of the door and physically controls the JM's static line.

(3) Due to the movement of the safeties and length of the aircraft, it is advisable to use a USAF BA-18 emergency parachute when using one safety. When using the safety harness, two safeties are required. Safety number 1 is positioned forward to inspect jumpers. Safety number 2 is positioned aft to assist the JM.

(4) The JM must exit first when jumping this aircraft. When the JM gives the command STAND IN THE DOOR, he moves to the door and assumes a proper door position. On the command GO, the JM looks at the number 2 man, gives the command GO, and then exits the aircraft. Once the JM exits, the safety controls the static lines and the parachutists' flow out of the aircraft.

c. **Parachute Fouling.** To prevent fouling of the T-10-series or MC1-series type parachute on the aircraft when jumping the C-46 or C-47, a 5-foot static line extension must be used. This extension has a snap hook at one end and a nondetachable connector link at the other end. The connector link on the extension is attached to the snap hook of the parachute static line. The snap hook on the parachute static line is safetied with a short piece of wire covered with a canvas duck sleeve and taped in place. The extension is stowed by using rubber retainer bands, continuing to stow the extension on the pack body. When the stow is completed, the static line and extension should have four stows on the right and three stows on the left.

NOTE: When jumping the C-46 utilizing the MC1-series, parachutists use only the left aft door.

d. **Anchor Line Cables.**

(1) The C-46 has two anchor line cables. Both anchor line cables must be detached from their floor fittings and anchored to the floor in the rear section of the cargo compartment. A wooden block is used to support the cable at the anchor point.

(2) The C-47 has one permanently installed anchor line cable that must be secured to the overhead attachment points provided in the center of the aircraft.

- The aft edge of the jump door is lined with a smooth metal tubular stripping, or it is padded and taped.
- The flooring of the jump door is made smooth by the insertion of an additional plywood section to butt against the tubing and existing flooring.
- If aft troop exit doors are installed, they are opened at the 20-minute warning.
- A slowdown from cruise airspeed (when applicable) is initiated in enough time to allow drop-speed cruise two minutes prior to drop time.
- Personnel and cargo drops are normally made from 105 to 125 knots.
- When possible during personnel drops, propeller RPM should be reduced to lessen the blast effect on the jumpers.

Section VI. DC-3 (CONTRACT AIRCRAFT/CIVILIAN SKYTRAIN)

The DC-3 is the civilian version of the C-47 Skytrain. Like the C-47, the DC-3 is a twin-engine, short-range transport aircraft.

19-35. SEATING CONFIGURATION

The seating configuration on the DC-3 is as follows:

a. A total of 24 parachutists can be jumped from the DC-3 using the aft troop door. This aircraft can accommodate two sticks: a 12-man stick sits on the starboard side, and a 12-man stick sits on the port side.

b. Contract DC-3 aircraft are not rigged with paratroop seats or with individual seat belts. Jumpers are required to sit on the floor. They are restrained by one safety belt over the entire stick of jumpers. USAF 10,000-pound tie-down straps or C-3A (NSN 1670-00-447-9504) modified safety belts must be supplied by the using unit.

19-36. JUMP COMMANDS AND PROCEDURES

Jump commands and procedures for the DC-3 follow.

a. **GET READY**. Jumpers respond in the same manner as for other fixed-wing aircraft.

b. **PORT SIDE PERSONNEL, STAND UP**. The odd-numbered jumpers seated on the left side of the aircraft stand up.

c. **STARBOARD SIDE PERSONNEL, STAND UP**. The even-numbered jumpers seated on the right side of the aircraft stand up.

d. **HOOK UP**. The open portion of the static line snap hook is toward the port side of the aircraft. The elbow of the arm holding the static line is kept close to the body. The static line is controlled by each parachutist in a reverse bight at waist level in the left hand. The odd-numbered jumpers hook up first, then the even-numbered jumpers hook up between the odd-numbered jumpers to form a staggered stick of jumpers.

e. **CHECK STATIC LINES**. Each jumper checks the man in front to ensure the static line snap hook cover on the 5-foot static line extension covers the snap hook.

f. **CHECK EQUIPMENT**. Each jumper checks the man in front to ensure that the man's elbow is close to his body and the static line extension hangs below and behind the arm.

g. **SOUND OFF FOR EQUIPMENT CHECK**. Jumpers respond in the same manner as with other fixed-wing aircraft.

h. **STAND IN THE DOOR**. A proper door position is taken by the parachutists.

i. **GO**. Personnel exit the aircraft at 1-second intervals.

19-37. SAFETY PRECAUTIONS

Safety precautions for the DC-3 are as follows:

a. **Parachutists**.

(1) Jumpers hook up using a reverse bight, with the elbow of the arm holding the static line kept close to the body.

(2) Upon exit from the aircraft, jumpers bring the feet and knees together to form a tight body position.

b. **Safety Personnel Duties**.

(1) When using DC-3 aircraft, the safety also performs duties as a loadmaster. Prior to takeoff, the safety ensures all jumpers are secured and prepared for takeoff.

(2) After he has checked each parachutist, the safety moves to the aft end of the aircraft, aft of the door, and physically controls the JM's static line.

(3) The safety maintains communications with the pilots through the ICS located in the aft of the aircraft, and relays all information to the JM.

(4) Due to the movement of the safeties and length of the aircraft, it is advisable to use an USAF BA-18 emergency parachute when using one safety. When using the safety harness, two safeties are required. Safety number 1 is positioned forward to inspect jumpers. Safety number 2 is positioned aft to assist the JM.

c. **Jumpmaster**. The JM must exit first when jumping this aircraft. When the JM gives the command, STAND IN THE DOOR, he moves to the door and assumes a proper door position. On the command GO, the JM looks at the number 2 man and gives the command GO, and then he exits the aircraft. Once the JM exits, the safety controls the static lines, the parachutists' flow out of the aircraft, and the jumper interval.

d. **Anchor Line Cables**. The permanently installed anchor line cable must be secured to the overhead attachment points provided in the center of the aircraft. On the C-47, the aft anchor point for the cable is located at the aft right side of station number 542. When the anchor line cable anchor point on the DC-3 is at station number 542, all procedures relevant to the C-47 (Section V) apply.

e. **Jump Door**.

(1) The aft DC-3 cargo door must be removed and rigged for jumping prior to takeoff.

(2) The aft edge of the cargo/jump door is rigged with a smooth metal tubular stripping or padded and taped.

(3) The two aft cargo door hinges, door hasp, and the door knob are padded and taped.

Section VII. C-212 (CASA 212)

The C-212 is a twin-engine, high-wing, multipurpose light transport designed for operations involving short, rough airfields (Figure 19-6). The aircraft can transport 24 personnel in a troop lift mode and 15 combat-equipped parachutists in the airdrop mode using the port door. Troops are loaded over the ramp.

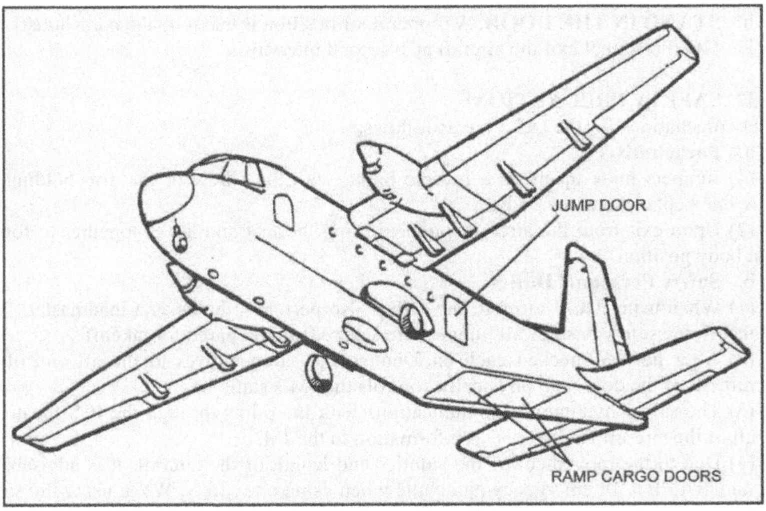

Figure 19-6. C-212 (Casa 212).

19-38. SEATING CONFIGURATION

Fifteen parachutists are seated in two sticks of jumpers (Figure 19-7). The odd-numbered personnel (eight) are seated on the starboard side and the even-numbered personnel (seven) are seated on the port side.

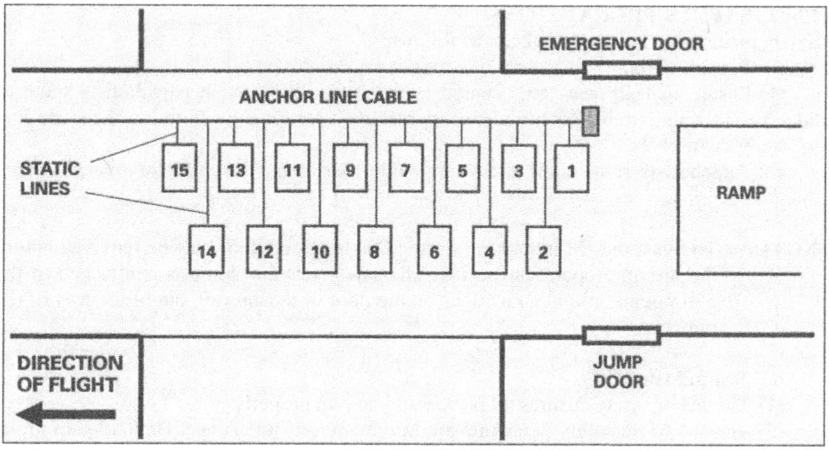

Figure 19-7. C-212 seating configuration.

19-39. ANCHOR LINE CABLE ASSEMBLY

There is one anchor line cable assembly in the C-212. It runs from the reinforced anchor line attachment plate on the forward bulkhead to the anchor line connector near the right side of the aft starboard emergency door.

19-40. SUPERVISORY PERSONNEL REQUIRED

The following personnel are required for airdrop operations from the C-212: one jumpmaster to perform standard aircraft check procedures, one safety, and one loadmaster/crew chief.

19-41. JUMP COMMANDS

The following jump commands are used with the C-212 aircraft.

a. **GET READY.**

b. **STARBOARD SIDE PERSONNEL, STAND UP.**

c. **PORT SIDE PERSONNEL, STAND UP.**

d. **HOOK UP.** On this command, the odd-numbered personnel hook up between the even-numbered personnel to form a continuous stick of parachutists, hooking the open portion of the snap hook facing inboard over the left shoulder. All parachutists take up a reverse bight.

e. **CHECK STATIC LINES, CHECK EQUIPMENT, and SOUND OFF FOR EQUIPMENT CHECK.** These commands are executed in the same manner as with other fixed-wing aircraft.

f. **STAND IN THE DOOR (Door)/STAND BY (Ramp).** A proper exit position is taken by the parachutist.

g. **GO.** Personnel exit the aircraft at one-second intervals.

19-42. SAFETY PRECAUTIONS

Safety precautions for the C-212 are as follows:

a. **Parachutists**.

(1) Ensure that all seats are secured in the **up** position when parachutists stand to hook up. During extreme air turbulence, parachutists take a short bight on the static line to steady themselves.

(2) Parachutists remain off the ramp while it is being lowered for over-the-ramp operations.

NOTE: To assist the JM in looking for the DZ, the troop door may be removed before the airborne operation begins. The safety restraint harness is attached to the 500-pound tie-down positions on the floor of the aircraft, out of the way of the jumpers.

b. **Jumpmaster**.

(1) The JM or safety ensures all personnel hook up properly.

(2) The JM (if no safety personnel are in the aircraft) jumps last. He hooks up to the anchor line cable, ensuring his static line does not become fouled.

NOTE: On aircraft that do not have a positive communication system, the following safety measure is recommended: one ring on the alarm bell signals the JM to look at the jump light or communicate with the cockpit.

c. **Equipment**.

(1) When adjustable individual weapons cases are jumped from the door, they must be reduced to 36 inches in length.

(2) When accompanying supplies and equipment are dropped from the door, the bundles must be standard air delivery containers no larger than 40 by 24 by 36 inches.

(3) When ramp bundles are dropped, either the 15-foot static line with drogue or the breakaway static line may be used. When door bundles are dropped, the 15-foot static line with drogue is used with cargo parachutes.

(4) When ramp bundles are dropped, troops may follow out the troop door. The JM and safety or loadmaster push the bundles out.

d. **Aircraft**.

(1) Aircraft speed during the jump is 90 to 110 knots.

(2) When parachutists are jumping from the troop door, the door may be opened or removed and set into the door recess provided on the ramp.

(3) When conducting bundle operations from the ramp, the JM must close the door. The door may be opened or removed before the ramp is lowered.

(4) The Omega antenna, located beneath the tail section, must be removed prior to ramp bundle operations.

19-43. TOWED PARACHUTIST PROCEDURES

Procedures outlined for other fixed-wing aircraft will be followed for the C-212.

19-44. AIRCRAFT CONFIGURATION FOR RAMP STATIC LINE PERSONNEL AIRDROP

The loadmaster configures the aircraft. The JM should verify the configuration. Static line ramp parachute operations are authorized only when the retrieval system is operational.

a. The aircraft is configured for a static line personnel airdrop. One of each of the following items of equipment is needed:

(1) Hand winch.
(2) Static line deflector block.
(3) Retrieval bar.
(4) Retrieval strap.
(5) Extended interphone cord.
(6) 2,500-pound tie-down strap.
(7) 5,000-pound tie-down strap.
(8) One 3-foot length of 1-inch tubular nylon.
(9) Cloth-backed adhesive tape.
(10) Anchor cable.
(11) Two restraint harnesses.

b. Install and preflight-inspect the following equipment:

(1) Attach hand winch to right tie-down row in zone 1 and check for security.
(2) Inspect cable for broken wires or kinks and check for operation.
(3) Ensure static line deflector block is attached to the right side of the ramp. Cover the bolt head with tape.
(4) Inspect retrieval base on board and attaching brackets.
(5) Install and check extended interphone cord for operation.
(6) Fit and adjust restraint harnesses.
(7) Ensure that the 3-foot length of 1-inch tubular nylon and the 5,000-pound tie-down strap are secured and available for immediate use.

19-45. C-212 JUMPMASTER CHECKLIST

The jumpmaster follows this checklist for the C-212:

a. **Seats**.
- Adequate seats are available for troop load.
- All seats have safety belts.
- Seat backs are secure.
- Seats are serviceable.
- There are no projections through seats.

b. **Floor**.
- Nonskid surface covering is in good condition.
- Floor is clean and safe to walk on.
- Loose equipment is secured and does not interfere with troops.

c. **Jump Door**.
- There are no sharp or protruding edges on door frame.
- Door opens and closes easily.
- Door sits in ramp recess properly.

d. **Jump Lights**. Check sets for operation.

- Set 1—above port aft jump door.
- Set 2—above starboard aft emergency door. Check alarm bell; it is the signal for exiting.

e. **Static Line Anchor Cable System.**

(1) *Forward support beam.*

- Bolts, nuts, and safety wire are present.
- Anchor cable is attached to centerline anchor point.
- Cable bolt, locking bolt, nut, and safety wire are present.
- Check anchor line tension indicator—red line indicator should not be seen.

(2) *Anchor cable.*

- Cable has no breaks, frays, or kinks.
- Cable is clean and free of rust.
- Swage is present.

(3) *Anchor line cable aft support.*

- Cable, locking bolt, nut, and safety wire are present.

f. **Emergency Equipment.**

- First aid kits are onboard (2).
- Fire extinguishers are onboard (2).
- Alarm system is operational.
- Sufficient emergency parachutes are available.

g. **Miscellaneous.**

- Lighting system is operational.
- Airsickness bags are available.
- JM kit (extra equipment) is onboard.
- Earplugs are available.

NOTE: Loose equipment and jump door (removed) are lashed to the cargo ramp or to the rear of the forward bulkhead.

PART FIVE
Drop Zones

CHAPTER 20
PROCEDURES ON THE DROP ZONE

A drop zone is any designated area where personnel and equipment may be delivered by means of parachute or free drop. The DZ is located where it can best support the ground tactical plan; it is selected by the ground unit commander. For tactical training, the USAF assault zone availability report should be checked for an approved DZ within the tactical area. If the selected DZ is not on the AZAR, a tactical assessment must be conducted.

Section I. DROP ZONE SELECTION AND METHODS

The GUC uses the tactical analysis to select an area that can best support his mission. This section discusses several technical selection factors that must be considered.

20-1. AIR DROP AIR SPEED

Table 20-1 provides recommended drop speeds for various aircraft.

TYPE OF AIRCRAFT	DROP SPEED
UH-1	50 to 70 knots—optimum 70 knots
UH-60	65 to 75 knots—optimum 70 knots
CH-46 (USMC)	80 to 90 knots
CH-47	80 to 110 knots—optimum 90 knots
CH-53 (USMC)	90 to 110 knots
CH/HH3 (USAF)	70 to 90 knots
C-5/130/141/17/KC-130	130 to 135 knots (personnel)
C-5/130/141/17/KC-130	130 to 150 knots—optimum 130 knots for all loads (door bundles, CDS, and heavy equipment)

Table 20-1. Aircraft drop speeds.

20-2. AIRCRAFT DROP ALTITUDES

See Table 20-2 for aircraft drop altitudes.

AIRCRAFT	DAY (AGL) (feet)	NIGHT (AGL) (feet)
Rotary Wing		
(All Services, Rotary Wing))		
Personnel < 90 KTs	1,500	1,500
> 90 KTs	1,500	1,500
Bundles	300	300
Fixed Wing (Troop Carrier)*		
Personnel		
Basic Airborne Training	1,250	1,250
*Training***	1,000	1,000
Tactical	800	800
Door Bundles	300	500
Heavy Equipment	1,100	1,100

* AWADS/SKE – Drop altitude is 500 feet above highest obstacle that falls within 3 miles either side of DZ run-in. CDS using G-13 or any other parachute, minimum drop altitude is 400 feet AGL. CDS using G-14 parachute can drop from 300 feet AGL.

** The training drop altitude of 1,000 feet AGL may be waived to 800 feet AGL by completing a mature risk assessment decision cycle.

Table 20-2. Aircraft drop altitudes.

20-3. TYPE OF LOAD

Type of load includes personnel (one-second drop interval between jumpers), CDS, CRRC, door bundles, heavy equipment (three-second exit interval), or heavy equipment followed by personnel (three seconds per CDS/CRRC/heavy equipment and one second for each personnel).

> **CAUTION**
> There must be a three-second interval between equipment drops and the exit of parachutists to avoid possible jumper entanglement. The DZSO or DZSTL must follow the procedures for heavy drop operations, but observe the jumpers as they exit the aircraft.

a. **Obstacles**. To ensure that the airdrop is safe and that equipment and personnel can be recovered or employed to accomplish the mission, the drop zone and adjacent areas should be free of obstacles. Examples of obstacles are—

- Trees 35 feet or higher impeding recovery of personnel or equipment.
- Water 4 feet deep within 1,000 meters from any edge of the DZ.

- High tension wire that is carrying active current of 50 volts or greater. (Should be turned off prior to drop.)
- Any other conditions that may injure parachutists or damage equipment (inactive electric wires, barbed wire fences, swamps, ditches, gullies, and so forth).

NOTE: See Appendix D for DZ risk assessment.

b. **Air Approach and Departure Routes**. Air routes to and from the DZ should not conflict with other air operations or restrictive terrain, or with manmade objects (television or radio towers).

20-4. METHODS OF DELIVERY

Different drop methods are described as follows:

a. **High Velocity**. A drogue chute stabilizes and keeps equipment upright but does not slow the descent (for example, a 12-foot, high-velocity [HV] parachute on a door bundle).

b. **Low Velocity**. A parachute slows the rate of descent for a soft landing (for example, personnel and cargo parachutes).

c. **Free Drop**. This is cargo that has no device to stabilize or slow the rate of descent (for example, durable items such as clothing bundles).

20-5. ACCESS TO AREA

The unit must have access to and from the DZ to recover equipment or conduct troop movement. DZs with no roads leading to them or next to a river with no bridges are examples of impeded access to areas.

20-6. SIZE

The following information provides minimum peacetime sizes when using fixed-wing aircraft and must be adhered to unless a waiver is issued. During contingency or wartime missions, DZ sizes may also be waived. However, size requirements remain a joint responsibility of the COMALF and the airborne commander.

NOTE: To convert yards to meters, multiply yards times .9144 (yards x .9144). To convert meters to yards, divide meters by .9144 (meters ÷ .9144).

a. **Personnel from Military Rotary-Wing and Small Fixed-Wing Aircraft**. The distance required for personnel is determined by the use of the formula D=RT.

b. **GMRS, VIRS, or WSVC DZ**. Minimum size is 275 meters by 275 meters.

c. **Personnel from Fixed-Wing C-130 and Larger Aircraft Using a CARP DZ**. The ground space required is 550 meters by 550 meters for one parachutist from a single aircraft (add 70 meters to the length for each additional parachutist).

(1) For drop altitudes above 1,000 feet AGL, add 28 meters (14 meters each side) to the width and 28 meters (trail edge) for each additional 100 feet.

(2) From official sunset to sunrise, add 90 meters to the width (45 meters each side) and length (both ends) for visual drops.

(3) For visual formation, add 90 meters (45 meters each side) to the width.

(4) For AWADS and SKE, add 370 meters (185 meters each side) to the width.

d. **Heavy Equipment Drops from Fixed-Wing Aircraft Using a CARP DZ**. The ground space required is 550 meters by 915 meters for one platform from a single aircraft.

(1) For a C-130, add 370 meters to the length (trail edge) for each additional platform.

(2) For a C-141, add 460 meters to the length (trail edge) for each additional platform.

(3) For drop altitudes above 1,100 feet AGL, add 28 meters to the width (14 meters each side) and to the length (trail edge).

(4) From official sunset to sunrise, add 90 meters to the width (45 meters each side) and to the length (45 meters each end) for visual drops.

(5) For visual formation drops, add 90 meters to the width (45 meters each side).

(6) For AWADS and SKE, add 370 meters to the width (185 meters each side).

e. **Containerized Delivery System (CDS) Drops for the C-130 Using a CARP DZ**. The ground space required is 370 meters by 370 meters for one container from a single aircraft (182.5 meters each side).

f. **Containerized Delivery System Drops for the C-141 Using a CARP DZ**. The ground space required is 410 meters by 540 meters for one container from a single aircraft.

(1) For altitudes above 600 feet, add 37 meters for each additional 100 feet to the width (18.5 meters each side) and to the length (trail edge).

NOTE: Altitudes above 1,000 feet are not recommended.

(2) From official sunset to sunrise, add 90 meters to the width (45 meters each side) and to the length (45 meters each end) for visual drops.

(3) For visual formation drops, add 90 meters to the width (45 meters each side).

(4) For AWADS and SKE, add 370 meters to the width (185 meters each side).

NOTE: The size of the CDS drop zone depends on drop altitude, number of bundles, formation, and type of aircraft. (Refer to AFI 13-217.)

Section II. AIRDROP RELEASE METHODS AND PERSONNEL

The number and type of aircraft that air-delivers personnel and equipment, using one of the four methods, usually dictate the type and composition of the ground support party, which can be tailored for a mission.

NOTE: The unit mission request for aircraft specifies the type of drop method to be used, such as CARP, GMRS, VIRS, or WSVC, and the composition of the ground support party (STT and DZSO; DZST and or DZSO).

20-7. METHODS

To ensure accurate delivery on the DZ, JMs use four different methods. Each method uses various input from the ground and air in the calculation formula.

a. **Computed Air Release Point (CARP)**. The CARP is the most often used method in aerial delivery for conventional airborne operations. The CARP is computed by the aircrew (navigator) and determines the release point from the air.

b. **Ground Marking Release System (GMRS)**. The GMRS is the method used mostly by special operations forces (SOF). The GMRS is computed by the DZSTL and determines the release point from the ground.

c. **Verbally Initiated Release System (VIRS)**. The VIRS is one of the two methods used by services (Army and USMC) having rotary-wing and fixed-wing aircraft for small DZs, for dropping a specified number of personnel. VIRS is computed by the DZSTL; the release point is indicated by an oral command to the aircraft.

d. **Wind Streamer Vector Count (WSVC)**. The WSVC is one of two methods used by services having rotary-wing and fixed-wing aircraft for DZs for dropping a specified number of personnel. The release point is JM-directed and is the only method not requiring markings on the DZ.

20-8. ORGANIZATION

To become operational, drop zones require key personnel to be located on the DZ for controlling, marking, medical evacuating, wind readings, and malfunctions.

a. **USAF STT and DZSO**. STT and DZSO personnel are normally used in joint airborne operations of more than four troop carrier aircraft. Reference MOA 87 airdrop operations without STT.

b. **DZST**. The DZST consists of trained military personnel. It is normally used in small joint airborne operations involving four or fewer troop carrier aircraft. CARP, GMRS, JSJR, WSVC, or VIRS is used.

c. **DZSO**. Acting alone, without STT support, the DZSO operates the drop zone, with a small number of aircraft dropping a limited number of personnel. The airdrop release method used is CARP or GMRS, or, if rotary-wing and small fixed-wing aircraft are employed, VIRS or WSVC.

20-9. DROP ZONE SAFETY OFFICER DUTIES

The USAF STT and Army DZSO have specific duties, which are discussed as follows:

a. When the USAF STT is supporting an airborne operation, the DZSO is the airborne commander's direct representative on the drop zone. He is responsible for the safe operation of the DZ. No personnel or equipment is dropped if the DZSO is not physically on the DZ.

NOTE: The prerequisites to perform the duties of the DZSO are outlined in Chapter 7.

(1) ***Special Duties***. The duties of the DZSO are—

(a) Coordinating with the USAF STT.

(b) Ensuring the drop zone is fully operational 1 hour before drop time.

(c) Opening the drop zone through range control and closing it when accountability of personnel, air items, and equipment is completed.

(d) Before the drop, conducting ground or aerial reconnaissance of the DZ for obstacles or safety hazards.

(e) Collocating with USAF STT and taking initial wind readings 1 hour before the scheduled drop time.

(f) Establishing communications with the DACO NLT 1 hour before drop time.

(g) Conducting continuous surface wind readings NLT 12 minutes before the scheduled drop. Giving the CLEAR TO DROP or NO DROP to the STT (to relay to aircraft) 2 minutes before the scheduled drop.

(h) Monitoring surface winds from the parachutists' point of impact and at the highest point of elevation on the drop zone. A NO DROP situation exists when surface winds exceed the maximum allowable limits within 10 minutes of the actual drop. (See paragraph 22-5.)

(i) Controlling all ground and air medical evacuations. Priority for airspace must be given to medical evacuations. This is particularly important when rescue or medical aircraft are involved, since they may be delayed if follow-on jumps continue.

(j) Ensuring that any water obstacle is covered by a boat detail. A water obstacle is water more than 4 feet deep and 40 feet wide that is within 1,000 meters from any edge of the surveyed DZ. Water that is deeper than 4 feet but less than 40 feet wide is not an obstacle that requires boats, but does require approved flotation devices for the jumpers.

(k) Submitting postmission reports (for example, MAC Form 168 or incident/accident forms) to the appropriate agency.

NOTE: The CLEAR TO DROP or NO DROP that is relayed at 2 minutes does not indicate the final wind reading. If surface winds increase beyond authorized limits, a NO DROP can be relayed at any time thereafter. If readings exceed the limits, the DZSO must reestablish a 10-minute window.

(2) ***Support Requirements.*** These apply to multiple aircraft formations (fixed-wing aircraft), personnel, and equipment, or to single aircraft operations on drop zones more than 2,100 meters in length or with 20 seconds exit time ("green light").

(a) The DZSO ensures the ground support team is in place on the DZ one hour before the drop. The support team includes:

- Assistant DZSO (not required for USN or USAF operations).
- Four medical personnel (with two FLA); USMC/USN/USAF require one qualified support person.
- Malfunction officer (with camera).
- Parachute recovery detail (with saw and tree-climbing equipment).
- Parachute turn-in detail (with vehicles).
- Radios—one for the DZSO and one for the assistant DZSO (minimum).
- Anemometers—Services should only use approved anemometers to measure surface winds during all personnel and cargo parachute operations. The approved anemometers are the DIC, DIC3, TurboMeter, and AN/PMQ-3A. The AN/ML433A/PM and the anemometers that use floating balls or small floating lightweight aluminum devices in a tube are not authorized for use during personnel or cargo airdrop operations. The DIC, DIC3, and

TurboMeter cannot be calibrated; they must be given an expedient check just before use.

- Ensure fresh batteries are installed in the anemometer.
- Check the anemometer in a no-wind condition such as in a vehicle cab or a building. Turn on the anemometer and, if any reading other than zero registers, the anemometer is not fit for use and must be discarded.
- Use a three-anemometer check by comparing the reading on three anemometers in identical conditions. Discard the one anemometer that doesn't read the same as the other two.
- The TurboMeter must be held within 20 degrees of wind line with the wind entering the rear of the meter to ensure accurate readings.
- Calibration requirements for the AN/PMQ-3A will be conducted IAW appropriate TMs. Other anemometers not tested and recommended for use should be employed only after a command-initiated risk assessment is completed. Regardless of the method or device used to measure DZ winds, the airborne commander is responsible for ensuring winds on the DZ do not exceed 13 knots during static line personnel airdrops.

- Compasses—two (one each for the DZSO and the assistant DZSO).
- Smoke grenades.
- Vehicles.
- Road guards.
- Military police (to control vehicles and spectators or appropriate crowd control).
- Boat detail.

NOTE: These requirements may be supplemented based on the type of drop, size of operation, number of aircraft, number of parachutists, and geographical location of the DZ.

(b) For single aircraft (no more than a 20-second exit time or no more than 2,100 meters of usable DZ), the composition of a partial control group is as follows:

- One DZSO (an assistant DZSO is not required).
- Wind reading from a single location on the DZ.
- Two medics (with FLA); USMC/USN require one qualified Navy corpsman.
- Malfunction officer (with camera, binoculars, and night vision goggles for night operations).
- Parachute recovery detail.
- Radio.
- Anemometers—Services should only use approved anemometers to measure surface winds during all personnel and cargo parachute operations. The approved anemometers are the DIC, DIC3, TurboMeter, and AN/PMQ-3A. The AN/ML433A/PM and the anemometers that use floating balls or small floating lightweight aluminum devices in a tube are not authorized for use during personnel or cargo airdrop operations. The DIC, DIC3, and

TurboMeter cannot be calibrated; they must be given an expedient check just before use.

— Ensure fresh batteries are installed in the anemometer.

— Check the anemometer in a no-wind condition such as in a vehicle cab or a building. Turn on the anemometer and, if any reading other than zero registers, the anemometer is not fit for use and must be discarded.

— Use a three-anemometer check by comparing the reading on three anemometers in identical conditions. Discard the one anemometer that doesn't read the same as the other two.

— The TurboMeter must be held within 20 degrees of wind line with the wind entering the rear of the meter to ensure accurate readings.

— Calibration requirements for the AN/PMQ-3A will be conducted IAW appropriate TMs. Other anemometers not tested and recommended for use should be employed only after a command-initiated risk assessment is completed. Regardless of the method or device used to measure DZ winds, the airborne commander is responsible for ensuring winds on the DZ do not exceed 13 knots during static line personnel airdrops.

- Compass.
- Smoke grenades.
- Boat detail and road guards.

b. The DZSO has operational responsibility for the drop zone. In addition to the DZSO's duties for drop zones, the DZSO must also—

(1) Be positioned at the point of impact 15 minutes before drop time. The assistant DZSO is at the highest point of the drop zone or at the opposite end. For combination airdrop operations, the DZSO/DZSTL must follow the procedures for heavy drop operations, but observe the jumpers as they exit the aircraft.

(2) Relay a ground weather decision and CLEAR TO DROP or NO DROP signal to the lead aircraft 2 minutes before the drop for each pass.

(3) During night drops, ensure all lights that are on or next to the drop zone and are not a part of the drop zone marking system are turned off 5 minutes before drop time and remain off during the drop (except those lights that mark obstacles).

(4) Contact the pilot of the aircraft immediately after the drop and ask if any personnel or equipment did not drop. He relays this information to the airborne commander on the drop zone.

c. When advised of the coordination appointment by the airborne commander or his representative, the DZSO is furnished the following information:

- Number of lifts.
- Type of aircraft.
- Drop zone requirements.
- Unit SOP.
- Station time.
- Drop time.
- Number of personnel for each pass.

20-10. DROP ZONE SUPPORT TEAM AND DROP ZONE SUPPORT TEAM LEADER DUTIES

In operations in which the STT is not present, the DZSTL has overall responsibility for the conduct of operations on the drop zone. He represents both the airborne and airlift commanders. The DZSTL assumes all the responsibilities normally associated with the USAF STT and DZSO. Refer to DZSO MOA between USAF and Army/USMC/Navy dated 6/26/87.

a. DZSTs have the primary mission of supporting wartime CDS airdrops for battalion-size units and below and peacetime airdrops of personnel, CDS, and heavy equipment for one- to four-aircraft operations. With some exceptions, these primary mission airdrops are limited to day and night visual conditions.

b. DZSTs also maintain the secondary mission of supporting other types of airdrops. The secondary missions may include wartime force projection and sustainment of personnel, equipment, and CDS; peacetime airdrops under AWADS and IMC conditions; and VMC formation drops with four or more aircraft.

c. The DZST consists of one or more individuals. More members may be required, depending on the complexity of the mission. The senior member of the DZST functions as the DZSTL. For combination airdrop operations, the DZSO/DZSTL must follow the procedures for heavy drop operations, but observe the jumpers as they exit the aircraft.

d. If an individual assumes the duties of both the DZSO and the DZSTL, he also is responsible for the following:

(1) Conducting premission coordination.

(2) Evaluating the DZ for suitability and safe operating conditions.

(3) Ensuring all DZ markings are properly displayed.

(4) Operating all visual acquisition aids.

(5) Ensuring NO DROP signals are relayed to the aircraft.

e. Once the DZSTL has been notified and assigned a mission, he must conduct accurate premission coordination. The recommended DZST crew mission briefing checklist (Figure 20-1) reflects the minimum essential information that must be addressed and confirmed by the DZSTL. Normally, peacetime drops should employ every acquisition aid and safety device available, including air-to-ground radio communications, PIBAL mean effective wind measurement, air traffic control light gun, and smoke or flares. During contingency or wartime operations, limited airdrop support equipment is available; therefore, it is important for premission coordination and briefings to be comprehensive with respect to visual signals (drop cancellation, postponement, and authentication procedures). The coordination must be timely to ensure the DZST has enough time for planning and for moving to and establishing the drop zone.

20-11. BRIEFING CHECKLIST

Figure 20-1 shows a recommended DZST/aircrew mission briefing checklist.

DZ name/location and JA/ATT mission sequence number verified. ________

TOT(s) block time (NO DROP procedures, for example, race track). ________

Valid DZ survey (date) verified. ________

Type drop (HE, PE, CDS). ________

Type release (VIRS, CARP, GMRS, AWADS, WSVC, visual). ________

Type parachutes. ________

Ground quick disconnects. ________

Number of jumpers/bundles. ________

Number and type of aircraft. ________

DZ information. ________

____

Markings/signals:

- Panels/lights. ________
- Block letter identification. ________
- Smoke, flares. ________
- Emergency NO DROP procedures. ________

  ____
- Mission cancellation indication. ________

  ____

DZ support capabilities:

- Radios available/frequencies. ________
- Visual acquisition aids available. ________
- NAVAIDS available. ________
- MEW equipment. ________

Airspace coordination verified. ________

Aircraft (mission) commander's name, unit of assignment, telephone number. ________

____

DZSTL name, rank, unit of assignment, telephone number. ________

____

Drop score/incident/accident reporting procedure. ________

____

Figure 20-1. DZST/aircrew mission briefing checklist.

20-12. EQUIPMENT

The DZSTL should maintain an inventory of basic equipment to support an airdrop mission:

- VS-17 panels.
- Smoke (red, yellow, green).
- White steady lights, white omni-directional.
- Amber rotating beacon.
- Air traffic control gun, SE-11 light gun, or 4-cell MAGLITE flashlight.

NOTE: ATC gun requires a special power source and plug to function properly.

- Signal mirror.
- Binoculars.
- Night vision goggles for night operations.
- Anemometers—Services should only use approved anemometers to measure surface winds during all personnel and cargo parachute operations. The approved anemometers are the DIC, DIC3, TurboMeter, and AN/PMQ-3A. The AN/ML433A/PM and the anemometers that use floating balls or small floating lightweight aluminum devices in a tube are not authorized for use during personnel or cargo airdrop operations. The DIC, DIC3, and TurboMeter cannot be calibrated; they must be given an expedient check just before use.
  - — Ensure fresh batteries are installed in the anemometer.
  - — Check the anemometer in a no-wind condition such as in a vehicle cab or a building. Turn on the anemometer and, if any reading other than zero registers, the anemometer is not fit for use and must be discarded.
  - — Use a three-anemometer check by comparing the reading on three anemometers in identical conditions. Discard the one anemometer that doesn't read the same as the other two.
  - — The TurboMeter must be held within 20 degrees of wind line with the wind entering the rear of the meter to ensure accurate readings.
  - — Calibration requirements for the AN/PMQ-3A will be conducted IAW appropriate TMs. Other anemometers not tested and recommended for use should be employed only after a command-initiated risk assessment is completed. Regardless of the method or device used to measure DZ winds, the airborne commander is responsible for ensuring winds on the DZ do not exceed 13 knots during static line personnel airdrops.
- Compass.
- Signal flares.
- PIBAL system with helium source.

NOTE: Other items of equipment/signals may be required by premission coordination and mission complexity.

CHAPTER 21

DROP ZONE COMPUTATIONS AND FORMULAS

Once the composition of the ground party and the selection of a drop zone have been established, several technical aspects must be considered and planned for marking the DZ. These technical aspects are critical because of the data that must be used (ground, winds, drift-distance formula, forward throw, release point).

21-1. DROP ZONE FORMULAS FOR GMRS AND VIRS

The following procedures for using the distance and time formulas apply to GMRS and VIRS.

a. **Distance Formula (D = RT).** Compute DZ length for a specific mission by using the D = RT formula. (D is the required length of the DZ in meters; R is the ground speed of the aircraft in meters per second; and T is the time required for the aircraft to release its cargo.) To use this formula, some conversions and mathematics are required.

(1) ***Airspeed Conversion to Ground Speed.*** To find the aircraft ground speed, convert aircraft airspeed (expressed in knots) to ground speed (meters per second). Do this by multiplying knots times .51 (knots x .51) (1 knot equals .51 meter per second). The following table from Chapter 20 is repeated to assist the estimation of aircraft airspeeds.

TYPE OF AIRCRAFT	DROP SPEED
UH-1	50 to 70 knots—optimum 70 knots
UH-60	65 to 75 knots—optimum 70 knots
CH-46 (USMC)	80 to 90 knots—optimum 90 knots
CH-47	80 to 110 knots—optimum 90 knots
CH-53 (USMC)	90 to 110 knots
CH/HH3 (USAF)	70 to 90 knots
C-5/130/141/17/KC-130	130 to 135 knots (personnel)
C-5/130/141//17KC-130	130 to 150 knots—optimum 130 knots for all loads (door bundles, CDS, and heavy equipment)

Table 21-1. Aircraft drop speeds.

(2) ***Time Over DZ Requirement.*** To determine the time over the DZ that is needed to release a parachutist or equipment, use the following factors:

(a) Allow 1 second for each parachutist to exit the aircraft; do **not** include the first parachutist (10 parachutists require 9 seconds). (Mathematically, this is represented as 10 x 1 - 1.)

(b) Allow 3 seconds per bundle to exit the aircraft; do **not** include the first bundle (3 bundles would require 6 seconds). (Mathematically, this is represented as 3 x 3 - 3.)

(c) Personnel jumping T-10-series parachutes may exit both doors simultaneously. The door with the most parachutists is used to calculate the time required.

EXAMPLE: D = RT

What length DZ would 8 jumpers require when jumping from an aircraft flying at a drop speed of 90 knots?

Step 1: Solve for R (answer is expressed in meters per second): airspeed x .51 (90 knots x .51) = 45.90 meters per second.

Step 2: Solve for T (answer is expressed in seconds): number of jumpers x 1 - the first jumper (8 x 1 - 1) = 7 seconds.

Step 3: Solve for D (answer is expressed in meters): 45.90 meters per second x 7 seconds = 321.30 meters. Always round **up** to the nearest whole number. Therefore, D = 322 meters, the required DZ length.

b. **Time Formula (T = D/R).** Solving this formula provides the seconds required to exit the jumpers over the DZ (time = meters divided by meters per second). If a DZ less than the required length must be used, compute the flight time over the DZ to determine how much of the load can be released in one pass. Use the T = D/R formula: T is the time the aircraft is over the DZ in seconds, D is the length of the DZ in meters, and R is the ground speed (rate) of the aircraft in meters per second.

(1) ***Airspeed Conversion.*** Convert the aircraft's airspeed (expressed in knots) to its ground speed (expressed in meters per second) as in the D = RT formula (knots x .51). Round **up** the answer to the next whole number.

(2) ***Determination of T.*** Divide the ground speed conversion number into D (the DZ length); this determines T. Any fractional answer is rounded **down** to the next whole number.

EXAMPLE: T = D/R

How many parachutists from a CH-47 (drop speed of 90 knots) can land on a 750-meter DZ each pass?

T = Number of parachutists.

D = DZ length is 750 meters (given).

R = Airspeed is 46 meters per second (90 knots x .51 = 45.9; round up to 46).

Solution: T = D/R (D ÷ R).

D/R = 750 meters divided by 46 meters per second = 16.3 seconds.

T = 16 seconds (round **down**).

16 seconds over DZ x 1 parachutist per second + 1 parachutist (the first parachutist exiting the aircraft does not affect the number of seconds spent over the DZ) = 17 parachutists. Thus, 17 parachutists per pass can land on the 750-meter DZ.

21-2. WIND DRIFT

Two means of determining wind drift are the WSVC method and the D = KAV formula.

a. **Wind Streamer Vector Count**. The WSVC method (Figure 21-1) is used when the release point is determined from the air. It is normally jumpmaster-executed and does not require markings to be placed on the DZ.

(1) ***Streamer Drop.*** On the first aircraft pass over the desired point of impact, a streamer is dropped from the aircraft. The aircraft then turns to allow the JM to keep the streamer in sight. The pilot adjusts his route so that the flight path is over the streamer on the ground and the desired impact point (in a straight line).

(2) ***Count.*** As the aircraft passes over the streamer, the JM begins a count, stopping the count directly over the impact point. He immediately begins a new count. When that count equals the first count, the aircraft is over the release point for the first parachutist.

(3) ***Aircraft Flight Adjustment.*** The pilot then maneuvers the aircraft to fly along the axis of the DZ and over the release point. Slight adjustments may be made by observing the parachutists as they land on the DZ.

NOTE: This method should not be used for tactical employment, since the aircraft is required to make multiple passes over the DZ.

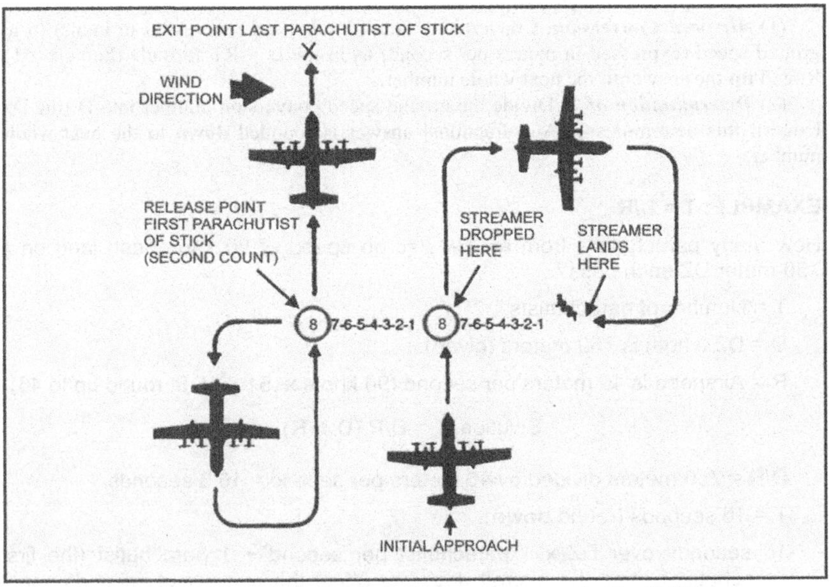

Figure 21-1. Determination of the release point by WSVC.

b. **D = KAV Formula**. This is another method for determining the effects of wind on a parachute: D = drift of parachute (in meters) from a given altitude; K = constant that represents the typical drift characteristic for a type of parachute. These constants are—

- 1.5 for cargo parachutes and heavy equipment (HD or HE).
- 3.0 for personnel parachutes.
- 2.4 for tactical training bundles.

A = drop altitude (expressed in hundreds of feet AGL); and V = velocity of wind. The mean effective wind (MEW) should be used for personnel. MEW is the preferred method for all other types of loads; however, the surface wind may be used during low-risk operations.

NOTE: If aircraft must be shut down for a long period, a wind drift indicator should be thrown at the last release point to ensure the release point is still valid.

EXAMPLE:

An aircraft is dropping cargo from 500 feet AGL with a surface wind of 10 knots. What is the calculated parachute drift? (The parachute drift is calculated using the D = KAV formula.)

D = Wind-induced drift in meters.

K = Wind drift constant for type of parachute.

A = Drop altitude expressed in hundreds of feet (500 feet would be expressed as 5).

V = Velocity of wind in knots (either MEW or surface wind measurement).

Step 1: K = 1.5 (cargo parachute or HE constant).

Step 2: A = 5 (500 feet).

Step 3: V = 10 (10 knots).

Step 4: D = 1.5 x 5 x 10 = 75.0, or 75 meters of drift. (Any fractional answer is rounded **up** to the nearest whole number.)

21-3. WIND VELOCITY

Two options are available for determining wind velocity.

a. **Mean Effective Wind**. The most effective option is the use of MEW. This reading is a constant wind speed average from drop altitude to the ground. The PIBAL system determines the MEW. This system should be used when possible; it is more reliable than the other option, which measures surface wind velocity only.

b. **Surface Wind Measurement**. Either the AN/PMQ-3A or commercial anemometers authorized by USAIS messages DTG 101000Z MAR 94, subject: Use of Anemometers During Airdrop Operations, and DTG 211200Z OCT 94, subject: Use of Turbometer During Static Line Airdrop Operations, are recommended for use. Other anemometers not recommended for use should be employed only after a command-initiated risk assessment is completed. Regardless of the method or device used to measure DZ winds, the airborne commander is responsible for ensuring winds on the DZ do not exceed 13 knots during static line personnel airdrops.

(1) ***Equipment.*** The equipment needed to compute the MEW by the PIBAL method is as follows:

- Helium source.
- Pilot balloons (10 or 30 grams).
- Clinometers or other devices for measuring from 0 to 90 degrees.
- Balloon measuring tape (to measure balloon circumference) (10-gram: 57 inches day, 74 inches night; 30-gram: 78 inches day, 94 inches night).
- PIBAL lighting units (Type 5) for night use (liquid-activated lights).
- Compass.
- Conversion charts (10- and 30-gram) (Tables 21-2 and 21-3).
- Watch with second hand.

(2) ***Procedure.*** The procedure for measuring MEW using the PIBAL is as follows:

(a) Fill the 10-gram or 30-gram balloon with helium to the required size.

(b) Check the conversion chart for drift time to drop altitude (Table 21-2, page 21-6, and Table 21-3, page 21-7).

(c) Release the balloon and begin timing.

(d) Keep the balloon in sight.

(e) Once the required time has elapsed, determine the azimuth to the balloon with the compass and read the degrees from the drift scale.

(f) Refer to the conversion chart and read down the angle column to the number closest to the angle on the scale.

(g) Read across the top of the chart (altitude in feet) to the drop altitude in use. Read down this column until the two lines (6 and 7) intersect.

(h) Where the two lines intersect is the MEW at drop altitude, in knots. The direction of the MEW is the back azimuth of the compass reading that was taken at the same time as the angle measurement.

(i) The MEW becomes the variable V in the D = KAV formula to determine the amount of drift in meters.

NOTE: A parachute's K-factor is based on the parachute's flight characteristics, not on its mode of use. IAW Chapter 2 of FM 10-500-3/TO 13C7-1-11, the K-factor for the T-10-series parachute used in the cargo mode is the same (3.0) as for personnel drops using the T-10-series parachute.

WIND SPEED IN KNOTS
10-GRAM HELIUM BALLOON
Inflate balloon to 57" circumference for day and 74" circumference for night.

ELEVATION ANGLE	DROP ALTITUDE IN FEET: 500	750	1000	1250	1500	1750	2000	2500	3000	3500	4000	4500	ASCENSION TABLE: TIME	ALT (FT)
70	02	02	01	01	01	01	01	01	01	01	01	01		
60	03	02	02	02	02	02	02	02	02	02	02	02		
55	03	03	03	03	03	03	03	03	03	03	03	03		
50	04	04	03	03	03	03	03	03	03	03	03	03	0:10	80
45	05	04	04	04	04	04	04	04	04	04	04	04	0:20	170
40	06	05	05	05	05	05	05	04	04	04	04	04	0:30	250
35	07	06	06	06	06	05	05	05	05	05	05	05	0:40	330
30	08	07	07	07	07	07	07	07	06	06	06	06	0:50	400
25	10	09	09	09	08	08	08	08	08	08	08	08	1:02	500
24	11	10	09	09	09	09	08	08	08	08	08	08	1:10	540
23	11	10	10	09	09	09	09	08	08	08	08	08	1:20	610
22	12	11	10	10	10	10	09	09	09	09	09	09	1:23	670
21	12	11	11	10	10	10	10	10	10	10	10	10	1:43	750
20	13	12	11	11	11	11	11	10	10	10	10	10	1:50	790
19	14	13	12	12	11	11	11	11	11	11	11	11	2:25	1000
18	15	13	13	12	12	12	12	12	11	11	11	11	2:44	1100
17	16	14	13	13	13	13	12	12	12	12	12	12	3:05	1250
16	17	15	14	14	14	13	13	13	13	13	13	13	3:49	1500
15	18	16	15	15	14	14	14	14	14	14	14	14	4:30	1750
14	19	17	16	16	16	15	15	15	15	15	15	15	5:11	2000
13	21	19	18	17	17	17	17	16	16	16	16	16	6:34	2500
12	22	20	19	19	18	18	18	18	17	17	17	17	7:58	3000
11	24	22	21	21	20	20	20	19	19	19	19	19	9:22	3500
10	27	25	23	23	22	22	22	21	21	21	21	21	10:44	4000
09	30	27	26	26	25	24	24	24	23	23	23	23	12:08	4500

Table 21-2. The 10-gram PIBAL chart.

21-4. FORWARD THROW

Forward throw is the effect that inertia has on a falling object. When an object leaves an aircraft, it is traveling at a speed equal to the speed of the aircraft. The parachutist (or bundle) continues to move in the direction of flight until the dynamics of parachuting takes effect.

a. **Forward Throw for Rotary-Wing Aircraft**. To determine the amount of forward throw for rotary-wing aircraft, divide the drop speed of the aircraft in half. This yields the forward throw in meters. (For example, an aircraft flying at 70 knots would have a forward throw of 35 meters.)

b. **Forward Throw for Fixed-Wing Aircraft**. To determine the forward throw for fixed-wing aircraft, the following distances apply (Table 21-4).

WIND SPEED IN KNOTS
30-GRAM HELIUM BALLOON
Inflate balloon to 78" circumference for day and 94" circumference for night.

ELEVATION ANGLE	DROP ALTITUDE IN FEET: 500	750	1000	1250	1500	1750	2000	2500	3000	3500	4000	4500	ASCENSION TABLE: TIME	ALT (FT)
80	01	01	01	01	01	01	01	01	01	01	01	01		
70	03	03	03	02	02	02	02	02	02	02	02	02		
60	04	04	04	04	04	04	04	04	04	04	04	04		
55	05	05	05	05	05	05	05	05	05	05	04	04	0:10	120
50	06	06	06	06	06	06	06	06	05	05	05	05	0:20	240
45	07	07	07	07	07	07	07	07	07	06	06	06	0:30	360
40	09	08	08	08	08	08	08	08	08	08	08	08	0:42	500
35	10	10	10	10	10	10	10	10	09	09	09	09	0:50	400
30	12	12	12	12	12	12	12	11	11	11	11	11	1:02	600
25	15	15	15	15	15	15	14	14	14	14	14	14	1:10	830
24	16	16	15	15	15	15	15	15	15	15	15	15	1:17	1000
23	17	17	16	16	15	15	15	15	15	15	15	15	1:46	1250
22	18	18	17	17	17	17	17	16	16	16	16	16	2:10	1500
21	19	19	18	18	18	17	17	17	17	17	17	17	2:34	1750
20	20	20	19	19	19	19	18	18	18	18	18	17	2:56	2000
19	21	20	20	20	20	20	19	19	19	19	19	18	3:43	2500
18	22	22	21	21	21	21	20	20	20	20	20	20	4:31	3000
17	23	23	23	22	22	22	22	22	21	21	21	21	5:21	3500
16	25	25	24	24	24	24	23	23	23	23	22	22	6:09	4000
15	27	27	26	26	25	25	25	25	24	24	24	24	7:00	4500
14	29	19	18	17	17	17	17	16	16	16	16	25		
13	31	20	19	19	18	18	18	18	17	17	17	27		

Table 21-3. The 30-gram PIBAL chart.

	C-130	C-141	C-5
PERSONNEL/DOOR BUNDLES	229 METERS/ 250 YARDS	229 METERS/ 250 YARDS	229 METERS/ 250 YARDS
HEAVY EQUIPMENT	458 METERS/ 500 YARDS	668 METERS/ 730 YARDS	668 METERS/ 730 YARDS
CONTAINER DELIVERY SYSTEMS	503 METERS/ 550 YARDS	686 METERS/ 750 YARDS	686 METERS/ 750 YARDS
TACTICAL TRAINING BUNDLE	147 METERS/ 160 YARDS	147 METERS/ 160 YARDS	147 METERS/ 160 YARDS
NOTE: To convert yards to meters, multiply yards by .9144. To convert meters to yards, divide meters by .9144.			

Table 21-4. Fixed-wing forward throw data.

21-5. DROP HEADINGS, POINT OF IMPACT, WIND DRIFT COMPENSATION, AND FORWARD THROW COMPENSATION

For CARP operations, the navigator onboard the aircraft determines when the load is to be released from the aircraft (when the green light is turned on). For GMRS and VIRS operations, ground personnel determine the release point (Figure 21-2).

a. **Drop Heading**. Drop heading on all DZs depends on three factors—the long axis, prevailing winds, and obstacles on approach and departure ends. The DZSO or DZSTL uses all three when the situation permits; however, the long axis is the primary concern. With a GMRS, WSVC, or CARP DZ, drop heading can be obtained from USAF Form 3823, DZ Survey (formerly MAC Form 339). A circular/random approach DZ does not have a set drop heading. The mission commander notifies the aircrew and the DZ commander of the drop heading to be used NLT 24 hours in advance of the airdrop operation.

NOTE: On some DZs, predetermined drop headings must be used.

b. **Point of Impact**. The location selected where the first bundle or parachutist should land is known as the PI. The PI should be located along the DZ centerline. However, due to the tactical situation, the PI may need to be located near a wood line. The DZSO or DZSTL uses a buffer zone of 100 meters on one end of the DZ for safety reasons. PI location for GMRS or VIRS is 100 meters in from the leading edge centerline for personnel. CARP PI is designated on USAF Form 3823, DZ Survey (formerly MAC Form 339).

c. **Wind Drift Compensation**. To compensate for wind drift, the DZSO or DZSTL moves from the desired PI into the wind the number of meters calculated using the D = KAV formula. (For example, if drift equals 350 meters from the PI, he faces into the wind and walks 350 meters in a straight line.)

d. **Forward Throw Compensation**. To compensate for aircraft forward throw, the DZSO or DZSTL faces the back azimuth of the drop heading and walks the appropriate forward-throw distance to the release point.

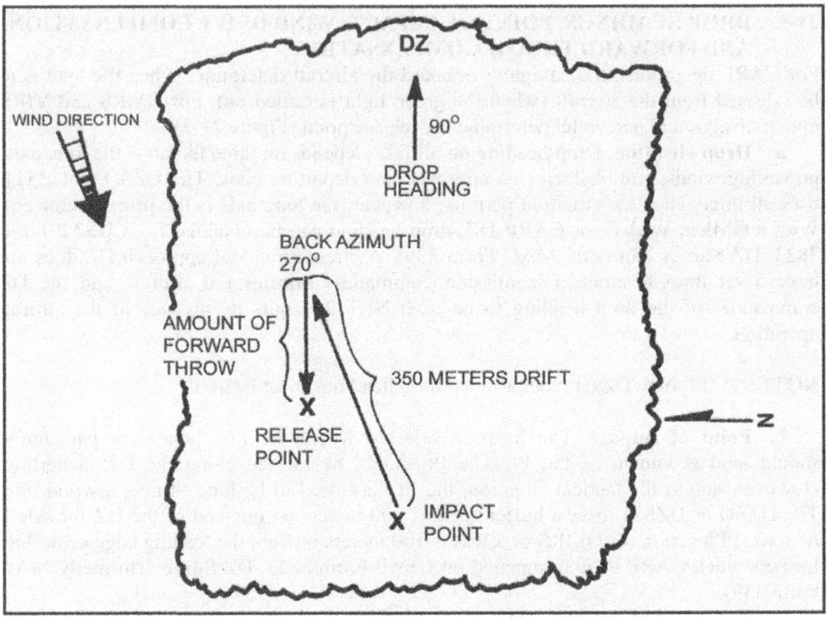

Figure 21-2. RP location for VIRS and GMRS.

CHAPTER 22

ESTABLISHMENT AND OPERATION OF A DROP ZONE

Five methods may be used to establish or operate a drop zone. Four of these require markings to be placed on the DZ: CARP, GMRS, and VIRS. The WSVC method requires no markings on the DZ.

22-1. COMPUTED AIR RELEASE POINT

CARP is used only by fixed-wing aircraft in conjunction with a CCT or qualified DZSTL.

a. **CARP Points of Impact** (Figures 22-1 and 22-2, page 22-2). The PIs for CARP operations are as follows:

(1) ***Personnel.*** For personnel, drops at the PI are 300 yards (day) or 350 yards (night) from the leading edge.

(2) ***CDS.*** For CDS bundles from a C-130, drops at the PI are 200 yards (day) or 250 yards (night) from the leading edge. For CDS drops from a C-141, the PI is 225 yards (day) and 275 yards (night).

(3) ***HE.*** For heavy equipment, drops at the PI are 500 yards (day) or 550 yards (night) from the leading edge.

NOTE: On surveyed DZs, the PI for a particular type load is predetermined. Its surveyed location can be found on AF Form 3823 or MAC Form 339. (Use of MAC Form 339 is authorized until supplies are exhausted. USAF is converting all MAC Forms 339 to AF Form 3823 when a DZ comes due for recertification.)

b. **No-Drop Communication to Aircraft**. No-drop conditions are relayed to the aircraft in the following ways: red smoke, red flares, forming the code letter into two parallel bars perpendicular to flight, or the absence of a planned signal. Forming the code letter into an X indicates mission cancellation.

NOTE: The type of marking used for no-drop conditions is coordinated in the premission briefing.

c. **Control Center**. Control center locations (location of DZSTL) are as follows:

(1) ***Personnel Drops.*** Personal drops are normally located at the PI.

(2) ***CDS Drops.*** CDS is located 150 yards to the 6 o'clock position of the PI.

(3) ***HE, Free Drops, High Velocity, AWADS.*** Heavy equipment, free drops, high velocity, and AWADS with a ceiling of less than 600 feet are all off the DZ. For this type of operation, the DZSTL uses the best vantage point off the DZ to observe the airdrop.

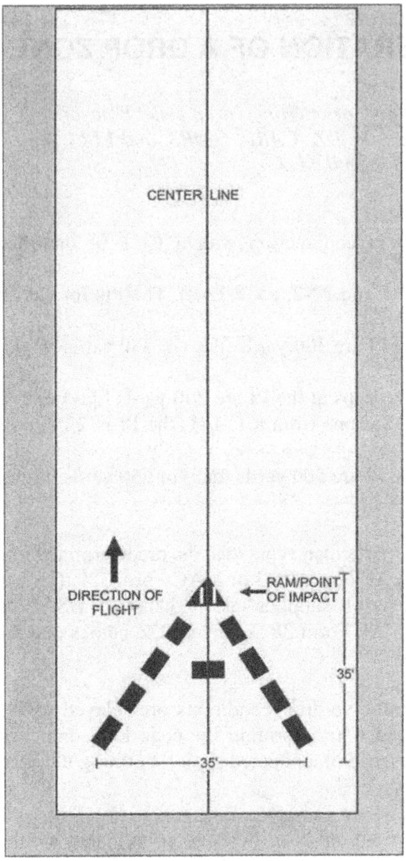

Figure 22-1. Day CARP drop zone markings.

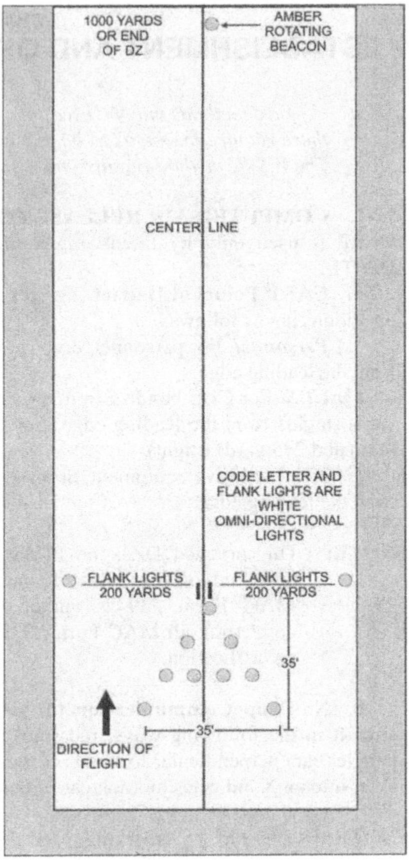

Figure 22-2. Night CARP drop zone markings.

22-2. DROP ZONE MARKINGS

A marked DZ has a PI or release point marked with a precoordinated visual or electronic signal. Standard DZ markings consist of raised angle markers (RAM), VS-17 marker panels, visible lighting systems, and light beacons. Virtually any type of lighting or visual marking system is acceptable if all participating units are briefed and concur. Night markings or visual acquisition aids may include a light gun, flares, fire pots, railroad fuses, flashlights, chem lights, and infrared (IR) lighting systems. Electronic NAVAID markings (ZM, SST-181, GAR-I, tactical aid to navigation (TACAN), and so forth) may be used for either day or night operations and placed as directed by mission requirements.

a. During day operations, the PI will be marked with a RAM or block letter. If authentication is required, a block letter will be used instead of the RAM. Authorized letters for PI markings are A, C, J, R, and S. The block letters H and O are authorized for random approach DZs. The block letters should be aligned with the surveyed DZ axis or with the aircraft line-of-flight, if different from the survey. The minimum size for block letters is 11 meters (35 feet) by 11 meters and consists of at least nine marker panels.

b. During night operations, the PI will be marked with a block letter and flanker lights. The apex of the block letter will be located on the PI. Flanker lights will be white and located 250 meters left and right abeam the PI. The minimum size is 11 meters by 11 meters and consists of at least nine white lights with a recommended minimum output rating of 15 candela. A trailing edge beacon will be used during actual personnel airdrops. When used, the amber trailing edge beacon will be placed along the surveyed DZ centerline 1,000 meters from the PI, or at the DZ trailing edge, whichever is closer to the PI. During premission coordination for personnel drops, aircrews will identify their trailing edge beacon requirements to STS or DZC. For all airdrops, the DZ identification must be coordinated and briefed to the ground party and aircrews.

c. When mission requirements dictate and aircrews are qualified and equipped, IR lights may be substituted for overt lights using the DZ marking patterns specified in paragraph 22-2b.

22-3. GROUND MARKING RELEASE SYSTEM

The GMRS uses markings known as the four-panel inverted L, six-panel T, or seven-panel H. The T or H pattern is recommended for C-141/C-5 airdrops due to aircraft's side angle vision limitations (Figure 22-3, page 22-4).

a. **Inverted L Marking**. When the drop aircraft is 100 meters directly to the right of the corner (A) panel, the drop is executed.

b. **Marking Placement for Inverted L**. Markings (four panels) are placed as follows (Figure 22-3, page 22-4):

(1) From the RP, move 100 meters to the left (90 degrees) of drop heading for the location of the corner (A) panel. Emplace a VS-17G panel with the long axis of the panel parallel with the drop heading. Elevate the panel at a 45-degree angle toward the approaching aircraft. This aids the aircrew and the JM in visual identification of the DZ.

(2) From the corner (A) panel, move in the same direction as above for 50 meters for the location of the alignment (B) panel. Emplace this panel as described above.

(3) From the alignment (B) panel, move 150 meters in the same direction as above for the location of the flanker (C) panel. Emplace this panel as described above.

(4) From the corner (A) panel, move 50 meters on a back azimuth of the drop heading for the location of the approach (D) panel. Emplace this panel the same as described above.

(5) At night, replace all panels with a white omni-directional light. Lights may be shielded on three sides or placed in pits.

(6) During day operations, smoke may be displayed at the RP. During night operations, a white air traffic control light may be used to mark the RP.

(7) NO DROP may be signaled to the aircraft by red smoke, red flares, scrambled panels, or the absence of a planned signal.

c. **Mask Clearance**. Since the aircraft is required to fly along the markings on the DZ, these markings must be visible to the aircrew. The markings are placed where obstacles will not mask the pilot's line of sight. As a guide, a mask clearance ratio of 1:15 is used, that is, one unit of vertical clearance for every 15 units of horizontal clearance. For example, if a DZ marker must be positioned near a terrain mask, such as the edge of a forest that is on the DZ track, and the trees are 10 meters high (33 feet), the markings would require 150 meters (492 feet) of horizontal clearance from the trees (Figure 22-4, page 22-6). This applies to static line jumps only.

d. **Code Letters**. If any portion of the inverted L falls within a 15 to 1 (15:1) mask clearance ratio of obstacles on the approach end of the DZ, a code letter (H, E, A, T) or far panel is required on the departure end of the DZ for CDS or bundle drop and should be coordinated during the DZST/aircrew mission briefing. This far marking is on line with the corner (A) panel to allow the aircrew to begin alignment on the release point until the inverted L comes into view. If a code letter is used, it can be used to distinguish the DZ from other DZs in the area.

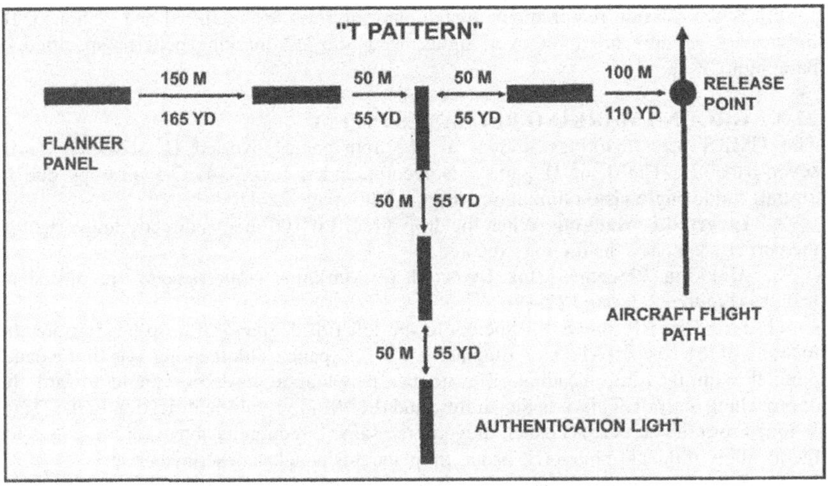

Figure 22-3. GMRS panel emplacement.

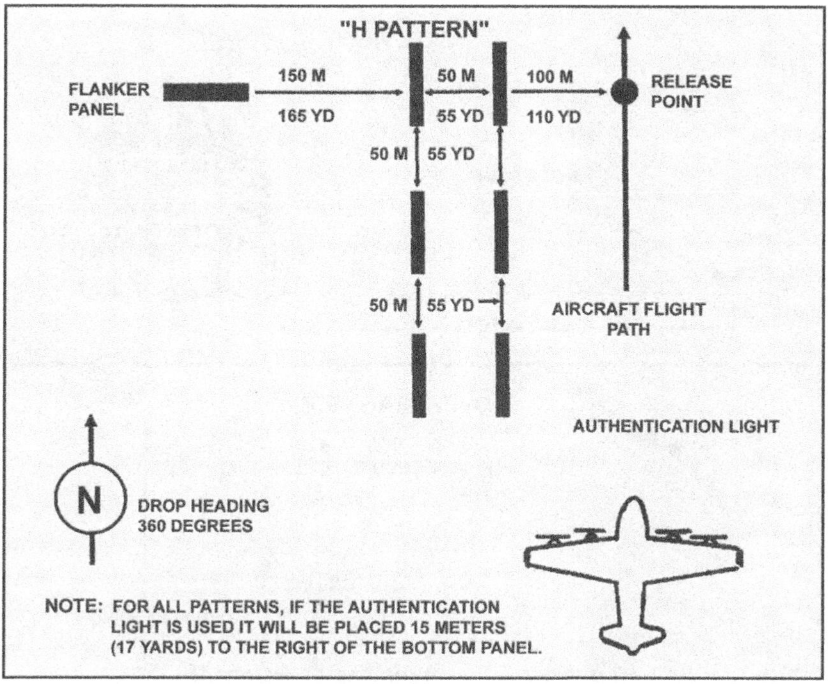

Figure 22-3. GMRS panel emplacement (continued).

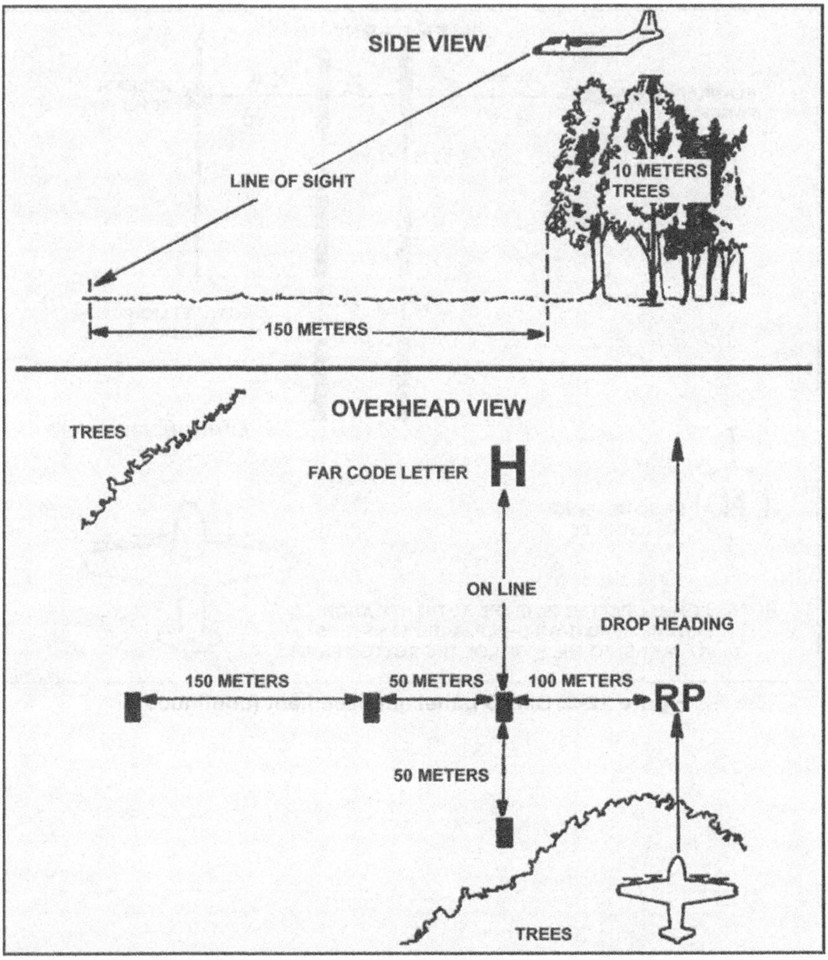

Figure 22-4. The 15:1 mask clearance ratio.

22-4. VERBALLY INITIATED RELEASE SYSTEM FOR ROTARY-WING AND FIXED-WING AIRCRAFT

VIRS is used to execute a drop over the RP by GTA verbal command. This method allows the conduct of the operation with a minimum amount of prior DZ information and coordination. The aircraft flies the given direction until the DZSTL sees the aircraft. A code letter (H,E,A,T) marks the RP. Once the crew identifies the DZ, the radio operator directs the aircraft over the drop heading RP. When the aircraft is directly over the RP, the command EXECUTE, EXECUTE, EXECUTE initiates the drop (Figure 22-5).

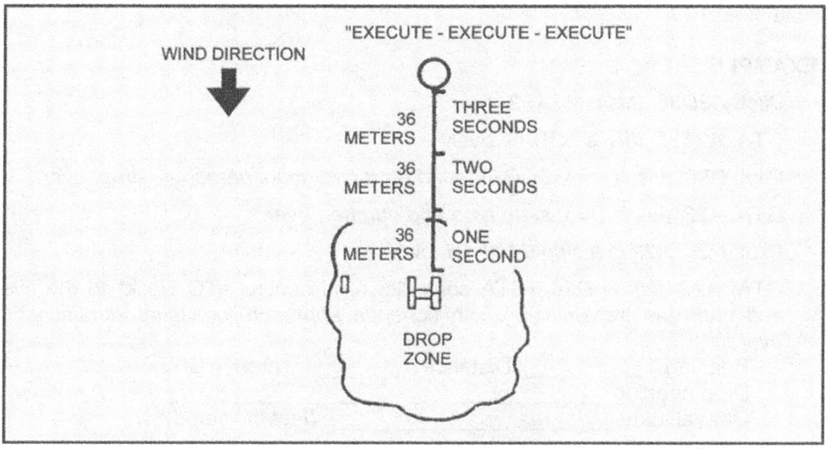

Figure 22-5. Army VIRS offset.

a. **Day DZ Markings**.

(1) ***Determination of Release Point.*** The DZSTL then places the code letter on the drop heading, with the base panel of the letter at the release point. The code letter is formed by VS-17G panels placed together. Each letter is two panels high and one panel wide.

(2) ***Placement of Flank Panel.*** The flank panel is placed parallel to the code letter and aligned with the base panel. It is placed 200 meters to the left of the code letter (or at the edge of the DZ, whichever is closer).

(3) ***Placement of Far Panel.*** The far panel is placed 500 meters from the base panel (or at the edge of the DZ, whichever is closer) and on line with the drop heading.

(4) ***Panel Construction.*** Both the far and flank panels consist of a single VS-17G panel. These panels may also be elevated at a 45-degree angle to improve visibility.

b. **Night DZ Markings**.

(1) ***Use of Lights.*** The procedures for establishing the DZ are the same for night operations except that white light is used for the code letter and far and flank markings.

(2) ***Code Letter Construction.*** Each code letter is four lights high and three lights wide. There is a distance of 5 meters between each light in the code letter. The far and flank lights are signal lights. Also, a white-and-red lens ATC (SE-11) light should be located at the RP.

(3) ***Security.*** Lights may be shielded on three sides or placed in pits to prevent enemy ground observation.

22-5. GUIDANCE PROCEDURES

During Army DZ operations, the GTA is responsible for guiding the jump aircraft to the DZ, over the DZ on the proper drop heading, and at the proper altitude and drop speed. He ensures the parachutists exit the aircraft at the proper release point. Once the parachutists have exited the aircraft, the GTA must then clear the aircraft from the control zone.

EXAMPLE:

Pilot: *C3D36, this is A2A22, over.*

GTA: *A2A22, this is C3D36, over.*

Pilot: *D36, this is A22, CCP inbound for a personnel parachute drop, over.*

GTA: *A22, this is D36, state type and number, over.*

Pilot: *D36, A22 is a single UH-1H, over.*

GTA: *A22, this is D36,* (GTA controller reads entire ATC block* to the pilot and ends the transmission with *continue approach for visual identification, over*).

Heading ____________ Distance __________ (from CCP)
Drop heading ____
Drop altitude _________________________ (feet indicated)
Drop speed ____
Number jumpers/bundles that can be accepted ________

Pilot: *Wilco.*

Upon sighting aircraft, the GTA tells the pilot:

GTA: *A22, this is D36, I am at your 11 o'clock, 500 meters, signal out, can you identify, over.*

Pilot: *D36, A22 identifies orange panel, over.*

GTA: *A22, D36 has visual contact, turn to drop heading, over.*

Pilot: *D36, A22 turning drop heading, over.*

GTA: *A22, this is D36, steer left/right, over.*

Pilot: *D36, A22 roger.*

GTA: *A22, this is D36, on course, over.*

Pilot: *D36, A22 roger.*

When aircraft is 8 to 10 seconds out from release point:

GTA: *A22, this is D36, with six jumpers, stand by, over.*

Pilot: *D36, this is A22, standing by, over.*

When aircraft is directly over release point:

GTA: *A22, this is D36, with six jumpers, execute, execute, execute.* (GTA must say EXECUTE or NO DROP at least three times, or until first load exits.)

At completion of operation, the GTA tells the pilot:

GTA: *A22, this is D36, I observe six jumpers away and clear, state intention and report when clear of my control zone,* (issue any advisories), *over.* (GTA must place aircraft into a closed traffic pattern with a reporting point if more than one pass is required.)

22-6. ACCEPTABLE WIND LIMITATIONS

Maximum allowable surface wind for static line parachute personnel airdrops is 13 knots (17 knots for WDZ). The maximum surface wind speed for static line heavy equipment airdrops is 17 knots with ground quick disconnects, 13 knots without ground quick disconnects, and 20 knots for CDS using G-13/14 parachutes. There is no wind speed restriction when HV parachutes are used with door bundles and CDS. There is no altitude wind limitation. Winds on the DZ are measured using the AN/PMQ-3A anemometer, or commercial anemometers authorized by USAIS messages DTG 101000Z MAR 94, subject: Use of Anemometers During Airdrop Operations, and DTG 211200Z OCT 94, subject: Use of Turbometer During Static Line Airdrop Operations)—two (one each for the DZSO and the assistant DZSO). Other anemometers not recommended for use should be employed only after a command-initiated risk assessment is completed. Regardless of the method or device used to measure DZ winds, the airborne commander is responsible for ensuring winds on the DZ do not exceed 13 knots during static line personnel airdrops.

22-7. THE 10-MINUTE WINDOW

On multiple aircraft operations or single aircraft operations using more than 2,100 meters of DZ, the surface wind is measured from the control center and the highest point of elevation on the DZ or the trail edge of the DZ. For single operations using less than 2,100 meters of DZ, the wind is measured from only one location, normally the control center. Beginning 12 minutes before TOT, the DZSO begins a constant monitoring of the surface wind using an anemometer.

a. **Surface Wind Exceeds Limits**. If the surface wind exceeds allowable wind limits, the aircraft is notified of a NO DROP, and a new 10-minute window is established. If the wind remains within limits during this new window, the drop takes place as planned. If the winds exceed allowable limits during the new window, NO DROP is relayed to the pilot and the entire procedure starts again.

b. **No-Drop Signal**. A NO DROP signal may be relayed to the aircraft by radio, red smoke, red flares, scrambled panels, or another planned signal.

22-8. POSTMISSION REQUIREMENTS

Immediately following the operation, several reports must be forwarded to higher headquarters.

a. **Required Reports**. Most of these reports are self-explanatory and require little time to complete. MAC Form 168 is used to record strike report information. All services will complete parachute malfunction and incident reports IAW AR 59-4/MCO 13480.1B/OPNAVINST 4630.24C/AFJ 13-210(I). All services may also be required to report malfunctions and incidents on the loss of equipment and or injury to personnel IAW service regulations, ground mishap reporting MCO P5102.1 and OPNAVISNT 5102.1C.

- DZSO report.

- Malfunction report.
- MAC Form 168, Airdrop/Airland/Extraction Zone Control Log (Figure 22-6, page 22-11).
- Incident reporting format (Figure 22-7, page 22-12).

NOTE: MAC Form 168 will become AMC Form168 in the next printing.

b. **MAC Form 168 Completion**. Complete the MAC Form 168 as follows:
(1) DATE box—date of airdrop.
(2) LOCATION box—name of DZ.
(3) SST AND UNIT box—SST name and unit.
(4) DZ/LZ/EZ CONTROL OFFICER AND UNIT box—DZSTL name and unit.
(5) DROP ZONE SAFETY OFFICER AND UNIT box—enter names.
(6) LINE NO column—mission sequence number of each aircraft.

NOTE: Every aircraft has a mission sequence number (entered under LINE NO column). Subsequent passes by that same aircraft will all be scored on separate lines, in the order that they occur, immediately below the line for the first pass.

(7) TYPE ACFT column—type of aircraft.
(8) UNIT column—unit of aircraft.
(9) CALL SIGN column—call sign of pilot.
(10) TYPE MSN column—type of mission; refer to LEGEND for abbreviations.
(11) ETA column—estimated time of arrival, estimated TOT, S3 air brief.
(12) ETA column—actual arrival time of every pass.
(13) STRIKE RPRT columns—

- YDS column—distance first jumper/container lands from PI in yards; if within 25 yards, it is scored a PI.
- CLOCK column—using direction of flight as 12 o'clock and its back azimuth as 6 o'clock, estimated direction from PI to first jumper/bundle.

(14) SURF WIND column—surface wind; direction in degrees and velocity in knots.
(15) SCORE METHOD column—refer to LEGEND.
(16) MEAN EFFECTIVE WIND columns—

- TIME column—time taken.
- ALT column—what altitude taken to (should be drop altitude).
- DIR & VEL column—wind direction in degrees and velocity in knots.

c. **MAC Form 168 Routing**. The DZSTL forwards the completed MAC Form 168 to his air operations officer, who in turn submits it through the chain of command to the USAF representative.

AIRDROP / AIRLAND / EXTRACTION ZONE CONTROL LOG

DATE	4 MAR 89
LOCATION	FALCON DZ
CCT AND UNIT	1721 CCS SSGT EVANS SRA GILL
DZ/LZ/EZ CONTROL OFFICER AND UNIT	
DROP ZONE SAFETY OFFICER AND UNIT	2FSSO CAPT STARKEY

LEGEND

AH–Airland (Heavy)
AL–Airland
CD–CDS
ED–Extraction (Drogue)

EX–Extraction
GM–GMRS
HE–Heavy Equipment
HO–HALO

IL–Inverted "L"
LS–Instrument Landing System
PE–Personnel
RB–Radar Beacon Drop

TC–TT B CDS
TH–TT B Heavy
TP–TT B Personnel
WD–AWADS

SCORE METHOD
E – Estimated
P – Paced
M – Measured

LINE NO	TYPE ACFT	UNIT	CALL SIGN	PILOT/ NAVIGATOR	TYPE MSN	ETA	ATA	STRIKE RPRT		AL/EX		SURF WIND	SCORE METHOD	MEAN EFFECTIVE WIND			REMARKS (Continue on Reverse)
							ATD	YDS	CLOCK	S	U			TIME	ALT	DIR & VEL	
1 140	C-130	317	27 COUNTY		HE	1000	1000	250	5			0209	E	0945	1100	158/09	
2					PE		1012	DRY	PASS								
3					PE		1025	50	6								
4					PE		1029	600	5								
5																	
6 653-7	C-141	437	52 BASKO		CD	1100	1100	250	6			0210	E	1045	600	150/11	
7			54				1105	100	8								
8			55				1110	200	4								
9			59				1115	200	6								
10																	
11																	
12																	
13																	
14																	
15																	

MAC FORM FEB 75 168 PREVIOUS EDITION IS OBSOLETE

Figure 22-6. Example of completed MAC Form 168.

A. GENERAL

- (1) JA/ATT Sequence Number
- (2) Date (of Operation)
- (3) TOT (Local Time)
- (4) Type Mission
  - (a) Number of Aircraft
  - (b) Type Aircraft
  - (c) Type Assault Zone
  - (d) Type of Delivery (CARP, VIRS, GMRS)

B. PERSONNEL INVOLVED

- (1) Flying Unit
- (2) Unit Supported
- (3) DZSTL (Name/Rank/Unit)
- (4) Medics (In Place)
- (5) POC for Further Information

C. ASSAULT ZONE

- (1) Name/Type
- (2) Location
- (3) Any Deviations from Survey
- (4) Marked IAW the Survey

D. COMMUNICATIONS WITH AIRCRAFT

- (1) Type Radios
- (2) Frequency Used
- (3) Problems

E. WEATHER PASSED TO AIRCRAFT

- (1) Time of Observation
- (2) Time Weather was Passed to Aircraft
- (3) MEW
- (4) Surface Wind
- (5) Remarks

F. POST-INCIDENT WEATHER OBSERVATION

G. NARRATIVE

Figure 22-7. Example of suggested format for incident reporting.

22-9. SURVEYS

USAF DZs are surveyed by qualified CCT/DZST (for SOF only—JM qualified at minimum). However, CCT are not required to be JM qualified. All DZs will be surveyed or tactically assessed by qualified STT/DZSTL/JM/Pathfinder personnel prior to use. Procedures can be found in FM 3-21.38(FM 57-38) and AFI 13-217. All information concerning the DZ is placed on MAC Form 339, Drop Zone Survey (Figure 22-8, page 22-14 and Figure 22-9, page 22-15), or AF Form 3823, Drop Zone Survey (Figure 22-10, page 22-16 and Figure 22-11, page 22-17). These forms provide the user the essential information needed to operate the DZ. Section 4 of the forms states what type of missions may be conducted on the DZ.

NOTE: When supplies of MAC Form 339 are exhausted, it will be replaced with AF Form 3823.

a. **Contingency/Wartime Operations**. During contingency/wartime and major exercises, DZSTLs may be expected to tactically locate, inspect, and approve a potential DZ for follow-up airdrop of resupply or reinforcements.

b. **Tactical Assessment**. All services will conduct airdrops on approved DZs. The tactical assessment is an approved means to certify a DZ for airdrop on both fixed-wing and rotary-wing aircraft. All tactical DZ assessments that can be used for future operations and that meet the standard for USAF aircraft will be forwarded to the USAF for inclusion in the AZAR. This will be accomplished using the following checklist, AFI 13-217, and FM 57-38. The JM/DZSTL will ensure all DZ requirements are within standards for the type of personnel, parachutes, and equipment airdropped. Once a DZ has been tactically assessed, it must be approved using AF Form 3823 (Drop Zone Survey).

- DZ name or intended call sign.
- Topographical map series and sheet number.
- Recommended approach axis magnetic course.
- Point of impact location (eight-digit grid).
- Leading edge centerline coordinates (eight-digit grid).
- DZ size in meters or yards.
- Air traffic restrictions/hazards.
- Name of surveyor and unit assigned.
- Recommended approval/disapproval (cite reason for disapproval).
- Remarks (include a recommendation for airdrop option, CARP, GMRS, VIRS, or blind drop).
- All other requirements must be met for a nontactical drop zone.

NOTE: Airdrop operations on tactically assessed DZs are made only under the following conditions:

- During training events, using rotary-wing or fixed-wing aircraft.
- The airdrop is located within a military reservation or on US government leased property.
- The supported service accepts the responsibility for any damage that occurs as a result of airdrop activity.

- There is adequate time for safe, effective planning.
- All hazards and obstacles are identified.
- A detailed risk assessment is completed.
- The tactical assessment is documented on AF Form 3823 and is approved by the first O6 in the chain of command.

DROP ZONE SURVEY

DROP ZONE SURVEY	1. DZ NAME: ARKMAN CIRCULAR	2. LOCATION: Ft. Benning , GA
	3. MAP SERIES / SHEET NUMBER / EDITION / DATE OF MAP: Ft. Benning Reservation Map AFPP Map # 39 (1973)	

4. SURVEY APPROVAL/DISAPPROVAL DATA

4A. DATE SURVEYED: 30 Jan 90	TYPED NAME AND GRADE OF SURVEYOR: John P. Werkheiser SSgt	PHONE NUMBER (AUTOVON): 486-2535	UNIT: 1721 CCS , Pope AFB,NC
4B. DATE REVIEWED: 5 Feb 90	TYPED NAME AND GRADE OF REVIEWER: Montgomery, Paul J. Capt, USAF; UNIT AND LOCATION: 317 TAW/DOXT, Pope AFB, NC 28308	PHONE NUMBER (AUTOVON): 486-4300	SIGNATURE: [signature]
4C. DATE APPROVED: 5 Feb 90	TYPED NAME AND GRADE OF APPROVING AUTHORITY: THOMAS P. QUANCE, Colonel, USAF; UNIT AND LOCATION: 317 TAW/DO, POPE AFB, NC 28308-5000	PHONE NUMBER (AUTOVON): 486-4414	SIGNATURE: [signature]

4D. DROP ZONE APPROVAL/DISAPPROVAL	FOR	COVCRS	PE	HE	HALO/HAHO	SATB	CRRC	HSLLADS		
A. APPROVED	DAY	A	A	B	A	A	B	D		
B. DISAPPROVED	NIGHT	A	A	B	A	A	B	D		

5. COORDINATING ACTIVITIES

DZ CONTROLLING AGENCY OR UNIT: Directorate of Plans and Training , Ft. Benning, GA	PHONE NUMBER (AUTOVON): 835-5245
RANGE CONTROL: SKYWATCH: 277.5(UHF) TAMPA NESTOR: 38.60(VHF/FM)	PHONE NUMBER (AUTOVON): 835-5186

6. DZ DIMENSIONS (Yds / Mtrs) (For Circular DZ, Enter Radius Only)

A. LENGTH: N/A	B. WIDTH: N/A	C. RADIUS: 500 yards	TAPING POINT DISTANCES	D. T/L FROM DZ TRAILING EDGE: N/A	E. T/P FROM DZ CENTERLINE: N/A
POINT OF IMPACT (PI) DISTANCES FROM DZ LEADING EDGE		F. CDS PI: N/A		G. PE PI: N/A	H. HE PI: N/A

7. DZ AXIS DATA

A. MAGNETIC: 360	B. GRID (UTM): 360	C. TRUE: 359	D. DATE OF VARIATION DATA: 1970

8. GROUND POINT ELEVATION	A. CDS PI: 540'	B. HE PI: -N/A-	C. PE PI: 540'	D. HIGHEST: 540'

9. DZ COORDINATES

A. SPHEROID: CLARKE 1866	B. GRID ZONE: ZONE 16	C. EASTING: 7	D. NORTHING: 35
E. POINT OF ORIGIN: 0750 7287			

F. POINT	UTM COORDINATES	LATITUDE (DMS)	LONGITUDE (DMS)
DZ CENTERPOINT	07347360	32 16' 54.1" N	84° 47' 54.3" W
CDS PI	07347360	32 16' 54.1" N	84° 47' 54.3" W
PE PI	07347360	32 16' 54.1" N	84° 47' 54.3" W
HE PI	N/A	N/A	N/A

G. DZ CORNERS UTM COORDINATES	
LEFT LEADING EDGE	RIGHT LEADING EDGE
LEFT TRAILING EDGE	RIGHT TRAILING EDGE
LEFT TIMING PT	RIGHT TIMING PT

10. ZONE MARKER (ZM) POSITION DATA (Enter Additional ZM Points in Remarks)

A. PRIMARY POINT OF IMPACT	B. DISTANCE PRIOR / AFTER PI (Yards)	C. DISTANCE LEFT / RIGHT OF PI (Yards)

MAC Form 339, FEB 89 PREVIOUS EDITION IS OBSOLETE

Figure 22-8. Example of completed MAC Form 339 (front).

11. DZ DIAGRAM / 12. REMARKS

DZ NAME

ARKMAN CIRCULAR DZ, GA

NOT TO SCALE

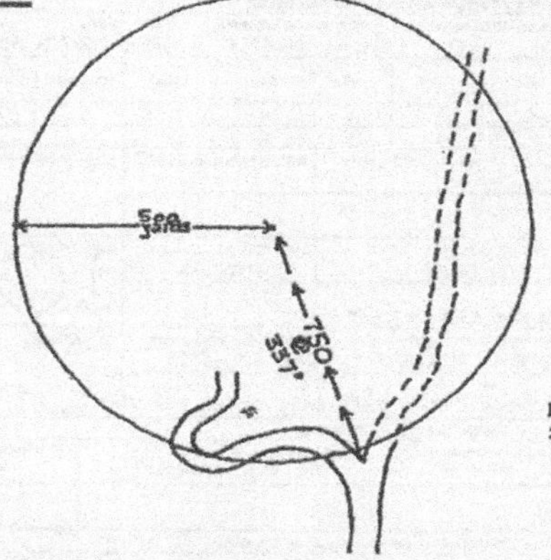

PI is located 750 yds at 337° from intersection.

1. DZ is for single ship operations only.
2. User accepts responsibility for injury to personnel and damage of equipment.
3. DZ is approximately 30% trees from 15' to 20' tall.
4. For pathfinder use, run in heading and PI will be pre-briefed.
5. Aircraft must contact Range Control at least 10 minutes prior to entry into the restricted area (R3002).
6. Ground party must establish radio communications with Range Control 30 mintes prior to TOT and maintain contact for duration of mission.
7. Recommend all turn outs to the west to avoid over flying the town of Cusseta, GA located 026° at 2000m.
8. Cellular One Telephone tower located 026° at 2600m. Height of tower is 795'MSL.
9. Dirt roads located on the south and east side of DZ.
10. TV tower 2249' MSL located at LAT/LONG 32° 19' 25"N, 84° 46'46"W; UTM grid coordinates 091784.

13. PHOTOGRAPHY AVAILABLE

☐ YES ☒ NO

LOW LEVEL ROUTES

☐ NONE AVAILABLE

☐ ROUTE NAME/DESIGNATOR

Figure 22-9. Example of completed MAC Form 339 (back).

AIRBORNE UNIT ASSUMES RESPONSIBILITY FOR PERSONNEL INJURY AND EQUIPMENT DAMAGE ON DZ

DROP ZONE SURVEY	1. DZ NAME LEDO I	2. LOCATION FT. BENNING, GEORGIA
	3. MAP SERIES/SHEET NUMBER/EDITION/DATE OF MAP AFPP MAP #39 FT. BENNING RESERVATION 1973/TPC G-21D/ATLANTA SECTIONAL	

4. SURVEY APPROVAL/DISAPPROVAL DATA

4A1. DATE SURVEYED	4A2. TYPED NAME AND GRADE OF SURVEYOR	4A3. PHONE NUMBER	4A4. UNIT
15 Aug 95	EDDIE D. KIDD, SSG	(DSN) 835-3218	HHC 1/507TH PIR

4B. DROP ZONE APPROVAL/DISAPPROVAL A = APPROVED D = DISAPPROVED	FOR	CDS/CRS	PER	HE	MFF	SATB	CRRC	HSLLADS	HVCDS	
	DAY	A	A	D	A	A	D	A	A	
	NIGHT	A	A	D	A	A	D	A	A	

4C. DATE APPROVED FOR GROUND OPERATIONS	NAME, GRADE AND SERVICE OF APPROVAL AUTHORITY	PHONE NUMBER (DSN)	SIGNATURE
	UNIT AND LOCATION		
4D. DATE SAFETY OF FLIGHT REVIEW APPROVED 31 Jan 95	NAME AND GRADE OF REVIEWING OFFICER JAMES B. GOOD, JR., Capt, USAF	PHONE NUMBER (DSN) 731-3988	SIGNATURE
	UNIT AND LOCATION 3134 OSS/OSTX, Little Rock AFB AR 72099-5043		
4E. DATE OF MAJCOM APPROVAL 17 Feb 95	NAME AND GRADE OF APPROVING AUTHORITY RONALD L. MOREY, COL, USAF	PHONE NUMBER (DSN) DSN 576-5007	SIGNATURE
	UNIT AND LOCATION HQ AMC/DOT, 402 Scott Drive Unit 3A1, Scott AFB IL		

5. COORDINATING ACTIVITIES

A. DZ CONTROLLING AGENCY OR UNIT	B. MEMORANDUM OF UNDERSTANDING/LAND USE AGREEMENT	C. PHONE NUMBER (DSN)
1/507th Para Inf. Reg. S-3, Ft. Benning GA	YES ☐ NO ☒ ATTACHED ☐	835-3012
D. RANGE CONTROL Range Control FM 38.60 UHF 277.5 (Skywatch)		E. PHONE NUMBER (DSN) 835-2152

6. DZ DIMENSIONS (YDS/MTRS) (FOR CIRCULAR DZ, ENTER RADIUS ONLY)

A. LENGTH	B. WIDTH	C. RADIUS	TIMING POINT DISTANCES	D. T/P FROM DZ LEADING EDGE	E. T/P FROM DZ CENTERLINE
925 yds	700 yds	N/A		N/A	N/A

POINT OF IMPACT DISTANCES FROM DZ LEADING EDGE	F. CDS PI	G. PE PI	H. HE PI
	350 yds	350 yds	N/A

7. DZ AXIS DATA (OPTIONAL FOR CIRCULAR DZ)

A. MAGNETIC	B. GRID (UTM)	C. TRUE	D. DATE OF VARIATION DATA
061 deg	058 deg	059 deg	1991

8. GROUND POINT ELEVATION	A. CDS PI	B. HE PI	C. PE PI	D. HIGHEST
	480'	N/A	480'	500'

9. DZ COORDINATES

A. SPHEROID	B. DATUM	C. GRID ZONE	D. EASTING	E. NORTHING
CLARKE 1866	1927 NA	16	7	35

F. GPS DERIVED COORDINATES	G. POINT OF ORIGIN
YES ☒ NO ☐	GL 1530 7913 Intersection of roads PI is 432m @ 224 deg

H. POINT	UTM COORDINATES	WGS84 LATITUDE (D-M.MM)	WGS84 LONGITUDE (D-M.MM)
DZ CENTERPOINT	GL 1509 7880	N 32d 19' 37.4"	W 84d 42' 54.1"
CDS PI	GL 1500 7874	N 32d 19' 35.7"	W 84d 42' 57.5"
PE PI	GL 1500 7874	N 32d 19' 35.7"	W 84d 42' 57.5"
HE PI	N/A	N/A	N/A

I. DZ CORNERS UTM COORDINATES

LEFT LEADING EDGE GL 1456 7884	RIGHT LEADING EDGE GL 1490 7830
LEFT TRAILING EDGE GL 1527 7929	RIGHT TRAILING EDGE GL 1562 7875
LEFT TIMING POINT N/A	RIGHT TIMING POINT N/A

AF FORM 3823, FEB 94 (EF-V1) (PerFORM PRO)

Figure 22-10. Example of completed AF Form 3823 (front).

DZ NAME

LEDO 1

10. DZ DIAGRAM/ REMARKS

1. User accepts responsibility for injury to personnel and damage to equipment and property that may result from an airdrop by AMC/ACC aircraft.
2. Drop zone is approximately 40% trees 35' to 50' tall.
3. TV tower 270 deg, 3.5 sm 2249' MSL
4. Radio tower 238 deg, 4.8 sm 795' MSL; tower 3 sm west of DZ 1010' MSL.
5. Prior coordination is required for access to R-3002.
6. Ft. Benning MOA located 3 NM past DZ. Contact DSN 835-5186 for info.

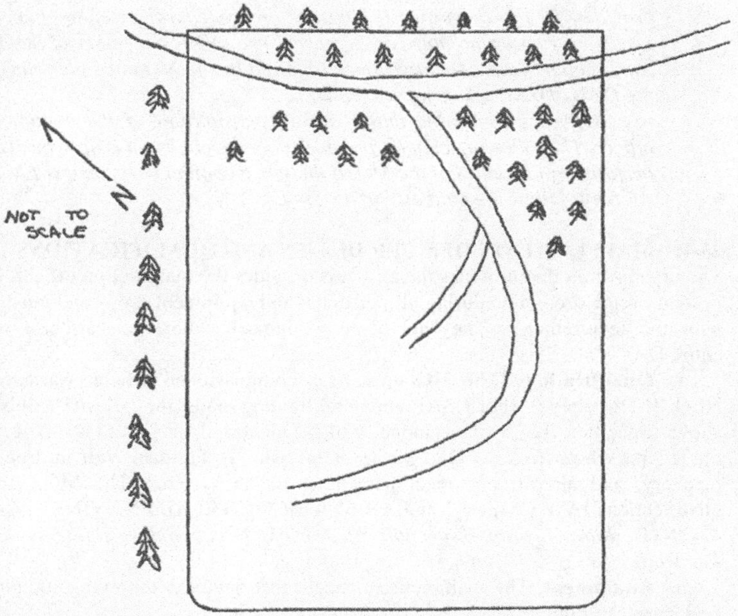

11. PHOTOGRAPH AVAILABLE

YES ☒ NO ☐

12. LOW LEVEL ROUTES

☒ NONE AVAILABLE

☐ ROUTE NAME/DESIGNATOR

AF FORM 3823, FEB 94 (REVERSE) (EF-V1) (PerFORM PRO)

Figure 22-11. Example of completed AF Form 3823 (back).

CHAPTER 23

MALFUNCTIONS REPORTING AND DUTIES OF THE MALFUNCTION OFFICER

The investigation of personnel, parachutes, and equipment malfunctions receives the highest priority and is secondary in priority only to medical aid for the injured. It supersedes all other aspects of the operation to include ground tactical play. Prompt and accurate investigations and reporting could save lives and equipment. The report provides data to determine if a system or procedural training change is necessary to prevent future occurrences. The MO is a member of the drop zone support team. Any assistance required by the MO must pass through the DZSO/DZST, who controls the DZ.

AR 59-4 governs the duties and responsibilities of the malfunction officer. The following chapter provides only the minimum requirements for performing the duties of the MO. If there is a conflict between this FM and AR 59-4, follow the guidance in AR 59-4.

23-1. MALFUNCTION OFFICER DUTIES AND QUALIFICATIONS

The organization that provides the air items provides the malfunction officer. He must be present on the drop zone during all personnel and equipment drops and must be familiar with the requirements. The MO positions himself where he can best observe the entire DZ.

a. **Qualifications**. The MO must be a commissioned officer, warrant officer, or NCO (E-5 or above). (For USAF unilateral training loads, the DZ MO will be an E-4 or above. Specific AFSCs of personnel will be identified in AFI 11-410.) He must be a trained parachute rigger (92R, 921A, 92D) who is familiar with airdrop, parachute recovery, and aircraft personnel parachute escape systems. The MO will meet the qualifications IAW Chapter 7 and AR 59-4/MCO13480. 1B/PNAVINST 4630.24C/AFJ 13-210(I), *Joint Airdrop Inspection Records, Malfunction Investigations, and Activity Reporting*.

b. **Equipment**. The malfunction officer must have the following equipment during performance of duty:

- A communication capability with the DZ control party.
- A good quality camera to take photographs of malfunctions or incidents, airdrop equipment, and impact sites. Photographic equipment is essential for the proper performance of MO duties. A video camera is optimal. Pictures recording a malfunction or the equipment involved greatly assist in the investigation.
- The forms and clerical supplies necessary to tag equipment and initiate reports.
- Binoculars (day) or night vision devices (night).
- Transportation to move around the DZ.

c. **MO Duties**. The MO performs the following actions in case of a malfunction:

(1) ***Partial or Total Malfunction with No Serious Injuries.***

(a) Secure and guard the impact site. Conduct an on-site investigation to determine, if possible, the cause of the malfunction, and to ascertain whether a criminal or malicious act may have caused the malfunction. (Refer to AR 59-4, Appendixes B and C, for checklists.)

(b) Photograph the malfunction as it happens (if possible), the malfunctioned equipment, and the malfunction or impact site. Sketch the impact site in relation to the DZ.

(c) Collect, properly identify, tag, and maintain chain of custody for items of physical evidence involved in the malfunction (such as clothing, equipment, air items, and personal property).

(d) Conduct a detailed component-by-component examination of all equipment.

(e) Collect statements from any personnel present who witnessed the malfunction. Take statements from JM/AJM, safeties, previous jumper, follow-on jumper, ground observers, and aircraft personnel who are able to provide facts.

(f) Appropriately tag and bag the collected equipment and evacuate it to an appropriate area to conduct the TM-10-1670-series rigger type investigation.

(g) Ensure all air items and evidence are retained until the investigating authority releases them.

(h) Prepare and submit the appropriate reports IAW AR 59-4, Chapter 5.

(2) ***Partial or Total Malfunction with Serious Injury or Death.***

(a) Place the impact area off limits and post a guard to ensure the integrity of the scene and to limit access to authorized personnel only. Take immediate possession of the log record book (DA Form 3912), and limit access to the appointed investigating officer. In the event of a death, ensure the DZSTL notifies the military police immediately. Access to the site is limited to the malfunction officer, the SME assisting the MO, and responding CID and medical personnel only.

(b) Photograph the parachutist, the impact site, and any obvious defects in the equipment (to include damage done by the impact). Sketch the impact site in relation to the DZ and mark the impact location of the parachutist and his equipment.

(c) Collect **all** items of equipment involved and maintain chain of custody.

(d) Immediately initiate an investigation prior to the arrival of the CID and ensure the scene is not altered. If determination of tampering or intentional sabotage is made, the entire investigation is turned over to the CID. If there has been no tampering or sabotage, continue with the investigation.

(e) Request that medical personnel secure and preserve the parachutist's clothing and equipment that is removed from the DZ.

(f) Gather statements from previous jumper, follow-on jumper, JM/AJM, safeties, and from any other ground or aircraft personnel who may be able to provide facts.

(g) Record the name and unit of any personnel who observed the incident even if the information is not new to the investigation.

(h) Secure a copy of the jump manifest. Reconstruct the jump stick from personnel present if required.

(i) Conduct a detailed component-by-component examination of all equipment after the parachutist has been evacuated.

(j) Request the DZSO/DZSTL notify the DACO so the aircrew can inspect the aircraft for any defects or damage that may have contributed to or caused the malfunction. Request DACO ensure the parachute deployment bags from that aircraft be identified and segregated from those of other aircraft.

(k) Obtain the deployment bag serial number from the log record book or parachute pack sheet and secure it with the remainder of the parachute assembly until the investigation is completed.

(l) When removing the parachute and related air items to a suitable area to perform the TM-10-1670-series rigger type investigation, ensure the parachute is loosely rolled, leaving any turns, tangles, or twists in the suspension lines. Appropriately tag and bag the air items and maintain chain of custody. Release these to the investigating SME only.

(m)Ensure that the DZSO/DZSTL signs for any organizational and personal clothing, weapons, and equipment not collected as part of the investigation. This equipment will be returned to the jumper's unit, and the unit will preserve the equipment until the completion of the investigation.

23-2. MALFUNCTION OFFICER RESPONSIBILITIES DURING INVESTIGATIONS

The depth of any investigation varies according to the severity of the malfunction and resultant injuries. In cases apparently not involving death or serious injury, the MO conducts the on-sight investigation solely to determine the cause of the malfunction and actions required to prevent future occurrences.

a. **MO Follow-On Investigation**. In cases involving misconduct, serious incident or injury, or death, the MO conducts a follow-on investigation according to service directives and AR 59-4. His investigative notes, insights, reports, and physical evidence are available to these investigations. The MO must ensure that after the component-by-component examination is completed, the equipment involved in the malfunction is secured and accessible only to the appointed investigating SME.

b. **Confidentiality of Investigation**. During the investigation, the MO gathers items of information and evidence that are sensitive in nature. He ensures the information pertaining to the investigation is given only to authorized personnel on a need-to-know basis. The MO must exercise great care so that the rights of involved personnel are not compromised and the government is not placed in an unfavorable position.

23-3. REPORTING DATA

The MO discusses as much data as allowable with the DZSO/DZSTL prior to transmitting it to the control group. The MO and the DZSO/DZSTL normally discuss this feeder information immediately after the jump is complete.

NOTE: The MO uses DD Form 1748-2 to report all airdrop personnel malfunctions IAW AR 59-4.

PART SIX
Special Airborne Procedures

CHAPTER 24
ADVERSE WEATHER AERIAL DELIVERY SYSTEM

AWADS is a navigational system installed in some USAF C-130 aircraft. It enables the aircraft to fly to a DZ during reduced visibility, and provides flexibility to the airborne force commander in the accomplishment of all airborne missions. AWADS is effective in large, joint operations, tactical reinforcements, and specialized missions.

24-1. MULTIPLE MISSION SUPPORT

AWADS operations facilitate rapid and continuous aerial deployment or resupply in adverse weather or darkness. Units can execute a parachute assault without a pre-positioned AFCCT or an Army DZST. As a result, time (length) of the air formation is shortened, and the air corridor must be cleared only once by tactical air to heighten the element of surprise. AFCCTs are introduced with the assault elements and assist the GUC with additional CDS, airland, or HD missions. AWADS enables a commander to conduct a rapid vertical reinforcement during instrument meteorological conditions or visual meteorological conditions for units threatened by enemy penetration.

NOTE: A minimum ceiling of 200 feet and a minimum visibility of 1/2 mile for personnel and equipment are imposed for tactical training.

24-2. TRAINING AND PREPARATION

AWADS operations demand detailed planning, rehearsal, training, and coordination between USAF and Army units to be effective. Due to limited visibility in the air and during assembly on the ground, AWADS requires both technical training and psychological preparation of the parachutists. AWADS sustained prejump training requires modification to normal jump conditions.

24-3. MODIFIED JUMPMASTER DUTIES

The JM relies on the loadmaster to obtain and relay en route information. Under AWADS conditions, the JM still attempts to perform all of the required checks. He may not be able to observe safety hazards beyond the immediate area of the door.

24-4. MODIFIED PARACHUTIST ACTIONS

Parachutist actions during descent under normal conditions must be modified in AWADS training. Modifications are made when using the T-10-series and MC1-series parachutes.

a. **T-10-Series Points of Performance.**

(1) Check body position and count.

(2) Check canopy and gain canopy control. Any malfunction under these conditions requires activation of the reserve since the parachutist cannot effectively judge the rate of descent.

(3) Keep a sharp lookout during descent.

(4) Do not slip, except to avoid collisions, until after breaking through the clouds.

(5) Give way. Give lower canopies the right of way; higher canopies slip to avoid them.

(6) Recheck the canopy. Make this check after breaking through the clouds if the canopy could not be checked while in the clouds.

(7) Prepare to land. Do not release equipment until the ground can be seen and it is clear below.

(8) Land. Execute a proper PLF.

b. **MC1-Series Points of Performance.**

(1) Check body position and count.

(2) Check canopy and gain canopy control. If the TU modification is to the front, there is a complete inversion. Do **not** activate the reserve since the canopy can still be controlled. For any other malfunction, activate the reserve immediately.

(3) Identify steering toggles. Grasp each one and bring them both to chest level to reduce lateral movement and perform braking; this helps to eliminate midair collisions and extreme dispersion due to excessive drift.

(4) If either or both toggles are broken, steer the canopy by pulling a slip with the rear riser on the side of the intended turn. In the clouds, anyone with a broken toggle and not applying "brakes" automatically has a greater lateral drift than anyone else and must keep alert during descent.

(5) Keep a sharp lookout during descent. During reduced visibility prepare to take immediate evasive action.

(6) Give way. Give lower canopies the right of way. If parachutists see the possibility of converging at any altitude and from any direction, they immediately turn away from each other by pulling the toggle that is away from the other parachutist.

(7) Recheck canopy. Perform a thorough canopy inspection after breaking through the clouds. Release the toggles when the ground is in sight and prepare equipment for landing. Use the lowering lines.

(8) Prepare to land. Turn into the wind 75 to 150 feet above the ground. Obstacles spotted when coming out of low clouds normally require rapid preparation for the appropriate emergency landing.

(9) Land. Execute a proper PLF.

CHAPTER 25

DELIBERATE WATER DROP ZONE OPERATIONS

Units conducting water drop zone operations should use the following procedures as a guide to ensure the mission is conducted safely. Conduct all jumps under the following conditions:

- *Daylight; warm weather conditions.*
- *Surface winds do not exceed 17 knots.*
- *Drop control—visual marking system.*
- *Helicopter—flying under visual flight conditions.*
- *Wind conditions permit operation on the WDZ.*
- *Water should be more than 10 feet deep with no underwater obstacles that depth.*
- *Briefing held for WDZ operations by DZSO/DZSTL and WDZ control personnel.*
- *All parachutists are classified as swimmers or strong swimmers. (USMC S-1 or higher).*
- *All parachutists have completed drownproofing training within the preceding 12 months.*
- *All parachutists have been trained on activation procedures for the life preserver in use, to include manual inflation procedures.*
- *All parachutists have attended prejump training on deliberate water DZ procedures according to unit SOP.*
- *All combat equipment is float checked.*

NOTE: The parachute operation OIC briefs parachutists, aircrew, JM, and WDZ control party regarding WDZ operations.

WARNING

The MC1-1B should be used, if available. The porosity of the MC1-1C is extremely limited and may create a suffocation hazard if the jumper is trapped under the canopy.

25-1. PERSONNEL AND EQUIPMENT

The following personnel and equipment are required for deliberate WDZ operations:

a. **Personnel.**

(1) JM, AJM, safety personnel, as required for the type of aircraft used.

(2) DZSO/DZSTL, JM qualified and current IAW Chapter 7.

(3) Medic/corpsman with resuscitator.

(4) Safety vehicle driver.

(5) Boat commander/coxswain for each boat.

(6) Malfunction officer.

(7) Safety swimmers—minimum of one safety swimmer is required to be on board each recovery boat. The safety swimmer must have fins, facemask, knife, and an inflatable life preserver. The safety swimmers will be used to recover personnel and equipment, and assist parachutists as needed. The safety swimmer cannot be additionally assigned as the boat coxswain or corpsman. The safety swimmer should be a qualified parachutist and, as a minimum, will be qualified as a US Army Class 1 advanced survival swimmer (IAW TC 21-21), USMC CWS1, Navy 1st class, or lifeguard certified. The safety swimmer will attend prejump training for intentional water jumps with parachutists and wet silk training.

b. **Water Drop Zone Equipment**.

(1) Equipment for safety swimmers.

(2) Serviceable boats, motors, and fuel (RB-15 boats, 14-foot engineer assault boats, combat rubber raiding craft, or civilian equivalent boats with motors). If available, hard-body boats should be used.

(3) Equipment for recovery and command boats.

(4) Required panels and smoke.

(5) Required communications equipment for each boat to include spare batteries and one complete spare radio set for the operation. Boat-to-boat and boat-to-air communication should be checked.

(6) Floating, nonflammable container with suitable anchor for smoke grenades.

(7) First aid equipment to include resuscitator and backboard.

(8) Bailing cup per boat.

(9) Motor tie-down rope per boat.

(10) Sheath knife (boat commander and safety swimmer) per boat.

(11) Pliers per boat.

(12) One extra life preserver for emergencies per boat.

(13) Life jackets or life preservers for all personnel onboard each boat.

(14) Oars or paddles per boat.

(15) Boat hook per boat.

(17) Approved anemometers. The approved anemometers are the DIC, DIC3, TurboMeter, and AN/PMQ-3A. (The AN/ML433A/PM and the anemometers that use floating balls or small floating lightweight aluminum devices in a tube are not authorized for use during personnel or cargo airdrop operations. See Chapter 20, paragraph 20-9 for use and calibration information.)

c. **Parachutist Requirements.** Parachutists, at a minimum, must be classified as a US Army Class 3 basic survival swimmer, USMC CWS-1, or Navy 1st class swimmer before making a water parachute drop. Parachutists will be current static line jumpers before making water jumps. Wet silk training for intentional water jumps will be conducted at a minimum of once every six months by all parachutists and safety swimmers involved in water jumps.

a. An individual's first water jump must be performed during the day and without combat equipment. To be classified as a Class 1 advanced water survival swimmer, personnel will be tested and documented on an annual basis and must pass the requirements contained in TC 21-21.

b. Wet silk training is conducted at the unit level by certified personnel (Class 1 advanced water survival swimmer [IAW TC 21-21], USMC CWS1, Navy 1st class, or lifeguard-certified current JM) by putting an unserviceable parachute in a pool, in a controlled environment, with safety swimmers. The jumpers, one at a time, demonstrate the actions to take during a water landing, jump into the water, and swim under the canopy to experience what it is like to be under a parachute in the water. The jumper demonstrates his ability to follow a radial seam to get safely out from under the canopy, make an air pocket under the canopy to breathe from, and breathe from the apex of the canopy. If a jumper is trapped under the canopy during wet silk training, the safety swimmers pull the jumper out of the water. A minimum of two safety swimmers with mask, fins, snorkel, and dive knife and a medic/corpsman will be available during this training. A standby diver on SCUBA is also recommended.

25-2. ORGANIZATION AND EQUIPMENT OF DROP ZONE DETAIL

The organization and equipment of the drop zone detail include the following:

a. **DZSO Command Boat**. The command boat will be separate from the recovery boats. It will be used by the DZSO and medic/corpsman and will not be used to recover parachutists or equipment except in the case of an emergency. The command boat (boat number 1) includes the following personnel and equipment:

NOTE: All boats required for the operation will be on station with required personnel and the engines running before the release of parachutists.

(1) ***Personnel.***

- Drop zone safety officer.
- Boat commander/boat coxswain.
- Safety swimmer with equipment.
- Medic/corpsman with equipment.
- Malfunction officer.

(2) ***Equipment.***

- Marker panels.
- Medical aid kit.
- Backboard.
- Resuscitator.
- Boat hook.
- FM radio (complete) with extra battery for boat-to-boat, boat-to-air, and surface communications.
- Approved anemometers. The approved anemometers are the DIC, DIC3, TurboMeter, and AN/PMQ-3A. (The AN/ML433A/PM and the anemometers that use floating balls or small floating lightweight aluminum devices in a tube are not authorized for use during personnel or cargo airdrop operations. See Chapter 20, paragraph 20-9 for use and calibration information.)
- Bailing cup.
- Motor tie-down rope.
- Sheath knife (safety swimmer and boat commander).

- Pliers.
- Life jackets or life preservers for all personnel on board.
- One extra life preserver for emergencies.
- Oars or paddles per boat.

b. **Recovery Boats for Equipment Drops**. A minimum of one power driven boat is required for two equipment platforms dropped on the same pass. Equipment recovery boats are to be used in the recovery of equipment parachutes and platforms. Recovery boats assigned to recover jumpers doe not meet this requirement when jumpers and equipment are on the same pass. Equipment recovery boats must be large enough to recover cargo parachutes and platforms. The boat coxswain's only duty is to navigate the boat. The boat coxswain cannot act as the safety swimmer or corpsman.

(1) Use extreme caution when recovering CRRC and other equipment in the water.

(2) The jumpers are the recovery personnel. They will have assisted in the rigging of the CRRC and container equipment before the airdrop. This will help they know what straps to cut without losing the equipment or injuring other personnel.

(3) Ensure a safety tie is rigged between the platform and boat and is cut last after the recovery boat has secured the platform.

(4) Tow the platform back to shore with the recovery boat. Place sandbags on the back of the platform to raise the front to ease towing.

WARNING

Ensure that all personnel are clear of the platform and equipment and not fouled in any lines before cutting the equipment loose from the platform. Failure to do so may cause serious injury or death if the platform sinks.

b. **Recovery Boats for Personnel.** A minimum of one power driven recovery boat is required for every three parachutists being dropped on the same pass. Parachutist recovery boats must have an inflatable boat or ladder rigged along side if they have a freeboard of more than 3 feet and or the boats do not provide an easy platform for recovery of personnel. Boats assigned as personnel recovery platforms may only be used to assist in the recovery of airdropped equipment after all parachutists have been recovered. The boat coxswain's only duty is to navigate the boat. The boat coxswain cannot act as the safety swimmer or corpsman.

c. **Equipment and Personnel Requirements for Recovery Boats.** Recovery boats require the following personnel and equipment:

(1) ***Personnel (Personnel and Cargo Recovery Boats).***

- Boat commander/coxswain.
- Safety swimmer with equipment.

(2) ***Equipment (Personnel and Cargo Recovery Boats).***

- Boat hook.
- Bailing cup.
- Motor tie-down rope.

- Sheath knife (safety swimmer and boat commander).
- Pliers.
- Life jackets or life preservers for all personnel onboard.
- Extra approved life preserver for emergencies.
- Oars or paddles per boat.
- Radio (complete) with extra battery for boat-to-boat surface communications.

(3) ***Safety Swimmer Equipment.***

- Swim fins—one pair.
- Face mask.
- Snorkel.
- Sheath knife.
- Life vest.
- Wet suit—optional.

(4) ***Safety Vehicle Operator.***

- Driver (on standby).
- Radio (communicate with command boat).

NOTE: The JM briefs the parachutists, aircrew, and WDZ control party regarding WDZ operations.

(5) ***DZSO (Additional Duties).***

(a) Briefs boat crews, safety swimmers, medic, and safety vehicle driver on the following:

- Overall organization of WDZ.
- Number of lifts and personnel to be dropped.
- Drop altitude and aircraft heading.
- Surface winds.
- Water depth.
- Turnaround time between drops.
- Recovery procedures.
- Communications plan.
- Emergency recovery and evacuation plan.
- Drop and abort signals.
- Applicable special instructions.

(b) Maintains visual observation of all parachutists until safely recovered.

(6) ***Jumpmaster.***

(a) Performs JM duties.

(b) Helps recover personnel from the water in an emergency.

(c) Keeps parachutists under observation. Does not release the next pass until all parachutists have been recovered.

(7) ***Boat Commander/Coxswain.***

(a) Ensures all personnel and equipment are on board.

(b) Ensures that all equipment is operational.

(c) Ensures that the safety swimmer has been briefed and understands instructions.

(d) Maintains visual observations of parachutists from time of exit to safe recovery.

(e) Controls actions of safety swimmer.

(f) Ensures all personnel are trained and rehearsed in their duties.

(8) ***Safety Swimmer.***

(a) Maintains visual observation of parachutists from time of exit to safe recovery. Is alert at all times for parachutists in trouble and is prepared to enter the water to assist parachutists.

(b) Enters water (on order of boat commander) and assists parachutist recovery.

(9) ***Medic/Corpsman.***

(a) Ensures the resuscitator is complete and in operational condition.

(b) Ensures that safety personnel understand lifesaving techniques.

(c) Uses applicable medical equipment and provides required first-aid treatment and medical evacuation.

25-3. SAFE CONDITIONS

Units conducting WDZ operations should use the following to ensure the mission is conducted safely. Ensure all WDZ parachute jumps are conducted under the following conditions for all services.

- Water will be more than 10 feet deep.
- The WDZ will not be in or near the surf zone.
- Surface winds will not be in excess of 17 knots.
- Sea state is no more than 2. (IAW MCWP 2-15.3, Table 7-1, or Joint Pub 4-01.6, sea state 2 indicates that wave height will not exceed 1 2/3 feet (20 inches), as measured from a wave crest and the preceding trough.)
- Parachutists jumping in cold water (60 degrees or lower) will wear wetsuits or dry suits.
- All parachutists are classified as strong swimmers, USMC CWS-1, or Navy 1st class swimmers.
- All safety swimmers should be parachutists and will be classified as Class 1 advanced water survival swimmers (IAW TC 21-21), USMC CWS-1, Navy 1st class swimmer, or lifeguard certified.
- All parachutists and swimmers have completed wet silk training in the past six months.
- All parachutists and swimmers have been trained on activation procedures for the life preserver in their use to include manual inflation procedures.
- All floatation devices are approved, serviceable, and properly maintained.
- All parachutists have attended prejump training on deliberate water DZ procedures according to this chapter and unit SOP.
- All combat equipment is waterproofed and float checked.
- All jumpers and equipment have chem lights and strobes attached for night jumps.
- Night vision devices will be used on the WDZ for all night drops.
- Drop control—visual marking system: mirror, smoke, VS-17 panels.
- Release method: CARP or WSVC.
- Helicopter or fixed-wing—flying under visual flight conditions.
- Only one to three parachutist jumps for each recovery boat for each pass. (Boat-to-jumper ratio depends on unit and jumper experience with water jumps.

One boat to one jumper is strongly recommended for units that rarely conduct WDZ operations.)

- Sufficient boats are properly manned with the engines running during all live passes.
- Clear to drop is by radio or by all the boats driving in one big circle.
- Abort the pass is by radio or by all the boats being scrambled.

25-4. JUMP RECOVERY PROCEDURES

After parachutists enter the water, recovery boats (boat numbers 2 and 3 and any additional recovery boats) proceed to the location of designated parachutists and begin recovery operations.

a. **Boat Commander/Coxswain.** The boat commander/coxswain of each recovery boat identifies his assigned parachutists as soon as possible.

(1) The boat coxswain determines whether the parachutist's life preserver has been inflated. If the life preserver has not inflated, the recovery boat coxswain immediately proceeds to the impact point, taking care to stay out of the way of the other parachutists. Upon reaching the assigned parachutist, the boat coxswain treats the situation as a parachutist in distress and takes appropriate action.

(2) The DZSO command boat (boat number 1) is located so the DZSO can observe the landing of parachutists and reinforces recovery boats with additional safety swimmers and a resuscitator as required.

(3) The following procedure is used for recovery operations.

(a) When possible, recovery boats approach parachutists from alongside. This will help avoid getting the parachute stuck in the engine crop. Recover parachutes by the apex.

(b) If the parachutist experiences no difficulties after impact, he signals "All OK" by raising one arm straight up (without waving).

NOTE: **Any other signal** or **no signal** given by the parachutist is considered as a distress situation and immediate action is taken.

(c) The boat coxswain directs the boat alongside the parachutist; the safety swimmer extends the boat hook so that the parachutist can grasp it. If the parachutist is unable to grasp the hook, the safety swimmer secures a portion of the parachutist's equipment with the hook.

b. **Safety Swimmers.** On instruction from the boat commander/coxswain, the safety swimmer enters the water alongside the parachutist to assist the parachutist and to recover the parachute canopy and other equipment. If a parachutist has gone underwater, the following action is taken.

(1) Red smoke is displayed to indicate an emergency and a message is transmitted to the command boat by radio. All other activities cease. (A red smoke grenade may be activated and dropped in an ammunition can bolted to a wood plate, or a smoke machine may be used.)

(2) The safety swimmer dives to recover the parachutist and cuts him free of equipment as required.

(3) The DZSO moves his boat to the scene and the medic/corpsman prepares to use the resuscitator and backboard.

c. **Recovery Helicopter (if used).** The jump aircraft is alerted to assist in emergency recovery operations.

(1) When directed by the DZSO, the pilot stands-by the location to provide medical evacuation by air, if necessary.

(2) After normal recovery operations, parachutists and equipment are unloaded at the assembly area. All boats are repositioned for the resumption of jump operations.

25-5. WATER DROP ZONE PREJUMP TRAINING

Prejump training for water jumps should be conducted within 24 hours before the scheduled jump. This training includes demonstrations and practical exercises for all parachutists; use of the suspended harness is recommended. Prejump training includes, but is not limited to, the following:

a. Six points of performance for water jumps.
  - Check body position and count.
  - Check canopy and gain canopy control.
  - Prepare to land with a life preserver (B-7 or UDT, see below).
  - Keep a sharp lookout during descent and maneuver to indicated impact area.
  - Prepare to land by turning and facing into the wind, maintaining position until just before landing.
  - Prepare to make a PLF (if water is shallow or ground contact is made).

b. Emergencies in the air.

c. Emergencies in the water.
  - Ensure the life preserver is fully inflated.
  - If trapped under the canopy, the jumper finds a radial seam and follows it to the edge.
  - If trapped under the canopy, the jumper lifts his hand slowly and creates an air pocket to breath, then cuts a hole in the canopy material. (If the jumper lifts his hand too quickly, he will create a suction and the canopy will stick to his face and arm.)
  - Wet silk training will have been completed and documented within six months prior to the jump.

d. Recovery procedures.
  - Recover parachutes by pulling them in the boat by the apex.
  - Ensure the boat is kept free of gasoline and oil, which would ruin the parachutes.

e. Orientation at the WDZ for the WDZ crew.

f. Any special instructions.

25-6. PROCEDURES FOR DELIBERATE WATER LANDINGS WITH A LIFE PRESERVER

The procedures for deliberate water landings with a life preserver are as follows:

- Parachutists may wear combat equipment only after it has been waterproofed and float checked. To waterproof equipment, use bags and rigger tape inside equipment containers. Float check by submerging in a tank of water until saturated and ensure the equipment is neutral or positive buoyant (floats).

- Parachutists will wear service-approved life preservers.
- Parachutists will wear the ballistic helmet or a service-approved helmet during WDZ operations.
- Parachutists jumping in cold water (60 degrees or below) will wear wetsuits or dry suits and may wear a wetsuit/dry suit hood under the ballistic helmet.
- Parachutists may wear wetsuit booties, coral booties, or tennis shoes.
- Parachutists may jump with mask and fins. Fins to be worn on the feet will be taped or tucked under the left arm and should be dummy-corded to the jumper with 1/4-inch cotton webbing.
- All jumpers will wear a knife for emergencies.
- Jumpers may follow equipment, such as door bundles, CRRC, and RAMZ, on a single-ship operation. There will be three seconds between the cargo exit and the first jumper's exit to ensure the jumpers do not become entangled with the cargo.
- The cargo and jumpers are hooked to the same anchor line cable on the same side when jumping the ramp.

a. **Deliberate Water Landing with the B-7 Life Preserver.** The B-7 is worn under the parachute harness with the inflatable portions under the jumper's armpits (Figure 25-1, page 25-10). At altitude, the jumper releases all equipment tie-downs and lowers equipment, but does not jettison it.

(1) During the descent, the parachutist inflates the life preserver by discharging the attached CO_2 cartridges. If necessary, the life preserver can be inflated by blowing air into the inflation valve hose.

(2) After entering the water, the parachutist activates both canopy release assemblies. He pulls the cable release safety clip out and away from his body (exposing the cable loops) and activates the canopy release assembly using one of the two methods (hand to shoulder or two-hand assist) as his feet touch the water.

(3) Since the B-7 life preserver will provide 500 pounds of positive buoyancy, the parachutist does not remove the harness and equipment.

(4) The parachutist swims upstream or upwind away from the parachute to avoid becoming entangled, and signals "All okay" to the recovery boat.

WARNING

Do not release the canopy release assemblies until the feet make contact with the water. Altitude is hard to judge over the water. If the jumper activates his canopy releases at even moderately high altitude (for example, 50 feet), serious injury or death may result.

CAUTION
The jumper must stay clear of the suspension lines to avoid entanglement. If the parachute and equipment are to heavy and hard to hold on to, the jumper lets go of the equipment to avoid being pulled under.

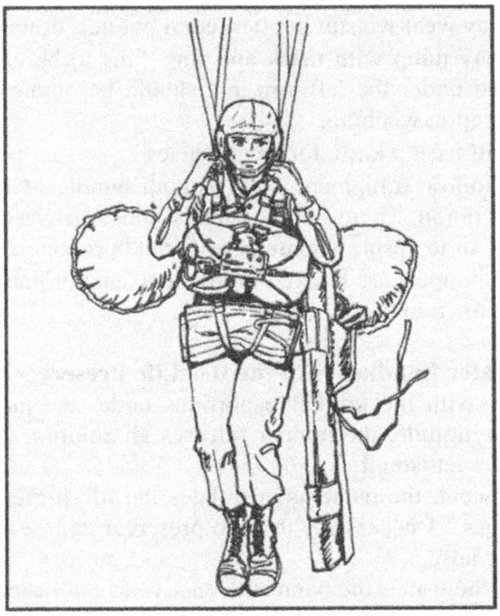

Figure 25-1. Landing with the B-7 life preserver.

b. **Deliberate Water Landing with UDT Life Preserver.** Parachutists wearing a UDT vest route the chest strap **under** the UDT vest so that the chest is not crushed if the UDT is inflated.

(1) Upon exiting the aircraft, the jumper checks the canopy and steers away from other jumpers. At altitude, the jumper releases all equipment tie-downs and lowers equipment, but does not jettison it.

(2) The jumper activates the quick-release on the waistband and unsnaps the left connector snap on the reserve parachute. He rotates the reserve to the right side of the parachute harness and seats himself well into the saddle.

(3) The jumper activates the ejector snap on the chest snap and inflates his life preserver with air by blowing into the inflation tube or activating the CO2 cartridge.

(4) Sitting back in the harness and using the right hand to hold onto the left main lift web, the jumper steers with his left hand. Just before entering the water, he places his hands on the ejector snaps on the leg straps.

(5) The jumper activates the ejector snaps on the leg straps, throws his arms up, and arches out of the harness when entering the water. He prepares to execute a PLF if the water is shallow.

(6) The jumper swims upstream or upwind away from the parachute to avoid becoming entangled, and signals "All okay" to the recovery boat. He recovers the parachute by grabbing the apex.

CAUTION

The jumper must stay clear of the suspension lines to avoid entanglement. If the parachute and equipment is too heavy to hold on to, the jumper legs go of the equipment to avoid being pulled under.

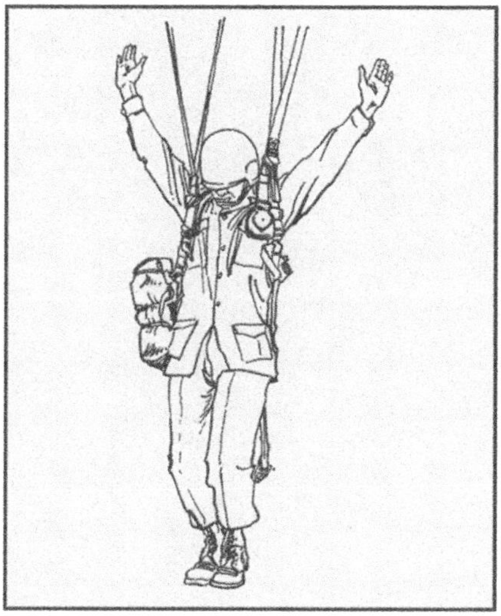

Figure 25-2. Landing with a UDT life preserver.

c. **Landing Procedures.** Upon landing, personnel must be prepared to perform the following procedures.

(1) If being dragged in the water, parachutists activate one (or both) canopy release assemblies after entering the water.

(2) Parachutists prevent fouling in the canopy or suspension lines during severe wind conditions.

(3) Parachutists signal "All okay" by extending one stationary arm overhead (DO NOT WAVE) or "Help" (any other signal or no signal) to the recovery boat, and stand by for pickup, remaining calm.

d. **OIC Responsibilities.** The parachute operation OIC ensures that—

- All personnel scheduled for a water jump are not less than Class 3 basic water survival swimmer qualified IAW TC 21-21 (Army), USMC CWS-1 (USMC), or Navy 1st class swimmers (USN).
- All jumpers and safety swimmers have completed wet silk training within the past six months.
- All equipment has been waterproofed and float checked.
- All jumpers are using approved serviceable floatation.
- All jumpers have completed the required prejump training.

CHAPTER 26

EXIT PROCEDURES

Two types of exit procedures can be used on fixed-wing aircraft: ADEPT exit and mass exit.

DANGER

ONLY ONE TYPE OF PARACHUTE (T-10- OR MC1-SERIES) MAY BE USED DURING ANY ONE GIVEN PASS OVER A DROP ZONE. THE MIXING OF PARACHUTES THAT HAVE DIFFERENT OPENING PERFORMANCE CHARACTERISTICS, SUCH AS NONSTEERABLE (T-10-SERIES) VERSUS STEERABLE (MC1-SERIES), CAN CAUSE HIGH-ALTITUDE INCIDENTS DURING ADEPT OPTION 2 OR MASS EXIT OPERATIONS.

26-1. ALTERNATE DOOR EXIT PROCEDURES FOR TRAINING (ADEPT) OPTIONS 1 AND 2

The ADEPT options are used when jumping the T-10- and MC1-series parachute. This training safety measure allows the maximum number of parachutists to exit the aircraft with a minimum risk of high-altitude entanglements.

a. **ADEPT Option 1**.

(1) During a single pass over the drop zone, only one stick of parachutists on one side of the aircraft jumps. After issuing the eighth jump command, the JM turns toward the paratroop door and regains control of his static line. He is positioned close enough to the door to control the flow of jumpers but far enough back not to impede the jumpers' movement to the door. The JM controls the number 1 jumper by grasping the jumper's saddle. When the jump caution lights turn green, the JM (on the active door for the pass) issues the ninth jump command, GO, to the first parachutist and taps him on the thigh. The JM controls the flow of parachutists (performing a visual inspection of each parachutist, his static line, and his equipment as he approaches the door) and observes the jump caution lights; the safety takes the static lines.

(2) Once the last parachutist exits the active door, the safety visually clears to the rear, gives the loadmaster a thumbs-up signal, and, with the assistance of the loadmaster and or static line retriever, pulls in the static lines and deployment bags.

(3) During the pass, the JM in the inactive door performs outside air safety checks and then observes the actions in the active door, since (under this option) he will not have any parachutists exiting his door.

(4) Subsequent passes alternate from door to door until all parachutists have exited. The AJM is the last parachutist on his side of the aircraft. The JM is the last parachutist onboard the aircraft; he exits from his door.

b. **ADEPT Option 2**.

(1) During a single pass over the drop zone, one stick of parachutists exit from the JM's door, followed by a stick of parachutists from the AJM's door. After issuing the

eighth jump command, the JM turns toward the paratroop door and regains control of his static line. He is positioned close enough to the door to control the flow of jumpers but far enough back not to impede the jumpers' movement to the door. The JM controls the number 1 jumper by grasping the jumper's saddle. When the jump caution lights turn green, the JM (on the primary door for the pass) issues the ninth jump command, GO, to the first parachutist and taps him on the thigh. The JM controls the flow of parachutists (performing a visual inspection of each parachutist, his static line, and his equipment as he approaches the door) and observes the jump caution lights; the safety takes the static lines.

(2) The AJM observes the actions in the JM's door. When he sees only three parachutists remaining in the JM's stick, he faces his parachutists and issues the eighth jump command. When the AJM sees the last parachutist clear the jump platform in the JM's door, he turns and rechecks his jump caution lights. (The JM is now observing the actions in the AJM's door.)

(3) If the jump caution lights are still green, the AJM verbally issues the ninth jump command, GO, to his first parachutist and taps him on the thigh. He controls the flow of parachutists (performing a visual inspection of each parachutist and his equipment as he approaches the door) and observes the jump caution lights; the safety takes the static lines.

(4) When the last parachutist exits from his side of the aircraft (last pass), the AJM gives his static line to the safety, checks the jump caution lights, and, if they are green, exits. The JM, seeing the AJM exit, passes his static line to the safety, checks the jump caution lights, and, if they are green, exits.

(5) Safety personnel visually clear to the rear of the aircraft and give the loadmaster a thumbs-up signal. They help the loadmaster recover static lines and deployment bags.

26-2. MASS EXITS

This exit procedure is used only when jumping the T-10-series parachute. During this type of exit, parachutists may exit from both doors at the same time. The JM gives the command GO, and the AJM then turns and gives his jumper the command GO to create a staggered effect.

CHAPTER 27
BUNDLE DELIVERY SYSTEM (WEDGE)

The bundle delivery system (referred to as a wedge because of its shape) is constructed with two Type II, 8-foot, modular airdrop platforms; a lightweight aluminum frame; four sections of skate wheel roller conveyors; and bundle release assemblies. The wedge weighs 930 pounds. It is designed to lock into the ramp restraint rails and to provide adequate tie-downs to secure various A-7A and A-21 configurations for flight. When positioned on the raised cargo ramp in flight, the wedge provides an inclined platform so that bundles can be released by rolling them out and off the ramp immediately before or after parachutists exit the doors.

27-1. APPLICATION
The wedge provides minimum dispersion between parachutists and equipment bundles on a single pass over the drop zone. The bundles can be released individually or in any group combination from the wedge when the loadmaster pulls the release pins from one or more release plates, depending on the number of bundles to be released at one time. Each release plate system employs a 1/8-inch-diameter steel release pin, a restraint release plate, and a 1-inch-wide tubular nylon restraint strap that secures each bundle, or each combination of bundles, to the wedge. A 1-inch-wide tubular nylon strap lanyard is attached to each steel release pin. Releasing the restraint strap allows the bundle(s) to roll off the aircraft ramp under the force of gravity.

27-2. RESTRICTIONS
The following restrictions apply when using the wedge:

a. **General Restrictions**. The wedge accommodates six A-7A or A-21 aerial delivery bundles, or a combination thereof, aboard the C-130 for cargo ramp airdrop. It limits parachutists to 20 for each door. (The anchor line cable stops are positioned about 26 inches forward of the center anchor line cable supports.)

NOTE: Wedge bundles can be dropped only on the first pass across the drop zone.

b. **Bundle Sizes**. Bundle sizes are limited to 27 inches long by 42 inches wide by 48 inches high, to include parachutes. The total rigged weight for each bundle must not exceed 538 pounds. Maximum allowable weight is 3,228 pounds for each six bundles. For example, maximum dimensions for mortar bundles are 27 inches long by 60 inches wide by 27 inches high (height dimension does not include the cargo parachute, paperboard honeycomb, and skid board). Mortar bundles must weigh between 320 pounds and 538 pounds (weights include skid board and cargo parachute). These dimensions allow up to three mortar bundles to be configured on the wedge, or mortar bundles can be mixed with smaller (27 inches by 42 inches) bundles to facilitate cross-loading. Two configurations are possible when mixing bundles: one mortar and four normal-size bundles (27 inches by 42 inches) or two mortar and two normal-size bundles.

c. **Aircraft**. During tactical training (visual or instrument flight regulations), bundle drops are restricted to the first three aircraft.

27-3. RIGGING PROCEDURES

Aerial delivery units usually rig the wedge; however, the system may be rigged in the airborne unit area. The areas of responsibility overlap; therefore, the JM must be familiar with the rigging procedures. However, the wedge must have a final inspection by a qualified rigger before the wedge leaves the rigging site.

27-4. BUNDLE DROP SEQUENCE

Bundles may be dropped from the wedge in the following sequence:

- A single bundle before and after the parachutists exit.
- A single unit of two to three bundles before or after the parachutists exit.
- Two units of two or three bundles for each drop with one unit dropped before and one unit after the parachutists exit.

The bundle sequence is determined before loading the wedge. If single bundles are to be dropped, a restraint strap and release lanyard are required for each bundle. If multiple bundles are to be dropped in units of two or three, a restraint strap and release lanyard are required for the aft bundle in each unit. Three exceptions are as follows:

a. **A-7 and A-21 Bundles.** The skid board size is 27 inches by 42 inches by 3/4 inch. Drill the skid board and center a 1-inch hole on the 42-inch side, 1 inch from the edge. Place two layers of honeycomb the size of the bundle on the skid board.

b. **Bundle Restraint Strap.** A restraint strap is formed by cutting a 12-foot length of 1-inch-wide tubular nylon webbing (requirement for maximum-size bundle). Form a 6-inch loop, with an overhand knot, 3 feet from one end of the strap. Lay the tie across the honeycomb with the loop next to the center hole in the skid. The 3-foot length of strap should be at the front of the skid (the front is the side with the 1-inch hole nearest the edge). Place the bundle on top of the honeycomb and rig the bundle (FM 10-500-3/ FMFM 7-47/TO 13C7-1-11, which supersedes FM 10-501/TO 13C7-1-11). Place the running ends of the restraint strap on top of the rigged bundle in preparation for loading on the wedge.

c. **Cargo Parachutes.** Use G-14 cargo parachutes, rigged with nonbreakaway static lines without drogue for A-7A or A-21 bundles dropped from the wedge. Attach the parachutes to the bundles after they are loaded aboard the wedge and restraint straps are secured.

27-5. INSPECTION

Before bundles are loaded, the wedge is inspected for completeness and serviceability and to ensure that the roller conveyors are properly positioned and secured to the platform. The bundle release plates are correctly installed as follows:

a. **Left-Side Release Plates.** Install left-side release plates (three each) between roller conveyors 1 and 2 at rollers 1, 2, 11, and 12, 21, 22, counting from the front to the rear of the wedge.

b. **Right-Side Release Plates.** Install right-side release plates between roller conveyors 3 and 4 at roller positions 1, 2, 11, and 12, 21, 22.

c. **Release Lanyard.** Install a 1-inch-wide tubular nylon webbing release lanyard with a 5 3/4-inch by 1/8-inch steel pin and ensure it is serviceable, marked, and secured to the correct release plate with a 20-inch length of Type III nylon cord. (The 20-inch cord forms a safety to prevent rebound of the lanyard and possible injury to personnel releasing the bundles.)

NOTE: If a lanyard is unserviceable, make a new one with a length of 1-inch tubular nylon webbing and a 5 3/4-inch by 1/8-inch steel pin. Using a bowline knot, attach the pin to one end of the webbing. Insert the pin in the appropriate release plate hole and extend the webbing forward to a point 30 inches from the forward edge of the wedge. Cut the webbing and form a 6-inch loop in the running end to provide a secure handhold for releasing the bundles.

27-6. LOADING, RIGGING, AND RESTRAINING BUNDLES TO WEDGE

Bundles are loaded, rigged, and secured to the wedge at the rigging site, and then transported to the aircraft for loading. The following procedure is for loading and rigging bundles to be dropped individually.

a. Load bundle number 3, left unit, with the loop of the restraint strap forward.

b. Position the loaded skid board with the 1-inch hole centered over the hole in the forward release plate. Pass the 6-inch loop of the restraint strap down through the hole in the board and release plate. Insert the release lanyard pin through the restraint strap loop and the hole in the release plate.

c. Form a loop in the forward running end of the restraint strap. Pass the forward running end of the restraint strap through the loop, cinch the strap tight around the bundle, and secure with two half hitches and an overhand knot.

d. Secure the release lanyard pin with one complete turn (attach cotton ticket number 8/7). Pass the cord down through the hole in the release plate and through the pin loop. Tie the running end with a square knot and a locking knot.

NOTE: Repeat the above procedure for each bundle in the following sequence: bundles 2 and 1, left side; bundles 3, 2, and 1, right side.

e. Position, attach, and secure G-14 cargo parachutes to the bundles. Using tie-down straps, secure the bundles to the wedge to prevent movement and possible damage to the release pin during transport and to the wedge when loading aboard the aircraft.

NOTE: If bundles are to be dropped in multiples of two or three, only the aft bundle requires a restraint strap and release lanyard. Bundles are loaded on the wedge in reverse order—left to right, front to rear—with the loop of the restraint loop strap forward.

27-7. JUMPMASTER PROCEDURES

JM procedures include managing anchor line cables and performing inspections before and after loading.

a. **Anchor Line Cables.** The JM ensures that—

(1) Anchor line cables are disconnected from the center anchor cable supports and secured to preclude obstruction to personnel.

(2) Stops are installed and taped on the inboard anchor line cable at fuselage station 749 for the bundle static lines, and about 26 inches forward of the center anchor cable support for parachutist static lines.

NOTE: Static line retrieval spools are forward of these stops. The stops and spools are adjusted so that the distance from the forward edge of the anchor cable supports to the forward edge of the spool is 26 inches.

b. **Inspection Before Loading**. The JM is responsible for inspecting, with a qualified parachute rigger, the following items on the wedge before loading. (He also inspects them with the loadmaster after loading.)

- Bundles A-7A or A-21 are present and are properly secured to plywood skid board (FM 10-500-3/TO 13C7-1-11).
- Quick-release assemblies are properly seated and safety clip is inserted (A-21 bundle).
- Correct size and layers of honeycomb are present.
- Bundles are properly secured to plywood skid board.
- A one-inch tubular nylon restraint strap is routed vertically around bundles, with a six-inch lanyard overhand loop at the release point.
- G-14 cargo parachute(s) is rigged for nonbreakaway and without drogue.
- Parachute is properly attached and secured to bundle(s).

c. **Inspection After Loading**. The JM is responsible for inspecting the wedge for correct positioning and locking into the aircraft restraint rails when the following actions are accomplished:

- Wedge is positioned so that aft restraint rail lock in platform number 6 indent is in place (counting from the rear of the platform).
- Cargo parachute static lines are connected to the inboard anchor line cables forward of the stops at station 749.
- Bundle release lanyards are correctly routed.
- Release pins are seated in the release plates.
- Safety tie is in place.
- Bundles are secured for flight.

27-8. BRIEFING RELEASE PROCEDURES

Briefing release procedures include the following:

a. Identify the jumpmaster, loadmaster, safety personnel, and assistants; brief each individual on his responsibilities.

b. Coordinate type signals to be used to conduct the drop—for example, thumbs-up for DROP, thumbs-down for NO DROP.

c. Coordinate bundle release and parachutist exit sequence to be used—for example, on the green light, the number 1 loadmaster releases the bundle(s) on his side of the aircraft. All parachutists will exit after the last bundle clears the aircraft ramp. After the last parachutist exits the aircraft, safety personnel give a thumbs-up, signaling the number 2 loadmaster to release the bundle(s) on his side of the aircraft.

d. Review procedures for a bundle hung up in the aircraft—for example, the loadmaster will signal NO DROP and notify the pilot; the pilot turns on the NO DROP red light. The drop is canceled if the problem cannot be immediately identified and safely corrected.

27-9. LOADMASTER AND JUMPMASTER DUTIES DURING FLIGHT

Airdrop sequence of bundles and personnel include:

a. **Loadmaster**. At the 3-minute slowdown, the loadmaster raises the cargo door and installs the anchor line cables in the center anchor line cable supports.

b. **Jumpmaster**. The JM hooks up, issues the 6-minute warning, and begins the jump commands.

NOTE: All commands are the same (whether using the wedge or not) unless all bundles are to be released before the exit of the parachutists. Then, the command STAND BY is given. The number 1 parachutist in each door assumes a position at a 10-degree angle to the forward edge of the doors and observes the JM.

(1) ***Bundles Precede Jumpers.*** On the green light, the loadmasters release the bundles. When the JM observes that the last bundle has cleared the aircraft ramp, he gives the command GO. All parachutists then exit the aircraft.

(2) ***Jumpers Precede Bundles.*** If all bundles are to be released after the parachutists exit, the JM gives the command STAND BY for the first two parachutists. On the green light, the command GO is given. When the parachutists have exited, the static safety personnel give a thumbs-up signal for the loadmaster to release the bundles.

(3) ***Bundles Released Before and After the Jumpers.*** On the green light, the number 1 loadmaster releases the required bundles. When the JM observes that the last bundle has cleared the aircraft ramp, he gives the command GO. All parachutists then exit, and the static safety personnel signal thumbs-up to the number 2 loadmaster to release the remaining bundles. Static lines are retrieved, and the cargo and troop doors are secured by the loadmaster.

CHAPTER 28
COMBAT AIRBORNE OPERATIONS

This chapter describes allowable combat techniques for airborne operations, which may deviate from those for peacetime operations. Commanders should conduct a detailed risk assessment before modifying normal peacetime techniques to which their soldiers have grown accustomed in training. Failure to do so can result in unnecessary loss of combat power and additional casualties that must be cared for on the drop zone. Commanders should ensure any changes are thoroughly briefed, understood, and rehearsed during troop leading procedures and pre-jump training. Within the constraints of this chapter, MACOMs should develop their own combat procedures based upon their mission, organizational structure, and level of training.

28-1. MODIFICATIONS TO PERSONNEL AND EQUIPMENT PROCEDURES

Personnel and equipment can be dropped in any sequence that best supports the ground tactical plan. Commanders may modify peacetime aircraft, serial, and formation interval rules based upon METT-T.

a. All aircraft are loaded to the maximum.

b. The static line snap hook safety wire is not required to be inserted in the snap hook.

c. The right connector snap of the reserve parachute does not require a safety wire.

d. The waistband, if not used, is rolled and secured with a piece of masking tape to keep it out of the way.

e. Jumpers should only carry minimum mission-essential equipment to make room for ammunition.

f. Commanders should determine, based upon METT-T, when jumpers should hook up. If faced with a well-developed enemy air defense threat, jumpers should be stood up prior to entering the threat zone.

g. The green light should go on to achieve the desired PI. There is no requirement for safety buffers on the lead or trail edge of the DZ, provided there are no obstacles. The red light should not go on unless there is a significant hazard (such as a large body of water) or the aircraft is scheduled to continue to another DZ.

h. Weapons should be rigged in the M1950 weapons case. (A magazine is loaded, weapon placed on SAFE, and no round is chambered.) Jumping weapons exposed does not significantly decrease the time required to place the weapon into action. More importantly, with jumpers and jumpmasters executing procedures that they have not rehearsed, jumpers are more likely to become entangled or towed. The M1950 weapons case is equipped with a quick release zipper. The jumper should not attempt to unzip the zipper but should grasp the tab thong and pull upward sharply (as if to finish zipping completely). When the slide fastener disengages, the zipper teeth will uncouple the length of the M1950.

i. Leaders and JMs should consider the heavy and nonstandard loads paratroopers carry. An excessively heavy load or protruding object may prevent a jumper's strong exit.

More importantly, the load may become an obstacle and may result in the failure of a stick to exit.

j. Jumpmasters should either lead or follow their sticks out on combat jumps. Commanders should load-plan their jumpmasters to come from the lead or trail edge units, thereby landing closer to their objectives. Jumpmasters may jump in the stick closest to their exit points. If safeties are designated to jump, they should come from the trail edge unit. They should rig as late as possible in order to conduct their required safety checks unencumbered. Safeties must continue to control static lines until all jumpers have exited the aircraft.

CAUTION

Do not abandon your training techniques just because you are in combat. Do a detailed risk assessment to ensure changes are necessary. Ensure changes are well briefed, understood by all, and rehearsed during pre-jump training.

k. Parachutists may be briefed to "pop and go," leaving all their air items where they land. If the DZ is under enemy attack and it is necessary to find cover and start fighting, they may drop their weapons cases and reserves, activate both canopy release assemblies, and move out, leaving their harnesses on to be discarded later. METT-T may require a follow-on airland. Jumpers should be briefed to "bag and go" if they land near a runway or taxiway. This means the jumper places his parachute in the kit bag and leaves it upside down no closer than 10 meters from the active runway or taxiway. METT-T determines whether units designate a "clean side" or "dirty side" of the active runway or taxiway.

l. Drop altitude is determined by coordination between the airborne commander and the air mission commander. The JTF commander makes the final jump altitude decision. When conducting risk assessment, commanders are advised of the altitude chart located in Chapter 26 and are reminded of the 135-foot altimeter error. If the decision is made to jump at 600 feet AGL or lower, pre-jump training should include the proper modifications (such as lowering equipment after the second point of performance).

m. Leaders should jump their own radios. Radio frequencies should be accessible, short whip antennas screwed in, and handsets plugged in. Leader radios are ready to go immediately after landing.

28-2. MOVEMENT FROM ASSEMBLY AREAS

The quicker assault objectives are secured, the better the odds for a successful mission. Units should proceed to the assault objective IAW unit SOP.

a. Determine minimum force required to seize and secure the assault objective. When that force has assembled (to include any "lost" troopers from other units), the senior leader of the group may begin the assault on the objective IAW the commander's intent.

b. Leaders should never bypass a soldier on the DZ. Every soldier "wandering around" ought to be either pointed in the right direction or integrated with that leader's unit.

CAUTION
If you train as you fight, you will fight as you train. Do not let heavy loads exceed the capabilities of your jumpers just because you are in combat.

28-3. LANDING PLAN

The development of the landing plan must be thorough. Assembly areas must be selected. A heavy drop plan with multiple HD impact points along the length and across the width of the DZ may be necessary. Offset HDs as required.

a. Vehicle drivers and equipment operators should land in the same sector of the DZ where the impact point for their equipment is located. Commanders must spread the HD impact points along the entire DZ; they do not saturate one small sector of the DZ, thus making a lucrative enemy target. They must consider the time and distance factors where load number 1 and load number 2 from the same aircraft will land. For example, delivering a howitzer and its prime mover from the same aircraft ensures they both land on the same DZ but physically separated. The crew should be cross-loaded accordingly.

b. GUCs adjust loads and HD impact points so howitzers and prime movers land at or near the same point. If the crew cannot locate their assigned prime mover right away, they should take the nearest one they can find and move out.

c. Crew members should not waste time looking for a particular bumper number on the DZ. If the vehicle belongs to the unit, recover it.

d. GUCs carefully plan CDS delivery. Rather than dropping the CDS in the center of the DZ, they must drop it along the edge of the DZ or on the unit assembly area. If a CDS is needed later, GUCs deliver it on or near the assault objective. Regardless of the decision as to where and when the CDS is needed, they must plan carefully so that backhauling is reduced.

28-4. HEAVY DROP LOADS

Minimum-essential HD loads should precede the personnel drop. Other important HD equipment should follow the assault personnel drop as soon as possible. Units cannot fight without heavy weapons, vehicles, and ammunition bundles. Peacetime safety rules require a 30-minute to one-hour separation between personnel drops and HD unless the HD is delivered first. These rules **do not** apply in combat. Door bundles or wedge loads can be dropped from any aircraft. Bundles should be exited on the green light with troops following immediately.

28-5. INJURED PERSONNEL

Mission completion is paramount in combat. Soldiers will want to provide aid to their buddies injured during the jump, but **speed is critical**, and **every fighter is needed to**

seize and secure the assault objective. Therefore, injured personnel are cared for as follows:

a. Medics must quickly join and move to the assault objectives with their assigned units. Medics should not stop on the DZ to care for injured parachutists (non-life-threatening casualties); injured jumpers will be cared for later.

b. Jumpers must be briefed and understand that they cannot stop on the DZ to help their buddies who were hurt during the jump and who received non-life-threatening injuries.

NOTE: The most effective way to aid an injured paratrooper on the DZ is to eliminate the enemy threat.

28-6. SUPPLIES

Units should obtain more critical items than are actually needed for mission accomplishment to compensate for those items that are damaged or destroyed during the airdrop.

a. Resupply and ammunition replenishment are the parachutist's most serious problem. Commanders should devise alternate methods of insertion for items such as LAWs, antitank mines, RAAWs, mortar rounds, or radio batteries. They should **not** rely on the individual paratrooper.

b. Keep door bundles rigid and light if you are exiting jumpers after them. Litters provide an excellent platform to build door bundles on. They provide rigidity and a means to move the load quickly while on the ground.

CAUTION

If you train as you fight, you will fight as you train. Do not let heavy loads exceed the capabilities of your jumpers just because you are in combat.

APPENDIX A

AIRBORNE REFRESHER TRAINING

Airborne refresher training is required for personnel who have not jumped within a 6-month period. The length of the refresher training depends on the proficiency of the parachutist. The minimum requirements are in Table A-1. Airborne refresher training will be instructed and documented by a qualified and current JM. Individual service components may modify these requirements depending on training aids and equipment availability.

PERIOD	HOURS	LESSON	TRAINING AIDS/EQUIPMENT
1	1	Sustained airborne training (pre-jump) to include PLFS (front, side, and rear) and methods of recovery.	Two-foot PLF platforms, sawdust pit, and mock door.
2	1	Fitting and wearing the T-10-series or MC1-series parachute and B-5/B-7 life preservers; rigging individual equipment in the HSPR; and packing individual weapons/equipment containers.	TOE and TDA equipment, ALICE pack and HSPR, T-10-series and MC1-series parachute assemblies, B-5/B-7 life preservers, and individual weapons/equipment containers.
3	1	Actions in the aircraft rehearsal, proper exit procedures, sequence of jump commands, all jump commands, first two points of performance.	Mock door structures and 5-foot static lines with snap hooks.
4	1	Control of canopy, turns, slips, entanglements, emergency landings, landing attitude, and activation of the reserve.	Suspended harness apparatus (if available) and swing landing trainer apparatus (if available).
5	1	Door exit procedure rehearsal, sequence of first three points of performance, releasing equipment containers, and activation of reserve.	34-foot tower (if available) and troop parachute harness.
6	1	Parachute jump.	Aircraft, parachutes, DZ, unit equipment.

Table A-1. Minimum requirements, airborne refresher training.

APPENDIX B
JUMPMASTER TRAINING COURSE

The Jumpmaster Training Course is the standard US Army course of instruction designed to qualify individuals as jumpmasters for conventional static line parachuting. It is the standard for all airborne unit activities. The course is approximately 2 weeks long and contains 97 hours of instruction.

SUBJECT	SCOPE	HOURS
In-processing	Administrative in-process of class.	1.0
Orientation and Administration	Orientation on the course requirements that the student must meet to include student handouts, questionnaires, the entire training schedule, and the grading system used to evaluate students.	1.0
Duties and Responsibilities of the Jumpmaster and Safety	Discussion of the duties and responsibilities of the jumpmaster and safety personnel from the time of notification until completion of the airborne operation.	2.0
Individual Equipment Containers	Discussion of the characteristics and nomenclature of individual equipment containers to include a demonstration on the correct rigging, attachment, and lowering procedures for the ALICE pack, M1950 weapons case, and AT4JP.	2.0
Army Aircraft Orientation	Familiarization with the preparation, inspection, and jump procedures for the UH-1H, UH-60A, and CH-47 helicopters.	1.0
Jumpmaster Personnel Inspection	Demonstration of the correct method of inspecting a parachutist and attached combat equipment. Remaining hours are spent on practical exercise using two-man buddy teams, where the student is required to conduct a personnel inspection and find and report major and minor rigging deficiencies that have been placed in the parachute assembly and attached equipment. Students are changed over often to ensure all receive the same amount of inspection time.	37.0

SUBJECT	SCOPE	HOURS
USAF Aircraft	General aircraft descriptions, jumpmaster procedures, and aircraft preparations. Critical elements of airborne operations are the aircraft and drop altitude. These aircraft are service tested and approved for troop drops. Minimum jump altitude and considerations that apply to basic airborne jumps, tactical training jumps, and combat jumps are discussed.	1.0
Prejump Training	Discussion of the five points of performance and methods of activating the reserve, towed parachutist procedures, collisions, entanglements, and the three types of emergency landings.	1.0
A-Series Containers	Discussion of the characteristics, capabilities, and methods of packing, rigging, and inspecting the A-series containers, and how to attach and inspect the cargo parachute.	2.0
Duties and Responsibilities of the DZSO/DZSTL	Discussion of the prerequisites to perform duties as the drop zone support team leader (DZSTL), drop zone safety officer (DZSO), assistant drop zone safety officer (ADZSO). Discussion of tactical DZ assessment and composition of the team for different airdrop scenarios. Scoring procedures using MAC Form 168 are included.	1.0
Jump Commands, Door Procedures, and Door Bundle Ejection Procedures	Demonstration and practical exercise of the proper sequence of jump commands and time warnings with proper hand-and-arm signals; the door procedures used by a jumpmaster; and door bundle ejection procedures using aircraft mock-ups. Remaining time is spent for practical exercise in the mock-ups (or actual aircraft).	4.0

SUBJECT	SCOPE	HOURS
CARP Drop Zones	Additional discussion includes drop zone surveys (USAF Form 3823, DZ Survey, formerly MAC Form 339). Discussion of the methods of marking drop zones for computed air release point (CARP). Discussion of drop zone selection factors. Practical exercise on determining the mean effective wind and use of authorized anemometers. Familiarization with drop zone marking requirements for day and night airborne operations. Discussion of ADEPT options (1, 2, and mass exits).	2.0
Nomenclature Examination	Written nomenclature examination.	0.5
Review and Critique	Brief review of the previous week's instruction to ensure all questions and tasks are clear.	1.0
Nomenclature Examination Retest	Personnel who failed the initial test are retested.	0.5
Written Examination	Written comprehensive examination which covers all instruction that has been presented.	1.0
Jumpmaster Personnel Inspection Pre-Examination	Each student inspects three parachutists—two with no combat equipment and one combat-equipped parachutist (5-minute time limit).	4.0
Jumpmaster Personnel Inspection Examination	Each student inspects three parachutists—two with no combat equipment and one combat-equipped parachutist (5-minute time limit).	3.0
Jumpmaster Personnel Inspection Retest	Personnel who failed the initial test are retested.	2.0
Prejump Training Examination	Oral presentation by each student to determine his ability to effectively conduct prejump training.	6.0

SUBJECT	SCOPE	HOURS
Jumpmaster Briefing	A briefing on all airborne operations and grading procedures.	.5
Aircraft Inspection Class	Practical exercise on aircraft inspection procedures using the C-130 or C-141.	1.0
Written Examination Retest	Personnel who failed the initial test are retested.	1.0
Day Practical Work in Aircraft Examination	Graded practical exercise jumpmastering personnel with attached equipment. Half the students are graded while the other half act as their jumpers. A second lift is required to test the second half of the class.	10.0
Maintenance of Air Items	Inspection, maintenance, and turn-in of all air items used during the airborne operations.	1.0
Out-processing	Administrative out-processing the class.	2.0
Graduation		1.0

APPENDIX C

JUMPMASTER REFRESHER COURSE

The Jumpmaster Refresher Course is designed to update qualified jumpmasters who are not in a current status IAW Chapter 7. It ensures that standardization of course content is maintained. The unit designated to conduct this course ensures that equipment normally used is available for this training. The course will be instructed and documented by a current and qualified JM. All JMs should attend a refresher course on a yearly basis, no matter the level of experience or currency. This will ensure all personnel are familiar with current changes, procedures, and standardization. Any unit having special requirements, such as nonstandard aircraft or special items of equipment, may add periods of instruction to the course as needed. Individual service components may modify these requirements depending on training aids and equipment availability.

SUBJECT	SCOPE	HOURS
Individual Equipment	Characteristics and nomenclature of individual equipment containers. Demonstration of the correct rigging, attaching, and lowering procedures for approved and tested equipment in use by the training unit.	1.0
Jumpmaster Personnel Inspection	Demonstration of the correct method for inspecting a parachutist with attached combat equipment. Practical exercise uses two-man buddy teams, where the individual is required to conduct a personnel inspection and find/report major and minor rigging deficiencies that have been placed in the parachute assembly and attached equipment. Personnel changed often to ensure all receive the same amount of inspection time.	2.0
Pre-Jump Training	Discussion of the five points of performance and methods of activating the reserve, towed parachutist procedures, collisions, entanglements, and the three types of emergency landings.	0.5

SUBJECT	SCOPE	HOURS
Drop Zone Support Teams, DZ Procedures, and DZ Formulas	Discussion and familiarization of the prerequisites to perform duties as the drop zone support team leader (DZSTL), drop zone safety officer (DZSO), assistant drop zone safety officer (ADZSO); and composition of the team for different airdrop scenarios. Scoring procedures using MAC Form 168 are included.	1.0
Jump Commands, Rigging Procedures, Spotting Procedures, Door Procedures, and Door Bundle Ejection Procedures, CRCC	Discussion and practical exercise (PE) of the proper sequence of jump commands and time warnings with appropriate hand-and-arm signals, door procedures used by the jumpmaster, and door bundle ejection procedures. A demonstration is given using aircraft mock-ups. The remaining time is spent for practical exercise in the mock-ups (or actual aircraft).	1.0
CARP Drop Zones	Additional discussion of methods of marking drop zones for computed air release point (CARP). Students are familiarized with drop zone marking requirements for day and night airborne operations.	1.0

APPENDIX D

PLANNING CONSIDERATIONS FOR WATER, WIRE, AND TREE EMERGENCY LANDINGS

An airborne airdrop is an inherently high-risk operation. When water, wire, or tree obstacles are on or close to the intended drop zone, the jumpers face an even higher risk of injury. This appendix assists the commander in conducting a DZ risk assessment analysis. This appendix also provides the commander, DZSO/DZSTL, and JM with operational and logistical planning measures to lower the risks jumpers may encounter during airborne operations. Individual service components may modify these requirements depending on training aids and equipment availability. Individual service components will follow their own service risk assessment procedures and regulations.

D-1. WATER OBSTACLES

A water obstacle is any body of water (for example, a lake, pond, river, stream, or canal) that has a depth of 4 feet or more, is 40 feet wide or wider, and is located within 1,000 meters of any edge of the surveyed or tactically assessed DZ.

a. **Risk Assessment Analysis for a Drop Zone with Water Obstacles**. When making a training parachute jump DZ risk assessment, the commander should consider the proximity of the water obstacle to the DZ, the depth of the water obstacle, and the width of the water obstacle. Additionally, the following factors may enter into the water obstacle risk assessment: the condition of the water obstacle bottom, the current of a free-flowing water obstacle, water temperature, the number of obstacles, the equipment available to reduce the risk level, jumper experience levels, jump time (day or night and percent of illumination), and whether or not the selected DZ is critical to mission success. The following risk categories are assigned to DZs with water obstacles:

(1) ***High Risk.*** A high-risk condition exists if a water obstacle is within 1,000 meters of any edge of the DZ, water depth is 4 feet or more, and water is 40 feet wide or wider. If a high-risk condition exists, it will be necessary to use a boat detail and have approved life preservers on the jumpers. If the water is 4 feet deep or more, but not over 40 feet wide, a boat detail is not required. However, approved life preservers are still required for the jumpers.

(2) ***Medium Risk.*** A medium-risk condition exists if a water obstacle is more than 1,000 meters but less than 1,500 meters from any edge of the DZ, water depth is 4 feet or more, and water is 40 feet wide or wider.

(3) ***Low Risk.*** A low-risk condition exists if a water obstacle is more than 1,500 meters from any edge of the DZ, water depth is 4 feet or more, and water is 40 feet wide or wider.

b. **Planning Considerations**. The commander, DZSO/DZSTL, and JM perform the following actions to reduce the risks associated with water obstacles:

(1) ***Commander.***

(a) Ensure a risk assessment analysis has been conducted to determine the unintentional water landing risk level for jumpers. If a high risk exists, select (if possible) an alternate DZ that allows mission conduct at a lower risk level.

(b) Ensure a follow-on assessment has been made to determine whether the jumpers' risk level has changed.

(c) Ensure that key leaders, jumpmasters, and jumpers have been informed of the water obstacle risks and the risk level (high, medium, or low).

(d) Ensure USAF Form 3823, DZ Survey (formerly MAC Form 339), is current and available. Ensure the DZSO/DZSTL and JM have read it and completely understand the unintentional water landing risk level and the safety measures that are to be used.

(e) If a boat detail is used, ensure the DZSO/DZSTL and the unit providing the detail have properly conducted initial or refresher training. Ensure the OIC/NCOIC of the boat detail knows where and when the detail begins its duties and how to contact the DZSO/DZSTL.

(f) Ensure that approved life preservers are coordinated for if used.

(g) Ensure the DZSO/DZSTL, JM/AJM, safeties, and jumpers are informed of all water obstacle risks and that the DZSO/DZSTL and JM complete their duties.

(2) ***DZSO/DZSTL.***

(a) Determine if a follow-on assessment of the DZ has been conducted to confirm the current status.

(b) If the risk assessment indicates high risk and a boat detail is necessary, ensure the OIC/NCOIC is fully briefed on the plan. Ensure all boat detail personnel have been trained and have all necessary equipment available to conduct the mission.

(c) Read all applicable regulations, FMs, and SOPs. Ensure copies are present throughout mission.

(3) ***Jumpmaster.***

(a) If approved life preservers are to be used, ensure they have been inspected within the last 180 days and are serviceable, and that all jumpers have been trained on life preserver wear, fit, and use (to include manual inflation).

(b) Ensure all personnel have received prejump training within 24 hours prior to drop time, with special emphasis on unintentional water landings.

c. **Water Obstacle Coverage**. Each water obstacle may require a different type of coverage. The following is an example composition of a boat detail. Equipment should be altered to best accomplish the mission.

(1) OIC/NCOIC (qualified as a boat operator) and assistant boat operator.

(2) Qualified boat operators - 1 primary and 1 assistant for each boat.

(3) Recovery personnel - 2 for each boat (one may be lifeguard qualified and combat lifesaver certified). All boat detail personnel should be strong swimmers.

(4) Each recovery boat team may need the following equipment:

- Boat (Zodiac RB-10 or solid-bodied boat of comparable size) with operable outboard motor.
- Enough fuel/oil to complete the mission.
- Life vest/floatation device for each boat detail member and additional floatation devices for jumpers.
- Life ring with attached rope—1.
- Radio with spare battery—1.
- Shepherd's crook—1.
- Grappling hook—1.
- Long backboard to facilitate CPR—1.

- Aid bag with resuscitation equipment—1.
- Rope, 120 feet long—1.

(5) For night recovery operations, each recovery boat must have operational night vision devices with spare batteries - 2 for each boat.

d. **Optional Training**. The following optional training is suggested:

(1) Suspended harness training on second through fifth points of performance may be given. Step-by-step training on the procedures jumpers will take for an unintentional water landing may be conducted.

(2) An optional dunk tank training device may be constructed to allow the lowering of jumpers (wearing parachute harness and B-5 or B-7 life preserver) into the water. This training familiarizes the jumpers with the proper emergency water landing procedures.

D-2. WIRE OBSTACLES

A wire obstacle is a wire or set of wires (regardless of height or type) located within 1,000 meters of any edge of the surveyed or tactically assessed DZ. The types of wire obstacles that could pose a risk to jumpers are power, telephone, or cable television wires. Wire fence can be regarded as an obstacle if it will pose a hazard to jumpers. Power line capacity (voltage or amperage) is **not** a factor when determining the risk to jumpers who may come in contact with a wire obstacle on or near the DZ.

CAUTION

Regardless of voltage or current-carrying capacity, if a power line is located within 1,000 meters of any edge of the surveyed or tactically assessed DZ, the power should be cut off before using the DZ, if possible.

a. **Risk Assessment Analysis for a Drop Zone with Wire Obstacles**. When making a training parachute jump DZ risk assessment, the commander should consider the proximity of the wire obstacle to the DZ, the height of the wire obstacle, the number of obstacles, the equipment available to reduce the risk level, jumper experience levels, drop time (day or night and percent of illumination), and whether or not the selected DZ is critical to mission success. The following risk categories are assigned to DZs with wire obstacles:

(1) ***High Risk.*** A high-risk condition exists if a wire obstacle is within 1,000 meters of any edge of the DZ. If a high-risk condition exists, it may be necessary to have a recovery detail at the DZ.

(2) ***Medium Risk.*** A medium-risk condition exists if a wire obstacle is more than 1,000 meters but less than 1,500 meters from any edge of the DZ.

(3) ***Low Risk.*** A low-risk condition exists if a wire obstacle is more than 1,500 meters from any edge of the DZ.

b. **Planning Considerations**. The commander, DZSO/DZSTL, and JM perform the following actions to reduce the risks associated with wire obstacles:

(1) ***Commander.***

(a) Ensure a risk assessment analysis has been conducted to determine the unintentional wire landing risk level for jumpers. If a high risk exists, select (if possible) an alternate DZ which allows mission conduct at a lower risk level.

(b) Ensure a follow-on assessment has been made to determine whether the jumpers' risk level has changed.

(c) Ensure that key leaders, jumpmasters, and jumpers have been informed of the wire obstacle risks and the risk level (high, medium, or low).

(d) Ensure USAF Form 3823, DZ Survey (formerly MAC Form 339), is current and available. Ensure the DZSO/DZSTL and JM have read it and completely understand the wire landing risk level and the safety measures to be used.

(e) If a recovery detail is used, ensure the DZSO/DZSTL and the unit providing the detail have properly conducted initial or refresher training. Ensure the OIC/NCOIC of the recovery detail knows where and when the detail begins its duties and how to contact the DZSO/DZSTL.

(f) Ensure the DZSO/DZSTL, JM/AJM, and jumpers have been informed of all wire obstacle risks. Ensure the DZSO/DZSTL and JM complete their duties.

(2) ***DZSO/DZSTL.***

(a) Determine if a follow-on assessment of the DZ has been conducted to confirm the current status.

(b) If the risk assessment indicates high risk and a recovery detail is utilized, ensure that the OIC/NCOIC is fully briefed on the plan. Ensure all recovery personnel have been trained and have all necessary equipment available to them to conduct the mission.

(c) Ensure that coordination with the power company has been made to cut off the power not later than 1 hour prior to drop time, if possible.

(d) Read all applicable regulations, FMs, and SOPs. Ensure copies are present throughout mission.

(3) ***Jumpmaster.***

(a) Ensure all personnel have been briefed on the wire obstacles.

(b) Ensure all personnel have received prejump training within 24 hours prior to drop time with special emphasis on unintentional wire landings.

c. **Wire Obstacle Coverage**. Each wire obstacle may require different types of coverage. The following is an example composition of a recovery detail. Equipment should be altered to best accomplish the mission.

(1) OIC/NCOIC and assistant.

(2) Enough personnel to recover jumpers who may become entangled in the wire obstacles.

(3) A recovery team may need the following equipment:

- Radios with spare batteries - 2 (1 for OIC/NCOIC and 1 for recovery team).
- Grappling hook—1.
- Tree-climbing kit—1.
- Long backboard to facilitate CPR—1.
- Aid bag with resuscitation equipment—1.
- Ropes, 120 feet long—2.
- Wood poles 15 feet long—2.
- **Wood** extension ladder, 20 feet long—1.

- Snap links—4.

(4) For night recovery operations, the following equipment should be added:

- Night vision devices (with spare batteries) for each team.
- Operational flashlights (with spare batteries) for each team.

d. **Optional Training**. Optional training includes suspended harness training on the second through fifth points of performance. Step-by-step training on the procedures jumpers take for unintentional wire landings may be conducted.

D-3. TREE OBSTACLES

A tree obstacle is any tree or group of trees that are on, around, or within 1,000 meters of any edge of the drop zone.

a. **Risk Assessment Analysis for a Drop Zone with Tree Obstacles**. When making a training parachute jump DZ risk assessment, the commander should consider the proximity of the tree obstacles to the DZ, the number of obstacles, the equipment available to reduce the risk level, jumper experience levels, drop time (day or night and percent of illumination), and whether or not the selected DZ is critical to mission success. The following risk categories are assigned to DZs with tree obstacles:

(1) ***High Risk.*** A high-risk condition exists if a tree or group of trees are within 1,000 meters of the DZ or are on any edge of the DZ and have a height of 35 feet or more. If a high-risk condition exists, it may be necessary to have a recovery detail present at the DZ.

(2) ***Medium Risk.*** A medium-risk condition exists if a tree or group of trees are on or within 1,000 meters of any edge of the DZ and have a height of 20 to 35 feet.

(3) ***Low Risk.*** A low-risk condition exists if a tree obstacle having a height of less than 20 feet is on or within 1,000 meters of any edge of the drop zone.

b. **Planning Considerations**. The commander, DZSO/DZSTL, and JM perform the following actions to reduce the risks associated with tree obstacles.

(1) ***Commander.***

(a) Ensure a risk assessment analysis has been established to determine the tree landing risk level for jumpers. If a high risk exists, select (if possible) an alternate DZ that allows mission conduct at a lower risk level.

(b) Ensure a follow-on assessment has been made to determine whether the jumpers' risk level has changed.

(c) Ensure that key leaders, jumpmasters, and jumpers have been informed of the tree obstacle risks and the risk level (high, medium, or low).

(d) Ensure USAF Form 3823, DZ Survey (formerly MAC Form 339), is current and available. Ensure the DZSO/DZSTL and JM have read it and completely understand the tree landing risk level and the recovery measures to be used.

(e) If a recovery detail is used, ensure DZSO/DZSTL and the unit providing the detail have properly conducted initial or refresher training. Ensure the OIC/NCOIC of the recovery detail knows where and when the detail begins its duties and how to contact the DZSO/DZSTL.

(f) Ensure that the DZSO/DZSTL, JM/AJM, and jumpers have been informed of all tree obstacle risks and that the DZSO/DZSTL and JM complete their duties.

(2) ***DZSO/DZSTL.***

(a) Determine if a follow-on assessment of the DZ has been conducted to confirm the current status.

(b) If the risk assessment indicates high risk and a recovery detail is used, ensure that the OIC/NCOIC is fully briefed on the plan. Ensure all recovery personnel have been trained and have all necessary equipment available to conduct the mission.

(c) Read all applicable regulations, FMs, and SOPs; ensure copies are present throughout mission.

(3) ***Jumpmaster.***

(a) Ensure all personnel have been briefed on the tree obstacles.

(b) Ensure all personnel have received prejump training within 24 hours prior to drop time, with special emphasis on unintentional tree landings.

c. **Tree Obstacle Coverage**. Each tree obstacle may require a different type of coverage. The following is an example composition of a recovery detail. Equipment should be altered to best accomplish the mission.

(1) OIC/NCOIC and assistant.

(2) Enough personnel to recover jumpers who may become entangled in the tree obstacles.

(3) A recovery team may need the following equipment:

- Radios with spare batteries - 2 (1 for OIC/NCOIC and 1 for recovery team).
- Grappling hook—1.
- Tree-climbing kit—1.
- Long backboard to facilitate CPR—1.
- Aid bag with resuscitation equipment—1.
- Ropes, 120 feet long—2.
- Wood poles 15 feet long—2.
- **Wood** extension ladder 20 feet long—1.
- Snap links—4.

(4) For night recovery operations, the following equipment should be added:

- Night vision devices (with spare batteries) for each team.
- Operational flashlights (with spare batteries) for each team.

d. **Optional Training**. Optional training includes suspended harness training on the second through fifth points of performance. Step-by-step training on the procedures jumpers will take for unintentional tree landings may be conducted.

D-4. DROP ZONE RISK ASSESSMENT DECISION MATRIX AND LEADER'S CHECKLISTS

Following are Table D-1, drop zone risk assessment decision matrix, and Figures D-1 through D-3 (pages D-7 through D-8), leaders' checklists.

	HIGH RISK	MEDIUM RISK	LOW RISK
WATER	Within 1,000 meters of any edge of DZ, more than 4 feet deep, and more than 40 feet wide.	More than 1,000 meters but less than 1,500 meters from any edge of DZ, more than 4 feet deep, and more than 40 feet wide.	More than 1,500 meters from any edge of DZ, more than 4 feet deep, and more than 40 feet wide.
WIRE	Within 1,000 meters of any edge of DZ.	More than 1,000 meters but less than 1,500 meters from any edge of DZ.	More than 1,500 meters from any edge of DZ.
TREE OR TREES	Within 1,000 meters of any edge of DZ and 35 feet tall or taller.	Within 1,000 meters of any edge of DZ.	Within 1,000 meters of any edge of DZ, but less than 20 feet tall.
OBSTACLE TRAINING AND EQUIPMENT	Required.	Recommended, but at commander's discretion.	Not required.

Table D-1. Drop zone risk assessment decision matrix.

COMMANDER

_____ Risk assessment and follow-on assessment have been conducted.

_____ Key leaders, jumpmasters, and jumpers have been informed of water obstacle risks and the risk level (high, medium, or low).

_____ USAF Form 3823, DZ Survey (formerly MAC Form 339), is current and available; the DZSO/DZSTL and JM have read it and completely understand the unintentional water landing risk level and the safety measures that are to be used.

_____ The DZSO/DZSTL and OIC/NCOIC of the boat detail have been briefed and understand their mission.

_____ B-5/B-7 life preservers are coordinated for if used.

_____ The DZSO/DZSTL, JM/AJM, safeties, and jumpers have been informed of all water obstacle risks.

_____ The DZSTL/DZSO and JM have completed their duties.

DZSO/DZSTL

_____ Risk assessment analysis has been conducted.

_____ If a boat detail is used, the OIC/NCOIC of the boat detail is fully briefed on the plan.

_____ All personnel have been trained and have all necessary equipment available to conduct the mission.

_____ All applicable regulations, FMs, and SOPs have been read.

_____ The boat detail maintains communications throughout the mission. Communications are established 1 hour prior to drop time and checked 15 minutes prior to drop time.

JUMPMASTER

_____ All personnel have been briefed on the water obstacles.

_____ B-5/B-7 life preservers have been inspected within the last 180 days and are serviceable.

_____ All jumpers have been trained on wear, fit, and use (to include manual inflation) of life preservers.

_____ All personnel receive prejump training within 24 hours of drop time, with special emphasis on unintentional water landings.

Figure D-1. Leader's checklist for possible water landings.

COMMANDER

_____ Risk assessment and follow-on assessment analysis have been conducted.

_____ Key leaders, jumpmasters, and jumpers are informed of wire obstacle risks and the risk level (high, medium, or low).

_____ USAF Form 3823, DZ Survey (formerly MAC Form 339), is current and available. The DZSO/DZSTL and JM have read it and completely understand the wire landing risk level and the safety measures that are to be used.

_____ The DZSO/DZSTL and the OIC/NCOIC of the recovery detail are briefed and understand their mission.

_____ The DZSO/DZSTL, JM/AJM, safeties, and jumpers are informed of all wire obstacle risks.

_____ DZSTL/DZSO/JM complete their duties.

DZSO/DZSTL

_____ Risk assessment analysis has been conducted.

_____ If a recovery detail is used, the OIC/NCOIC of the recovery detail is fully briefed on the plan.

_____ All personnel have been trained and have all necessary equipment available to conduct the mission.

_____ All applicable regulations, FMs, and SOPs have been read.

_____ The recovery detail maintains communications throughout the mission. Communications are established 1 hour prior to drop time and checked 15 minutes prior to drop time.

JUMPMASTER

_____ All personnel have been briefed on the wire obstacles.

_____ All personnel have received prejump training within 24 hours of drop time with special emphasis on wire landings.

Figure D-2. Leader's checklist for possible wire landings.

COMMANDER

_____ Risk assessment and follow-on assessment analysis have been conducted.

_____ Key leaders, jumpmasters, and jumpers are informed of tree obstacle risks and the risk level (high, medium, or low).

_____ USAF Form 3823, DZ Survey (formerly MAC Form 339), is current and available. The DZSO/DZSTL and JM have read it and completely understand the tree landing risk level and the safety measures that are to be used.

_____ The DZSO/DZSTL and the OIC/NCOIC of the recovery detail are briefed and understand their mission.

_____ The DZSO/DZSTL, JM/AJM, safeties, and jumpers are informed of all tree obstacle risks.

_____ The DZSTL/DZSO and JM have completed their duties.

DZSO/DZSTL

_____ Risk assessment analysis has been conducted.

_____ If a recovery detail is used, the recovery detail OIC/NCOIC is fully briefed on the plan.

_____ All personnel have been trained and have all necessary equipment available to conduct the mission.

_____ All applicable regulations, FMs, and SOPs have been read.

_____ The recovery detail maintains communications throughout the mission. Communications are established 1 hour prior to drop time and checked 15 minutes prior to drop time.

JUMPMASTER

_____ All personnel have been briefed on the wire obstacles.

_____ All personnel receive prejump training within 24 hours of drop time, with special emphasis on unintentional tree landings.

Figure D-3. Leader's checklist for possible tree landings.

GLOSSARY

ADEPT	alternate door exit procedure for training
ADP	airdrop personnel
ADZSO	assistant drop zone safety officer
AF	Air Force
AFCCT	Air Force combat control team
AFJMAN	Air Force joint manual
AFR	Air Force regulation
AGL	above ground level
AJM	assistant jumpmaster
AIRPAC	all-purpose weapons and equipment container system
ALICE	all-purpose, lightweight, individual carrying equipment
AMC	Air Mobility Command
AMCR	Air Mobility Command regulation
ANGLICO	air and naval gunfire liaison company
APFT	Army Physical Fitness Test
AR	Army regulation
ARS	automatic release system
ASOP	airborne standing operating procedure
ASP	ammunition supply point
ATC	air traffic control
AWADS	adverse weather aerial delivery system
AZAR	assault zone availability report
BCU	battery coolant unit
CARP	computed air release point
CCP	circulation control point
CCT	combat control team
CDS	container delivery system
CFM	cubic feet per minute
CO	commissioned officer
CO2	carbon dioxide (gas)
COMALF	commander of the airlift forces
CPR	cardiopulmonary resuscitation
CWIE	container, weapon, individual equipment
DA	Department of the Army
DACO	departure airfield control officer
DC	District of Columbia
DMJP	Dragon missile jump pack
DOD	Department of Defense
DTG	date-time group
DZ	drop zone
DZSO	drop zone safety officer

DZST	drop zone support team
DZSTL	drop zone support team leader
ETLBV	enhanced tactical load-bearing vest
FHT	field handling trainer
FLA	front line ambulance
FM	field manual
FPLIF	field pack, large, internal frame
FS	fuselage station
GLO	ground liaison officer
GMRS	ground marking release system
GTA	ground-to-air
GUC	ground unit commander
HAHO	high-altitude, high-opening
HALO	high-altitude, low-opening
HD	heavy drop
HE	heavy equipment; high explosive
HEAT	high-explosive antitank
HPT	hook-pile tape
HQ	headquarters
hrs	hours
HSPR	harness, single-point release
IAS	indicated airspeed
IAW	in accordance with
ICS	intercommunication system
IFF	identification, friend or foe
IMC	instrument meteorological conditions
intercomm	intercommunication (radio)
JA/ATT	Joint Airborne/Air Transportability Training
JACRS	code letters used on CARP drop zones
JM	jumpmaster
JMPI	jumpmaster personnel inspection
JPI	joint preflight inspection
JSJR	jumpmaster spotted, jumpmaster released
JTF	joint task force
KT	knot
LAPES	low-altitude parachute extraction system
LAW	light antitank weapon
lb	pound

LCE	load-carrying equipment
LDA	lateral drift apparatus (training apparatus)
MAC	Military Airlift Command
MACOM	major Army command
MCWP	Marine Corps Warfighting Publication
METT-T	mission, enemy, terrain, troops and time available
MEW	mean effective wind
MFF	military free fall
MG	machine gun
MIRPS	modified improved reserve parachute system
MO	malfunction officer
MOD	modified
mph	miles per hour
MSL	mean sea level
MTT	mobile training team
MTOE	modified table of organization and equipment
NAVAIDS	navigational aids
NCO	noncommissioned officer
NET	not earlier than
NLT	not later than
No.	number
NOPO	nonporous
NVIS	night vision imaging systems
OD	olive drab
OIC	officer in charge
OPCON	operational control (the authority delegated to direct those personnel needed to accomplish a specific mission)
PASGT	personnel armor system, ground troops
PE	personnel
PI	point of impact
PIBAL	pilot balloon
PIE/R2	parachutist's individual equipment rapid release
PIL	parachutist's impact liner
PIR	Parachute Infantry Regiment
PLD	personnel lowering device
PLF	parachute landing fall
port	left side of aircraft (direction of flight)
PP	parachute personnel
PRB	parachute recovery bag
RAAWS	ranger antiarmor/antipersonnel weapon system
RAM	raised angle marker

reps	repetitions
RMP	reprogrammable microprocessor
RP	release point
SAW	squad automatic weapon
SCUBA	self-contained underwater breathing apparatus
SFQC	Special Forces Qualification Course
SH	suspended harness (training apparatus); student handbook
SKE	station-keeping equipment
SLT	swing landing trainer (training apparatus)
SMJP	Stinger missile jump pack
SOF	special operations force
SOLL	special operations low level
SOWT	special operations weather team
SOI	signal operation instructions
SOP	standing operating procedure
SSN	social security number
STABO	a system for extracting personnel by helicopter (combined first letters of the surnames of the five tactical airdrop personnel who designed the system)
starboard	right side of aircraft (direction of flight)
station time	time the aircraft is to be loaded and prepared for flight (normally, 35 minutes prior to takeoff)
STT	special tactics team
TACSAT	tactical satellite
TALCE	tanker/airlift control element
TAP	tactical airdrop personnel
TC	training circular
TDA	table of distribution and allowances
TM	technical manual
TO	technical order (USAF)
TOE	table of organization and equipment
TOT	time on target
TP	training practice
TTB	tactical training bundle
US	United States
USA	United States Army
USAF	United States Air Force
USAIS	United States Army Infantry School
USASOC	United States Army Special Operations Command
USMC	United States Marine Corps

VIRS	verbally initiated release system
VMC	visual meteorological conditions
VS	visual signal
WDZ	water drop zone
WDZSO	water drop zone safety officer
WDZSTL	water drop zone support team leader
WO	warrant officer
WSVC	wind streamer vector count

REFERENCES

Documents Needed

None required.

Readings Recommended

These readings contain relevant supplemental information.

AF Form 3823	Drop Zone Survey. February 1994.
AFI Reg 13-217	Assault Zone Procedures, 1 June 1999.
AR 59-4/AFJI 13-310/ OPNAVINST 4630.24C/ MCO 13480.1B	Joint Airdrop Inspection Records, Malfunction Investigations and Activity Reporting. 1 May 1998.
AR 350-1	Army Training. 1 August 1981, with Change 1, 1 August 1983.
AR 350-2	Opposing Force Program. 15 June 1983.
DA Pam 351-4	U.S. Army Formal Schools Catalog. 31 October 1995.
DA Form 1306	Statement of Jump and Loading Manifest. May 1963.
DD Form 1574	Serviceable Tag - Materiel. October 1966.
DD Form 1748-2	Joint Airdrop Malfunction Report (Personnel-Cargo). November 1997.
FM 10-500-2/TO 13C7-1-5	Airdrop of Supplies and Equipment: Rigging Airdrop Platforms. 1 November 1990, with Change 1, 14 December 1992, and Change 2, 23 December 1996.
FM 10-500-3/TO 13C7-1-11	Airdrop of Supplies and Equipment: Rigging Containers. 8 December 1992, with Change 1, 26 September 1996.
FM 10-550/TO 13C7-22-71	Airdrop of Supplies and Equipment: Rigging Stinger Weapon System and Missiles. 29 May 1984, with Changes 1-3, 14 February 1989 - 28 June 1996.

FM 10-542	Airdrop of Supplies and Equipment: Rigging Loads for Special Operations. 07 October 1987.
FM 21-20	Physical Fitness Training. 30 September 1992, with Change 1, 1 October 1998.
FM 3-21.32(FM 57-38)	Pathfinder Operations. 1 October 2002.
FM 31-19/MCWP 3-15.6 AFI 11-411(I)/ NAVSEA SS400-AG-MMO-010	Military Free-Fall Parachuting Tactics, Techniques, and Procedures. 1 October 1999.
FM 31-71	Northern Operations. 21 June 1971.
FM 100-27/AFM 2-50	US Army/US Air Force Doctrine for Joint Airborne and Tactical Airlift Operations. 31 January 1985, with Change 1, 29 March 1985.
MAC Form 168	Airdrop/Airland Extraction Zone Control Log. 1 December 1992.
MAC Form 339	Drop Zone Survey. February 1989 (obsolete but still in use; replaced by AF Form 3823).
TC 31-24	Special Forces Air Operations. 9 September 1988.
TC 31-25	Special Forces Waterborne Operations. 3 October 1988.
TM 1-1520-237-10	Operator's Manual for UH-60A, UH-60L Helicopters, and EH-60A Helicopters. 31 October 1996, with Changes 1-6, 30 June 1997 - 3 April 2000.
TM 9-1425-429-12	Operator's and Organizational Maintenance Manual for Stinger Guided Missile System Consisting of Weapon Round Basic, Round Post, Weapon Round RMP, Trainer Handling Guided Missile Launcher M60, Interrogator Set AN/PPX-3A/3B, Interrogator Set Programmer AN/GSX-1/1A. 21 April 1992, with Change 1, 30 October 1992.
TM 10-1670-201-23/ TO 13C-1-41/ NAVAIR 13-1-17	Organizational and Direct Support Maintenance Manual for General Maintenance of Parachutes and Other Airdrop Equipment. 30 October 1973, with Changes 1-7, 7 March 1974 - 17 May 1990.

TM 10-1670-251-12&P	Operator and Unit Maintenance Manual for Personnel/Cargo Lowering Device 500-lb Capacity. 11 June 1991, with Change 1, 30 June 1994.
TM 10-1670-262-12&P	Operator and Unit Maintenance Manual Including Repair Parts and Special Tools List Personnel Insertion/Extractions Systems for STABO, Fast Rope Insertion/Extraction System, and Anchoring Device. 25 September 1992, with Change 1, 24 November 1993, and Change 2, 30 June 1994.
TM 10-1670-298-20&P	Unit Maintenance Manual Including Repair Parts and Special Tools List for Container Delivery System A-7A Cargo Sling; A-21 Aerial Deliver Cargo Bag; A-22 Aerial Delivery Cargo Bag; A-23 Aerial Delivery Cargo Bag; Capsule, Cargo, CTU-2/A; Strap Connector, 60 Inches Long; Strap Connector, 120 Inches Long. 15 September 1995.
TM 38-250/ AFJMAN 24-204/ NAVSUP PUB 505/ MCO P4030.29G/ DLAI 4145.3	Preparing Hazardous Materials for Military Air Shipments 1 March 1997.
US Army Special Operations Command Regulation 350-2	Training, Airborne Operations Manual. 1 June 1995.

INTERNET WEB SITES

U.S. Army Publishing Agency
http://www.usapa.army.mil

Army Doctrine and Training Digital Library
http://www.adtdl.army.mil

INDEX

*FM 7-85

FIELD MANUAL
No. 7-85

HEADQUARTERS
DEPARTMENT OF THE ARMY
Washington, DC, 9 June 1987

FM 7-85
RANGER UNIT OPERATIONS

IMPORTANT

U.S. Army Infantry School Statement on
U.S. NATIONAL POLICY CONCERNING ANTIPERSONNEL LAND MINES

TABLE OF CONTENTS

DISTRIBUTION RESTRICTION: Approved for public release; distribution is unlimited.

PREFACE

This publication contains doctrinal guidelines for the employment and training of ranger units. The intended audience is intermediate and higher level commanders, staffs of those headquarters, and newly assigned officers and NCOs of the ranger regiment. It is to be used with all other published US doctrine concerning NBC operations and the climatic or geographic influences on combat. The fundamentals apply to the employment of ranger units on worldwide operations and across the spectrum of conflict from low to high intensity warfare. This includes all aspects of the integrated battlefield: nuclear, biological, and chemical. It is not intended to apply to the employment of long-range reconnaissance patrol (LRRP) companies, even though such units carry the designation "ranger." These LRRP companies, as well as the long-range surveillance company (LRSC) (corps level) and the long-range surveillance detachment (LRSD) (division level), are organized, trained, and equipped for long-range reconnaissance in enemy territory.

The scope of this publication addresses the mission, organization, equipment, capabilities, limitations, planning, training, operations, and logistical support of ranger units. It describes how the ranger regiment and its subordinate battalions train, receive taskings, plan missions, task-organize, deploy, and integrate supporting elements to perform the mission. Once in combat, the ranger regiment maneuvers to accomplish the mission in the manner of other light infantry units. Ranger units make full use of published light infantry doctrinal guidance. This manual will repeat procedures or techniques contained in other publications only if necessary for understanding or continuity.

This publication does not contain everything the ranger needs to know to successfully execute a unit mission. Rather, it supplements unit training and the ranger unit commander's guidance; it does not supersede it.

The proponent of this publication is HQ, TRADOC. Submit changes for improving this publication on DA Form 2028 (Recommended Changes to Publications and Blank Forms) and forward it to Commandant, US Army Infantry School, ATTN: ATSH-B-ID, Fort Benning, GA 31905.

Unless otherwise stated, whenever the masculine gender is used, both men and women are included.

CHAPTER 1

Introduction

The ranger regiment is a major component of the US Army's special operations forces. It is a unique light infantry unit tasked to conduct special military operations in support of national policies and objectives. These operations require highly trained, well-disciplined units capable of employment in any environment, either alone or in concert with other military forces. Within this publication, the term *ranger force* describes any size force consisting mainly of members of the ranger regiment and led by a member of the ranger regiment's chain of command. A ranger force may be a TOE unit or it may be a specially organized task force for a specific mission.

1-1. Mission.

a. The mission of the ranger regiment is to plan and conduct special military operations. These operations are conducted by specially trained, equipped, and organized forces against strategic or tactical targets in pursuit of national military, political, economic, or psychological objectives. They may support conventional military operations or they may be performed independently when conventional forces cannot be used. (See Figure 1-1.)

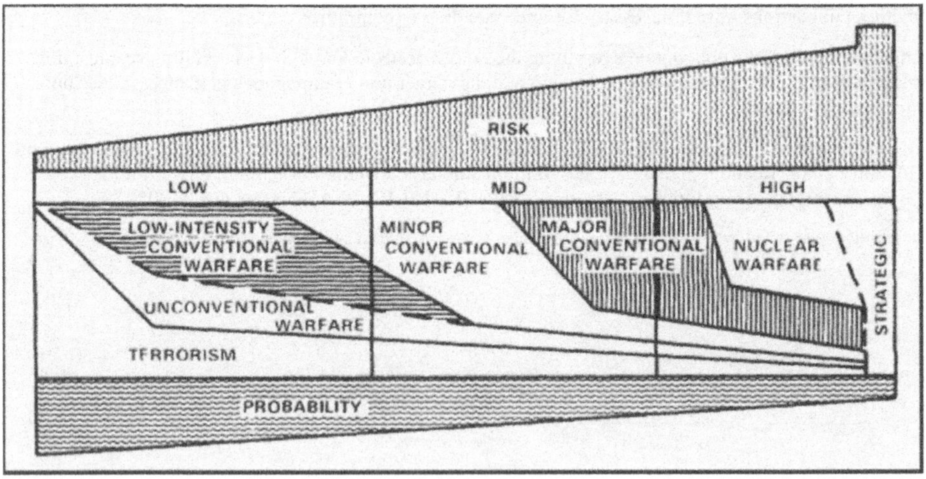

Figure 1-1. Spectrum of conflict.

b. Special military operations conducted by the ranger regiment include strike operations, usually deep penetration, and special light infantry operations. Strike operations include raids, interdiction, and recovery operations. Special light infantry operations include many of the light infantry missions assigned to airborne, air assault, or light infantry battalions and brigades. These operations are conducted in support of the AirLand Battle at all levels of intensity.

1-2. Purpose and function.

The ranger regiment provides the national command authority (NCA) the ability to move a credible military force quickly to any region in the world. The regiment uses the entire spectrum of intelligence support, from national systems to organic assets. Ranger units maintain a readiness posture that supports their immediate commitment to battle once deployed. They are often tailored for specific missions and may require augmentation from external sources. Tactical mobility may be augmented by USAF or Army special operations aviation (SOA) aircraft.

a. The ranger regiment conducts light infantry operations on the integrated battlefield as well as in low intensity conflicts. Special light infantry operations are conducted to accomplish certain contingency missions during peacetime or as part of the echelons above corps (EAC) battle.

b. The regiment conducts both strategic and operational missions. Its efforts are combined into an overall plan to destroy, delay, and disorganize the enemy, or to cause him to divert his attention and combat forces to rear area security. The limited number of ranger units and the diverse targets dictate a careful assignment of missions.

c. The mission, enemy, terrain, troops and time available (METT-T) factors govern the command and control arrangement under which the regiment operates. The ranger regiment or separate ranger battalions are normally assigned to the headquarters whose area of responsibility includes the regiment's operational area(s). The strategic or operational value of potential targets means that ranger units are normally employed at no lower than corps level. Employment at EAC is routine. The EAC headquarters employing ranger units could be the commander in chief's (CINC's) special operations command or field army. The controlling headquarters could also be that of a joint task force (JTF) or the Army component command (ARFOR) of a JTF. The ranger regiment's organization, communications equipment density, and training programs do not support operational level reconnaissance missions.

d. Ranger battalions are not oriented to a specific theater. Current force structure, contingency plans, and training needs (see Appendix G) preclude committing battalions to one region.

1-3. Fundamentals of ranger operations.

a. The success of an operation by a ranger unit depends on the observance of the fundamentals of the US Army's AirLand Battle doctrine (see FM 100-5). The ranger regiment fights the enemy as a light infantry force. It follows infantry doctrine and observes all the basic rules of infantry operations.

b. Along with initiative, depth, agility, and synchronization, ranger operations require--

(1) Detailed planning and coordination that allow the ranger unit to discern and exploit the enemy's weaknesses while avoiding his strength. Both permit the evaluation and use of information gathered from all sources. Detailed coordination integrates all supporting units and services. During planning, the ranger unit conducts pre-mission training, briefings, and rehearsals for all personnel. The use of special mission equipment is practiced and perfected.

(2) Decentralized execution IAW mission orders and the commander's intent. Special operations forces must use individual and unit initiative. Mission-type orders give the ranger force commander the flexibility to take advantage of opportunities on the battlefield.

(3) Surprise, achieved through the ranger unit's ability to move by uncommon means, along unexpected routes, and over rough terrain. Ranger units normally conduct operations during poor weather and reduced visibility, aided by night vision devices. This adds to the attainment of surprise.

(4) Survivability, achieved by using the classic infantry combat techniques of stealth and concealment. The ranger unit engages the enemy at the time and place of its own choosing. It takes full advantage of terrain, and destroys or suppresses enemy weapons. The ranger unit seeks to destroy or neutralize the enemy's command and control systems, his surveillance assets, and his mobility assets. Survivability is enhanced by rapid mission accomplishment and a prompt departure from the objective area.

(5) Mobility, speed, and violence of execution that allow the ranger unit it to close quickly on the objective area and complete the mission before the enemy can react. The speed at which events take place confuses and deceives the enemy as to the intent of the ranger unit. This forces the enemy to react rather than to take the initiative. Tactical mobility lets the ranger unit break contact and withdraw from the objective area.

(6) Shock effect, which is a psychological advantage achieved by the combining of speed and violence with the precision of the ranger attack. The ranger unit strives to apply its full combat power at the decisive time and place, and at the point of the greatest enemy weakness. The ranger unit strives to achieve maximum physical and psychological effect on the enemy by exhibiting aggressiveness and reasoned audacity.

(7) Multiple methods of insertion and attack, trying not to repeat operations thus decreasing the chance the enemy will detect a pattern. This is achieved through imaginative training and planning.

(8) Deception, practiced by the ranger unit during all phases of its operations, from deployment through the insertion phase, to the actions in the objective area and extraction. The enemy is kept unaware of the ranger unit's intentions, is fed conflicting and wrong information, and is kept from knowing the size or mission of the ranger force. The ranger force commander makes full use of ruses, feints, false insertions, electronic countermeasures, and dummy transmissions. Concurrent military operations by US or allied forces may be used as cover for ranger operations. Ranger units are careful to hide every possible aspect of an operation and to disguise those that cannot be hidden. While in the operational area, or when departing any location, ranger units ensure that no material is left that could provide the enemy with information concerning the ranger force.

(9) Audacity, achieved by a willingness to accept risk. The ranger force commander considers what the enemy expects the unit to do and the actual ranger abilities, He then chooses a course of action that may confuse the enemy, while remaining within the capability of the ranger unit. This fundamental is combined with deception and surprise to disrupt the enemy command and control.

c. Detailed planning and coordination finds the enemy. Surprise, mobility and speed, variety, deception, and audacity combine to shock and disorient the enemy, fixing him in place. The violence and precision of the ranger attack finishes the enemy while ensuring the ranger force survives.

1-4. Capabilities and limitations.

a. The ranger regiment has the following capabilities:

(1) Deploying quickly to conduct operations on all types of terrain and in all kinds of weather.

(2) Establishing a credible American presence in any part of the world to show US interest or resolve.

(3) Infiltrating and exfiltrating an area of operations, and assaulting an objective by land, sea, or air.

(4) Conducting strike operations to include raids, personnel and equipment recovery operations, and interdiction of key areas.

(5) Conducting special light infantry operations to include seizing and securing airfields, communications centers, command and control facilities, and key bridges; and other special light infantry operations.

(6) Performing short-duration reconnaissance of assigned ranger objectives for the ranger force commander.

(7) Operating for up to three days without resupply, and for longer periods when provided with accompanying or airdropped supplies.

(8) Providing liaison, communication, and coordination personnel and equipment to integrate the deployed ranger force into the logistical, intelligence, and operational system of the theater or joint task force (JTF) commander.

(9) Assuming operational control, for a limited time, of other US military forces such as engineers or infantry, airborne, or air assault battalions.

(10) Providing the focal point for all-source intelligence support to attached and assigned units of the regimental task force.

(11) Conducting limited combat operations under conditions of chemical, nuclear, or biological contamination.

b. The ranger regiment has the following limitations:

(1) Limited capability against armored or motorized units in open terrain.

(2) No organic transportation.

(3) Limited sustained combat capability due to the shortage of organic combat support and combat service support elements.

(4) Limited organic air defense weapons.

(5) Limited organic indirect fire support.

(6) No casualty evacuation capability.

(7) Reconstitution and retraining needed to replace combat losses.

1-5. Employment considerations.

a. Ranger units are characterized by the quality, motivation, training, and individual skill of their members. This produces units with superb collective abilities, able to adapt well to changing, complex situations.

b. Ranger units can conduct either deliberate or quick-response operations.

(1) Deliberate operations rely on careful planning, reconnaissance and surveillance of the target area, deception, secrecy, thorough preparation and rehearsals, and violent execution. A deliberate operation aims to complete the mission even though the enemy may have heavy forces on or near the objective area. Deliberate operations allow for detailed planning, evaluation, rehearsal, and coordination before insertion. A deliberate operation is likely to succeed against targets that the enemy has protected in depth, that have strong natural defenses, or that need a detailed and long insertion process.

(2) Quick-response operations rely on the high level of training and readiness of the ranger regiment to execute a mission before the enemy can react. These operations are conducted when there is little time for long, detailed planning. They rely on set procedures set forth by the ranger regiment and its supporting elements. A quick-response operation aims to complete the mission before an enemy can react. This type of operation may be chosen due to the time-sensitive nature of the target, political or military goals, the time frame of other operations, or the increased chance of enemy detection.

(3) Whether an operation is to be deliberate or quick response is often a difficult and time-sensitive decision by a high-level command authority. Decision-makers must consider the enemy's strength in the area, his intentions, his ability either to reinforce or to alter the target area, and the consequences of success or failure of diplomatic or military initiatives in related areas. The ranger unit commander tries to use the existing time, manpower, and resources to complete a detailed and coordinated plan. He refines that plan up to the insertion into the objective area.

c. Ranger units train to operate in any environment or weather condition. They regularly perform operations during periods of limited visibility. Ranger units maintain a high state of physical fitness and often train in close quarters combatives. In addition to completing advanced marksmanship training with standard US weapons, each member of a ranger unit trains with many foreign weapons. Ranger units are trained to operate on urbanized terrain, becoming specialists in entry and clearing techniques and quick-fire methods, especially during periods of limited visibility.

d. Because ranger units have limited vehicles, logistics operations capability, indirect fire support, and heavy weapons systems, they are not designed for continuous operations. During all phases of operations and training, ranger units need responsive external support.

e. Ranger units are normally employed against targets and under conditions that need their unique skills. Although targeting priority is set by the overall commander, ranger units are not normally assigned missions that can be done by conventional aerial bombardment or by other units.

f. Ranger units are oriented toward offensive operations. They are not normally employed as a rear area protection force. Although the ranger regiment has a small reconnaissance unit, ranger units do not normally conduct long-range reconnaissance missions. The structure, communications, and training of the ranger unit do not prepare it for LRRP missions. Ranger units engaged in strike and special light infantry operations have a secondary mission to collect and report combat information.

g. Ranger units can be deployed worldwide when US military presence or participation with a host national military activity would serve US interests. This deployment shows a readiness to commit forces into a threatened area or proves US national resolve. After the deployment, other activities include staging operations, rehearsing combat operations, securing base areas for use and deployment of other forces, and so on that provide a clear signal of US intent. Ranger units are not trained or organized to provide mobile training teams (MTT) to train indigenous forces. The US Special Forces or other special operation forces are trained to conduct such security assistance operations.

h. Ranger units can serve as an example to a host country and provide limited military advice and training. They will normally be augmented with linguists and technicians to increase their abilities. The regiment would normally still function as a unit. Most of its assistance would be through short-term, high-impact, unit-oriented operations. They would not be long-term individual efforts associated with advisory-type activities.

i. Ranger units may be deployed to engage in combined training exercises with allies. This enhances US national image by demonstrating the outstanding abilities of the American ranger. These activities may include ranger, light infantry, airborne, air assault, or amphibious operations.

j. Ranger units use standard US nuclear, biological, chemical (NBC) warning; detection; protection; and decontamination equipment and doctrine when operating on the integrated battlefield.

1-6. Command and control.

a. **External.**

(1) The ranger regiment is a key component of the US Army's special operations forces. The other elements of special operations forces are special forces (SF), psychological operations (PSYOP), civil affairs (CA), and special operations aviation.

(2) The 1st Special Operations Command (1st SOCOM) is responsible for command of all active component special operations forces elements, to include the ranger regiment, in peacetime. The 1st SOCOM has command and control of all assigned special operations forces, less operational command (OPCOM) of special operations forces units forward deployed. (See Figure 1-2.) The ranger regiment, while based in the continental United States (CONUS), is under the control of 1st SOCOM. During peacetime, the OPCOM of deployed ranger units is through channels chosen by the NCA through the Joint Chiefs of Staff (JCS) to the unified command.

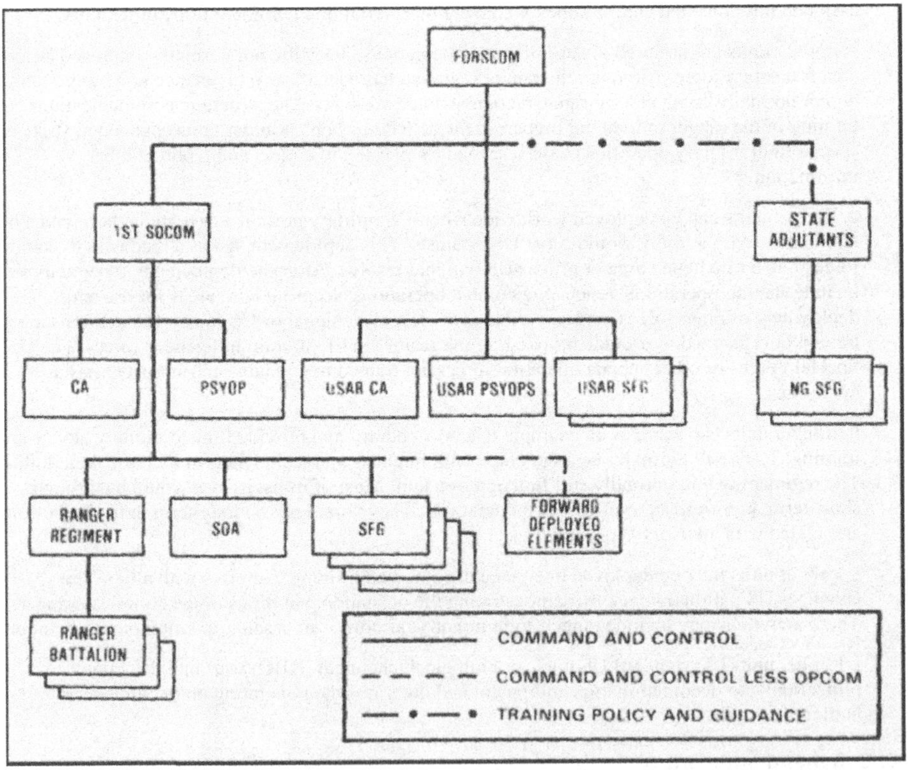

Figure 1-2. Command and control of Army Special operations forces during peacetime.

(3) In support of national policy, the NCA may direct contingency operations involving US

forces. These are politically sensitive military operations normally characterized by the short-term rapid projection or employment of forces in conditions short of conventional war. One aim of the NCA is to conclude them without going to war. The flexibility and effectiveness of the ranger regiment make it possible to use military force quickly to complete the mission and limit the spread of conflict. Command and control of special operations forces in a contingency operation is usually by a special operations task force (SOTF). (See Figure 1-3.) The SOTF and designated special operations forces are under the OPCOM of the highest level of command responsible for the operation. This command may be the unified command responsible for the contingency area (Option A), or a JTF designated by the NCA through the JCS (Option B). If no SOTF is formed, the ranger regiment could be OPCOM to the Army component commander (Option C). The ranger regiment or its battalions are not normally placed OPCOM to a division. A division's area of influence rarely contains appropriate targets. The division also lacks the assets to adequately support ranger operations. Employment of a ranger unit by a corps-level command is normally on a case-by-case basis. The theater or JTF commander retains centralized control over the mission assigned to the ranger unit.

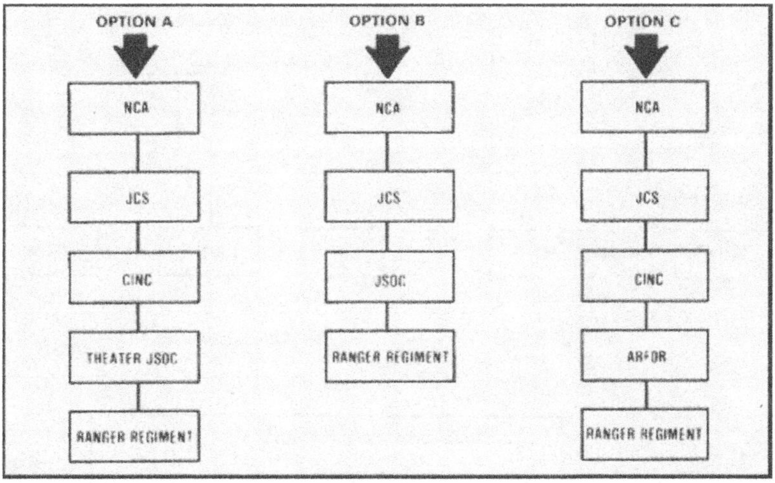

Figure 1-3. Command and control of Army Special operations forces during contingency operations.

(4) The 1st SOCOM is responsible for the deployment of the ranger force to the theater area. The headquarters having OPCOM provides the all-source intelligence, secure communications, insertion and extraction assets, and logistical support needed. The responsible commander in chief, through the Army component commander, provides support during the contingency operation.

(5) During wartime, ranger units are deployed as strategic assets to the theater of operations.

Strategic mission support is provided to the unified commander (CINC) from the NCA through the Joint Chiefs of Staff. It may involve the execution of sensitive operations. The employment of ranger units is controlled at the highest level of command directly responsible for the mission. Ranger units are assigned to a REMAB based on mission, force protection, and OPSEC considerations. The employing command and other elements provide timely and accurate intelligence, communications, logistical, and administrative support. Command and control of these units conforms to the provisions of JCS Pub 2. (See Figure 1-4.)

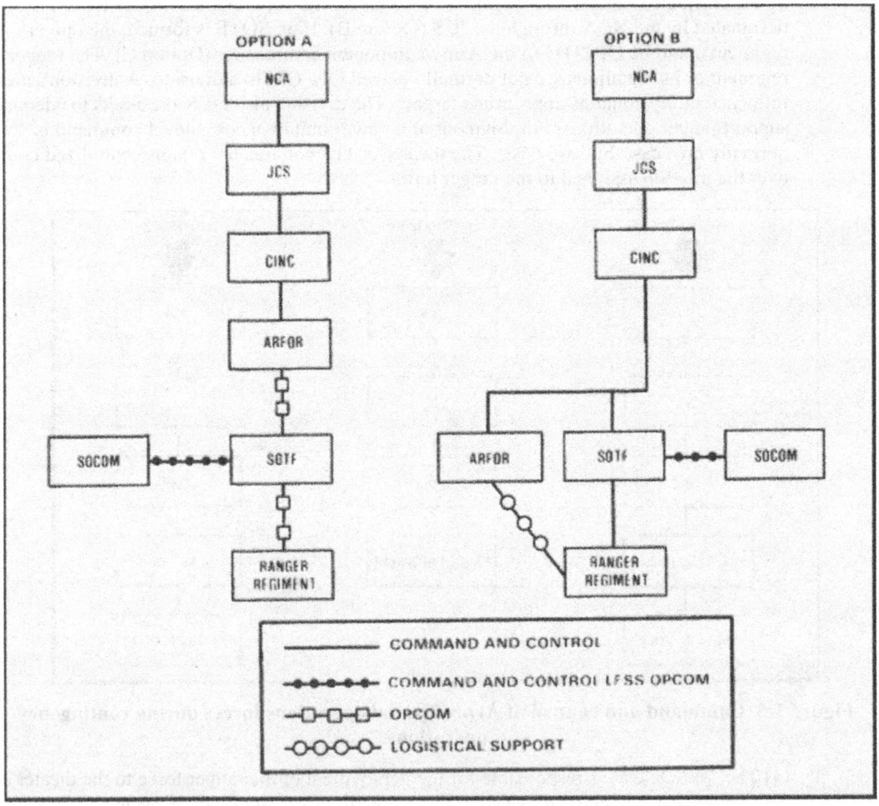

Figure 1-4. Command and control of Army Special operations forces during wartime.

(a) A theater joint special operations command (JSOC) is subordinate to the unified command to supervise the command and control, employment, and support of all US and allied special operations forces. The JSOC is a key element in the command and control of ranger units performing strike missions or special light infantry operations (Option A).

(b) The JSOC may be placed on a lateral line with other in-theater service components (Option B). This can occur when ranger units are -employed far from conventional forces and against strategic targets. It can also occur when the JCS or a CINC forms a JTF to conduct a limited-duration operation in an area where ranger units and conventional forces may operate near each other, or where operations must be coordinated on a regional level.

(c) Under the Army special operations command (ARSOC) concept, 1st SOCOM deploys as a command and control headquarters within the theater. When this occurs, the command relationships will be as shown in Figure 1-5.

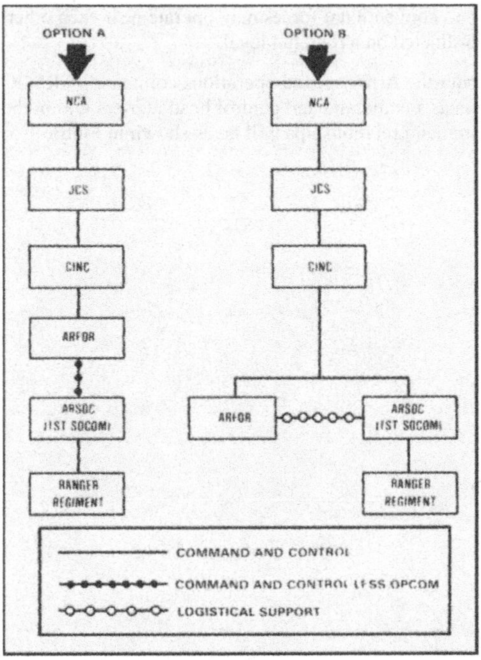

Figure 1-5. Command and control of Army Special operations forces under the ARSOC concept.

(d) Recently passed legislation will modify existing high level SOF command and control arrangements. All US Army, USN, and USAF special operations forces will be consolidated into joint commands at both national and theater level. The ranger regiment will continue to operate under the command of the 1st SOCOM, while increasing its ability to conduct joint operations.

(6) Ranger units conducting strike or special light infantry missions are not normally offered for transfer of authority (TOA) to allied forces. However, this will not preclude allied commands and US elements from combined planning and targeting for ranger missions.

(7) Requests for ranger units, from either US or allied commands, go through normal command channels to the CINC. The CINC, through the JSOC, is responsible for the employment of all special operations forces, including ranger units. This provides flexible response to JCS-directed missions, ensures quick response to the CINC's needs, enhances the joint employment of special operations forces, and lessens the chance of overcommitting and degrading the ranger force.

(8) Command and control of the ranger regiment and its battalions is IAW AirLand Battle doctrine. Before commitment to a mission, command and control is kept at a level where the unit's unique skills can be applied worldwide. Once committed, the ranger unit is placed under the operational control of the command that is responsible for the mission. This normally corresponds to the command whose area of influence includes the target area. Ranger task forces may operate directly under the authority of the senior American representative in the country. Command and control arrangements may vary according to the theater involved and METT-T considerations.

(9) The command and control of special operations is centralized and begins at the highest level of the military and civilian decision-making process. The specific command relationships are normally based on the mission.

b. **Internal.**

(1) The ranger regimental headquarters is organized to operate like that of a brigade headquarters. In addition to commanding and controlling all three ranger battalions, the regimental headquarters can assume operational control (OPCON) of conventional combat and combat support units, and other special operations forces for limited periods. It provides the regimental commander with a well-balanced responsive staff. It is structured as an operational headquarters that can deploy to the mission area and act as the ground tactical headquarters.

(a) The regimental headquarters prepares for combat by forming two small command and control groups. Each of these can deploy to an objective area and control combat operations. Other elements of the staff operate the tactical operations center (TOC) and the logistical operations center (LOC). These may be collocated depending on the situation. The TOC and LOC are normally located well behind the FEBA, at the REMAB or at the intermediate staging base (ISB). The ranger regiment also has two small liaison teams consisting of LNOS, staff representatives, and communications elements. (See Appendix B.)

(b) When a single ranger battalion is committed to an operation, the regimental commander normally exercises control from the TOC. A liaison and communications team is detached from the regimental headquarters to augment the ranger battalion. This team is attached to the deployed battalion until the end of the mission.

(c) If two or more ranger battalions are used in a single operation, the regimental commander normally deploys both command and control groups. He commands from the objective area, The use of two small, mobile control elements makes up for the possible loss of the regimental commander and the primary control element.

(d) Whenever the regimental commander is controlling an operation from the REMAB or the objective area, he normally provides a liaison team with a communications element to the higher headquarters exercising operational control. This cell can provide ranger representation to the controlling staff, operations interface, logistics interface, fire support planning and coordination, intelligence analysis and dissemination, and communications between the higher headquarters and the ranger battalion.

(2) The ranger regimental headquarters can serve as a ranger or special operations force task force headquarters. With augmentation, it can function as the Army special operations force command element. The regimental headquarters can also function as the Army component command of a JTF.

1-7. Communications.

a. Ranger missions need unique and dedicated communication support assets. Along with the ranger regiment's organic communications means, communications support assets are organized within the 1st SOCOM that-can deploy in support of the ranger regiment. This ensures that there are reliable communications to all elements of the SOTF. A special operations communications battalion (SOCB), deployed by the 1st SOCOM, provides more communications systems to support the special operations plan. The SOCB does not replace or duplicate existing in-theater communications. It augments the CINC's communications with the US Army's portion of the critical special operations force command, control, and communications link unique to the SOTF.

b. The ranger regiment's communication platoon can provide secure, long-range communication links from the CINC to a deployed ranger force. It can also enter the SOCB and the joint communications support element (JCSE) nets to provide a direct link with the NCA. Its equipment consists of four light vehicles equipped with various multiple communications systems. The two regimental liaison teams use one of these vehicles each. The TOC and LOC each use one.

c. Secure satellite communications (SATCOM) and radios with AM or FM capability are the main means of communication within the ranger regiment. These are also used for communications within the ranger battalions. The regimental commander has two light vehicles equipped with multiple communications systems. One of these is used to support each of the two mobile command and control elements of the regimental headquarters. Some of the radio systems can be dismounted and man-packed.

d. The ranger regimental headquarters has both man-packed and vehicular tactical satellite communication systems. Each of the ranger battalions has one base station and three man-packed systems.

(1) The man-packed radio is battery-operated. It provides two-way communications in both line-of-sight (mobile) and satellite (at-halt) modes. The digital message device group (DMDG) is the input and output device for data transmission. Secure voice communications is achieved using various encryption devices.

(2) The vehicular net control station uses the basic receiver/ transmitter unit installed on a vehicle. Each can serve up to 15 terminals in a communications net.

e. The reconnaissance platoon is equipped with multiple, man-packed communications systems. These let its teams operate in most of the communications nets within the regiment. Each five-man reconnaissance team has this communications ability, as does the platoon headquarters.

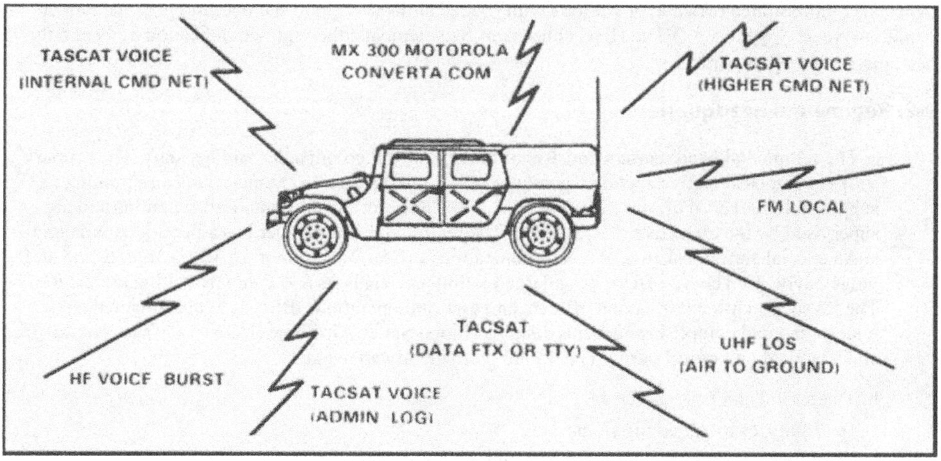

Figure 1-6. Communications capabilities within the regimental headquarters.

CHAPTER 2

Organization and Equipment

The ranger regiment is a specially organized and trained light infantry force consisting of about 2,000 personnel. It includes a regimental headquarters and headquarters company and three ranger battalions. (See Appendix A.) The regiment can operate as a single unit, separate battalions, or separate companies on specific missions. Independent missions are not normally assigned to ranger platoons.

2-1. Headquarters and headquarters company.

The HHC of the ranger regiment consists of 130 personnel, the majority being senior noncommissioned officers, commissioned officers, or soldiers with special skills to support the operations of the ranger battalions. (See Appendix A.) The HHC consists of two elements: the regimental headquarters and the headquarters company (minus).

2-2. Regimental headquarters.

a. The regimental headquarters consists of the regimental commander and his staff. The primary staff is composed of five sections: personnel (S1), intelligence (S2), operations and training (S3), supply (S4), and civil affairs (S5). These staff sections perform the standard functions and are supervised by the executive officer (XO). The regimental commander has a deputy commander and a special staff consisting of a communications officer, fire support officer, surgeon, and staff judge advocate. The S2 officer has a large intelligence analysis and counterintelligence section. The S3 section has extra liaison officers and assistant operations officers to coordinate the widespread and complex ranger missions. A United States Air Force (USAF) staff weather officer and a tactical air control party (TACP) are permanently attached.

b. The regimental headquarters:

- Analyzes assigned missions.
- Plans, coordinates, and controls training and operations.
- Coordinates the infiltration, resupply, exfiltration, and debriefing of the reconnaissance teams.
- Provides liaison support.
- Provides administrative, medical, NBC, and logistical planning and coordination support to the committed ranger battalions.
- Coordinates special support requirements (rigger, transportation, communications security [COMSEC]) for the ranger battalions.

2-3. Regimental headquarters company.

The regimental headquarters company (minus) consists of the company commander and his administrative personnel, plus the limited combat support (CS) and combat service support (CSS) elements assigned to the regiment. These are the fire support element, the communications platoon, the reconnaissance platoon, and the medical treatment team.

a. The company headquarters:

(1) Provides command and control of assigned elements.

(2) Provides administrative, personnel, logistical, and maintenance support to the assigned elements.

b. The fire support element:

(1) Plans fires to support the commander's ground tactical plan.

(2) Coordinates the integration of all available fire support means (mortars, artillery, armed helicopters, close air support, naval gunfire) to support mission execution.

c. The communications platoon (see Appendix A):

(1) Provides secure SATCOM teletype (TTY) and facsimile transmissions (FAX) abilities at the remote marshalling base or intermediate staging base (REMAB or ISB).

(2) Provides secure SATCOM, TTY, and FAX to the regimental headquarters and ranger battalions in the objective area.

(3) Provides cryptographic support to the regimental headquarters.

(4) Provides limited repair of radio and cryptographic equipment.

d. The reconnaissance platoon (see Appendix A):

(1) Trains, equips, and controls the employment of three reconnaissance teams.

(2) Infiltrates the objective area by parachute (high-altitude, low-opening [HALO]; high-altitude, high-opening [HAHO]; static line), helicopter, fixed-wing aircraft, self-contained underwater breathing apparatus (SCUBA), small boat, or other means.

(3) Remains undetected in the objective area for up to five days.

(4) Performs reconnaissance.

(5) Performs demolition target analysis.

(6) Uses small boats and inflatable rafts.

(7) Emplaces and monitors unattended ground sensors.

(8) Uses night image enhancement devices.

(9) Uses photographic equipment.

(10) Emplaces electronic target acquisition and designation devices.

(11) Collects specific combat information to satisfy priority intelligence requirements (PIR).

(12) Reports information using secure, long-range communication equipment.

Figure 2-1. Reconnaissance platoon communications.

(13) Performs drop zone (DZ) and landing zone (LZ) selection, marking, and reception duties.

(14) Escapes and evades enemy areas.

(15) Performs limited attacks or ambushes when tasked by the regimental commander.

(16) Reports weather in the objective area.

e. The medical treatment team:

(1) Provides medical support to the regimental headquarters.

(2) Coordinates required medical support and evacuation.

2-4. Ranger battalion.

a. The ranger battalion is organized and equipped to conduct both special and conventional combat operations. It contains organic combat, CS, and limited CSS elements. It consists of a battalion headquarters, a headquarters company, and three ranger rifle companies. (See Appendix A.)

b. The ranger battalion is the combat element of the ranger regiment. It can conduct independent combat operations or subdivide into its separate ranger rifle companies and their subordinate elements who can conduct semi-independent operations.

c. The ranger battalion headquarters and headquarters company (see Appendix A) includes:

(1) **Battalion headquarters.** The battalion headquarters consists of the commander; the executive officer; the command sergeant major; the S1, S2, S3, S4, and S5 officers; the chaplain; the communications officer; the fire support officer; and two radio operators.

(2) **Battalion headquarters section.** The battalion headquarters section consists of the S3 air and assistant S8, an assistant S2, an intelligence sergeant and a senior intelligence analyst, two operations sergeants and an operations assistant, a personnel staff noncommissioned officer (PSNCO), the chaplain's assistant, two clerk typists, two radio operators, a chemical officer and NCO, a personnel administration center (PAC) NCO, three PAC clerks, and a legal specialist. There is a USAF tactical air control party permanently attached.

(3) **Communications section.** The communications section consists of one telecommunications technician, one tactical communications chief, two combat signal radio team chiefs, one tactical communications systems supervisor, three radio operators, one switchboard operator, one single channel radio operator, and two equipment mechanics. The section provides communications support and associated maintenance.

(4) **Battalion medical section.** The battalion medical section consists of one general medical officer, one physician's assistant, two medical specialists, and twelve aidmen. The section provides unit-level medical support and trains selected members of the battalion in advanced medical procedures.

(5) **Battalion support platoon.** The battalion support platoon consists of a battalion support platoon leader, an ammunition NCO, and a food service section. The platoon provides limited CSS for assigned and attached units. This platoon does not have organic transportation. The battalion food service section consists of a food service sergeant (E7), a first cook (E6), and eight cooks.

(6) **Fire support team headquarters.** The fire support team (FIST) headquarters is supervised by the battalion fire support officer. It consists of the fire support sergeant and two fire support specialists. The headquarters controls three 9-man fire support teams. These teams are normally with the ranger rifle companies in a habitual support relationship. When operations are conducted near water, the fire support team headquarters may be augmented by spotter teams from the naval shore fire control party.

(7) **Company headquarters.** The HHC company headquarters consists of a commanding officer, an executive officer, a first sergeant, a supply sergeant, and an armorer.

d. The ranger rifle company (see Appendix A) includes a company headquarters, three rifle platoons, and a weapons platoon. Each company can conduct semi-independent operations for a short time when augmented by assets from the ranger battalion headquarters company.

(1) **Ranger rifle company headquarters.** The ranger rifle company headquarters consists of a commanding officer, an executive officer, a first sergeant, a supply sergeant, a tactical communications chief, two radiotelephone operators, an armorer, and an NBC NCO. It also has a two-man sniper team equipped with a sniper weapon. It normally has one of the 9-man FISTs attached from the ranger battalion headquarters company. Elements of this FIST may be further attached to the ranger rifle platoons.

(2) **Ranger rifle platoon.** The ranger rifle platoon consists of a headquarters section, three rifle squads, and a machine gun squad. The headquarters section personnel include a platoon leader, a platoon sergeant, and a radiotelephone operator. Each rifle squad consists of a squad leader and two balanced fire teams of four men each. The machine gun squad consists of a squad leader and three M60 machine gun crews of three men each. The machine gun squad is normally employed as a single element to provide quick, accurate, long-range supporting fire. One or more crews may be attached to other squads within the platoon.

(3) **Ranger weapons platoon.** The ranger weapons platoon consists of a headquarters section, a mortar section, and an antiarmor section. The headquarters section consists of a platoon leader, a platoon sergeant, and a radiotelephone operator. The mortar section has two 3-man 60-mm mortar squads. The mortar squads are normally employed as a section. They may be attached by section or squad to other platoons. The antitank section normally consists of a section sergeant and three 3-man 90-mm recoilless rifle or Dragon missile teams. They are normally employed as a section. The antiarmor section may also be attached by section or team to other platoons.

2-5. Special teams and elements.

The ranger regiment forms special teams to do critical tasks. These teams are made up of members of the unit and are used when needed. Examples of these special teams and elements are--

a. SCUBA teams whose members are graduates of the Special Forces Underwater Operations School. They conduct special insertion and reconnaissance duties.

b. HALO teams whose members are graduates of the Special Forces Freefall School and who conduct high-altitude, low-opening parachute operations.

c. HAHO teams whose members are also graduates of the Special Forces Freefall School and who conduct high-altitude, high-opening parachute operations, allowing offset parachute insertions.

d. Scout swimmer teams whose members are specially trained by the Landing Force Training Command (United States Marine Corps [USMC]) to conduct operations along coastlines, waterways, and inland waters. These teams conduct special insertion, reconnaissance, and security tasks often in support of amphibious operations.

e. Sniper teams, whose members are authorized by the table of organization and equipment (TOE), are used for both conventional and special operations. Along with the two-man sniper teams in the rifle company headquarters, each rifle platoon has two trained snipers.

f Demolition teams whose members are trained in advanced demolition and incendiary techniques and the use of special explosives,

g. Air defense teams trained in the use of man-portable air defense system (MANPADS) weapons. Such teams give the ranger battalion limited self-defense against air attack.

h. Liaison teams that are used by the ranger regimental commander to augment the existing command and control links from the employing unit headquarters to the ranger regiment. These teams are placed where they are most needed and may assist the higher headquarters staff in planning and target selection, as well as performing other liaison duties. See Appendix B for a detailed listing of the personnel and equipment included in the ranger liaison teams.

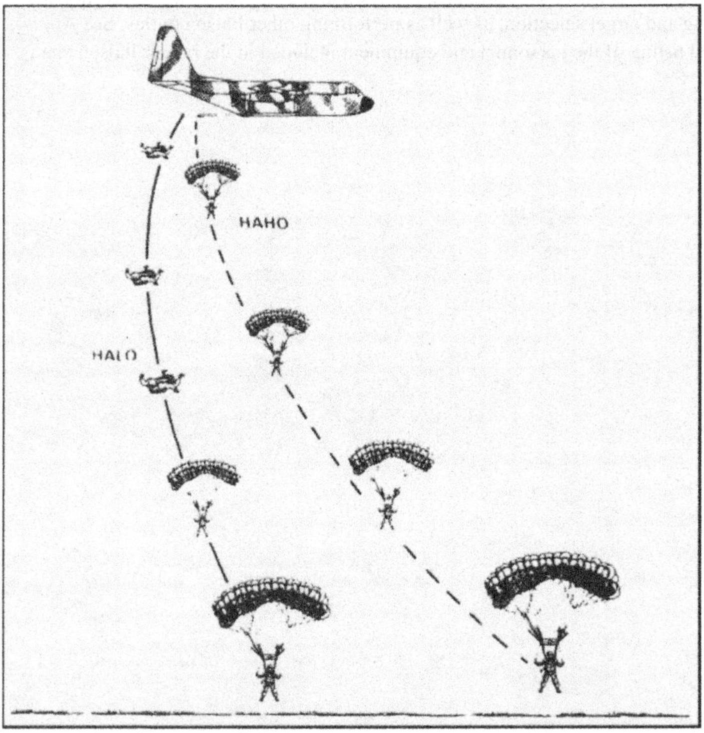

Figure 2-2. HALO and HALO parachute operations.

i. The ranger support element (RSE).

(1) The RSE is a special support element provided by host installation commanders at CONUS posts. It consists of elements and teams from either TOE units stationed at the installation or from the table of distribution and allowances (TDA) organizations under the command of the installation commander.

(2) The RSE provides the necessary support to ensure a timely and efficient deployment of the ranger battalion. During CONUS deployments, the RSE is responsible for both the departure and the reception support provided the battalion. During OCONUS deployments, the RSE is responsible for initial deployment and continuing support until the deploying ranger force is OPCON to the supported unified command or JTF. The responsible theater commander assumes support of the ranger battalion at the REMAB or the ISB if outside continental United States (OCONUS). On deployments that require the ranger battalion to move directly to the objective area from CONUS, the RSE continues support of the ranger battalion until released, and is prepared to accept considerable augmentation assets. The RSE provides the following support:

(a) Furnishing meals that are either served by a supporting unit in an established facility or taken to the ranger battalion at an isolated location.

(b) Issuing supplies from existing stocks, or from contingency items stored for use by the ranger battalion.

(c) Transporting supplies, food, personnel, and equipment.

(d) Palletizing equipment and ammunition for air movement and airdrop.

(e) Rigging vehicles and equipment for airdrop.

(f) Refueling vehicles and equipment.

(g) Issuing barrier and construction material for building of rehearsal sites.

(h) Helping in the construction of rehearsal sites and targets.

(i) Securing, storing, moving, and issuing ammunition and other Class V items.

(j) Maintaining ranger weapons and equipment, to include communications equipment. This is normally done by maintenance contact teams.

(k) Setting up and operating the RSE emergency operations center.

(l) Securing the REMAB if it is in CONUS.

CHAPTER 3
Deployment

Ranger units can quickly deploy by any means of transportation to any location in the world.

3-1. Readiness.

One ranger battalion is always in an advanced readiness condition as the Ranger Ready Force (RRF) available for immediate worldwide deployment. A second battalion is prepared to deploy later and a third battalion to follow. The regimental headquarters maintains command and control, liaison, communications, and reconnaissance elements immediately available for deployment. Higher status of readiness in response to specific world situations can be achieved.

3-2. Ranger ready force.

In order to maintain advanced readiness conditions, ranger battalions are rotated as the RRF.

a. Before taking over the duties of the RRF, battalions must do the following:

(1) Complete preparation for oversea movement (POM) qualification, to include an individual records check and required inoculations.

(2) Test fire all weapons, confirm battlesight settings, and repair or replace faulty weapons.

(3) Assemble and load on pallets the unit basic load (Classes I and V) and other supplies and equipment.

(4) Check unit-derived packing list components for accountability and serviceability.

(5) Prepare special operations troop listings.

b. During an RRF period, ranger units:

(1) Have a positive recall system for all personnel.

(2) Store enough air items for a battalion-size parachute assault.

(3) Conduct no off-post exercises, unless approved by the regimental commander.

c. Each unit keeps an alert roster and a telephone notification system for recalling personnel. The rosters ensure that the duty officer and the unit charge of quarters make a minimum number of calls. Each commander briefs personnel assigned to his unit on the recall system and tests it often.

d. Ranger units can maintain an increased readiness condition for long periods, but they should not be kept in that condition longer than needed. The effort to maintain this level of readiness will degrade the unit's training and effectiveness. If the needs of the operations dictate long periods at the highest readiness condition, provisions must be made for sustainment training. Also, added support will be required from the RSE.

e. The ranger regiment maintains a strong family support system when the unit is training or deployed on combat operations. Each member of the chain of command is responsible for family support.

3-3. N-hour sequence.

The N-hour sequence is a departure time sequence designed to help the ranger force complete all required deployment actions within a certain time. The length of the N-hour sequence depends on the readiness condition of the battalion before deployment. The N-hour sequence begins when the battalion is alerted and ends when the first elements depart. The N-hour sequence does not include travel time to the objective area.

3-4. Intelligence and security.

a. **Essential elements of friendly information.** Items of sensitive information and subject areas, if not secured, can be collectively analyzed by a hostile force. This analysis can reveal the intentions of a unit or compromise a mission. These items of information are designated as essential elements of friendly information (EEFI). They are never transmitted without encryption. The EEFI associated with a ranger unit alert normally are the mission, the use of staging bases, nature of the target, special terrain or weather conditions, and time limits. The ranger force commander must ensure that EEFI are not discussed outside of secure areas or made available to personnel without a security clearance and a need to know.

b. **Security plans.** Security plans for billets, marshalling areas, and out-loading areas must remain flexible to adapt to any situation. Certain items are always addressed in security plans. These are access control and physical security.

(1) **Access control.** Ranger units control access to their respective areas by allowing only authorized personnel to enter or leave. All entrances are secured and an armed guard with the access roster is posted. Access to telephones is restricted.

(2) **Physical security.**

(a) Equipment is open to theft and sabotage during the early phases of deployment. Units must provide guards to secure all unit and personal equipment.

(b) Weapons, ammunition, and sensitive items must be stored in a restricted area and guarded.

(c) Classified documents are open to compromise during deployment. Steps must be set up for maintenance, transportation, turn-in, and destruction of classified material. The most secure facility available must be used.

c. **Rear detachment security.** Security after deployment (maintained by the rear detachment) is as important as security before deployment. No information should be given to the news media from other than official sources. The rear detachment may continue to play a role in the unit cover and deception plan.

d. **Tactical operations center security.** Only personnel listed on TOC access rosters are allowed free entry into the TOC area. Personnel wanting to enter the TOC, who are not listed on the access rosters, are detained until their identity and purpose are proven. Such personnel are escorted at all times while within the TOC area.

e. **Marshalling and out-loading area security.** The ranger unit normally marshalls its personnel and equipment in the unit area, providing its own security force. The USAF security police provide this support at the departure airfield. On Army airfields, Army military police provide security. En route security is provided by the RSE.

f. **Personnel and unit equipment sanitization.**

(1) Before deployment, personnel are relieved of non-mission-essential material having intelligence value. This reduces the intelligence available to hostile forces if they capture a ranger force member.

(2) The ranger force commander sets the degree of sanitization IAW mission requirements.

(3) If strict sanitization is imposed, guidelines are issued as follows:

(a) The only personal items retained are identification card and tags.

(b) Unit insignia and markings are removed from uniforms and equipment.

(c) Personal items, such as wallets and photographs, are secured by the rear detachment commander.

(d) Personnel are briefed on how to compose and retain combat diaries. No personal diaries are allowed.

(4) The commander decides how much US currency is to be taken with the unit. '

g. **Area sanitization.** To avoid the compromise of classified or sensitive mission-related material, searches are conducted before and after leaving the host installation, REMAB, or ISB. These searches are held to locate and remove any material, indicators, litter, or other items that could compromise the ranger unit's mission. Unit commanders ensure that a thorough sanitization of their areas is conducted, to include dismantling rehearsal areas, destroying terrain models, erasing chalkboards, and emptying all classified and nonclassified waste containers. During their searches, commanders must pay close attention to emergency operations centers, tactical operations centers, and marshalling and loading areas. The ranger unit's intelligence officer coordinates support from local counterintelligence elements to search the area again after the ranger unit leaves.

3-5. Operations.

a. **Actions upon receipt of the OPORD.**

(1) Task-organize and equip units for special mission(s) IAW the warning order.

(2) Allocate/reallocate mission-essential equipment.

(3) Distribute augmentee personnel.

(4) Redistribute key personnel,

(5) Prepare a deployment time sequence.

(6) Prepare a movement plan.

(7) Prepare and maintain a current operation estimate.

b. **Advance elements.** A planning element is normally deployed to the REMAB/ISB or JTF headquarters in advance of the main body if the ranger force does not deploy directly to the objective area.

c. **Emergency operations center.** An EOC is setup by the ranger battalion headquarters. The EOC is responsible for relaying all reports to higher headquarters. It remains operational until ordered to cease.

d. **Air movement.** The ranger force always deploys by USAF aircraft. This is normally an administrative air movement to a REMAB or ISB. It may be followed by tactical air deployment from there. For some missions, tactical deployment begins at the unit's home station. Even if the movement is administrative, key personnel are cross-loaded. This is done under the direction of the operations section air officer based on aircraft allocations and load plans.

3-6. Logistics.

The following describes logistics readiness needs and steps for deployment.

a. **Supply.**

(1) **Class I.**

(a) Combat rations are normally palletized when the unit is RRF. These rations are to be used during the tactical operation. If the mission calls for deployment directly into the objective area, the palletized stock of rations is off-loaded in the unit area and issued as needed.

(b) Each ranger deploys with emergency combat rations.

(2) **Classes II and IV.**

(a) **Individual clothing and equipment.**

- Individual A bags are packed IAW a predetermined packing list, tagged, and marked. They are palletized, moved to the departure airfield by the RSE, and loaded IAW the air movement plan.
- Individual B bags are packed, assembled, and forwarded on order by the rear detachment.

- Individual environmental equipment is issued. Equipment shortages are replenished out of stocks maintained by the RSE.

(b) **Air items.**

- While the unit is on RRF, a contingency load of main and reserve parachutes is maintained by the RSE.
- Individual and unit air items are palletized by the RSE and shipped in bulk to the REMAB or ISB.
- For direct-deployment missions, air items and packing material are taken to the unit by the RSE.

(c) **General supplies.** The RSE maintains pallets of common contingency items and special equipment. Pallets are picked up by the out-loading support unit and moved to the unit area for issue or moved with the other pallets to the departure airfield.

(d) **Bulk supplies.** Barrier material is identified, relocated, palletized, and out-loaded IAW the air movement plan.

(3) **Class V.**

(a) The unit basic load of ammunition is secured, palletized, and loaded by the RSE IAW the air movement plan.

(b) For direct deployment missions, selected pallets are taken to the unit area and ammunition is issued. The convoy route is coordinated through the post EOC, and secured by the RSE.

(c) Designated items are rigged by the RSE for airdrop, if enough time exists before movement to the REMAB or ISB begins.

(4) **Class VI.** Personnel deploy with comfort items as needed.

(5) **Class VII.**

(a) All weapons and sensitive items not deployed are assembled and secured by the rear detachment personnel.

(b) Unit vehicles are prepared for airland or airdrop.

(c) Inoperable mission-essential equipment is repaired or replaced.

(d) Special equipment requests are forwarded to the logistics section as they are identified.

(6) **Class VIII.**

(a) Individuals deploy with one set of individual Class VIII items.

(b) Unit medics deploy with modified aidbags.

(c) Controlled medical items (less drugs) are palletized and deployed with section supplies. Controlled drugs are secured and issued to unit aidmen on order.

(d) Medical supply chests are maintained in the unit aid station.

(7) **Class IX.**

(a) Units deploy with initial prescribed load list (PLL) items.

(b) Critical PLL shortages are identified to the RSE for follow-on issue.

b. **Maintenance.**

(1) Upon alert, priority maintenance or replacement of key items of equipment is coordinated by the logistics officer. Only minimum mission-essential equipment is replaced.

(2) Items to be returned from supporting maintenance activities are identified, and pickup is coordinated by the property book officer.

c. **Out-loading coordination and support.**

(1) Upon alert, the RSE provides a liaison officer to the ranger logistics section.

(2) The support platoon leader (SPI,) attends the air movement conference and coordinates the following:

(a) Movement times (for both troops and equipment) to the departure airfield.

(b) Steps for weighing and joint inspection of the pallets and vehicles upon arrival at the departure airfield control group (DACG).

(c) Limitations on the pallet height and weight due to aircraft type, allowable cargo load (ACL), and availability.

(d) Changes in the planned number of pallets due to the mission requirements or airframe availability.

(e) Rigging and loading of the heavy drop and supply bundles.

(3) Empty pallets are placed in position for loading by the RSE in coordination with the SPL.

(a) Companies and staff sections prepare pallets.

(b) For airdrop operations, vehicles are prerigged at the unit location. On order, vehicles are driven to the heavy drop rigging site. The SPL and the officer in charge of the rigger detachment coordinate for a rigging site and the delivery of the required air items to that location. Rigging is done by the RSE, and supervised by a member of the ranger unit. Loading aboard the aircraft is IAW the air movement plan and supervised by RSE and USAF aerial port personnel.

(4) The DACG is provided by the host installation. Upon arrival at the DACG area, unit personnel and equipment are under the control of the DACG. As the deployment begins, a

member of the ranger unit is present at the DACG to ensure an orderly loading of personnel and equipment.

(5) The DACG plans for troop-holding areas in the event of delayed or aborted aircraft.

(6) Unit movement data are maintained by the air officer in the operations section.

3-7. Communications-electronics.

a. **Alert notification communications.** Alert notification of the ranger battalions by the regiment is done by secure communications means. Units in local training areas are informed by radio or through their assigned range field telephone. Units operating out of normal FM radio range during RRF periods set up a high frequency net, or other communications, to their higher headquarters.

b. **Telephone service.** Strict control is maintained over key telephones being used during unit alert. Care is taken to ensure that COMSEC and OPSEC violations do not occur. All telephones are assumed to be monitored; classified, sensitive, or official information is not discussed.

c. **Radio communications.**

(1) **Voice communications.**

(a) A secure radio net is set up early in the deployment. The ranger unit EOC, the installation EOC, and the RSE EOC operate in this net. A liaison officer from the ranger unit is placed at the installation EOC.

(b) The ranger support element liaison officer, attached to the logistics section, operates a secure voice station on a separate logistics net to the RSE EOC and the RSE commander.

(2) **Cryptographic material.** During RRF, COMSEC codes and enabling equipment are kept available to set up the secure voice net.

3-8. Remote marshalling base and intermediate staging base.

a. Ranger units can deploy from their CONUS base directly to the objective area. A more common method would be for the ranger unit to deploy first to either a REMAB or an ISB before insertion into the area of operations. (See Figure 3-1.)

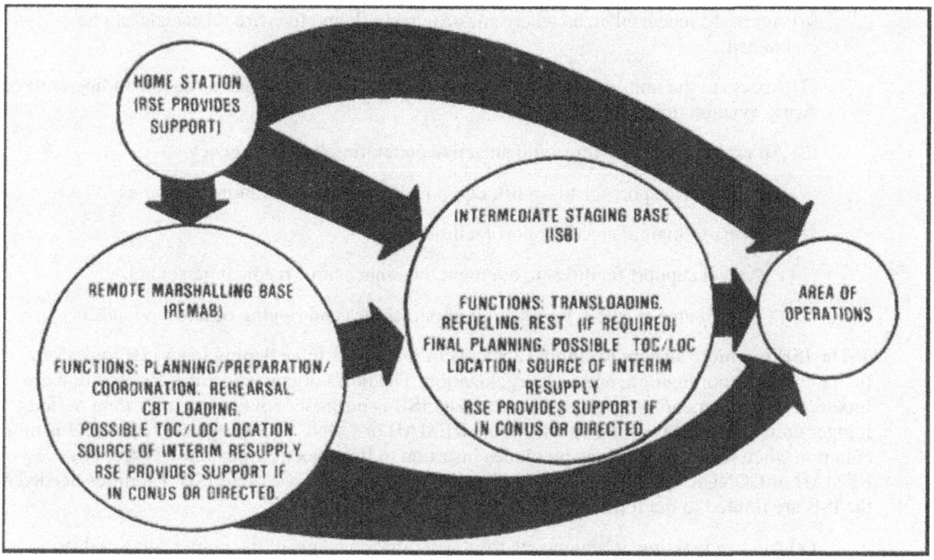

Figure 3-1. Deploying ranger units.

b. The REMAB is a secure base to which the entire ranger unit, including organic and attached support elements, deploys and where mission planning continues. The REMAB is normally within the geographical area encompassed by the command authority of the theater or JTF commander. This ensures that the CSS elements providing support to the ranger unit are operating within their normal area. It prevents or lessens out-of-sector support requirements for CSS elements.

c. The REMAB also provides a secure location for the ranger unit to conduct detailed planning and coordination with the controlling headquarters staff. In the REMAB, the ranger force commander conducts rehearsals, refines and modifies plans, determines PIR, and coordinates with the proper intelligence source to receive that information. It is in the REMAB that specialist augmentees to the ranger force are integrated into the unit, if they have not already joined. Specially trained supporting units, such as aviation and communications elements, also join the ranger force at the REMAB. The REMAB must provide--

(1) Access to the controlling headquarters staff.

(2) Physical security of billeting, planning, maintenance, and communications areas.

(3) Mess, billeting, latrine, and shower facilities for the ranger force and its supporting elements.

(4) Access to a C-141 or C-130 capable airfield, possibly with all-weather operations.

(5) Access to secure communications and processed mission intelligence.

(6) Access to rehearsal areas where sites can be built and live-fire rehearsals can be conducted.

(7) Access to the unit locations of major supporting elements, such as naval landing craft or Army aviation units.

(8) An external security force and an active counterintelligence agency.

(9) Vehicular transport for troop lift, equipment transfer, and administrative use.

(10) Access to maintenance support facilities.

(11) Medical support facilities to augment the ranger force medical personnel.

(12) Covered area to allow for packing of parachutes and rigging of airdrop loads.

d. The ISB is a more austere location. Elements of the ranger force deploy to an ISB to perform final planning, coordination, and task organization. The unit's organization and composition are finalized for movement to the objective area. The ISB is normally not occupied for long periods. Ranger units may deploy to the ISB from the REMAB or CONUS. Deployment to the ISB is more common when terrain or distance precludes insertion to the objective area directly from the REMAB or CONUS. Quick-reaction missions often involve the use of an ISB. Facilities needed at the ISB are limited to the following:

(1) Austere messing arrangements for the combat elements of the ranger force and its attachments.

(2) Austere billets or rest areas.

(3) Water point.

(4) Fuel for aircraft and vehicles.

(5) Areas for test firing of weapons.

(6) Covered and concealed areas for assembly of the ranger force and rigging of parachutes and door bundles.

(7) Austere airfield support facilities, possibly capable of all-weather operations.

(8) Remote location far away from civilians or traffic routes.

(9) Security and counterintelligence elements.

(10) Secure communications.

e. The REMAB and ISB should be located in an area similar in terrain and climate to the objective area. Time spent at the REMAB or ISB lets the ranger force begin acclimatization. Ranger units train yearly in all climates and emphasize physical conditioning. This allows acclimatization quicker than units not as well trained or conditioned. However, ranger units still need an acclimatization period if the terrain and climate of the objective area are different from that of their CONUS base. Not enough acclimatization means that the ranger force could be less efficient, especially when the objective area is hot, dry, or at a great altitude.

CHAPTER 4

Insertion, Extraction, Escape, and Evasion

Insertion is the entrance of a ranger force into a hostile area to execute a particular mission. It is the first critical phase of a ranger operation. These operations often involve deep penetrations of enemy territory by air, water, or land. This chapter emphasizes fundamentals and techniques employed by ranger units when their mission requires insertion and extraction. Extraction is conducted quickly after the mission to avoid casualties.

4-1. Intelligence.

a. Insertion plans are based on timely and accurate intelligence. The headquarters directing the operation must provide current intelligence information to the ranger force. The ranger regiment normally takes action to solicit required information from the controlling headquarters and from other agencies, if security permits. The type of intelligence needed varies depending on the specific mission, but normally includes the following:

(1) Current photographs from aerial or overhead reconnaissance.

(2) Detailed maps.

(3) Data on the population, terrain, and weather from area studies.

(4) Current enemy order of battle, enemy troop locations, and enemy capabilities.

(5) Details on the target area and infiltration routes.

b. To meet its need for current intelligence, the ranger force also relies on aerial reconnaissance and surveillance of the terrain over which it will move. Emphasis is on getting information on the enemy's ability to detect and engage the inserting forces. The location and abilities of air defense radar and weapons systems are critical. Terrain analysis is detailed and focuses on areas suitable for the insertion. Weather information must be detailed and current.

c. In getting information and intelligence data, OPSEC measures are stringent and must be enforced. Coordination with other agencies is kept within security constraints approved by the directing headquarters.

4-2. Deception.

There are always plans to deny the enemy knowledge of the ranger unit's insertion, or to deceive him as to the location or intent of the operation. Feints, false insertions, and other deception operations all add to ranger deception plans. Selection of unexpected means, time, place, and routes of insertion, as well as use of speed and mobility, give less reaction time to the enemy. Planning may include the use of diversionary fires to direct the enemy's attention away from the inserting unit. Deception techniques that may be used include--

a. Multiple airdrops, water landings, or both. This precludes the entire force from becoming engaged at the same time and place, if detected.

b. Dispersion of insertion craft (air or water) both in time and location.

c, Landing a force in an area closer to other potential targets than to the actual targets.

d. Leaks of false information and false messages.

e. False insertions or radio traffic to disguise actual insertions.

f. Diversionary actions such as airstrikes in other than the target areas.

g. Increased reconnaissance flights, either by manned aircraft or remotely piloted vehicles, over false targets.

4-3. Speed and Mobility.

a. Speed of action must be emphasized when the enemy can react quickly to the ranger force presence or when there are limited means of withdrawal. The operation must be conducted swiftly, allowing the ranger force to withdraw before the enemy can react.

b. Major problems facing ranger units are the need for speed and mobility, and the heavy loads carried by soldiers for sustained operations. Ranger units are required to be flexible and to be able to react to any problem within the objective area. However, if equipment is carried to cover every encounter, the load will become too heavy. This will reduce the speed and mobility of the ranger unit. The ranger force commander must always have a balance between the equipment carried and the mobility of his unit. He does this by making three different types of loads: fighting, mission, and existence. He decides what items will make up these loads, where they are carried, and what will be done with them upon contact with the enemy.

(1) The fighting load consists of those items of equipment, weapons, ammunition, food, and water that are common to all soldiers and are necessary for immediate use in combat. These items are normally carried and are retained at all times while in contact with the enemy.

(2) The mission load consists of those items of equipment or ammunition that are required by the mission, but are not normally carried by all rangers in a unit. The load may include spare batteries, demolitions, radios, antitank weapons, or mines. This load is normally spread out among members of the unit and is rotated often. It may be dropped upon enemy contact, but is normally carried to the objective and either used or emplaced there.

(3) The existence load consists of any items designed to provide protection from the elements: sleeping gear, changes of clothes, spare rations, or tentage. This load may or may not be brought into the objective area, depending on the tactical situation. If it is brought, it is normally dropped upon enemy contact or left in an assembly area and retrieved later.

4-4. Stealth.

Methods and steps chosen for insertion must emphasize stealth and try to avoid detection by the enemy. Multiple routes or methods of insertion may be used to preserve the ranger force if some soldiers or units are seen.

4-5. Suppression.

Effort must be made to suppress enemy detection devices, weapons systems, and command and control facilities. This may include electronic countermeasures or direct attack. This helps prevent the enemy from finding and engaging the ranger force en route to the objective area. Deception tactics add to suppression activities. If the ranger force is inserting by infiltration, suppression must not disclose friendly activities.

4-6. Security.

Emphasis on OPSEC prevents compromise of an impending operation. The operation must not be disclosed by overt rehearsals or training activities, the open use or procurement of special items of equipment, or the location of the marshalling area. Several measures can help in maintaining security:

a. Guarded and secure facilities for the headquarters and staff during planning.

b. Issue of operation orders only after units are moved to secure marshalling areas.

c. Alerts, rehearsals, and training to get the local people used to conditions that will exist during actual marshalling.

d. A secure facility set up to isolate units selected for an impending operation.

4-7. Night vision and electronic devices.

a. Rangers use night vision devices to detect and avoid the enemy's forces and his detection devices. These devices also assist in controlling and speeding movement and allow units to traverse almost impassable terrain.

b. The use of electronic homing devices permits precise navigation during reduced visibility. These devices may also be used with visual marking systems.

c. Passive night vision devices are used to assemble rapidly and to reorganize.

d. The use of active night vision devices in assembling, and the degree to which drop zones or assembly areas are "sterilized," depends on the following:

(1) Mission success may depend on not being seen during insertion. When a clandestine insertion is made, the time needed to get to the objective will normally be greater, the use of passive devices will prevail, and the sterilization of the area will be vital. Each ranger must be briefed on sterilization plans and techniques for erasing signs of the insertion.

(2) When detection is likely and the mission depends on getting to the objective before the

enemy can react, speed becomes crucial to the success of the insertion. Insertion may be near or on the objective. Rangers must get in and out quickly. They do minimum sterilization of the area and can use active RSTA devices.

4-8. Rehearsals.

Rehearsals should be close to the actual conditions of insertions and extractions. Security of rehearsals and rehearsal sites is important to the success of the operation. Rehearsals are held on terrain and structures like those in the objective area. If such structures are not available, they are built by the ranger force and the RSE with its supporting engineers.

4-9. Sand tables and terrain models.

Sand tables are useful to acquaint personnel with insertion sites and surrounding terrain. The use of sand tables and terrain models during the issuance of prejump orders and briefings enhances an orderly and speedy tactical assembly.

4-10. Communications.

Reporting during insertion and extraction missions is normally by exception, using an operations schedule.

4-11. Planning insertions.

a. **Centralized planning.** Planning for insertion of ranger units is centralized at the lowest level responsible for the overall operation. This is normally at the joint task force level, with detailed coordination required among all supporting forces. These forces may include other Army elements, USAF, United States Navy (USN), United States Marine Corps (USMC), other US agencies, and allies. Preparation must include detailed contingency planning for use of firepower, shock effect, speed to break contact (if detected), control measures, procedures, and planning. When planning takes a long time, intelligence must be continually updated through all available sources. This includes overhead photography and ground sources to ensure that the initial planning data are still current. Planning stresses the use of deception. Escape and evasion plans, external communication nets, and the location of caches must be known to leaders at all levels.

b. **Reverse planning sequence.** Insertion planning is done in the reverse planning sequence with the ground tactical plan being prepared first. Actions at the objective serve as the basis for the landing and assault plan, the plan for insertion and movement, the loading plan, and the plan for marshalling. All plans are interrelated. Some may be developed at the same time. Consideration of the factors of METT-T guides all planning. Special considerations that apply to these operations are:

(1) **Mission.** The nature of actions at the objective and requirements for special items of equipment and nonorganic personnel impacts on how the unit inserts. When the mission requires rapid deployment, the fastest method of insertion may be needed. In other operations, where success of the mission depends on maintaining secrecy, speed may be less important.

(2) **Enemy situation.** Enemy disposition, strength, and security measures along the route(s) near the objective area affect the method of insertion to be used. Planners must take into consideration whether the enemy:

(a) Uses RSTA detection devices or not.

(b) Has a strong air defense system or possesses air superiority.

(c) Has gaps between his defensive positions.

(3) **Terrain and weather.**

(a) **Topography.** Terrain such as mountains, swamps, or heavy forests favors land infiltration. Deserts, arctic regions, and open prairies favor air or water insertion. Areas not inhabited or with a friendly but scattered population favor land infiltration. A heavily populated area, or one which contains large enemy forces, favors air or water insertion. Hydrography--tides, currents, surf, reefs, and sandbars--must be considered when planning water insertion.

(b) **Weather.** Bad weather with reduced visibility and high winds favors land infiltration, and adversely affects air or water insertion. Limited visibility favors land infiltration; good visibility does not. Adverse weather aerial delivery systems (AWADS) lessens the impact of reduced visibility as a limiting factor on air infiltration. High winds may preclude airborne insertion.

(4) **Troops.** The number of men to be inserted and their level of training may be limiting factors. A need for special skills may call for use of nonranger unit personnel whose physical stamina and ability to perform many infiltration techniques may not equal that of the rangers. Whether supporting forces and special items of equipment are available must also be considered.

(5) **Time.** The main concern is whether the ranger operation is quick-response or deliberate-response. During a quick response mission, planning, preparation, and rehearsal time is short. The timing of the insertion is crucial with respect to weather, the enemy, and if an insertion craft is available.

4-12. Air insertion.

The most rapid form of insertion is by air. Rangers and equipment can be delivered by parachute (static-line or freefall technique), fixed-wing aircraft (airlanded), or helicopter (airlanded, parachute, or rappeling).

a. Suppression of enemy air defense along the insertion corridor may be necessary. The USAF will normally assume the major responsibility for suppressing enemy air defense artillery (ADA) capabilities that may interfere with the inserting forces. This is done by a variety of sophisticated electronic countermeasures applied against enemy equipment and by strikes against known enemy positions. Army fire support elements, special electronic mission aircraft (SEMA), and ground jammers may participate, as well as Navy aircraft, jammers, and gunfire.

(1) A main danger area is the forward edge of the battle area (FEBA) or frontier areas where

the enemy employs most of his sophisticated weapons systems and air defenses. Other dangerous areas are population or troop concentration areas behind enemy lines protected by point air defense systems.

(2) Since the majority of the enemy's detection devices and air defense weapons may be located near the point of entry, fire support, smoke screens (even at night), and suppressive measures may be critical. The use of special items of equipment may be needed to counter the enemy's RSTA effort.

(3) Long-range fires (artillery and naval gunfire) should be planned on known and suspected enemy antiaircraft locations and on prominent terrain features along the route(,s). Once beyond the FEBA, inserting units may be beyond the range of conventional indirect fire. More reliance will then be placed on air assets for fire support and electronic countermeasures.

b. There must be close coordination between the ranger unit and the transporting unit. What the transporting unit does during the insertion has a great impact on the success of the mission. To decrease the chance of detection, maximum use is made of periods of reduced visibility and tactical cover and deception. Drop zones and landing zones are chosen behind tree lines, in small clearings in forests, or on other types of terrain not easily seen.

c. All flight routes over enemy territory should be planned in order to avoid occupied areas and enemy air defenses. Routes should complement the cover and deception plan. They should avoid giving away final ranger objectives.

d. In-flight emergencies, mainly in deep-pedetration flights, should be considered, The ranger force receives a preflight briefing on the route to be flown and is informed of checkpoints along the route. Simple ground assembly plans for contingencies are set up before enplaning. In an emergency, the leader of the unit involved decides whether to continue the mission or abort. This is based on METT-T factors, contingency plans, and distance to the objective as compared to the distance to friendly territory. Contingency provisions for air and water rescue are made.

e. Deciding the DZ location is a joint effort. After receiving input from the air and ground commanders, the joint task force commander makes the final decision. During air assault operations, the ground commander is responsible for the overall process. After reviewing MEYF-T factors, it may be determined that surprise is more important than speed, or vice versa. This may determine the number, size, and locations of DZs and LZs.

f. Planning and preparation continue during the marshalling phase of any operation using air delivery for insertion. A review must be made of items of equipment that are crucial to the ranger force reaching its target area, accomplishing the mission, and returning safely. The need for special supplies and equipment, their transport, the distance to be traversed, the delivery means, and external support are all factors that enhance or reduce the chance for success. Once the means of insertion has been decided, logistics planning continues with emphasis on requesting and coordinating external support. The following must be considered:

(1) Detailed planning for mission logistical needs.

(2) Extra sets of key items of equipment to ensure redundancy and self-sufficiency of

inserting units.

(3) Plans for accompanying, planned, and emergency resupply. The ranger force should prepackage loads to meet specific contingencies. It should consider the use of supplies and equipment already in the operational area.

(4) In quick-response operations, only manportable items of equipment needed to accomplish the mission are carried. Plans and signals must be established for resupply of key items.

(5) Specific plans to evacuate casualties.

(6) All attached personnel and augmentation equipment must be properly prepared for movement.

(7) Provisions must be made for evacuating captured enemy equipment and personnel.

(8) All inserting units should be cross-loaded. Items to be considered in a cross-loading plan are the ground tactical plan, the assembly plan, the configuration of the DZ, the light route and formation of the aircraft, and the number of aircraft available. Cross-loading ensures that key leaders and equipment--for example, door bundles--are evenly distributed throughout the assaulting force. If one or more aircraft abort or are shot down, some key leaders and equipment will still arrive at the DZ. This permits mission accomplishment. Cross-loading aids rapid assembly on the DZ in support of the ground tactical plan. If a ranger is to assemble near the leading edge of a DZ, he is loaded on the aircraft to exit near the front of his stick.

(9) A bump plan is used if an aircraft should abort while still on the ground and there are no spare aircraft. The bump plan ensures that key leaders and equipment are not left behind when the main body takes off. If an aircraft aborts on the ground, key personnel move to another aircraft and replace nonessential personnel. This concept promotes success on the objective.

4-13. Airborne insertion.

a. During the planning phase of an airborne operation, there is no room for error in judgment. It can be planned and conducted well only if there is a constant exchange of information. All personnel involved must be informed of all changes in plans, times, and locations.

b. The reverse planning sequence is the key to the success of an airborne insertion. Most of the planning is based on the ground tactical plan, followed by the air movement plan to support it.

c. The USAF is responsible for flight planning, initial point (IP) selection, and crew procedures throughout the flight. The senior ranger in the aircraft must remain oriented and keep abreast of any last-minute changes. He must coordinate with the aircraft commander.

d. Most airborne insertions are made at either very high altitudes (HALO/HAHO) or very low. For combat operations, drop altitudes may be less than 500 feet above ground level (AGL), depending on the type of parachute used.

e. In airborne insertions, emphasis is on the use of special delivery or navigational techniques since normal insertion is during periods of limited visibility.

(1) **Adverse weather aerial delivery system.**

(a) A self-contained navigational system in some USAF C-130 aircraft is AWADS. The system allows the aircraft crew to fly a specific route to a DZ during darkness or bad weather. The system works with a Doppler radar that determines the ground speed of the aircraft, and an AWADS computer programmed with special flight route information (start point, route, checkpoints, wind speed and direction, and air speed). These two instruments show where the aircraft is in relation to the programmed checkpoints. This eliminates the need for airdrops under visual weather conditions.

(b) Only certain aircraft in formation are equipped with the actual AWADS device. Other aircraft are equipped with station-keeping equipment (SKE). This SKE consists of a computer that tells the aircrew where their aircraft is in relation to other aircraft in the formation.

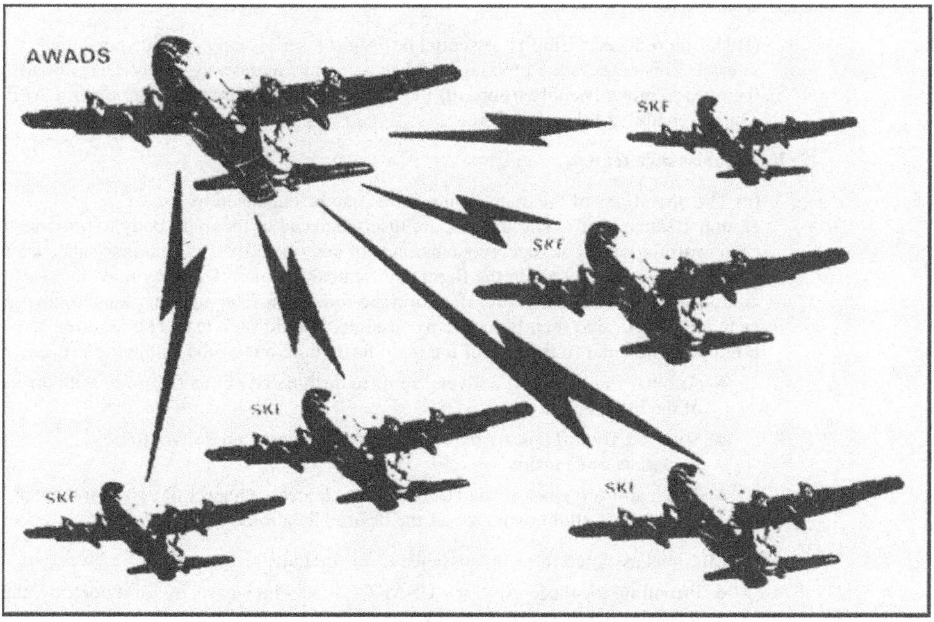

Figure 4-1. AWADS abd SKE aircraft formation.

(c) Airborne insertions using AWADS give the ranger commander the ability to airdrop personnel and equipment during low or zero visibility. This increases the safety of USAF aircraft. It protects them from visually tracked enemy antiaircraft guns and non-all-weather fighters. It also uses adverse weather as a combat force multiplier. Using AWADS, personnel and equipment may be inserted day or night without a pre-positioned USAF combat control team (CCT) or an Army assault team. Time length of the air train is shortened. The air corridor has to be cleared only once. This enhances the element of surprise. The CCTs are brought in with the assault elements. They are able to direct other low altitude parachute extraction systems (LAPES), container delivery systems (CDS), and air-land or heavy drop missions. Use of AWADS enables the ranger commander to conduct rapid, vertical reinforcement of units threatened by an enemy attack during bad weather.

(d) The use of AWADS requires detailed planning, rehearsal, and close coordination among the forces involved. Intense training of both USAF and Army personnel is required.

(e) Personnel must be briefed on terrain through use of photographs, maps, and terrain models. Limited visibility relies mainly on compasses and audible and visual assembly aids.

(f) During reduced visibility, essential heavy-drop serials may precede personnel aircraft. This reduces the time lag between personnel and heavy-drop chalks due to the need to move friendly troops off the DZ. Remaining heavy-drop serials or LAPES can be scheduled following P-hour.

(2) **Reconnaissance teams.**

(a) The operations of the main ranger force may be enhanced by use of reconnaissance teams. These teams are inserted ahead of the main body to provide early warning or to conduct reconnaissance of key objectives. If reconnaissance teams are required at a point along the flight path or near the main DZ, they may be cross-loaded on board any aircraft within the formation. They exit at planned times so as to land as close to their objective as possible. Should such teams be required at a point perpendicular to the axis of the main formation, one of the following is used:

- Another flank aircraft delivers teams to designated points before or with arrival of the main body.
- Selected aircraft deliver the main body, turn, and then deliver the reconnaissance teams.
- The teams are cross-loaded as complete elements on aircraft delivering other units whose flight path crosses the desired locations.

(b) Missions assigned to reconnaissance teams include:

- Providing local security for a USAF CCT, in which case, by joint doctrine, they form the Army assault team (AAT).
- Establishing surveillance or blocking positions on likely avenues of approach into the objective area or DZs.
- Locating and marking targets or assembly areas for the main assault force.

(c) Reconnaissance teams may be inserted by static line or HALO/HAHO operations.

(d) If the reconnaissance element is to locate a specific target or objective, it may be inserted up to 24 hours early. The longer the reconnaissance team remains near the objective area, the greater the danger of it being seen.

(e) In order for the information to be used by the ranger force commander, secure communications and a linkup plan must be prepared and rehearsed.

(3) **Assembly.**

(a) The ranger airborne force can conduct a quick assembly of combat power (weapons, personnel, and command and control elements) on the DZ. Quick assembly and reorganization is crucial because the unit is open to attack. Assembly must be fast and precise. Assembly areas and plans are set forth after considering all the factors of METT-T, mainly the location of the enemy, visibility, terrain, drop information, dispersion pattern, and cross-loading.

(b) The number of assembly areas depends on the location of the DZ, size of the assembly areas, and enemy's detection ability. Unit members move directly to the assembly areas without breaching security. Dissemination of information to all jumpers is the key to a successful assembly plan.

(c) For battalion-size or larger operations, general assembly areas are prescribed in the ranger force OPORD. Company commanders select and designate the exact location of their assembly areas. Although assembly area locations are terrain-dependent, they are generally on the DZ if the drop is at night and off the DZ if the drop is during daylight.

(d) Personnel chosen to carry assembly aids are cross-loaded in the center of their unit sticks. Strong, aggressive leaders should jump first and last in each stick. Upon landing, the leaders roll up their sticks toward the center, locate and move to the proper assembly aid. They may also carry or wear assembly aids. Individual rangers then follow these key personnel. During daylight operations, radiotelephone operators (RATELOS) can display color-coded pennants on their radio antennas to facilitate linkup with leaders. Key personnel may carry and jump with their own radios, and also display color-coded pennants on the radio antennas. Key personnel and RATELOs put their radios into operation before moving to assembly areas. For drops occurring near dawn or dusk, both daylight and night assembly plans are required.

(e) When tactical vehicles or door bundles are dropped, they are normally marked with panels, chemical lights, or strobe lights. The delivery parachutes are often color-coded.

(f) A ranger force should be able to assemble on an unfamiliar DZ, at night, within 45 minutes of the final aircraft pass. Because most ranger operations are time-sensitive, the unit normally moves from the assembly area to the objective when enough of the force has assembled to accomplish the mission. A straggler control point remains at the assembly area under the control of a designated leader. When everyone assembles, or at a specific time, the leader moves to link up with the unit. If available, a radio is left with the straggler control point.

(g) Assembly area security plans are prepared and disseminated. These plans must address hasty defense, sectors of responsibility, observation posts, fire support, and chain of command.

(4) **Contact on the DZ.** During the conduct of parachute insertion, rangers must be ready for enemy engagement at all times. Rangers are open to attack on the DZ before assembling. Just as contingency plans and chance immediate-action drills are developed, briefed, and rehearsed for other chance enemy contacts, rangers must be prepared to execute immediate-action drills for enemy contact on the DZ. The following actions are taken when there is enemy contact on the DZ:

(a) Rangers immediately assemble in fire team-size groups and fight off the DZ to designated assembly areas. Elements attempt to break contact and assemble.

(b) Assembly areas designated by commanders in the mission OPORD serve as primary rally points.

(c) Alternate assembly areas (normally 2 to 5 kilometers from the DZ) are designated in the mission operation order (OPORD). They are used by elements that cannot link up or assemble in the primary assembly area. These elements continue with the mission once assembled.

(d) During limited visibility, personnel should move directly and quickly off the DZ. They must avoid mistaking other friendly elements for enemy, which could cause firefights among friendly forces.

(5) **Reorganization.**

(a) The unit reorganizes according to plans, using designated assembly areas, aids, and identification markings for personnel and equipment. Replacements are named for key leaders or personnel who are lost or injured.

(b) Security teams assemble and move to their positions. Remaining units move quickly to their assembly areas, carrying supplies and equipment.

(c) Subordinate leaders keep the commander informed during assembly and reorganization.

(d) Runners may be used to report assembly status. This maintains radio silence. Designated personnel quickly recover supplies and equipment and move to the assembly areas.

(e) Reorganization is complete when all units have assembled or have been accounted for, and control has been established.

(6) **Pop-up technique.** This involves close terrain flying until the DZ is reached, then the aircraft quickly climbs to jump altitude. Special navigational equipment and aircrew training

are needed but may not always be available.

(7) **HALO/HAHO.** This involves high-altitude, low-opening and high-altitude, high-opening jumps with high performance parachutes designed to maneuver the jumpers to a specific point on the ground. During these operations, midair assembly procedures may be used.

(8) **Arctic.** Airborne insertion in extremely cold conditions is difficult. Times allocated for all parachute and equipment rigging must be increased. Rigging of individual equipment must be IAW FM 57-220. Ahkios can be rigged, packed inside a double A22 container, and dropped in platoon bundles. Because of the increased amount of individual equipment needed, allowable cargo loads for USAF aircraft are reduced.

(9) **Water.** Because of a lack of suitable land DZs, or to avoid detection, rangers may conduct airborne operations onto water DZs close to land. Procedures and techniques are similar to land DZ operations.

(10) **In-flight rigging.** On long-range operations involving extended flight time to the objective area, personnel may rig in flight. In-flight rigging requires detailed planning, coordination, of all jumpers. A rehearsal of these procedures should be conducted before deployment. Aircraft loads must be reduced to carry out in-flight rigging procedures. (See FM 57-220.)

4-14. Air-land insertion (fixed-wing).

a. **Planning factors.** Most of the planning factors that apply to airborne insertion operations also apply to air-land insertion operations. Most USAF aircraft cannot conduct air-land operations in total instrument weather conditions. The AWADS equipment is accurate enough for airdrop navigation, not air-land. Air-land operations are conducted, however, under conditions of extremely reduced visibility. Specially trained USAF crews, using sophisticated navigational techniques, can locate and land on small airfields in bad weather. As in airborne operations, ranger reconnaissance teams, as well as specially trained CCT elements, can jump before the air-land aircraft arrive. They emplace navigational and location markers to assist the pilots. Rangers may rely solely on air-land operations for insertion. This method does not provide for delivery of the force in the shortest time possible. Air-land insertions are normally in concert with an insertion of airborne elements.

b. **Time.** Since turnaround time is crucial, loads in aircraft should be unchained as the aircraft taxies. This also applies to personnel and seat belts. Aircraft should stop as close as possible to the assembly areas where personnel and equipment are off-loaded.

c. **Method.** Air-land insertions normally begin after a ranger Army assault team (AAT) and a USAF CCT have parachuted into the initial objective area and conducted reconnaissance, clearing, marking, and security operations. The ranger ground force commander is the arrival airfield commander. He has control over all arriving, taxiing, and departing aircraft (fixed-wing or rotary) through the CCT, which locates near him. The ranger force command and control element is usually on the first aircraft in and the last aircraft out.

d. **Security.** Air-land operations concentrate large numbers of men and equipment, which presents an open target to the enemy. The ranger ground force commander must ensure rapid off-loading, assembly, dispersion, and movement toward the objective.

e. **Communications.** The ranger ground force commander must have secure radio communications with the ground force, CCT, fire support elements, and higher headquarters. Wire should be laid to connect ground command and control locations.

f. **Operation duration.** Depending on the factors of METT-T, the aircraft may remain on the ground for extraction while the mission is being accomplished or take off at once. They may then orbit, pending recall, or return at a prearranged time.

g. **Extraction.** Extraction of the air-landed ranger force is normally planned and accomplished like an insertion, only in reverse order. A simple, yet effective, way must be found to see that no ranger personnel are left in the objective area.

4-15. Air assault insertion.

Ranger air assault (air-land, rappel, or combination) insertions normally take place during limited visibility. They quickly place the ranger force on, or close to, the objective. The ranger force (depending on the mission) normally plans on fighting when it hits the ground, completing the mission, and then being extracted. Air assault insertions need detailed planning, surprise, flexibility, speed, shock effect, concentration of combat power, and precise timing.

a. **Ground tactical plan.** The reverse planning process is important. The ground tactical plan, developed from the mission assessment, is the first planning area. All other planning begins from this point. The ranger commander and his staff decide stated and implied tasks, and request the proper helicopters for the mission.

b. **Fire support plan.** The fire support available to the ranger commander may be artillery, naval gunfire, mortars, attack helicopters, and USAF or USN tactical aircraft. Because of the timing of ranger air assault operations, preparatory fires are often not used. If they are used, enemy ADA sites normally have priority.

c. **Landing plan.** This plan puts the ranger force into the objective area at the time, place, and proper sequence to support the ground tactical plan. The ranger commander selects primary and alternate LZs and the priority for landing. The exact location of an LZ is determined after a detailed analysis of METT-T. Rehearsals are important. The ranger force must quickly assemble, reorganize, and accomplish the mission. The senior ranger on each aircraft ensures that the personnel on his aircraft offload at the correct LZ.

d. **Air movement plan.** The air movement plan provides for the coordinated movement of the ranger force into the zone of action in support of the 'landing plan. Flight routes, air movement tables, flight formation, altitude, and air speed should be considered. The ranger commander must have secure in-flight communications with his subordinate commanders.

e. **Loading plan.** This plan supports the air movement plan and provides for the pickup zone (PZ) setup and control. It sets priorities for the commander for loading personnel, equipment, supplies, and cross-loading. Air-loading table, aircraft loading formations, bump plan, and rehearsals should be considered.

f. **Staging plan.**

(1) This plan includes PZ selection and operation. The selection of PZs and LZs needs planning and coordination for best use of air assault assets. Site selection in or near the objective must be coordinated in person between the supported ranger commander and the aviation commander. The tactical situation is the key factor in PZ and LZ selection. Other factors are size of landing zones, surface conditions, ground slopes, approach and departure directions, prevailing winds, obstacles, communications, cross-loading, aircraft command and control, PZ and LZ markings, and rehearsals.

(2) The PZs are set up and secured about 15 minutes before touchdown of first aircraft. The fire support plan must support the other plans. All supporting fires must be coordinated with the air mission commander (AMC). Because of the depth of most objective locations, USAF tactical aircraft will be the main fire support method. Orbit locations for fire support aircraft must be planned.

g. **Other planning considerations.** These include escape and evasion, actions at the last LZ, assembly plan, downed aircraft procedures, control measures, forward area rearm and refuel point (FARP), weather delays, and deception plans.

4-16. Amphibious insertion.

Insertion by water may be by SCUBA, surface swimming, small boats, submarines, surface craft, helocasting, or a combination of these methods.

a. **Planning.** Planning for waterborne operations must be thorough. The plan is decided upon between the ranger commander, the transporting unit, and the joint headquarters directing the operation. Plans must be made for all types of enemy action and weather while on board the transporting craft. The transporting unit is given information on a need-to-know basis. Information that could endanger the operation may be withheld until the mission is under way. Initial planning includes:

(1) Time schedule of all events, from the beginning until the end of the joint operation, to be used as a planning guide. Timing for each event must be exact; the success of the operation depends on it. (See Figure 4-2)

Figure 4-2. Waterborne insertion.

(2) Embarkation site, where troops and equipment are loaded on the transporting craft.

(3) Drop site, where the ranger force will depart the primary craft and load into smaller boats.

(4) Handing site, where the unit will beach its boats, or land directly from amphibious craft.

(5) Loading, when loads are lashed and waterproofed IAW unit standing operating procedures (SOPs). Inspections are important. A bump plan for key personnel and equipment must be developed.

(6) Cross-loading, which will help preserve the ranger force. location of leaders, crew-served weapons, and needed equipment are stressed.

b. **Landing site.**

(1) **Beaches.** Beaches are checked to find one that allows an undetected approach. Landing sites that cannot be approached from different directions are avoided. Stretches of defended or patrolled coastline should be avoided. If sand beaches are used, tracks and other signs of

operation that may compromise the mission should be erased. Rural, isolated areas are best. The area behind the landing site should provide an assembly area and concealed routes for exit.

(2) **Barriers.** Reconnaissance may find that good landing sites are not suitable because of barriers between the site and the objective. Such barriers may be patrolled roads, large areas giving little or no concealment, enemy obstacles or defensive positions, areas populated with enemy sympathizers, or an area used for enemy troop maneuvers and training. Unless a means of avoiding such barriers is devised, landing sites should be limited to those without barriers.

(3) **Other factors.** Beach selection should also consider:

(a) The scheme of maneuver.

(b) Surprise.

(c) Enemy dispositions.

(d) Distance to the target.

(e) Landing and exit sites.

(f) Trafficability of the beach.

(g) Cover and concealment.

c. **Routes.** The route the transporting craft takes to the drop site should be planned to deceive the enemy. A major change in the route immediately after the unit's debarkation could compromise the mission. The route should be the same as a route that would be used in some other naval operation--for example, minelaying or sweeping, or a naval patrol. An alternate route must be planned.

d. **Tactical deception.** There should be other plans to deny the enemy knowledge of the insertion besides the water approach route plan. Plans may include electronic countermeasures (ECM), diversions, feints, ruses, or demonstrations.

e. **Actions at the drop site.** Some operations may permit landing directly from the transporting craft. If not, primary and alternate drop sites must be set. The drop site should be at least 1,500 meters offshore to cover any noise in the loading and launching. If the enemy has a surface radar capability, a drop site should be chosen to avoid detection. The drop site may have to be several miles offshore or may need the use of ECM. The manner in which men, small boats, and equipment are loaded must promote fast action at the drop site.

f. **Navigation.**

(1) Ship-to-shore navigation (to the landing site) may be done by dead reckoning. The course may also be maintained by compass navigation, reference to a shoreline silhouette, radar, or prearranged signals from the shore.

(2) Boat handlers, navigators, and other experts are relied on in various phases of the

transport operation.

(3) Lead reconnaissance teams or scout swimmer teams may be used to plant radar, radio, or infrared buoys, or distinct, shielded lights. Boats may be equipped with radar reflectors or beacons to permit accurate tracking.

g. **Actions on the beach.**

(1) Boats hold offshore while ranger scout swimmers move to a covered and concealed position. After reaching the beach, they make a short listening halt. They check the beach and the area immediately beyond the beach for enemy activity. Swimmers signal the force "all clear," mark the landing site, and provide local security.

(2) When the boats land, chosen personnel move directly into covered and concealed security positions. They prepare to defend the landing site.

(3) If the ranger unit is not carrying equipment, boats may be moved off the beach and buried or camouflaged. If heavy equipment makes it hard to move the boats, chosen personnel start unloading at once. The remainder of the force secures the landing site. Each boatload moves into position under the control of a designated leader. Teams are positioned with the best cover, concealment, and fields of fire available.

(4) The ranger force leader sets up a command post from which he can control the operation. He may establish secure communication with his headquarters.

(5) Boats may be deflated and buried or camouflaged near the landing site or away from it. If the boats are to be hidden near the landing site, a team may be chosen to dig holes or cut brush for camouflage along with the unloading of the boats. A small team may be chosen to sweep the beach to erase tracks and drag marks.

(6) The ranger force should be assembled and prepared to move within minutes after the last boat lands. Every ranger must be briefed on the unit's plan should enemy contact be made upon landing.

h. **Insertion by air from ship.**

(1) Helicopters from a ship can launch the ranger force from over the horizon and move at a low level to the LZ.

(2) Helicopters may be vectored from ships to a landing zone. The aircraft may return to the ship while the ranger force conducts the mission, they may orbit offshore, or they may stay at the LZ. Other aspects of landing and assembling are the same as discussed for helicopter insertions.

i. **Helocasting.** This form of insertion combines helicopters and small boats in the same operation. It is planned and conducted the same as air assault operations, except that the LZ is in the water. While helicopters are moving at low levels and low speeds, rangers drop small boats and jump into the water. They then assemble, climb into the boats, and move to shore. Ranger elements are usually assembled and loaded into the small boats within 20 minutes of the last ranger entering the water. This type of operation is normally conducted during limited visibility.

j. **Fire support.** Fire support may be ship-to-air, ship-to-shore, air-to-ground, or ground-to-ground. Specific missions include:

(1) Suppressing enemy air defense.

(2) Neutralizing enemy field artillery within range of boat or helicopter routes.

(3) Employing on-call fires against shoreline targets.

(4) Destroying enemy aircraft.

k. **Contingency planning.** The following must be planned for:

(1) Actions on enemy contact.

(a) En route.

(b) Helocast site.

(c) Flares.

(d) Aerial attack.

(e) Small arms fire.

(f) Indirect fire.

(2) Bump plan with priority to key leaders, crew-served weapons, and radios.

(3) Downed craft procedures.

(4) Escape and evasion.

(5) High surf.

(6) Adverse weather.

(7) Separation.

l. **Rehearsals.** The ranger force must rehearse all aspects of the amphibious insertion. This includes boat launching, paddling, beaching and assembly, boat commands, and capsize drills.

4-17. Land infiltration.

Land infiltration to the objective may be the best way to complete a mission. This is normal when the enemy has total air control or effective air defenses in the operational area to be penetrated, or when the objective is close to friendly territory. Rangers can infiltrate over any type of terrain, in any climate. However, thick forests, swamps, and broken or steep terrain offer the best chance for success.

a. **Planning.** There are many methods of infiltrating by land, and each has its good and bad points. Plans are designed to get the infiltrating force to the objective area without being seen. Plans should include:

(1) Selecting concealed routes for surprise and protection. Primary and alternate routes are based on detailed map and aerial photo study, ground reconnaissance, and data from agent reports.

(2) Avoiding obstacles, populated areas, silhouetting (when forced to cross obstacles or ridge lines), enemy positions, main avenues of approach, and movement along often used routes and trails.

(3) Selecting the time of infiltration to make use of reduced visibility, such as darkness, rain, snow, and fog, and periods of reduced enemy alertness, such as early morning, holidays, and times following protracted movement or combat. The time of the infiltration is important during critical phases (crossing borders or passing through enemy troop concentrations or populated areas).

(4) Using speed and stealth. All rangers must know routes, linkup points, alternate linkup points, time schedules, danger areas, and actions on enemy contact and at linkup points.

(5) Centralizing planning and decentralizing execution. Centralized coordination must ensure that units are acting IAW cover and deception plans. Control should be given to the leaders of the infiltrating groups. Within their designated zones, these leaders select their own routes and formations. Leaders control the rate of advance. When the enemy has electronic detection devices, radio communications must be kept to a minimum. Key leaders normally go with the group that is using the fastest lane or the lane least likely to be detected.

(6) Seeding infiltration routes to the assembly area with unattended ground sensors. These sensors can help the infiltrating force by reducing the chance of surprise contact during the infiltration, and by helping in controlling the forward progress of the units.

(7) Planning for fire support, which must include coverage of routes, danger areas, assembly areas, objectives, and PZs.

(8) Rehearsing actions at the objective, actions on enemy contact, actions at danger areas, and movement techniques.

b. **Movement.** Land infiltration is conducted using one of three techniques:

(1) **Movement in small groups along one axis.** All members of the force use the best route. Small groups are harder to detect, easier to control, and do not compromise the total force if detected. This technique may take too long and it needs an assembly area or linkup point. If the lead group is detected, other groups may be ambushed.

(2) **Movement in one group.** This technique has no reassembly problem. The same route is used by all, making navigation easier. A large force can fight out of a dangerous situation easier than a small one. It has no coordination problems between other infiltrating units. This form of movement is easier for the enemy to detect. If the force is detected, the mission may be endangered.

(3) **Movement in small groups along several routes at the same time.** This technique avoids putting the total force in danger, is less likely to be seen, forces the enemy to react in

many locations, and makes it harder for the enemy to determine the size of the force or its mission. Some groups travel over poorer routes and have problems of assembly, control, and support; the capture of personnel may endanger the mission. Regardless of which technique is used, ranger units normally move to reduce the chance of enemy contact.

c. **Assembly area.** Following movement into the general area of operations, infiltrating units converge on the assembly area or linkup point. Assembly may begin at once after the passage of lines, or it may take place at any time before reaching the objective. The infiltrating force must remain undetected before assembly. If an infiltrating unit is seen at this time, the enemy may learn the location of the assembly area, and the objective. It may take action to destroy the infiltrating force.

d. **Linkup point(s).**

(1) Selection characteristics include terrain that--

(a) Is free of enemy troops.

(b) Offers cover and concealment from enemy observation and patrols.

(c) Does not restrict movement of friendly units.

(d) Accommodates the unit.

(e) Is easily identified.

(f) Offers escape routes to alternate linkup points.

(g) Offers concealment from air attack.

(2) Action sequence for occupying the linkup point.

(a) Halt short of the point.

(b) Establish local security.

(c) Conduct a reconnaissance of the linkup point.

(d) Occupy the linkup point.

(e) Emplace observation posts and listening posts.

(f) Link up with other groups.

(g) Conduct final preparations.

(h) Continue the mission.

(3) Security may be enhanced with the aid of devices such as night vision goggles and night sights. Unattended ground sensors may be placed on likely avenues of approach into the linkup point. Normally, only passive observation devices are used to assist in avoiding detection.

(4) Plans must set a time schedule for actions at the linkup point. All unit members must move in quickly, consistent with security, so that the unit may proceed to the objective rally point (ORP) and on to the objective.

4-18. Stay-behind technique.

A ranger force can be positioned behind enemy lines by employing the stay-behind technique. In this method, a ranger force allows itself to be bypassed as the enemy advances. This is done to allow the ranger force to perform a specific mission. Stay-behind operations may require the concealment, or cache, of supplies before the enemy bypass. Rangers may need to use captured enemy supplies. The location of the assembly area and the plan for extraction are important parts of stay-behind plans.

4-19. Extraction.

The ranger force is normally extracted after the mission is completed. The enemy may also force the unit to abort its mission and to call for extraction. Contingency plans must be made to extract a force under enemy pressure, as well as one that is not in contact. Primary and alternate extraction sites are planned and coordinated with all ,porting forces. Ranger operations are normally conducted against deep targets. Extraction may begin with land movement but normally it includes the use of boats or aircraft. Plans may include primary I alternate means of extraction.

a. **Terrain.**

(1) The terrain is important in choosing the point of extraction - for example, whether or not the extraction site offers good cover from enemy direct-fire weapons and positions from which the extracting force can suppress the enemy. Tidal conditions and PZ size must be considered.

(2) Unlikely locations such as swamps, jungles, or mountain areas are often used for extraction. Specially trained air and naval crews, using the latest extraction devices, make such terrain useable.

b. **Enemy.**

(1) Extraction is best conducted without enemy pressure. However, such pressure could develop during the operation. Detailed plans are made for emergency or contingency extractions forced by the enemy as well as clandestine extractions not under enemy pressure. If under pressure, units will have to use planned fire support to suppress enemy weapons. At the beginning of the extraction with security posted to provide early warning, the unit may start to thin out the same as with a night withdrawal. Ground security may be relied on until the last unit withdraws under cover of suppressive fires.

(2) If enemy contact is made during an extraction, the commander must decide whether to reinsert units or break contact and move to another extraction point. Fire support and suppression are important and include suppressing enemy air defense by deception, ruses, or decoys.

(3) The ranger unit may have special weapons systems or equipment to aid in the extraction. Shoulder-fired antiaircraft missiles may be used to defend against enemy aircraft threatening

the extraction.

c. **Extraction by air or water.** Extraction by air or water is preferred when the resources are available and their use will not endanger the mission. Other factors that favor their use are when--

(1) Long distances must be covered.

(2) Time of return is essential.

(3) The enemy does not have air and naval superiority.

(4) Heavily populated hostile areas block land movement.

(5) Ranger forces are burdened with prisoners of war,casualties, or critical items of equipment.

d. **Extraction by land exfiltration.** Stealth and evasiveness are key elements of exfiltration. This method of extraction is favored when--

(1) The enemy has air superiority.

(2) The enemy is able to prohibit air or water extraction.

(3) The distance to friendly lines is short.

(4) The terrain provides cover and concealment (for movement on foot) and limits enemy mobile units.

(a) Multiple routes of exfiltration may be used to prevent destruction of the force if soldiers or units are detected. Captured enemy vehicles and equipment may be used to assist in the exfiltration.

(b) The ranger force may exfiltrate in one body or in small groups. Exfiltrating in small groups avoids the time delay of assembly.

(5) The exfiltrating force is lightly equipped and not burdened with captured personnel or material.

(6) The exfiltration route passes through an area occupied by friendly civilians or guerrilla forces who can assist the movement.

(7) Areas along exfiltration routes are not inhabited.

(8) The enemy force is dispersed or is not able to concentrate against the exfiltrating force.

4-20. Escape and evasion.

a. A vital part of all premission planning is the development of en-route and post mission E&E plans. Such plans enhance survival of the ranger force, and the transporting air or boat crews. The ranger force commander is responsible for--

(1) Helping in the development of an E&E plan in coordination with all supporting agencies.

(2) Ensuring that all members of the ranger force and the supporting boat or air crew are briefed on the E&E plan.

b. Each E&E plan is unique. Each situation has unique problems associated with E&E. The E&E plan devised by the ranger force commander must address these problems, while profiting from the individual abilities and training of the rangers and their supporting air and boat crews. The following considerations apply to E&E plans devised during ranger operations:

(1) The plans enhance survival of personnel who can no longer accomplish their assigned mission. The senior ranger in a group must decide if there are any remaining missions that the group can accomplish. If not, then he must try to evade and escape enemy capture if unable to link up and be extracted with the rest of the ranger force.

(2) Because of the depth of penetration behind enemy lines, most successful E&E plans may involve either air or water movement away from enemy-held territory. The air and boat crews supporting ranger operations are highly skilled and dedicated. If there is any chance of a successful rescue, these forces continue to try to retrieve rangers escaping from an enemy-held area.

(3) Rangers possess stamina and endurance. They can move a great distance (especially at night) and over rugged terrain to reach an area where they can rendezvous with search-and-rescue (SAR) aircraft or boats. Rangers are also resourceful and highly trained in communication, stealth, and camouflage. This, coupled with their self-discipline, means that successful E&E can be conducted over a great distance, through heavy concentrations of enemy, and over a long period.

(4) Escape and evasion plans for ranger elements should include avoiding contact with the local civilians. However, the aid of friendly insurgent forces may be enlisted. The ranger E&E plan may also include the use of other special operations forces E&E networks that are in place behind enemy lines. These nets must not be compromised by the volume of evaders.

c. Escape and evasion plans should be based on three phases:

PHASE ONE: That portion of the insertion following entry into the hostile zone where personnel survival and recovery is the only thing considered. Should an aircraft or boat carrying rangers be disabled during phase one, the ranking survivor assumes command. He must consider E&E planning and the experience and skills of other survivors. He should be concerned with the security of the mission, caring for wounded, and moving away from enemy activity while contacting SAR elements for recovery.

PHASE TWO: That portion of the insertion that is near the objective and could permit the ranger element to successfully pursue its mission. Although part of the E&E plan, actions by the ranger force during this phase become a forced land infiltration rather than an E&E. The senior surviving ranger must decide if enough personnel and equipment can be moved to the objective area in time to accomplish the mission. If so, he must proceed, taking along any surviving air or boat crew members. If not, he may proceed directly to the extraction point and wait, or he may contact SAR elements for recovery.

PHASE THREE: That portion of the operation after mission accomplishment. It is a hard time to escape and evade since the enemy is alert and reacting to the ranger forces. The confusion caused by the violence of the attack, coupled with the rangers' ability to move quickly cross-country away from the objective area, offers the best chance of a successful escape and evasion. The surviving ranger elements that cannot be extracted can do one of the following:

(1) Move over land at once to a PZ or beach for extraction by SAR elements.

(2) Move to an area where reentry into friendly lines can be made.

(3) Move to a neutral or nonaligned country where either internment or return to friendly control can be effected.

d. Any plan for E&E should provide for the issue to each ranger of an area map, a covert means of recognition, and a small compass. An emergency ration may also be included and any items needed for the climate in that objective area. The ranger force commander must balance the need for other survival equipment against the added weight and bulk involved.

CHAPTER 5

Strike Operations

The ranger regiment conducts strike operations in support of the Air Land Battle. Ranger strike operations are integrated into an overall plan designed to destroy, delay, and disorganize the enemy. They may also cause him to divert a major part of his combat forces and power to rear area security. Strike operations also help create an environment in which other special operations forces can operate. Strike operations can complement air power in disrupting enemy forces in depth. They allow the theater commander to regain the initiative and lead to decisive action.

5-1. Application.

a. The ranger regiment can conduct strike operations either as deliberate-response or quick-reaction missions. It can be inserted into the area of operations by land, sea, or air.

b. Strike operations can be conducted throughout the depth of the battlefield in support of conventional or unconventional warfare. Multiple strike operations can have a synergistic effect and can create situations that help friendly forces seize and maintain the initiative.

c. Strike operations include raids, personnel and equipment recovery, and interdiction of lines of communication. They are best accomplished by ranger units who use conventional techniques and special skills. Strike operations vary in size. When well augmented and supported, the ranger regiment can conduct strike operations deep in enemy territory using up to three battalions. Strike operations normally are limited in time and may be conducted unilaterally or in conjunction with other special operations forces or allied forces. They are normally ended by the extraction or exfiltration of the ranger force.

5-2. Raid operations.

a. A raid is a strike operation conducted behind enemy lines against strategic objectives, targets of high tactical value, time-sensitive targets, or key personnel and facilities in enemy rear areas. It normally involves a swift penetration of hostile territory to confuse the enemy or destroy his installations. Ranger units can conduct deep penetration raids when given the appropriate assets for insertion.

b. A strategic raid by the ranger regiment would be directed by the NCA and conducted under the OPCOM of a unified or JTF commander. An operational raid would be directed by the unified or JTF commander and conducted by that command directly or under the OPCOM of a corps commander. Typical targets would include:

(1) Command, control, communications, and intelligence centers at front level or combined arms army level.

(2) Nuclear and chemical weapons storage sites and delivery means.

(3) Key logistical centers, warehouses, ammunition complexes, or fuel pumping centers supplying logistical support to a specific front.

(4) Air defense and air traffic control integrating centers and air defense weapons sites.

(5) Key power generating and transmitting facilities, hydroelectric dams, and irrigation systems.

(6) Key ports or rail complexes.

(7) Key installations or facilities, such as airstrips, buildings, bridges, dams, or tunnels.

c. The size of the ranger raid force is determined by a detailed analysis of METT-T. The unit committed to the ground phase of the raid mission is kept as small as possible. It may be as small as a ranger company or it could include all the battalions of the ranger regiment. The force must be tailored to complete the mission quickly, violently, and with few casualties. The size of the force may also be affected by the augmentation of specially trained personnel needed for that mission.

d. Raids are normally conducted in five phases. (See Figure 5-1).

Figure 5-1. Five phases of a raid.

PHASE 1. The ranger force inserts or infiltrates into the objective area.

PHASE 2. The objective area is then sealed off from outside support or reinforcement, to include the enemy air threat.

PHASE 3. Any enemy force at or near the objective is overcome by surprise and violent attack, using all available firepower for shock effect.

PHASE 4. The mission is accomplished quickly before any surviving enemy can recover or be reinforced.

PHASE 5. The ranger force quickly withdraws from the objective area and is extracted.

e. This paragraph discusses the raid after insertion/infiltration. All considerations discussed in Chapter 4 must be applied in order for the unit to get deep behind enemy lines. Upon receipt of a raid mission, the following basics are emphasized:

(1) **Maximum use of intelligence information.** The gathering and dissemination of information must be continuous and provided to the raid force even while en route to the target area. To ensure mission accomplishment, the ranger unit must be kept informed of the latest enemy developments in the objective area to prevent being surprised. Only then can the full combat power of the rangers be concentrated at the decisive time and place.

(2) **Plan development.** The reverse planning sequence and the planning steps set forth in FM 101-5 are followed in the plan development.

(3) **Coordination.** Coordination is through the operational headquarters with friendly units (such as higher and supporting Army or joint headquarters, intelligence agencies) for fire, RSTA, special equipment, and personnel and logistical support.

(4) **Rehearsals.** Rehearsals help validate all aspects of planning for the raid and ensure precision in execution-They allow changes to be made in the plan before it is carried out.

f. A successful raid is ensured by--

(1) Launching the raid at an unexpected time or place by taking advantage of darkness and other periods of limited visibility, and moving over terrain that the enemy may consider impassable.

(2) Avoiding detection through proper movement techniques and skillful camouflage and concealment to include taking advantage of natural cover of the terrain, and using sophisticated equipment to detect and avoid enemy forces.

(3) Timing the operation as close as possible.

(4) Using all available support, both organic and nonorganic, to include use of special weapons such as Air Force "smart bombs" and artillery cannon-launched guided projectiles, with the ranger unit using laser target designators.

(5) Performing quick, violent, precise, and audacious actions that focus full combat power at the decisive time and place.

(6) Disengaging quickly upon mission completion.

(7) Withdrawing swiftly using planned routes and including a deception plan.

g. Four functions are normally performed by ranger tactical formations when conducting a raid. Each subelement is organized and equipped to do a specific part of the overall mission. Depending upon the specific mission, nature of the target, enemy situation, and terrain, these four functions are as follows:

(1) The command group controls movement to and actions at the objective. This unit normally consists of the ground commander, other subordinate leaders in the raid organization, and communications to support these leaders.

(2) The security element, whose organization is determined by the mission of the raid force, size and type of enemy force and its mobility and state of alert, terrain and avenues of approach into the area, and the time needed to seal off the objective area. The security element may perform the following:

(a) Secure the ORP.

(b) Give early warning of enemy approach.

(c) Block avenues of approach into the objective areas.

(d) Prevent enemy escape from the objective.

(e) Provide overmatch for the units at the objective and suppressive fires for their withdrawal.

(f) Provide short-range air defense.

(3) The support element provides the heavy volume of fire needed to neutralize the objective. Because fires from this unit are violent and devastating, they must be closely controlled to ensure the precision needed. On order or as planned, fires are lifted and shifted to cover the maneuver of the assault element by suppressing enemy fire from the objective or aerial fires. The support element may also be given specific locations to cover by fire in support of the security element if an enemy quick-reaction force moves toward the objective area. These may include routes into and out of the objective site, key terrain features, or installations adjacent to the main objective. Once the assault has been completed, or on order from the raid force commander, the support element displaces to the next planned position. Organization of the support element is determined by the following:

(a) Size of the objective, the geography of surrounding area, and the enemy threat (to include air) in the area. This element should be able to neutralize the objective (when supported by air or naval gunfire) and to lift or shift fires either when the assault is launched or when ordered by the raid force commander.

(b) Mission of the assault unit.

(e) Suitable firing positions.

(d) Size and nature of the enemy force in the objective area and those enemy forces capable of reinforcement at the objective.

(e) Fire support from other units (air strikes, naval gunfire, surface-to-surface missiles, and artillery fire).

(4) The assault element seizes and secures the objective and protects demolition teams, search teams, prisoner-of-way teams, or other teams.

(a) The organization of the assault element is always tailored to the mission. Each target area must be examined carefully. To assault, seize, and destroy an installation, position, or equipment, the assault element could be organized into one or more assault teams. The element's mission is to overcome resistance and secure the objective and to destroy the installation or equipment. Other specialized teams may also be needed. For example, sniper teams could be needed to remove key sentries. To capture prisoners, liberate personnel, and seize or destroy equipment, the assault element could be organized into assault teams, prisoner teams, search teams, medical teams, demolition teams, or breach teams.

(b) To destroy a point target or installation in a heavily defended area where the USAF cannot get close en to be effective, the assault element might be organized with one small team equipped with laser target designators. From covered and concealed positions, members of the team could then provide guidance for USAF delivery of laser-guided munitions from a safe distance.

h. The site chosen for the raid LZ or DZ must support the planned actions at the objective. There are two options when choosing sites:

(1) The ranger force can land on or near the objective and seize it before the enemy can react. (See Figure 5-2.) This avoids forced marches over land carrying heavy combat loads. If there is no suitable landing area near the objective, or the enemy has a strong reaction force nearby, this option is not favored.

Figure 5-2. Airborne landing on the objective.

(2) The ranger force can land unseen far from the objective. It then assembles, reorganizes, and moves into an ORP near the objective. (See Figure 5-3.) The objective is seized after security and support elements are in place. This option makes coordinated action easier by setting up control of small units before engaging the enemy.

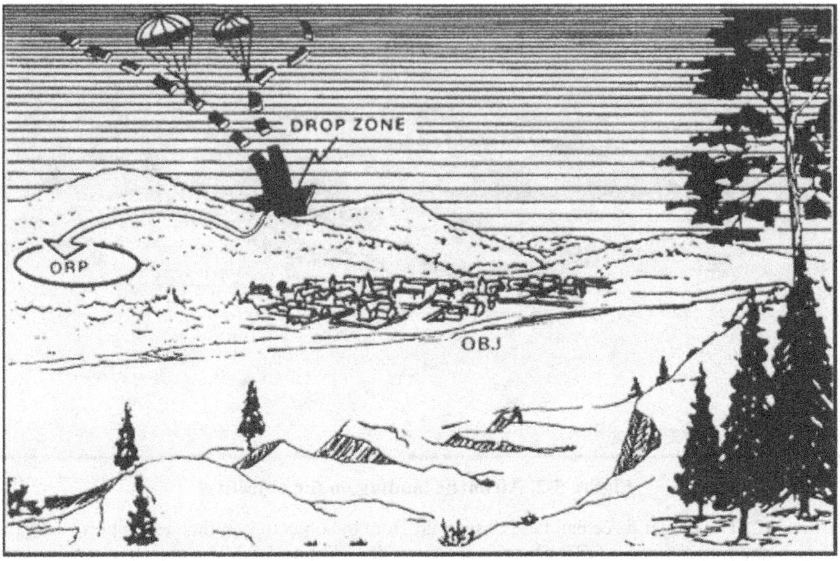

Figure 5-3. Airborne landing distant from the objective.

5-3. Interdiction operations.

a. AirLand Battle doctrine set forth in FM 100-5 is a shift in emphasis toward keeping the initiative in battle. The deep battle seeks out enemy elements separated from the main battle area and attacks them before they can be deployed. Interdiction of selected enemy forces by attacking them or blocking chokepoints hinders, delays, or interrupts the enemy's use of his lines of communication; it denies him the use of certain areas; it destroys industrial facilities, military installations, and equipment; and it stops or slows his advance along a particular route of approach. The ranger regiment's ability to conduct interdiction is vital to execution of the AirLand Battle doctrine.

b. Interdiction by ranger forces includes blocking or channeling approaches, holding key terrain for a limited time. and destroying facilities.

c. Ranger forces conduct interdiction missions that supplement those conducted by long-range Army and USAF weapon systems. Large-size ranger forces are normally used only when other systems fail or are not available, and the target is of such priority that it must be attacked. Small ranger laser target designator teams may work in concert with USAF or USN strike aircraft to direct terminally guided weapons systems. Rangers should be directed against crucial interdiction targets that disrupt, confuse, or delay the enemy and allow friendly forces to gain the initiative at the forward line of own troops (FLOT).

d. Typical interdiction operations are--

(1) Destroying rail lines, switches, rail yards, and rolling stock.

(2) Destroying bridges, overpasses, tunnels, and truck parks.

(3) Closing mountain passes or routes in restricted terrain.

(4) Cutting oil or natural gas pipelines and destroying pumping and distribution systems.

(5) Destroying dams, locks, spillways, and flood control gates.

(6) Destroying rail transload points, switches, and yard engines.

(7) Destroying shipyards, dry docks, wharfs, lighters, cranes, and stevedore facilities.

(8) Destroying electrical generation facilities, power lines, transformers, and grid monitoring centers.

(9) Destroying water pumping stations and filtration systems.

(10) Destroying radio and TV stations, phone lines, microwave terminals, satellite receiving stations, and telephone exchanges.

(11) Ambushing enemy forces and personnel moving along lines of communication.

e. Considerations when using rangers for interdiction are:

(1) Insertion and extraction.

(2) Support resources available.

(3) Ranger force vulnerability.

(4) Target identification.

(5) Threat environment.

(6) Other or supporting means of interdiction.

(7) Command and control.

(8) Time and magnitude of the effect on the enemy forces at the forward edge of the battle area.

(9) Collateral effects of target destruction.

f. Interdiction targets are not identified and attacked at random. They are part of an overall deep attack plan to destroy or cripple an entire system. Interdiction is based on the assigned deep campaign plan that directs which targets are to be attacked, by whom, the results desired, and the priorities of attack for specific systems. Based on his plan, the ranger force commander selects those ranger elements best suited to conduct the attack. Target selection, which requires detailed intelligence and planning, is based on:

(1) **Target criticality.** A target is critical when its destruction or damage would have a major influence on the enemy's ability to conduct or support operations. Each target is considered in relation to other elements of the particular target system designated for interdiction. The criticality of a target changes with the situation. For example, when the enemy has few locomotives, a railroad bridge is less critical.

(2) **Target accessibility.** The ranger force must be able to insert into the target area. Terrain, security controls, location, and means of insertion are considered when studying a target for accessibility. systems. Based on his plan, the ranger force commander selects those ranger elements best suited to

(3) **Target restorability.** The enemy's ability to restore a damaged facility to normal operating capacity affects target selection. The enemy's repair and maintenance ability and his reserve stocks must be considered.

(4) **Target vulnerability.** A target is vulnerable when it is open to attack by means available to ranger forces. Vulnerability is affected by the nature of the target such as type, size, disposition, and security. It is also affected by the means available to attack it such as explosives, incendiaries, and special devices.

g. Key to the interdiction process is the ability to locate and evaluate target systems for attack and the timing of the attack. A target system is a series of related functions that serve a common purpose.

(1) A target is one element, installation, or activity identified for attack such as a locomotive, train, bridge, or prison.

(2) A target complex is several targets in the same general area, such as a railway marshalling yard, an airfield, or port and dock facilities.

(3) A target system consists of an industrial system and its sources of raw materials; the rail, highway, waterway, or airway systems over which these materials are transported; the source of power and method of transmission; the factory complex; and the means by which the finished product is transported to the user. The targeting process is usually accomplished in the special operations force cell at EAC and the battle coordination element (BCE) at corps level.

h. Interdiction techniques include ambush, emplacement of conventional demolitions, target designation for laser-guided munitions, and emplacement of beacons or sensors.

NOTE: The enemy surface-to-air threat normally precludes the use of an airborne platform for direct observation and designation. This may be accomplished by the on-ground ranger force.

i. Through interdiction, ranger units are an ideal force to contribute to the Army's AirLand Battle operational concept. This includes taking the initiative, moving quickly to strike deep in the rear of the enemy, and defeating the enemy force beyond the FEBA. Planners and commanders must ensure that the ranger units are used on missions that are synchronized to contribute to the overall deep battle campaign. Strikes against critical targets must combine with efforts of other assets to take the initiative from the enemy, limiting his flexibility, initiative, and momentum.

5-4. Personnel and equipment recovery operations.

Recovery operations are specialized raids to liberate imprisoned or detained personnel or to return certain items of equipment to friendly control. These operations include recovering and extracting downed or hijacked aircrews and political or military leaders. Rangers use surprise and combat power to overwhelm resistance before harm can come to the detainees or prisoners.

a. **Planning.** The planning of personnel and equipment recovery operations is the same as for the raid. The main difference is that the ranger force commander must plan for the extraction of the recovered personnel and the loading and extraction of sensitive equipment. Personnel and equipment recovery raids are often executed as quick-response missions. They normally require the ranger force commander to plan and execute the recovery quickly.

b. **Augmentations.** This type of strike operation often requires that the ranger force be augmented with nonranger personnel with special skills or abilities. Examples are medical personnel, technical experts, mechanics or crew members trained in repair and retrieval operations, linguists or translators, and other special operations force elements.

c. **Organization.** The organization of the recovery force is the same as for a raid. Sometimes special teams are needed to perform certain missions involved in the recovery. This mission is conducted aggressively, usually ending in close combat during darkness or other limited visibility conditions. Reconnaissance teams may be inserted ahead of the ranger force to reconnoiter the objective and to locate the detained personnel.

(1) All planning and execution takes place as described for the raid. The emphasis must be on detailed, timely intelligence.

(2) Medical teams must be available to care for the detainees.

(3) The ranger personnel must be adept at searching and clearing of buildings.

(4) The ranger force must--

(a) Locate the specific building or area where the friendly detainees are being held.

(b) Gain entry into buildings, use stun grenades, and avoid obstacles and booby traps.

(c) Engage all enemy personnel using selective-kill techniques.

(d) Locate, identify, and secure detained personnel.

(e) Move personnel to a secured extraction site.

(f) Leave no detainees or rangers behind in the objective area.

d. **Execution.** The key to success is surprise and speed. The enemy must be assaulted when least expected. The ranger force must ensure that friendly detainees are not harmed during the assault - either by rangers or enemy personnel.

e. **Extraction.** The ranger force commander should plan for the extraction point to be as close as possible to the point of recovery. This prevents the detainees from having to move cross-country or for the ranger force to transport bulky equipment long distances. For small groups of detainees or small items of equipment, the extraction point can be further away. Recovery operations may use any method or combination of methods of insertion and extraction (see Chapter 4). Close planning and coordination is normally required with US Army, USAF, or USN aviation for both delivery and evacuation of the target area.

CHAPTER 6

Special Light Infantry Operations

Ranger units are normally employed on strike operations. However, broad US strategic needs demand a wide range of other military abilities. The ranger regiment may conduct special light infantry operations when larger conventional forces could not be used. The regiment conducts these operations using light infantry doctrine, methods, and techniques.

These operations include, but are not limited to, movement on foot, military operations on urbanized terrain (MOUT), passage of lines, ambush, and reconnaissance. Ranger units do not normally conduct planned operations against enemy armored forces. Battlefield surprises or METT-T considerations may cause ranger units to engage armored forces for short periods. The ranger regiment can conduct a conventional light infantry defense for short periods. A defense is not normally conducted adjacent to other friendly forces. It is normally a perimeter defense from which the unit then breaks out from encirclement and links up with friendly forces.

6-1. Movement.

a. Most missions involve movement on foot in tactical operations. Ranger units infiltrating behind enemy lines seek to avoid chance enemy contact. They choose the movement that allows them to retain security and control. To avoid loss of surprise and initiative, casualties, and mission failure, ranger units normally--

(1) Avoid chance enemy contact, if possible.

(2) Move on covered and concealed routes.

(3) Avoid likely ambush sites and other danger areas.

(4) Practice camouflage, noise, and light discipline.

(5) Maintain all-round security.

(6) Make unavoidable enemy contact with the smallest element possible.

(7) Retain the initiative to attack at the time and place of the ranger unit's choice.

b. In planning the movement of his unit, the ranger commander must consider all the elements of METT-T and determine how they will affect the movement. He must determine the route, formations, security measures, methods of command and control, location of key leaders and weapons, and action to be taken on chance enemy contact.

(1) **Routes.** The ranger force commander conducts a careful route analysis, using the factors of observation, concealment, obstacles, key terrain, and avenues of approach (OCOKA). He chooses a route that maintains the security of the force while ensuring surprise. He tries to choose a route that will avoid enemy contact and speed movement. The commander makes a detailed route reconnaissance using maps, aerial photographs, and reports from persons familiar with the area. If possible, he reconnoiters the route himself or uses a reconnaissance team. He may have that team act as guides or mark the route using covert marking devices.

(2) **Formations.** Ranger units use the movement techniques found in current light infantry doctrinal publications. The traveling, traveling overmatch, or bounding overmatch technique is chosen based on the likelihood of enemy contact. The ranger force normally tries to avoid detection and contact with the enemy while moving. The commander chooses a movement technique that ensures that if unexpected contact were made, it would be with the smallest element possible and not the entire formation. The distance between moving ranger elements depends on the terrain, visibility, and enemy situation. March interval is normally great enough to allow each succeeding element either to deploy or change direction if enemy contact is made. Distances are close enough that each element could quickly assist the element to its front.

(3) **Security.** Security is to the front, rear, and flanks when enemy contact is likely. A front security element is placed well forward with adequate communications with the main body. The security element is used to detect the enemy and warn the main body. The main body then either changes direction or hides until the enemy is past. If the enemy does detect the main body, the security element may be used to attack the enemy from the flanks or rear. Flank security elements are placed next to the route of march. They move forward either by alternate or successive bounds. An alternative is to have the security element moving adjacent to the column along routes paralleling the direction of march. Rear security is handled the same as front security. Plans are made for the rear guard to support the main body during chance enemy contact. This is done either by maneuver or by furnishing supporting fire.

(4) **Command, control, and communications.** All available communication is used, consistent with OPSEC and movement security, to assist in maintaining control during movement. March objectives, checkpoints, and phase lines may be used to aid in control. The number of reports is reduced, normally only exception reports are needed. The ranger force commander should be well forward in the formation but may move throughout as the situation demands. Communications with security elements are mandatory. Operations security often precludes the use of radios. Therefore, connecting files, runners, and visual signals are normally used. Detailed planning, briefing, rehearsals, and control are valuable if there is enemy contact. Alternate plans are made to cover all possible situations.

(5) **Weapons.** Rangers are placed within the formation to allow a mix of their individual weapons to cover the front, flanks, and rear of the formation. Rifles, grenade launchers, and light automatic weapons cover all sectors. Crew-served weapons such as machine guns, antitank weapons, and mortars should be given designated locations within the formation. This allows the element leader to employ them quickly. They should be placed near the lead element, but not so close as to be engaged along with it. Crew-served weapons are used to

provide overmatch, especially when employing the bounding overmatch technique of movement.

(6) **Rehearsals.** The ranger force, expecting movement as part of an operation, conducts the following combat drill rehearsals:

(a) Conduct movement.

(b) React to enemy contact.

(c) Maneuver under fire (fire and movement).

(d) Cross danger areas.

(e) Break contact.

(f) Conduct a hasty ambush.

(g) Make a temporary halt.

(h) Establish a perimeter.

(i) Establish a rally point.

(j) Take actions upon consolidation.

(k) Use arm-and-hand signals.

(l) React to flares and indirect fire.

c. After a ranger force is inside enemy territory, it must be alert to avoid detection before reaching the objective.

(1) Should the ranger force security element become aware of enemy presence, it alerts the main force. The main force tries to move away without alerting the enemy and continues the mission.

(2) If the enemy does make contact, the security element then engages the enemy, trying to deceive him as to the size of the ranger force and its direction of movement. After breaking contact, the security force divides into small groups and moves to a rally point to rejoin the main force.

(3) The ranger force does not fight en route to the objective unless there is no alternative, then it quickly breaks contact and moves away.

(4) Ranger units may need to set up a temporary defensive position for resupply, evacuation of wounded, planning, or extraction. One technique that can be used is the temporary perimeter defense.

(5) If the moving ranger force comes under indirect fire, it moves quickly out of the impact area and continues the mission.

(6) The most effective means of combating an ambush is through the immediate return of a

high volume of fire by all weapons. This requires alertness, discipline, dynamic leadership, and rehearsed plans.

(a) Regardless of the method of movement, all weapons should be positioned for immediate use. A well-executed ambush is hard to counteract. However, the effectiveness of an ambush can be reduced by the immediate return of fire. Enemy weapons may be silenced, thus creating weak points in the ambush position. Smoke and high-explosive (HE) grenades are effective in executing counterambush actions. They create confusion, provide a screen for movement, and disrupt the ambush plan.

(b) When all elements of a ranger force are not trapped within the ambush, the elements that are free to maneuver should initiate an immediate flank or rear assault against the ambush force. A flank assault permits better coordination with personnel in the ambush. Supporting fires are provided by those weapons that cannot be hand carried and fired in the assault.

(c) Specific actions against a near and far ambush are as follows:

- **Near ambush.** Elements caught in the kill zone return fire at once and move out of the kill zone. Elements not in the kill zone help by maneuvering and conducting a flanking attack. The unit assaults the ambush position.
- **Far ambush.** Those ranger elements not in the kill zone set up an overmatch position and lay down a base of fire for the elements in contact. The elements in contact move out of the kill zone and prepare to assault the enemy, using individual movement techniques and fire and movement.

(7) The reorganization after enemy contact involves the use of rally points, plans for local security, reorganization of the force, evacuation of dead and wounded, and movement based on the unit mission.

6-2. Ambush operations.

a. **Definition.** An ambush is a surprise attack, from a concealed position. It is used on a moving or temporarily halted enemy to destroy or capture him and his equipment. An ambush is a useful tactic because--

(1) Small, well-trained, disciplined forces, with limited weapons and equipment, can destroy much larger enemy forces.

(2) It reduces the enemy's overall combat effectiveness by destruction and harassment of his forces.

(3) Enemy morale and effectiveness suffer heavily at little cost to the force executing the ambush.

b. **Execution.** A successful ambush must be executed with precision, violence, speed, and audacity of execution. For success, ambush operations must emphasize the following:

(1) **Surprise.** Surprise, more than any other single aspect, enhances the value of an ambush. Surprise increases the potential for inflicting damage on the enemy with less danger to the

ranger force.

(2) **Coordinated firepower and shock effect.**

(a) Coordinated firepower is used for maximum shock effect. Massive volumes of accurate fire, explosives, and mines, coupled with an aggressive attack, breaks the enemy's spirit to fight back. Surprise increases shock effect and the chances for success. Shock effect can cover unexpected defects in an ambush--for example, ambushing a much larger force than expected.

(b) All weapons must be sited with interlocking fires in the kill zone and along likely avenues of entrance or exit. Mortars should be used if the terrain permits. Tripods and traversing and elevating mechanisms are normally used with machine guns to lock in fires. All riflemen use firing stakes to mark left and right limits, and elevation stakes. There is a tendency to shoot high in an ambush--especially at night. The M203 grenade launchers are sited to cover the dead space and routes of escape. When the rules of engagement permit, plastic bags containing CS crystals may be taped to the front side of Claymore mines.

(3) **Control.** Control is essential.

(a) Leaders must have contact with all members of their unit to alert them to the oncoming enemy. Leaders should not move around the ambush site during this crucial period. A method used to alert members can be to tie strings or vines to rangers' legs or arms. By a series of light tugs, all members of the ambush can be alerted to enemy presence.

(b) The leader must initiate the ambush with a casualty-producing device. A bank of Claymore mines on a double-ring main is an excellent device to spring an ambush. Other good techniques are to use a 90-mm recoilless rifle firing antipersonnel (APERS) rounds or a machine gun. All must be under the direct control of the ranger leader. Whistles or pyrotechnics must not be used. They will allow the enemy time to react.

- As soon as the enemy is hit, he reacts. The ambush force has only a few seconds to destroy the enemy in the kill zone. He will quickly recover from the initial shock and leave the kill zone--either with a direct counterattack or withdrawal, Subsequent fires and other banks of Claymore mines must be planned.
- The leader always executes the ambush, except when a member of the ambush knows he has been discovered. He then has the authority to execute--with killing fire, not by yelling.
- The cease-fire must be controlled by the ranger leader. A whistle or other device may be used to get attention, and then cease-fire is signaled.

(4) **Security.** The flanks and rear of an ambush site are open to counterattack. Flank and rear security may be enhanced by -

(a) Echeloning in depth.

(b) Designating sectors of observation.

(c) Positioning of RSTA devices.

(d) Enforcing noise and light discipline.

(e) Having a good withdrawal plan.

(f) Securing routes of withdrawal.

(g) Executing with speed and violence.

(h) Positioning security forces to sea] off the ambush area.

(i) Having good camouflage.

(5) **Simplicity.** A simple, direct plan improves the chance of success. The ambush plan must be clear yet concise to offer the greatest likelihood of success. For example:

(a) Mission statements for security, support, and assault elements must be clear, concise, and direct.

(b) Tasks to be performed by the ambush elements should be easy to understand.

(c) Contingency plans should be simple.

(d) Routes into positions and withdrawal routes should not cross. They should be the shortest, most secure routes.

(6) **Training and self-discipline.** Precise and violent execution is how ranger units win against enemy forces. All advantages must be exploited. Discipline must be strict. There must be no sleeping, talking, eating, or smoking in the ambush site. If an ambush is to be set up for long periods, then the elements of the ambush must be pulled back at set times to the ORP for rest. Extended ambushes of 24, 36, or 48 hours require six- or eight-hour shifts. It may take a company to man an extended platoon ambush position. Tired troops cannot man an ambush well. Other vigorous operations cannot be performed all day and have an alert ambush all night.

(7) **Maximum use of specialized equipment.** This includes the most sophisticated weapons, communications equipment, and RSTA devices. The effectiveness of any mission is enhanced by the use of this equipment.

(8) **METT-T.** Every ambush mission must be evaluated by the ranger commander IAW METT-T to design an ambush that works.

(a) **Mission.** Clearly state the mission--for example, stop enemy resupply, ambush enemy armor columns, or capture personnel or equipment.

(b) **Enemy.** State the type of enemy units--for example, guerrillas, main force, logistics carrying parties, or armor units. What is the size of the force, its habits when attacked, and type of support? What is the enemy night vision ability and expected time of movement?

(c) **Terrain.** Is it wooded, mountain, jungle, or urban? Some good ambushes can be set in a city or suburb. Are the routes available to the enemy suitable to his activity?

(d) **Troops.** Squads and platoons are the basic ranger ambush forces. An ambushing force should be able to engage a force two to three times its size--but evaluate enemy and terrain, A platoon ambushing an enemy tank company in daylight in open terrain is risky. But engaging that same column at night or in a forest may give the ranger platoon an advantage.

(e) **Time.** How long will it take to emplace the ambush? A complex, lethal ambush with plenty of "dirty tricks" takes hours to set up; a hasty ambush on a trail takes 15 to 30 minutes. How long is the ambush to be in place? What hours of the day, or at nighttime only? How soon must it be set up? Do not set up an ambush in darkness--even if it is to be executed in darkness. Try to allow every ranger to '6see" and "lay" in his weapon in daylight. Time is important.

c. **Organization of the ambush force.** A ranger unit conducting an ambush must be task-organized to perform the following functions: assault, security, and support. The ambush forces should be task-organized according to the TOE--by platoons, squads, and fire teams. The TOE should not be changed to create smaller elements for an ambush. The TOE formations may be reinforced with machine gun, recoilless rifle teams, or a 60-mm mortar squad.

(1) **Assault.** The elements assigned the assault mission either move directly into their positions or through a release point. The mission may include any combination of the following:

(a) Conduct the main assault.

(b) Halt an enemy's motorized column or any moving target,

(e) Kill or capture personnel.

(d) Recover supplies and equipment.

(e) Destroy vehicles and supplies.

Search teams are not always used. The ranger leader must decide how and when to use search teams. When rangers leave the security of their well-chosen, concealed ambush position, they are subject to the fires of the enemy who may also be hidden and ready. If it is at night, do not throw tripflares or shoot illumination to light the search area as this will illuminate and expose rangers to the enemy. Always assume there is hidden enemy--the ambush will not kill them all. Night observation devices (NODs) or a red filtered flashlight taped to M16s should be used to quickly search in ranger buddy teams. If the return fire from the enemy is great or if the ambush missed the main body, then the leader may choose to break contact and leave without searching the kill zone.

(2) **Security.** The elements assigned the mission of security may move to their positions

directly or by way of a release point. Their missions may include any or all of the following:

(a) Secure flanks, rear, or ORP.

(b) Provide early warning.

(c) Seal off the kill zone to prevent the enemy from escaping or reinforcing.

(d) Assist in executing the ambush.

(e) Cover withdrawal of main ambush force.

(3) **Support.** The units assigned a support mission provide fires that may include employment of--

- Heavy automatic weapon fires.
- Antitank fires.
- Mortar fires.
- Mines.
- Flame munitions.

d. **Ambush site.**

(1) **Choose an ambush site.** When choosing an ambush site, all sources of information must be used to enhance surprise, exploit the enemy's weak points, and take advantage of the terrain. An analysis of the terrain is also based on the other factors of METT-T to include maps, aerial photos, and, if possible, personal reconnaissance. Emphasis is on--

(a) Natural cover and concealment for the ambush force.

(b) Routes of entry and withdrawal (at least two) that should be direct and easy to reach.

(c) Good observation and fields of fire.

(d) Harmless-looking terrain.

(e) Few enemy escape routes.

(f) Limited enemy reinforcement ability.

(g) Nearby assembly or rendezvous area.

(h) Terrain that will canalize enemy into kill zones, and natural obstacles to keep him there.

NOTE: Try to select a site covered by friendly supporting indirect fires.

(2) **Take advantage of the terrain.** Emphasis must be on exploiting all natural cover and concealment afforded by the terrain. Site the ambush and individual positions based on the terrain rather than trying to adapt the terrain to a fixed geometric design.

(3) **Restrict enemy movement.** Restricting enemy movement by natural or man-made

obstacles should also be planned.

e. **Types of ambushes.**

(1) Ambushes have two categories for ranger purposes--area ambush and point ambush.

(a) Area ambushes may be set up by ranger platoons, companies, or battalions. They are used to interdict enemy movement in a given area or inflict casualties on his forces. Area ambushes consist of a series of point ambushes. The size and location of the ambushes are dictated by the METT-T analysis.

(b) Point ambushes are set at the most ideal location to inflict damage on the enemy. Such ambushes must be able to accept the enemy force from more than one direction. The ranger force levels may enable it to execute an ambush in two or three main directions. The other directions must be covered by security that gives early warning of enemy attack.

(2) The basic ambush is linear (Figure 6-1). An ambush is set on an expected avenue of approach. This ambush can accept contact from three basic directions--left, right, and front. The rear is secured by a security team.

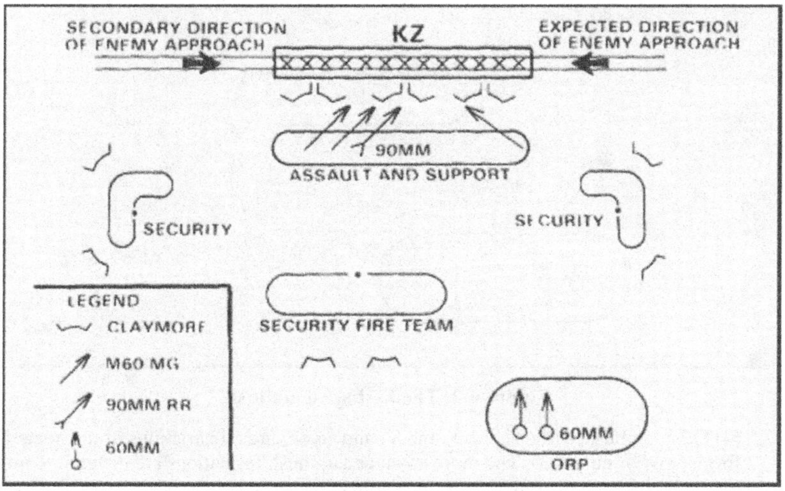

Figure 6-1. The linear ambush.

(3) A variation of the linear ambush is the L-shaped ambush (Figure 6-2). The L-shaped ambush is formed with the base (bottom) of the L perpendicular to the expected enemy direction of advance. This is a good ambush for a road, jungle trail, or an area where the enemy is canalized and his approach route is known. The L-shaped ambush can handle an enemy coming from the expected direction (toward the base of the L) from the front (stem

of the L). It is less effective against an enemy formation that comes from the opposite direction. The ambush must be executed when the enemy main body has cleared the base of the L. The flank security must protect the rear of the base of the L. In the L-shaped ambush, weapons must be sited to avoid direct or ricochet fire into friendly forces. The security elements must be considered as fire fans are being planned.

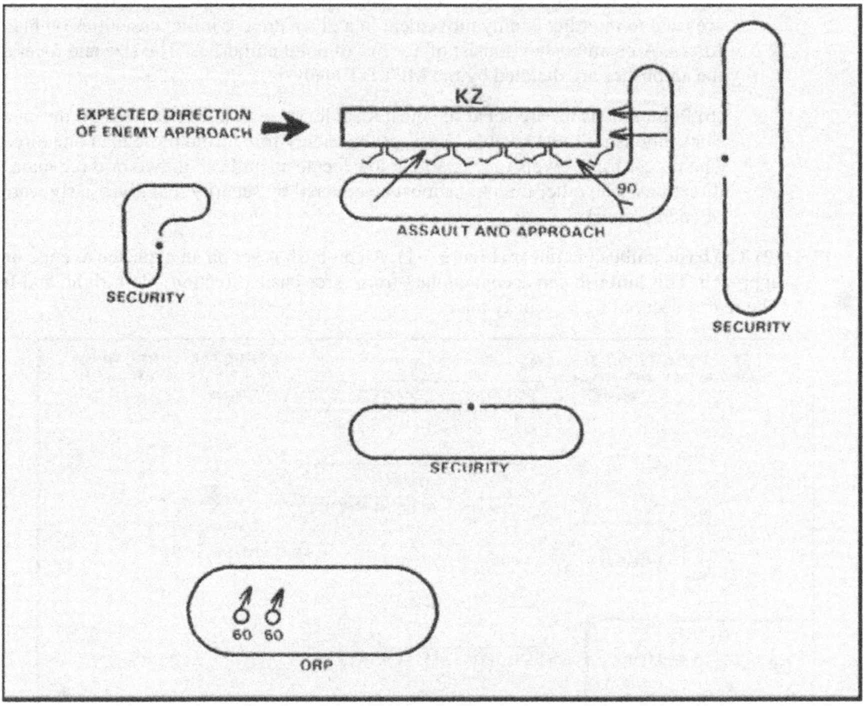

Figure 6-2. The L-shaped ambush.

(4) Other ambushes (the Z, the X, the V, and so on) are all variations of the linear and L formations (Figure 6-3). The more advanced ambush formations are difficult. Units must be expert in the basics before they are ready to employ such techniques.

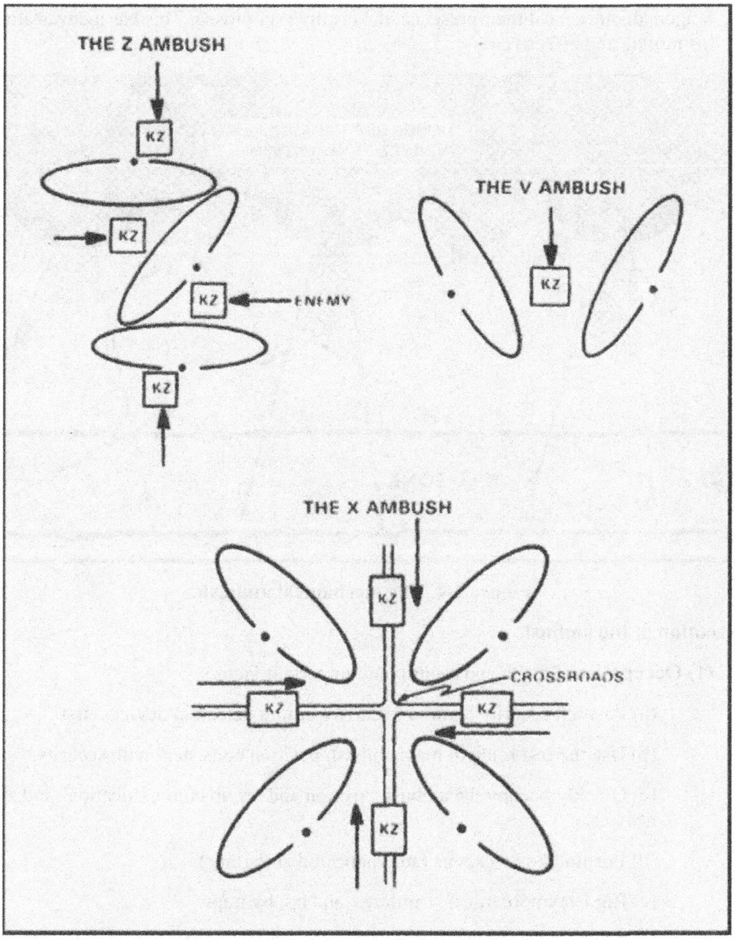

Figure 6-3. Other ambush variations.

(5) Another type of point ambush is the mechanical ambush (Figure 6-4). It consists of Claymore mines set in series with a double-ring main. It is command detonated or detonated by a triggering device activated by the enemy. Mechanical ambushes are normally manned. Rangers prepare to engage the enemy with direct fire if the mechanical ambush does not detonate or if it is wholly or partially ineffective. Mechanical ambushes are an effective way to interdict a large area using a small ranger force. If the mechanical ambush is effective and

rangers do not reveal their presence, the enemy is confused. This has a devastating effect on his morale and effectiveness.

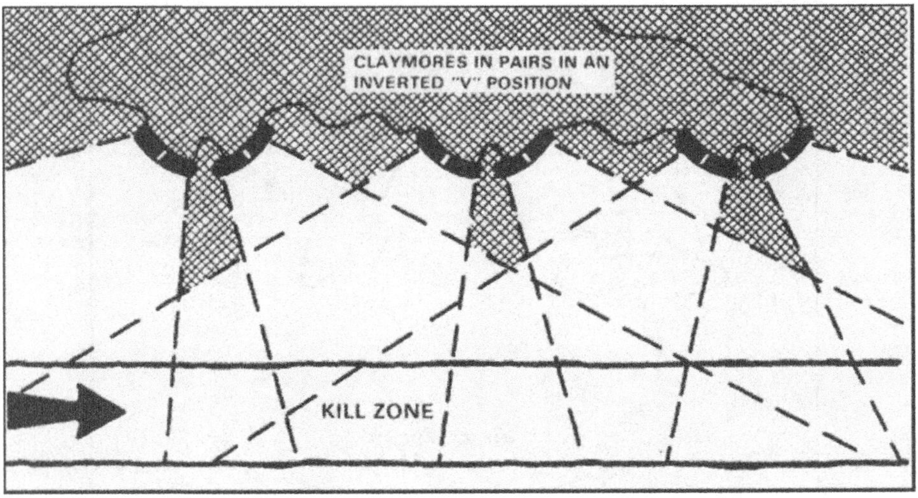

Figure 6-4. The mechanical ambush.

f. **Execution of the ambush.**

(1) **Occupation.** Stealth and security are important factors.

(a) Position security teams and early warning detection devices first.

(b) Use the rest route to main ambush position consistent with security.

(c) Quickly occupy the ambush position and set up communications and signaling devices.

(d) Position key weapons (automatic and antiarmor).

(e) Rig Claymore mines, tripflares, and booby traps.

(f) Ensure that all weapons are correctly positioned. Assign sectors of fire to provide mutual support and cover dead space.

(2) **Camouflage.** Each man must be hidden from the enemy, During preparation for the mission, each man camouflages himself and his equipment, and secures his equipment to prevent noise. At the ambush site, positions are prepared with minimum change in the natural appearance of the site. All debris resulting from preparation of positions is concealed to prevent evidence of occupation.

(3) **Movement.** Movement is kept to a minimum. The number of men moving at a time is closely controlled. Every man is as quiet as possible, especially at night. Light discipline is rigidly enforced at night. Smoking is forbidden.

(4) **Signals.** Audible and visual signals, such as whistles or pyrotechnics, must be changed often to avoid setting patterns and alerting the enemy. Three or four simple signals are needed to execute the ambush. Signals are used to--

(a) Provide early warning of an enemy approach. A signal by the security force to alert the patrol leader to the correct direction of enemy approach may be given. This includes arm-and-band signals, radio, or field telephone.

(b) Initiate the ambush. This may be the detonation of mines or explosives, Fire is then delivered at once in the heaviest, most accurate volume possible. Properly timed and delivered fires add to the achievement of surprise, as well as to the destruction of the target.

(c) Lift or shift fires, Signal for lift or shift fires, if the kill zone is to be assaulted. Voice commands, whistles, or pyrotechnics may be used. When the kill zone is assaulted, the lifting or shifting of fires must be as precise as when starting the ambush. Otherwise, the assault is delayed and the enemy has a chance to recover and react.

(d) Withdraw. A signal for withdrawal can be voice commands, whistles, or pyrotechnics.

(5) **Objective rally point.** The ORP is located far enough from the ambush site so that it will not be overrun if the enemy manages to attack the ambush unit(s). Existence loads are normally left in the ORP. Routes of withdrawal to the ORP are scouted (when possible by each man), Withdrawal routes should provide cover and concealment for the unit and hinder enemy pursuit. Withdrawal routes are a main consideration in the selection of the ambush site. They may be the key to survival after execution of the ambush. On signal, the ambush force quickly (but quietly) withdraws to the ORP. If pursued, the withdrawal may be by bounds, with mines or hasty ambushes used to delay pursuing forces.

(6) **Ambush variety.** If one method of ambush is used predominantly, the enemy will develop an effective defense against it and will be affected less by the shock of the ambush since he knows what to expect. More than one method of ambush should be used. No single method will fit all combinations of terrain, equipment, weather, and enemy capabilities. Variety should also be used when signaling - for example, use multiple signals, both audible and visual. Use weapons fire, mines, and RSTA when possible and vary signals to avoid compromise.

(7) **Swift action.** Speed in the execution of the ambush and the withdrawal should prevent enemy reaction forces from engaging the ambush force. Speed is often a shield against casualties and failure. When there is contact with reaction forces, speed may enhance quick disengagement. Ranger units must not only perform better than the enemy but also move faster than the enemy can react.

g. **Successful ambush.** Emphasize the following:

(1) Intelligence to ensure the enemy is ambushed at a time and place when he least expects or is least prepared to fight. This is preferred during reduced visibility to achieve surprise.

(2) Detailed planning and thorough training and rehearsing of all elements in all phases of the ambush. This ensures maximum shock effect through swift, precise execution.

(3) All available RSTA devices to permit daytime effectiveness at night when moving, shooting, or detecting enemy movement. To avoid detection, active RSTA devices should not be used until after the ambush has been triggered.

(4) All available firepower with emphasis on antiarmor, area and automatic weapons, and precision-guided munitions.

(5) Speed to achieve surprise and enhance security of the force.

(6) Cover, concealment, and overall protection afforded by the terrain when moving or when occupying ambush positions.

6-3. Antiarmor operations.

a. Antiarmor operations are not primary missions for ranger forces. They lack antiarmor firepower and have limited antiarmor ability. Ranger units can, however, perform limited antiarmor missions on close terrain by interdiction or ambush. The antiarmor weapons they do have are man-portable, but the ammunition is too heavy to carry many rounds cross-country. A ranger battalion has ten Dragon missile launchers and nine 90-mm recoilless rifles. It is also equipped with light antitank weapons (LAWs) and laser target designators (LTDs) that can guide Copperhead rounds.

b. Every ranger must be skilled with the limited antiarmor systems available. Proper selection, fortification, and camouflage of antiarmor ambush sites, route selections, withdrawal procedures, and initiative are important. Armor kill zones (AKZs) are normally set up after infiltrating the ranger force behind enemy lines under limited visibility.

c. A ranger antiarmor force is organized into four elements: command and control, support, security, and armor-killer teams (Dragons, LAWs, 90-mm recoilless rifles). These elements function the same as a raid force. The exception is that the assault force is the armor-killer team with the mission of destroying enemy armor, and it will probably not enter the AKZ. After the ambush occurs, the ranger force withdraws to its ORP. Indirect fires (if available) are then brought to bear on the AKZ.

d. Ranger units may conduct antiarmor ambushes and interdiction operation by use of laser-guided Copperhead rounds or air-delivered precision-guided munitions. Air-delivered family of scatterable mines (FASCAM) can be used to provide the ranger force security against attack by armored forces.

6-4. Military operations on urbanized terrain.

a. The ranger regiment normally tries to avoid the manpower-intensive, high-casualty battles associated with MOUT. However, due to the nature of strike operations and the types of targets selected, the regiment may conduct MOUT for limited periods. As in all ranger operations, the ranger force commander tries to surprise the enemy, attack him from unexpected directions, and destroy him with a violent, precise attack. The maze of cellars, rooms, sewer systems, rooftops, and alleys common to urban areas presents many opportunities for the ranger force.

b. Offensive operations by a ranger force exploit the centralized planning and decentralized execution aspects of ranger unit training. Individual action and initiative are always channeled to accomplish the commander's intent. The ranger unit's offensive tactics are based on quick action and boldness. Attacks by the ranger force are short and sharp. Execution is marked by swiftness, daring, timing, and surprise. The enemy's sleeping habits, eating habits, and relief times are watched. launched almost always at night or under smoke screens, attacks are often made without mortars or artillery, relying on surprise for success. Snipers and sharpshooters are used to shift the balance of battle.

c. Deliberate defensive operations are not good ranger unit missions. However, if a ranger unit must defend for a short period, the close terrain of urban areas allows it to function well against a stronger force. The buildings in a city break up advancing enemy formations and force them to move along streets. The ranger force executes an active defense anchored on strongpoints, such as a building or a group of buildings. It launches surprise counterattacks against the enemy's flank and rear. The aim of ranger defense is to give the enemy no respite and to make him feel threatened by a sudden, deadly attack. Ranger units should be withdrawn from defensive positions in urbanized areas as soon as possible. Prolonged MOUT and the associated casualties quickly degrade a ranger unit's combat effectiveness.

6-5. Perimeter defense.

The Perimeter defense may be conducted to control key terrain, gain time or economize forces.

a. **Planning.**

(1) **Warning order.** The order must be issued soon after mission receipt. It should contain the mission, time schedule (including rehearsal and inspection times, mission time, and place for OPORD), uniform and equipment, and special instructions.

(2) **Operation order.** The leader normally gives the OPORD on the ground to be defended. Locations and sectors of fire for squads and crew-served weapons should be designated. If time permits, leaders walk the terrain with subordinates as the OPORD is issued. The OPORD covers security, communications, signals, target reference points (TRPs), armor kill zones, and priority of engagement rules for antiarmor weapons. The OPORD also states the priority of work.

(a) Set up security and communications.

(b) Position crew-served weapons and personnel.

(c) Clear fields of fire.

(d) Prepare fighting positions.

b. **Organizing.**

(1) Squads are positioned with overlapping and interlocking sectors of fire and observation.

(2) The M60 machine guns are positioned either with an assigned final protective line (FPL) of fire or with a primary direction of fire (PDF). The PI)Fs are employed on covered avenues of approach such as gullies. The M60 FPLs should interlock across the unit front.

(3) The 90-mm recoilless rifles and Dragons are positioned with frontal cover to engage attacking armored vehicles from the flank.

(4) Riflemen are positioned to support and protect crew-served weapons. Riflemen are assigned FPLs that interlock at a point just beyond hand-grenade range.

(5) Dead space is covered with Claymore mines, antipersonnel mines, and M203 fires.

(6) Antitank mines, positioned to provide close-in protection from armored vehicles, are covered by LAW fires.

(7) Leaders plan indirect fire for long-range suppression of armored vehicles and short-range protection from dismounted assault. Mortar or artillery final protective fires (FPF) are placed on the most dangerous dismounted avenue of approach.

(8) Range cards are prepared for all crew-served weapons. Squad and platoon leaders prepare sector sketches and fire plans.

(9) The ranger force camouflages everything from ground and aerial observation.

(10) Leaders designate alternate and supplementary positions.

c. **Executing.** Leaders control the distribution and concentration of fires in their sectors. They constantly consider the overlap of fires with units to their left and right.

(1) Squads and crew-served weapons engage targets in sectors assigned by platoon leaders.

(2) Each weapon is used in its best role:

(a) Rifle, against enemy infantry within 300 meters.

(b) Machine gun, against enemy infantry within 1,000 meters.

(c) Grenade launcher, against infantry within 350 meters and armored vehicles within 200 meters.

(d) Light antitank weapon, against armored vehicles within 200 meters.

(e) Recoilless rifle, against armored vehicles within 400 meters.

(f) Dragon, against armored vehicles within 1,000 meters.

(3) Most dangerous targets are engaged first.

(4) Squads shift fires as needed or as directed by platoon leaders.

(5) The platoon masses fire on selected targets while maintaining coverage of the entire platoon sector.

(6) Platoon leaders control fires.

(7) After repelling an enemy assault, the platoon consolidates and reorganizes the defense.

6-6. Breakout from encirclement.

A breakout from encirclement is conducted when units, operating behind enemy lines, find themselves cut off from friendly forces and surrounded by superior forces. Given ranger unit missions, the chance of operating as an isolated force behind enemy lines is great. The breakout is characterized by determining enemy weak points, deception, massing of combat power, and a direct attack for a violent and timely breakout.

a. **Characteristics and fundamentals.** The breakout from encirclement involves the following characteristics and fundamentals:

(1) Set up a hasty defense. Adjust the perimeter as needed and make contact between subunits. Assign each a specific sector.

(2) Plan fire support. Plan FPFs on likely dismounted avenues of approach. Mass antiarmor weapons on the most dangerous mounted avenue of approach.

(3) Conduct reconnaissance. Identify enemy weak points. Locate enemy armored forces and the routes they must use. Locate enemy automatic weapons.

b. **Planning.**

(1) Organize all support personnel into maneuver elements--for example, company headquarters, mortars, and antitank. Organize the ranger force into a rupture force, reserve, support force, and a rear guard.

(2) Issue a fragmentary order. It should contain current local enemy situation and cover all pertinent areas of the five-paragraph field order.

(3) Select rally points.

c. **Execution.**

(1) Reduce the defense perimeter. This is done slowly until the unit is massed for the breakout attempt.

(2) Create a diversion. This is done either by fire from the rear guard force or by a limited attack in a direction other than that of the main breakout attack.

(3) Break through the defense with the rupture force. Hold the shoulders of the breakthrough.

(4) Move the reserve through to exploit the breakthrough.

(5) Move the support troops through with casualties. Have them collect casualties en route.

(6) Have the rear guard follow the support troops. It may have to fight a delaying action and use fire and movement to break contact with the enemy.

(7) Consolidate and reorganize following the breakout.

(8) Have the rupture force break contact with the enemy once the rear guard joins.

6-7. Linkup operations.

During operations behind enemy lines, ranger units may be required to link up with other friendly ground forces. Linkup may occur in airborne, air assault, amphibious, or shore-to-shore operations. It may occur during the breakout of an encircled force or during the convergence of separate forces. It may involve conventional or unconventional US forces, allied forces, or friendly guerrillas. Rangers may be required to link up with other rangers, partisans, or friendly ground forces that are conducting offensive operations.

a. **Planning.** Ranger commanders plan detailed linkup operations to avoid friendly forces firing on each other. Coordination and planning of communications, recognition signals, fires, command relationships and responsibilities, and control measures are essential. Coordination should be done before the start of the operation.

(1) Site. The linkup site should be easy to recognize, have cover and concealment, be located away from prominent terrain the enemy might use, be defensible, and provide multiple access and escape routes. An alternate site should be planned in case of compromise.

(2) **Communications and recognition signals.** Frequencies, call signs, codes, visual signals, and alternates for each of these should be planned before departure of friendly lines. Radios may be used to report the location of each unit to the site, and occupation and securing of the site. A system of mutual recognition must be devised to keep friendly patrols from firing on each other. This may include visual signals such as arm bands, panels, colored lights, or RSTA devices.

(3) **Fires.** The headquarters directing the linkup sets up fire control measures and priority of fires. The most commonly used control measure is the restrictive fire line (RFL), beyond which one force may not fire without coordination with another force. The RFL may be adjusted as two units draw closer together. Such an adjustment needs close coordination and should be planned before the operation. Successive phase lines between the two units can serve as on-order RFLS, provided that neither unit comes too close to the RFL before it is shifted.

(4) **Command relationships and responsibilities.** These are set up by the headquarters directing the linkup. Ideally, liaison personnel are exchanged before the operation. The key to success is liaison and guides.

b. **Execution.** Each member of the ranger force must understand the plan. Rehearsals are necessary in order to avoid confusion during the linkup. One uninformed ranger can start firing and cause enough confusion that friendly forces could be identified as enemy. Actions must be quick since units are open to enemy attack while they try to link up.

(1) **Moving units.** A linkup between moving units is hard. As the units move closer to each other, the chance of them engaging one another increases. Therefore, the linkup units must adjust their movements to each other and continuously coordinate on a designated secure radio net. If possible, one or both units should come to a short halt before linkup.

(2) **Stationary units.** Moving units must know the positions of the stationary units. They must keep stationary units advised of locations. Stationary units guide the moving units to the linkup point by radio. The stationary units must be ready to accept and position the moving units smoothly and quickly.

6-8. Passage of lines.

A passage of lines is an operation in which one unit moves through another unit that is stationary and disposed in a tactical formation on a FEBA. Ranger forces may conduct a passage of lines to get behind the enemy. Movement in forward unit areas must be controlled, coordinated, and kept to a minimum. This avoids conflict with friendly troops or the activation of RSTA devices. Ranger units treat the positions of forward units as danger areas. They must be assumed to be under enemy surveillance in all weather or visibility. The ranger force is open to attack during the passage. Personnel and units may be more concentrated than is desired. Fires of the stationary unit may be masked for a limited time. The passing ranger unit may not be in a good position to react to enemy contact. Detailed reconnaissance and coordination are crucial to ensure that the passage is conducted quickly and smoothly.

a. **Principles.**

(1) **Coordination.** The ranger force commander, or his representative, must coordinate departure and reentry of friendly front lines (FFL). This is done either directly with the units through which the rangers pass or through designated staff agencies. Items to be coordinated include fire support, passage lanes, control measures, guides, communications, intelligence, CSS, and actions on enemy contact.

(2) **Reconnaissance.** The ranger force commander should make a ground reconnaissance of the area through which the unit will pass. He should pick the initial rally point (IRP) at this time. The reconnaissance should include observing the area just forward of the friendly unit for routes, danger areas, and obstacles. During reconnaissance or coordination, rangers should wear the same uniform as the troops in the forward unit. This will avoid attracting the enemy's attention.

(3) **Security.** Maintain security to avoid contact with the enemy while departing or reentering. It is important that the ranger force make no enemy contact at this time. Control is hard if a fire fight starts just forward of an FFL.

b. **Departure of friendly front lines.**

(1) Designate an IRP.

(2) Maintain security. Use an appropriate movement formation when departing the FFL.

(3) Do not move within the friendly unit's area without a guide.

(4) Make final coordination with the friendly unit commander.

(5) Count the ranger force members as they depart.

(6) Make a security and listening halt.

c. **Departure of friendly units.**

(1) The ranger force arrives at the forward unit and is met by a guide. The guide leads the rangers to an assembly area chosen by the ranger commander during initial coordination with the forward unit commander.

(2) No one should move, either singly or as a unit, anywhere in the FFL without a guide.

(3) The ranger commander makes final coordination with the forward unit commander. He determines if any changes have taken place since initial coordination and learns of any recent enemy activity.

(4) Before leaving to coordinate, the ranger commander gives instructions (called a contingency plan) for what should be done while he is gone. These instructions state where he is going, who is going with him, how long he will be gone, what to do if he does not return, and actions to be taken if there is enemy contact.

(5) On the commander's return from final coordination, he may issue a fragmentary order (FRAGO) to cover any changes.

(6) Three common threats and techniques for countering them during departure from the FFL are:

(a) **Ambush and chance contact.**

- The ranger force commander must avoid enemy contact while departing.
- The ranger force goes to the friendly side of the departure point where a point team is dispatched to see that the area forward of the barriers is clear.
- The point element should check the area out to the first covered position or an area large enough to allow the unit room to maneuver if it is hit while departing.
- The distance varies with the size of the ranger force and the terrain.
- The point team notifies the ranger commander when the area is clear. The rest of the unit is brought forward.

(b) **Indirect fire.** If enemy indirect fire is falling, the ranger commander should not halt his unit after final coordination at the forward command post. Instead, he should move it quickly through the FFL. The point element should be far enough ahead of the main body to provide security from ambush or chance contact.

(c) **RSTA.** If the enemy has radar, sensors, or night vision devices, they may be offset by the following countermeasures:

- Use a well-hidden departure point, such as a reverse slope or a dense woods.

- Infiltrate the unit through the departure area and have the rangers rendezvous in a defilade.
- Pass through during rain, fog, snow, or darkness to help conceal the passage.
- Employ electronic countermeasures.

(7) Have a security and listening halt after the ranger force has moved out of sight and sound of the FFL. This is a short halt to see if enemy is in the area, and to accustom the unit to the sights and sounds of the battlefield. Halt in a position that covers the unit from chance friendly small-arms fire. If the security and listening halt must be in an exposed position, have each man lie flat on the ground.

(8) During movement, have frequent halts to see that the unit is not being followed and that no enemy is in the area.

d. **Reentry of friendly front lines.**

(1) Set up and occupy a reentry rally point (RRP).

(2) Maintain security at the RRP and at the reentry point.

(3) Use as few personnel as possible to locate the reentry point while the bulk of the unit remains in the RRP.

(4) Meet a guide at the reentry point. Normally, a forward-of FFL password is used since it may be overheard by enemy.

(5) Count the ranger force members to ensure no infiltrators follow the unit, especially during reduced visibility.

(6) Give the friendly unit commander a spot report, providing combat information of immediate tactical value.

e. **Reentry of friendly units.**

(1) The unit is moved into a rally point near the reentry point. This rally point should be on a prominent terrain feature where the ranger commander can pinpoint his location, with respect to the reentry point.

(2) The forward unit is alerted by radio that the ranger unit is ready to reenter. A code word is used for security and brevity. The code word must be acknowledged by the forward unit before the ranger force reconnoiters the reentry point. This shows that a guide has been sent to the reentry point and is waiting for the unit.

(3) If the ranger commander is certain of the reentry point location, he moves the entire unit directly to the reentry point. He has the point element coordinate the reentry.

(4) An effective method of locating the reentry point is through the use of a thermal imagery device. It detects the body beat of the reentry guide, even though he may be well camouflaged or his location masked by smoke. If ground surveillance radar is available, it can be used to vector the ranger unit to the reentry point.

(5) The reconnaissance party should never move parallel to friendly barriers or move around the wire. If the reentry point cannot be found by the initial reconnaissance or if the reconnaissance party finds only the barrier wire, the unit should notify higher headquarters and move to another rally point to wait until daylight. The unit should not stay in a rally point from which a radio transmission has been made.

(6) When the reentry point is found, the ranger commander can bring the unit forward. If the reentry point is easy to find, he can call by radio for the unit to come forward.

(7) The guide leads the unit through the barriers to the security position coordinated with the forward unit commander. The unit halts in the security position, remaining in movement formation. The ranger commander then gives a spot report to the forward unit commander. The ranger commander tells him anything of intelligence value or of immediate tactical use to the forward unit.

6-9. Reconnaissance platoon operations.

a. The reconnaissance platoon organic to the ranger regiment consists of a platoon leader, a senior reconnaissance sergeant, and three five-man reconnaissance teams that include a communications specialist in each team (see Appendix A). This platoon is responsible for short-duration reconnaissance of objectives for the ranger force commander. This reconnaissance is routinely in support of future strike or special light infantry operations. The teams can also be used by the ranger force commander to conduct deep target acquisition and designation, or to confirm or disprove information reported by other sources. Rarely will teams be used to attack enemy targets.

b. The reconnaissance teams of the ranger regiment do not replace (or perform the same reconnaissance tasks associated with) long-range reconnaissance patrol (LRRP) units or long-range surveillance units (LRSUs) that may be organic to the corps or division. These units are organized, trained, and equipped for long-term passive surveillance and reconnaissance in enemy territory. The ranger reconnaissance teams are trained and organized to support the immediate intelligence-gathering needs of the ranger force commander. They are not to act as a human intelligence (HUMINT) collection asset for other operational headquarters.

c. The reconnaissance platoon headquarters trains, equips, and controls the employment of the three reconnaissance teams. It operates the communications stations needed to receive reports from deployed teams. In the absence of the S2, it debriefs team members upon extraction or exfiltration. It can provide one or more teams to support each ranger battalion on independent operations. The reconnaissance teams can--

(1) Infiltrate the objective area by parachute (HALO, HAHO, or static lines), helicopter, fixed-wing aircraft, SCUBA, small boat, foot, or other means.

(2) Remain undetected in the vicinity of the objective area up to five days.

(3) Perform reconnaissance operations employing a full range of night observation devices, infrared detection devices, unattended sensors, and photographic equipment.

(4) Perform demolition target analysis.

(5) Operate small watercraft and inflatable boats.

(6) Emplace unattended ground sensors, omni-directional navigational beacons, hand-emplaced expendable jammers, and electronic target designation devices.

(7) Collect combat information to satisfy priority information requirements and mission-essential elements of information. Teams report that information by use of long-range, secure, burst-transmission communications equipment.

(8) Perform DZ selection, marking and reception duties.

(9) Report objective area weather conditions.

(10) Perform highly selective, limited attacks or ambushes when so tasked.

(11) Link up with the main body of the ranger force in the objective area, or escape and evade the enemy in order to return to friendly lines.

(12) Act as part of the AAT during airborne operations.

d. There are many limits of the reconnaissance teams.

(1) Mobility is normally restricted to foot movement in the area of operations.

(2) Use of radio and other active electronic or optical devices makes the teams open to enemy detection.

(3) All supplies and equipment needed by the team must be carried on the initial insertion, since resupply increases the probability of detection by the enemy.

(4) Conduct of insertion and extraction operations deep in enemy-held territory is difficult due to the need for secrecy, timeliness, security, and accuracy of location.

(5) Medical treatment of team casualties is limited to individual first aid. Casualty evacuation is difficult unless linkup with the main ranger force has been achieved.

e. The training of reconnaissance team members parallels that given to all members of ranger units with emphasis on infiltration, communications, and reconnaissance skills. Extensive training is also given in survival, evasion, resistance, and escape (SERE) and advanced first aid. Training emphasis is placed on actions at the reconnaissance site. The reconnaissance teams must conduct their mission and remain unseen. Every team member is trained to consider the need for information balanced by the likelihood of detection. Detection by the enemy compromises future operations.

6-10. Reconnaissance planning.

a. Detailed planning ensures that the reconnaissance mission is successful. It is also vital to the survival of the reconnaissance team. This planning is conducted by the ranger force planning group assisted by elements of the JTF staff. The supporting air or naval commander, navigator, and electronic warfare officer may aid the ranger force staff in the planning.

b. Upon receipt of the mission, the team selected begins an intense preparatory phase at the REMAB or the ISB. There the team receives its initial mission briefing. The items covered allow the team to complete its detailed planning. The reconnaissance platoon leader reviews the team plan. He ensures compatibility and integration into the ranger force commander's plan for employment of the main force. Staff specialists at ranger regiment and battalion--such as those concerned with intelligence, weather, or communications--provide help.

c. Planning for the use of reconnaissance teams is conducted within the framework of the ranger planning sequence for the entire mission. There are five planning considerations for the employment of reconnaissance teams:

(1) Operations security must be considered in all staff efforts to provide maximum protection for the reconnaissance teams. This includes intelligence, CE, logistics, administration, and maintenance. Operations security must be integrated throughout every ranger force operation. It consists of four main categories of security measures: signal security, physical security, information security, and deception. All are related and must be considered at the same time for each operation.

(2) Some missions may require rapid deployment to the operational area, thereby dictating the quickest method of insertion. However, others may depend on maintaining total secrecy, regardless of time.

(3) The transportation means selected for the delivery and recovery of the teams depends on the mission. This selection is also based on the abilities, limits, and availability of mission support platforms.

(4) The distance to and from the objective area must be considered, as well as the distance from the insertion point to the actual target. Reconnaissance teams can move great distances over rugged terrain. However, this movement does require time and limits the team's ability to carry bulky or sensitive equipment.

(5) Consideration must also be given to intelligence information available concerning the objective area which includes, but is not limited to--

(a) **Enemy situation.** The enemy threat--his abilities, disposition, security measures, detection, and air defense systems - affects the selection of the means for delivery or recovery.

(b) **Terrain.** Land formations must be considered in selecting the method of infiltration and exfiltration. Terrain affects the selection of altitudes, approach and exit routes, landing areas, drop zones, and beach landing sites. Air infiltration routes that provide terrain masking are desirable in static line parachute operations.

(c) **Weather.** Seasonal weather conditions affect team delivery and recovery. Factors to be considered are temperature, precipitation, visibility, clouds, and wind. If para-SCUBA techniques are used, high surface winds and their effect upon surf conditions or period of reduced visibility may preclude the use of parachutes, inflatable boats, or swimming as entry or recovery techniques. These same conditions favor land infiltration and exfiltration, The AWADS reduces the impact of visibility as a limiting factor for air infiltrations.

(d) **Astronomical conditions.** Periods of sunrise and sunset, moon phase, moonrise and moonset, and twilight must be considered.

d. A reconnaissance team mission must be specific and support the mission. Information concerning the target may not be gathered because the risk of detection by the enemy is so great. Detection may preclude the conduct of the planned operation. If more than one mission is assigned, priorities are set, All team missions must be coordinated to avoid other friendly forces operating near the target. Ranger forces conduct operations at such depth behind enemy lines that no other US or allied forces are normally present. There is a chance that a friendly indigenous guerrilla force may be operating near the objective, but ranger reconnaissance teams do not normally make contact with them.

e. The reconnaissance platoon leader joins with the intelligence and operations sections of the ranger force headquarters in the initial planning for reconnaissance team operations. The method of operation while moving, communication procedures, reporting, surveillance procedures, and actions on enemy contact should be included in the platoon's SOPS.

f. The ranger force intelligence officer normally prepares the reconnaissance plan IAW the guidance provided by the controlling headquarters. He is helped by the operations officer and the reconnaissance platoon leader. Selected team members and a member of the unit providing the insertion and extraction means are briefed on the mission early in the planning phase. They should participate in the detailed planning to follow. The essential details of a reconnaissance plan normally include--

(1) Areas to be observed and recommended areas or positions from which this can be done. These positions are determined in advance of the insertions of teams. They are based on a study of terrain, road and rail nets, enemy order of battle, delivery means, operations plans of the controlling headquarters, and the desires of the ranger force commander.

(2) Load plans and procedures. This includes both insertion and extraction. Alternate sites are chosen for possible use. Primary and alternate DZs are selected, if applicable.

(3) The flight plan. This includes insertion and extraction flight routes and alternate routes.

(4) The movement plan to and from a patrol base. This includes the times, routes,. and destinations when individuals or small groups are required to operate away from the patrol base.

(5) The fire support plan. This includes the location and times of planned air strikes in support of the main body, and all planned fires by the main assault force. The reconnaissance teams should know the frequencies and call signs for tactical air support.

(6) The timing and ground tactical plan for execution of the main force operation. This includes disposition of the main force, special mission teams from the ranger force, or any other special operations force elements.

(7) The communications plan. This includes frequencies, reporting schedule, emergency reporting procedures, and alternate communications plans. It also includes the ground-to-air

communications plan and the friendly recognition signal to be used during linkup with the main assault force or other friendly forces.

(8) Plans for use of technical specialists or special equipment.

(9) The plan for the treatment and evacuation of sick or wounded team members from the objective area.

(10) The E&E plan. This includes routes out of the objective area, passwords, and alternate extraction zones or methods far from the objective area.

(11) Mission-specific elements of information desired by the ranger force commander, and any special or on-order missions the team must conduct.

g. The team leader uses specific steps in planning, preparing, and executing missions. These procedures are comprehensive, yet adaptable to any situation. Three of the most important steps are:

(1) **Issuance of warning order.** This action gives the team maximum time to prepare for the mission. It should be issued as soon as the team leader has been notified of a mission and updated after he has made his tentative plan.

(2) **Issuance of OPORD.** The team leader conveys the information and instructions needed to accomplish the mission. He requires a briefback from each team member.

(3) **Rehearsals.**

h. The specific steps used when planning missions are as follows:

Step 1. Study the mission.

Step 2. Plan use of time.

Step 3. Issue initial warning order.

Step 4. Study terrain and situation.

Step 5. Make tentative plan.

Step 6. Organize.

Step 7. Update warning order.

Step 8. Coordinate (continuous throughout).

Step 9. Consult aerial photographs and map updates.

Step 10. Complete detailed plan. (To be combined with paragraphs 3, 4, and 5 of the operation order.)

Step 11. Issue operation order.

Step 12. Supervise (throughout), inspect, rehearse.

Step 13. Execute mission.

i. A workable E&E plan must be developed. Such a plan enhances the survival of the reconnaissance team members. It ensures that they are psychologically prepared to conduct the mission, by knowing there is a system and plan for returning them to friendly lines. The E&E plan must include a section describing the actions of the reconnaissance team members should their insertion craft be disabled en route to the objective area. It should also include what to do if the team is unable to link up with the main ranger force, or if the main ranger force has not been inserted into the objective area.

j. Rehearsals and briefbacks are best for finding flaws in procedures or planning. Thorough coordination of all procedures is essential. The exact type of aircraft or other means that will be used for insertion should be used for training. Rehearsals should be conducted under terrain, astronomical, hydrographical, and meteorological conditions close to those to be met on the operation. The team leader conducts a briefback with the platoon leader. This lets the platoon leader decide if the team is ready for the mission. It also allows him to make changes.

6-11. Execution of insertion.

a. Reconnaissance teams normally conduct blind insertions. Blind insertions are amphibious, heliborne, or parachute drops onto unmarked DZs without help from a DZ-marking party or a CCT. Blind insertions often depend on favorable light and weather conditions in the objective area. Favorable conditions for the reconnaissance team often means low-lying cloud cover, fog, and darkness. The teams are able to operate during reduced visibility by the extensive use of night observation devices. The team's extensive training and high level of land navigation skills allow it to quickly traverse rugged terrain and not be detected.

b. Team insertion into the objective area often needs a cover and deception plan. Team members can conduct HALO and HAHO operations. Teams can be inserted into areas not normally acceptable as a DZ for conventionally trained parachutists. By the use of HAHO techniques, reconnaissance teams can assemble in the air and glide to a landing at a location far from the drop point.

c. If reconnaissance teams need special or bulky equipment that cannot be carried on the individual team member, it can be airdropped, hidden near the DZ, and retrieved later. Such container drops are often used during low-level static line insertions. The high-altitude airdrop resupply system (HAARS) can be used to deliver such items at the same time as reconnaissance teams are inserted by HALO.

CHAPTER 7

Combat Support

Combat support is used to enhance the combat power of the maneuver companies. Combat support elements may be organic or nonorganic. They can conduct a variety of missions in providing indirect and direct fire support, security, intelligence, and maneuver and fire control.

7-1. Intelligence support.

a. The success or failure of a ranger mission often depends on the accuracy, detail, and timeliness of intelligence information. Both deliberate and quick-response missions need specific information that can be obtained only by a multidiscipline collection and analysis agency. An active interface with all levels of the intelligence system helps get the needed information quickly.

b. The theater or JTF commander is responsible for providing intelligence support. The 1st SOCOM also provides information and analyzed intelligence to the ranger force commander. The main source of processed intelligence is the Echelons Above Corps Intelligence Center (EACIC), similar to the division level military intelligence (MI) all-source production center (ASPC). The EACIC provides intelligence, security, and electronic warfare (EW) interfaces (see Figure 7-1). It is through the EACIC that ranger commanders can request support from national, joint, combined, or CONUS intelligence analysis systems.

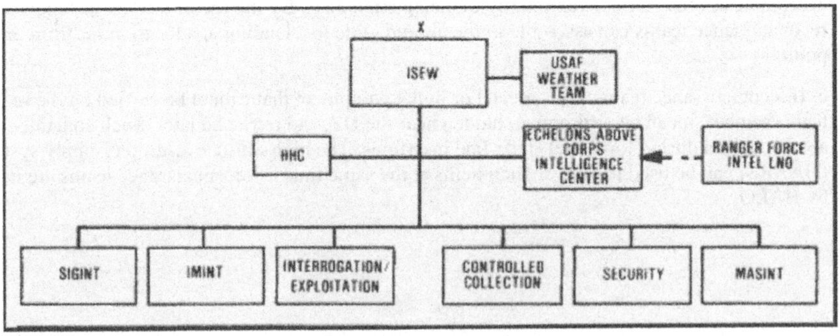

Figure 7-1. Intelligence at echelons above corps.

c. Intelligence requirements for ranger operations are unique, highly sensitive, and compartmented. The intelligence officer of the 1st SOCOM and the appropriate 1st SOCOM regional planning cell coordinate with theater intelligence agencies to provide needed information.

d. The ranger regimental staff and the ranger battalion staff have personnel in their intelligence sections to correlate and analyze information from many sources.

e. The reconnaissance platoon, organic to the ranger regiment, maintains surveillance of the objective area before the insertion of the main ranger force. It reports any last-minute developments that would affect the ground tactical plan. The platoon can also be tasked to determine any mission-specific items of essential information that the ranger force commander needs to complete his plan. This platoon is a highly specialized unit that is reserved for use by the ranger force commander. It is not a HUMINT collection agency for use by the theater or JTF commander. Reconnaissance teams normally link up with the ranger force in the objective area and are extracted with the main body. If this linkup cannot be made, they will exfiltrate enemy territory. They may use escape and evasion nets set up by other special operations forces or Department of Defense (DOD).

f. The ranger regiment normally places a liaison team at the headquarters of the command having OPCOM of a ranger force. This team provides command and control communications and liaison. It puts an intelligence liaison officer (LNO) at the EACIC or the appropriate ASPC (see Figure 7-2). This provides coordination to ensure that the intelligence needs of the ranger force are relayed to the intelligence processing center, and that the analysis is on the specific needs of the ranger force commander. This LNO coordinates with another located at the corps or EAC targeting center. Close coordination between the regiment, the intelligence analysis center, and the targeting center is essential to effective employment.

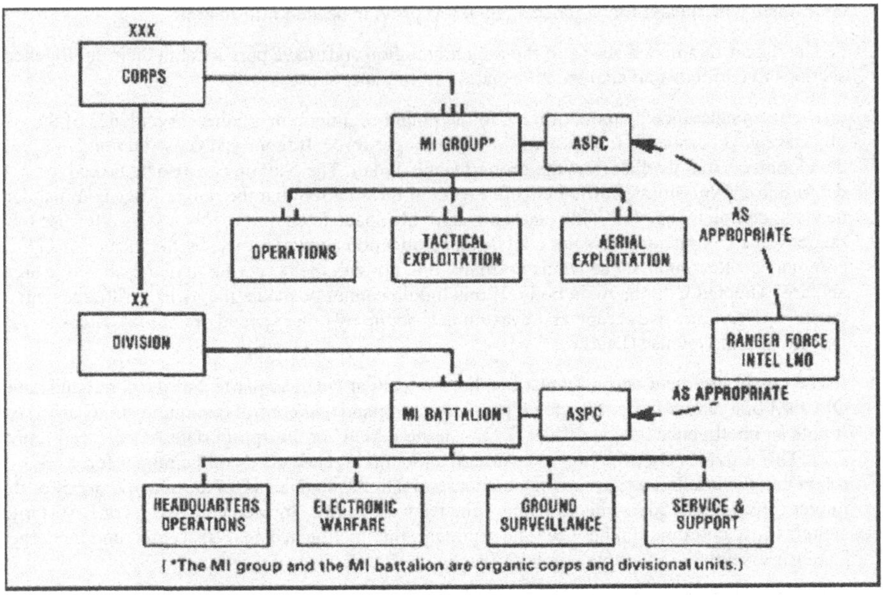

Figure 7-2. Intelligence source at corps and division.

7-2. Electronic warfare support.

a. Depending on the nature of the target and enemy abilities, a ranger mission may need support from EW units. This is mainly true during insertion. Electronic warfare operations disrupt or destroy the enemy's command and control of his forces and weapons systems, and retain friendly use of the electromagnetic spectrum. Electronic warfare is also used to support deception operations. They mislead the enemy by manipulation, distortion, or falsification of indicators, and persuade him to react in a manner prejudicial to his own interests.

b. Active jamming and chaff dispersal can prevent enemy early-warning radar from detecting the airborne force or determining its route. False transmissions can aid deception plans or feints used to help the ranger force's entry behind enemy lines. Selective jamming and imitative deception can disrupt the enemy command and control nets. This prevents the enemy from reacting to the presence of the ranger force in time to prevent mission accomplishment, or slows and interrupts his deployment of reaction forces.

c. Ranger elements, or other military forces, may conduct closely coordinated direct attacks against enemy EW sites during a ranger force's insertion or extraction. Not only does this reduce the enemy's ability to direct EW against the ranger force command, control, and communication system, but it misleads and confuses the enemy as to the true ranger objective.

d. The JTF or theater commander is responsible for planning the EW program to support a ranger mission. Planning and coordinating elements from the 1st SOCOM may also help. The EW program must be closely coordinated so as not to disrupt friendly air-to-ground or ground-to-ground communications, or to reveal the ranger force intentions.

e. The ranger regimental communications-electronics (CE) officer and those of the ranger battalions ensure that the required coordination concerning jammed frequencies, codes, authentication tables, and so on, is completed. The ranger regimental and battalion operations personnel ensure that the efforts of all EW units are combined to support the ground tactical plan.

7-3. Communications support.

a. Ranger operations are supported by secure long-range, lightweight, real-time, high-frequency (HF), and satellite communications. Effective long-range communications provide command and control links between deployed ranger units and the controlling headquarters. The JTF, ARFOR, or theater commander is responsible for communications between the controlling headquarters and the ranger regiment. The special operations force commander may also provide secure communications terminals to the ranger regiment or a deployed battalion.

b. Secure AM, FM, and SATCOM radios are the primary means of communication within the ranger regiment. Within the ranger battalion, AM and FM radios provide communications to company, platoon, and squad level.

c. During certain missions, specially trained and equipped quickreaction elements (QRE) deploy with the ranger force to provide secure communication links to the special operations force commander. These JCSE elements operate on either SATCOM or tactical satellite terminal (TACSAT) channels or through an airborne communications relay platform (see Figure 7-3). Depending on the mission, an airborne battlefield command and control center (ABCCC) aircraft or a joint airborne communications center/command post (JACC/CP) may be used. Their communications systems can operate at all levels of the national chain of command to permit a quick response to the tasking authority.

Figure 7-3. Airborne communications relay platform.

d. When only one ranger battalion is employed and the regimental headquarters is not the controlling headquarters, the ranger regiment provides a liaison cell to the controlling higher headquarters. This liaison cell includes a communications element from the regimental communications platoon. It can provide secure SATCOM, teletype, and facsimile transmission support to the ranger force in the objective area.

e. If two or more ranger battalions are employed, the ranger regimental headquarters deploys and acts as the command and control headquarters. The regimental communications platoon would then provide another communications link to the special operations force commander.

f. The ranger regimental CE officer ensures that the necessary communications links are set up and coordinated. The many communication means and channels available provide for effective control of a deployed ranger force. However, communication means must be closely coordinated at all levels of command to control the complex operations of a ranger mission. Planning and coordination with supporting aviation, transport, fire support, medical, and logistical elements before an operation are vital to efficient communications. The regimental CE officer must consider the communication systems linking the ranger force and other services. Air-to-ground and ship-to-shore communications are vital and must be set up early in ranger operations.

7-4. Fire support.

a. **Planning.**

(1) The ranger regiment has limited organic fire support assets. The six 60-mm mortars in each ranger battalion (two in each rifle company) provide the only organic indirect fire

available to the battalion commander. Each rifle company also has three 90-mm recoilless rifles and three Dragon missile systems that provide a small measure of organic direct fire support.

(2) The missions assigned to the ranger regiment and its subelements normally require fire support from organizations outside of the regiment. To plan, coordinate, and control these fires, the ranger regiment is organized with personnel dedicated to these tasks.

(3) The ranger regimental headquarters has a fire support element consisting of one fire support officer (FSO), two noncommissioned officers, and two fire support specialists. They prepare the fire support annex to the regiment's operation order. They ensure that all available fire support is planned to best support the regimental commander's ground tactical plan. They act as the primary coordinator for all external fire support, such as artillery, armed helicopter, close air support, naval gunfire, and short-range air defense. They also coordinate with each ranger battalion FSO to ensure unity of effort. The fire support element coordinates the efforts of air force, naval, and marine liaison officers when they are attached.

(4) The regimental commander may assign part of the fire support element to a committed ranger battalion as an attachment to help in the integration and planning of fires. The ranger battalions request through the fire support element that more or different fire support means be integrated into the ground tactical plan.

(5) The ranger battalion headquarters has a fire support team headquarters consisting of the battalion's FSO, two noncommissioned officers, and two fire support specialists. The FSO is the prime fire support coordinator for the ranger battalion commander. He maintains coordination with the regimental FSO and any attached fire support coordination elements, such as the USAF tactical air control party or naval shore fire control party. The FSO performs the following:

(a) Advises the ranger battalion commander on all fire support matters.

(b) Recommends allocation of fire support.

(c) Prepares fire support plans.

(d) Assigns target numbers.

(e) Processes target lists from the ranger company fire support teams to eliminate duplication.

(f) Monitors and functions as net control on the ranger battalion fire support coordination (FSCOORD) net.

(g) Reports changes in the status of fire support units to the ranger battalion commander, staff, and FISTS.

(6) When the ranger battalion is operating a single TOC, the FSO and his element are normally located there. When the ranger battalion is operating two TOCS, one TOC is normally manned by the FSO, the fire support sergeant, and a fire support specialist. The

senior fire support sergeant and the other fire support specialist would then monitor the FSCOORD net at the other TOC. The FSO ensures that any attached fire support elements also divide when needed to operate in a two-TOC configuration.

(7) The ranger battalion fire support team headquarters has three fire support teams assigned to it. These are normally allocated one to each of the ranger rifle companies. The company FSO and his team--

(a) Locate targets and request and adjust surface-to-surface fire support (mortar, field artillery, and naval gunfire).

(b) Plan fires to support the company ground tactical plan and prepare target lists.

(c) Coordinate fire support requests through the FSO for surface-to-surface and air-to-surface fires.

(d) Report battlefield information.

(e) Control air-to-surface fires of helicopters and prepare to assume control of fires from fixed-wing aircraft.

(f) Coordinate airspace use within the ranger rifle company operational area.

(g) Advise the ranger company commander on all fire support capabilities, limitations, and methods of employment.

(h) Inform all fire support units of target priority changes.

(i) Designate targets for laser-guided munitions.

(j) Assume operational control of, and administrative responsibility for, all augmentation fire support personnel (TACP, shore fire control party, and attack helicopter liaison teams).

(8) The company FSO is normally with the ranger rifle company commander. The FOs and their RATELOs are normally with the ranger rifle platoon leaders.

b. **Coordination.**

(1) Coordination of fires is a continuous process. It involves implementing the fire support plans of the ranger force commander and reacting to changing tactical situations. The success of a mission often hinges on how well fires are coordinated during the operation.

(2) The fire support planners within the many echelons of the ranger force become fire support coordinators during the actual execution of the operation. They use standard fire control coordination measures.

c. **Organic fire support.**

(1) **Indirect fires.**

(a) The 60-mortars constitute the only organic indirect fire support available within the ranger regiment. The mortar squads in each ranger rifle company normally operate as a section under the control of the mortar section sergeant. However, they can operate alone for short periods.

(b) The 60-mm mortar is used to destroy enemy equipment and light vehicles and to attack enemy personnel. Its main advantage is the ease with which it can be brought into operation against the enemy. This ease of employment allows the ranger force commander to concentrate combat power at the critical time and place. This helps to surprise and destroy the enemy.

(c) Mortar fires are employed by the ranger rifle company commander and coordinated by the company FSO. They are normally requested and adjusted by the FIST observers. The mortars are normally emplaced in a central position, and their fires are controlled by the fire direction center (FDC) using the indirect-fire method. The mortars may also use the direct-alignment or the direct-lay method.

(2) **Direct fires.**

(a) Each ranger battalion is authorized 90-mm recoilless rifles and Dragon missile launchers. These weapons constitute the heaviest organic direct-fire ability in the regiment. The antitank section in each ranger rifle company may operate as a complete section, but normally individual weapons are assigned throughout the company sector wherever their fires can best support the mission. These weapons can destroy most enemy armored vehicles and can penetrate the flank and rear armor of all enemy main battle tanks. The 90-mm recoilless rifle has a multiple projectile round that is effective against exposed enemy personnel. The HE round is highly effective against buildings and bunkers.

(b) The main disadvantage of direct-fire antiarmor weapons is their weight. These weapons, and their ammunition, are man-portable but place a strain on the unit when moving across rugged terrain.

(c) The ranger battalion is trained in the use of light antiarmor weapons, such as the US M72A2 LAW and the Warsaw Pact rocket-propelled grenade (RPG). The following provide protection for the ranger force against attack by enemy armored forces:

- The employment of light antiarmor weapons at all organizational levels.
- The placement of the 90-mm recoilless rifles or guided-missile systems on the most likely avenues of enemy-armor approach.
- An aggressive, offensively oriented employment.
- The ability to move into and across terrain impassable to armored vehicles.

d. **Field artillery fire support.**

(1) Many ranger unit operations are out of the range of supporting field artillery fires. Whenever such fires are available, they are planned for and integrated into the ranger force ground tactical plan.

(2) Coordination and exchange of call signs, frequencies, and target lists occur before insertion of the ranger force. Unless required by the tactical situation, ranger units do not normally receive augmentation forward observers from supporting field artillery units. The degree of mental and physical training necessary to execute certain ranger techniques requires more preparation than field artillery units are able to provide. The FIST from the ranger battalion requests and adjusts field artillery fires for the ranger force.

(3) Field artillery fires can be used to support the ranger force even if the objective area is out of range. Field artillery cannon fire and multiple rocket launch system (MRLS) fire can be used to suppress enemy air defenses. This helps the ranger force as it crosses the FEBA during airborne or air assault operations.

(4) Field artillery fires can be used to support the exflltration of ranger elements as they approach friendly lines. These fires can contribute to the deception plan and add combat power to feints used to support ranger operations.

(5) When in range of the objective area, field artillery units can be used to emplace field artillery delivered FASCAM to enhance the security of the ranger force.

(6) Copperhead rounds fired by 155-mm field artillery units can be terminally guided by the ranger FIST forward observers (FOs). They can attack hardened point targets or enemy armored vehicles by using a man-portable laser target designator (see Figure 7-4).

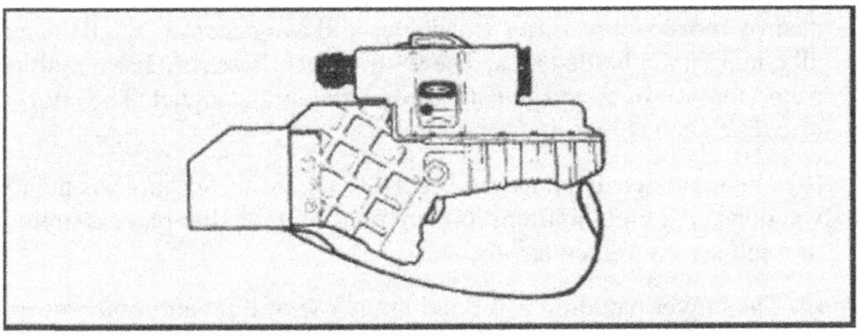

Figure 7-4. AN/PAQ-1 laser target designator.

e. **Aerial fires.**

(1) Aerial fire support is usually the prime means supporting the ranger force due to the distance behind enemy lines at which most ranger operations take place. Aerial fire support can be provided by either fixed-wing aircraft or helicopters.

(a) Fixed-wing aerial fire support may come from USAF, USN, or USMC units. The type of unit providing support, the aircraft, and the mix of ordnance carried all affect the fire support planning and coordination process. Some aircraft have a night and all-weather strike ability enabling them to support the ranger force during any level of visibility. Operations during poor weather that limits visibility to less than 3 nautical miles are still somewhat restricted. The ranger fire support coordinators must ensure that the correct aircraft are requested and employed effectively on the enemy. The

TACP directs and adjusts aerial fires in the objective area.

(b) The ranger force can use ground laser target designators to pinpoint targets for air strikes, as well as electronic navigation aids to permit nonvisual air strikes (beacon bombing). The ranger rifle company FIST or the TACP can control a laser-designated standoff air strike (see Figure 7-5).

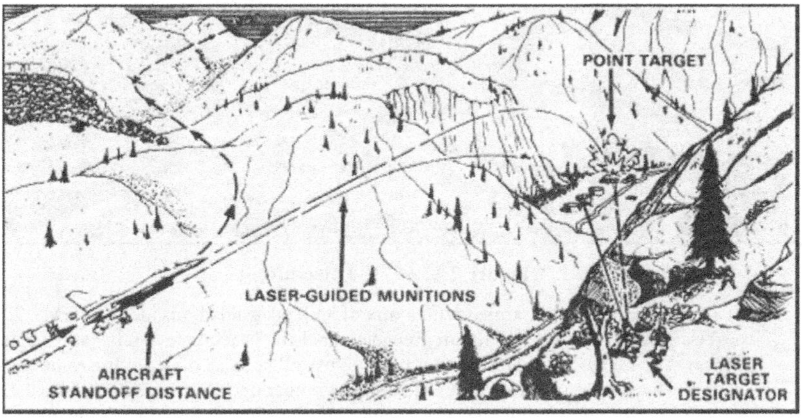

Figure 7-5. Standoff air strike.

(c) If the enemy ADA ability is not great or it can be degraded to a low level, the ranger force uses specially equipped and armed AC-130 aircraft (see Figure 7-6) for fire support. These aircraft provide an invaluable combination of firepower, night observation and illumination, communications, and long loiter time. A well-planned and executed suppression of enemy air defense (SEAD) program, coupled with ECM directed against enemy ADA units, normally permits the use of AC-130 aircraft.

Figure 7-6. AC-130 aircraft.

(2) The attack helicopter, armed with a mix of antitank guided missiles (ATGM), 2.75-inch rockets, 20-mm cannon, and 40mm grenade launchers, is accurate and responsive aerial fire support. The ranger force commander plans to use all sources of aerial fire to help accomplish his mission. However, the nature of ranger operations may preclude the extensive use of armed helicopters due to their limitations. Some attack helicopters are limited in range and lack an all-weather ability. They are restricted during operations at night by a lack of sophisticated night vision devices. They may lack the speed to go with the ranger force on some heliborne insertions. Attack helicopters may be used to escort and assist the ranger force as it crosses the FEBA. They may also be used to conduct feints and demonstrations to cover the insertion of the ranger force. As the MV-22 Osprey and the AH-64 Apache enter service, more use will be made of these aircraft to support ranger operations beyond the FERA, (See Figures 7-7 and 7-8).

Figure 7-7. MV-22 Osprey.

Figure 7-8. AH-64 Apache attack helicopter armed with Hellfire missiles.

(a) If attack helicopters are used to support a ranger operation, planned indirect fires are normally delivered along entry and exit corridors. Attack helicopters approach and depart the objective area using nap-of-the-earth flight profiles.

(b) Fires from armed helicopters are normally requested and controlled by the company FSO or one of his FOs, operating on a special ground-to-air net. The laser target designator may be used to precisely identify targets for the AH-64 Apache. Friendly unit locations may be marked by smoke, panels, lights, mirrors, or infrared sources.

f. **Naval gunfire.**

(1) During amphibious assaults or operations near a coastline, the ranger force may receive indirect fire support from naval gunfire. Normally, a ranger battalion would be supported by either a destroyer or a cruiser in a direct support role. However, the type and importance of the mission, the type of targets, the ships available, the hydrographic conditions, and the enemy capability determine how many and which type of ships are provided to support the ranger force.

(2) Naval gunfire is characterized by large volumes of highly destructive, flat-trajectory fire. Planned strikes in support of the ranger force may also include surface-to-surface missile fire. Some naval guns can fire a laser-guided projectile much like the Copperhead. This projectile, called a Deadeye round, is terminally guided the same as the Copperhead. Deadeyes can be guided either by an attached spotting party from the ANGLICO or by ranger LTD teams.

(3) When ranger units are being supported by naval gunfire, a shore fire control party is attached to the ranger force, This party normally consists of a liaison team and several spotter teams (forward observers). The liaison team is integrated into the operations of the fire support element (FSE) at the ranger battalion TOC. The spotter teams are attached to the ranger rifle companies. The shore fire control party LNO is the ship's representative to the ranger force commander, through the FSO. The spotter teams request and adjust fires from surface vessels. They can also request and control air strikes by carrier-based aircraft.

(4) The LNO and the spotter teams operate in the ground spot net, communicating with the ship by high frequency (HF) radio to request and adjust naval gunfire. The spotter teams communicate with the LNO using very high frequency (VHF) radios. The LNO also can communicate with aircraft using ultrahigh frequency (UHF) radios.

(5) Coordination and control measures that apply to naval gunfire are the same as for field artillery, with the addition of two terms, which are:

(a) Fire support area (FSA)--a sea area within which a ship may position or cruise while firing in support. It is labeled as FSA and numbered by a Roman numeral - for example, FSA VII.

(b) Fire support station (FSS)--a specified position at sea from which a ship must fire. This is very restrictive positioning guidance. It is labeled as FSS with a Roman numeral--for example, FSS VII.

g. **Air defense artillery.**

(1) Army air defense artillery units are rarely used to directly support the ranger force since

ranger operations normally take place deep behind enemy lines. However, it is possible that during the insertion or extraction phase of an operation, ADA units may support the ranger force as it crosses the FEBA.

(2) The ranger regiment's fire coordination element and the FSO in the ranger battalion are mainly concerned with adequate coordination of ADA fires. This prevents friendly units from being engaged. It allows the integration of ADA fires into the scheme of maneuver. The fire support planners ensure that enough coordination has been made so that ADA units are aware of the ranger force and know enough about its flight paths, routes, and altitudes to prevent engagement by friendly units. When conducting this coordination, care must be taken to pass only needed information to the ADA unit. To aid coordination, an ADA officer may be attached to the ranger force.

(3) The plan for a heliborne or airborne insertion and extraction of the ranger force must include the correct identification-friend-or-foe (IFF) codes, the flight profiles, the times of FEBA crossings, and the estimated number of aircraft making the crossing. There must also be a plan for crossing by straggler aircraft and battle-damaged aircraft that cannot reply to IFF interrogation. If possible, the routes selected should avoid concentrations of friendly ADA units.

(4) During an aerial crossing of the FEBA by a ranger force, the regimental headquarters may assign a liaison officer to the headquarters controlling the ADA fires in the crossing area. The LNO is then responsible for close and continual coordination. This ensures that the crossing is made without friendly ADA units engaging the ranger force while it is exiting or entering friendly lines.

(5) During ranger operations, the primary source of air defense fires is shoulder-fired antiaircraft missiles carried by specially trained individual rangers. These weapons provide the ranger force commander a limited defense against attack by enemy aircraft. The decision to take these weapons into the objective area depends on the factors of METT-T. Ranger operations should be completed before the enemy can react with ground forces. However, enemy aircraft may be able to react in time to threaten the ranger force. These lightweight missiles offer a highly effective, short-range counter to enemy air attack.

(6) The ranger force commander must define the weapons control status and rules of engagement for all short-range air defense (SHORAD) teams under his control. Priorities of target engagement must also be determined to prevent multiple engagements of a single enemy aircraft. The ranger battalion FSO and the attached ADA officer are responsible for the coordination of ADA fires within the objective area. Standard ADA weapons-control status and rules of engagement may be modified by the ranger force command.

7-5. Aviation support.

a. **Planning.**

(1) The ranger regiment may need extensive aviation support on some missions. The ranger battalion is organized and trained to move quickly to the objective area by air and to be resupplied by airdrop for short periods.

(2) Often the decision to execute a mission depends on the amount and type of available insertion and extraction aviation support. Normally, the theater commander or JTF commander allocates the type and amount of aviation assets to support a ranger operation. The theater or JTF commander's staff is responsible for the planning and coordination concerning aviation support. The ranger force commander must play an active part to ensure mission success.

(3) Due to the extensive training requirements and specialized operational techniques needed to support special operations, special air operations units in the Army and the USAF have been equipped and trained for the mission. These forces train with the ranger regiment and jointly develop tactics, techniques, and procedures of employment. These special units operate diverse aircraft that have unique capabilities. These help the ranger force commander to effectively employ his units. Command of these special units is maintained at a very high level in the chain of command. They are assigned to support a specific operation. Then they return to their parent unit's control.

b. **Coordination.** In addition to the staff of the theater commander or JTF commander, the coordination of aviation support occurs at all levels in the ranger force. These operations often are conducted in phases. A separate air mission commander is then designated. The AMC is responsible for the completion of each phase of the aviation mission, as well as the close and continuous coordination with the ranger force, commander. The ranger regimental air operations officer and the movements control officer are responsible for the planning, integrating, and controlling of aviation support.

c. **US Air Force.**

(1) The USAF provides special aviation support to the ranger regiment and other elements of the special operations force. The USAF has elements that are equipped with MC-130 Combat Talon aircraft (see Figure 7-9) and AC-130 Spectre gunships, as well as HH-53 Pave Low, UH-60, and UH-1 helicopters. These aircraft are specially equipped to conduct long-range infiltration, exfiltration, attack, and search-and rescue operations. The USAF also provides C-141B and C-130 aircraft for strategic and tactical airlift of the ranger force. There are specially trained USAF crews that can operate using AWADS and SKE. This gives them an excellent means of inserting the ranger force during reduced visibility.

Figure 7-9. MC-130 Combat Talon aircraft.

(2) The AFFOR commander provides the airlift control element (ALCE) support at the departure airfield. The ranger air operations officer ensures that coordination of aircraft load limits, air movement times, specific load plans, and manifests is conducted in an effective and timely manner. The air operations officer and the movements control officer ensure that a departure airfield control officer (DACO) from the deploying unit is designated to provide face-to-face coordination at the departure airfield.

(3) The USAF normally provides an airborne battle command and control craft. This is a specially equipped C-130-type aircraft that provides communications and data system down-links to ground stations.

(4) The terminal guidance needed for assault landings and air traffic control in the objective area is provided by a specially trained and equipped CCT. This element has habitual training and operations relationships with the ranger units. The USAF is also responsible for the specially trained and equipped weather teams that can be sent into the objective area with the CCT.

(5) The USAF Tactical Air Command (TAC) includes fighter, interceptor, ground attack, electronic countermeasure, early warning, reconnaissance, and airborne command and control squadrons, any of which may be used to support the ranger force. The Tactical Air Command provides the results of aerial reconnaissance to the AFFOR, who then processes it for use by the theater or JTF commander's staff. This reconnaissance may be by low-level photo, imaging infrared, side-looking airborne radar (SLAR), low-light-level television, or overhead photography.

(6) The Military Airlift Command (MAC) equips and trains special long-range SAR aircraft and crews. These aircraft, mainly helicopters, are equipped with sophisticated navigation, observation, and communications equipment, as well as extended-range fuel tanks and in-flight refueling capabilities (see Figure 7-10). The ranger regiment uses these aircraft for certain missions.

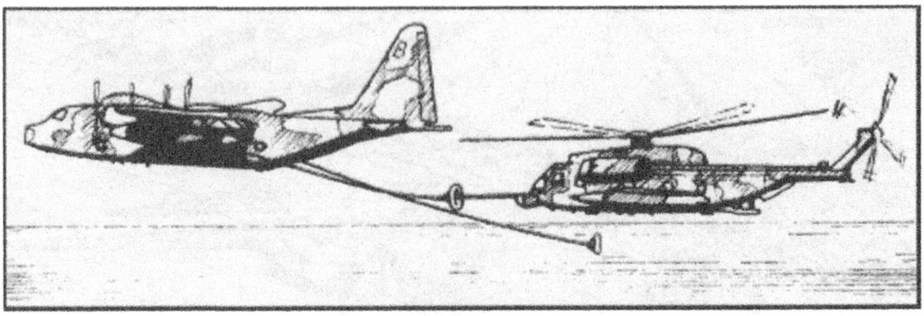

Figure 7-10. In-flight refueling.

d. **US Navy and US Marine Corps.** Both the USN and the USMC have units equipped and trained to support ground forces. The type of aviation support normally provided to the ranger regiment is close air support against targets in the objective area, or as suppression missions against enemy air defense installations. Long-range helicopters from the USMC may be used to support an amphibious operation or to insert special teams during ranger operations. The USN also has a search-and-rescue ability that may be used to support the ranger regiment.

e. **Special operations aviation units.**

(1) The US Army has specially equipped and trained SOA units tasked to support the ranger regiment, as well as other elements of the special operations force. These SOA units have a habitual training association with the ranger regiment and use dedicated aviation assets to -

(a) Insert, extract, and resupply the special operations force.

(b) Conduct armed escort, reconnaissance, surveillance, and electronic warfare in support of the special operations force missions.

(c) Provide airborne command, control, and communications for special operations force elements.

(d) Provide general support aviation during peacetime and contingency operations.

(2) Penetration of enemy airspace is normally done during limited visibility at very low altitudes. Penetrations are planned to maximize the use of existing terrain and exploit blind spots in enemy air defenses. On-board navigational aids are crucial to mission success. There normally is not any terminal guidance to the selected LZ or DZ. Deception operations of a tactical or operational nature help SOA units penetrate to the objective and reenter friendly lines. Advanced helicopters and short-takeoff-and-landing aircraft provide the help needed to do these missions.

(3) To survive enemy air defense systems and prevent mission compromise, SOA units avoid detection. Detected aircraft may reveal the location of supported special operations force ground forces or may be destroyed by enemy ADA fires. The SOA normally avoids detection by operating at night without illumination. This condition requires flight using sophisticated night vision devices. The SOA also operates in other conditions of reduced

visibility, such as clouds, fog, rain, snow, smoke, and dust storms.

(4) Before an operation, the ranger force commander may need to place reconnaissance teams near the objective. Army SOA assets can be used to transport ranger reconnaissance teams.

(5) Elements of SOA provide the ranger regiment with short and intermediate range penetration, resupply, and extraction capabilities. Long-range aircraft must be provided from USAF special operations assets.

(6) The ranger regimental or battalion commander is normally provided a liaison aircraft while the ranger force is in the REMAB. This ensures that he can travel quickly to the dispersed headquarters of the agencies providing support for the mission. This aircraft and crew are provided by the theater or JTF commander from EAC general support aviation assets. It may be a helicopter or fixed-wing aircraft, depending on the location of the REMAB to a suitable landing field.

(7) Army helicopters and fixed-wing aircraft may be used to provide SEMA aerial reconnaissance; radar, infrared and photographic imagery; and electronic intelligence support to the ranger force commander. This support is normally provided by the aerial exploitation battalion of the theater Army military intelligence group.

7-6. Engineer support.

a. The ranger regimental and battalion operations officers plan and request engineer support for ranger operations.

b. The most common engineer-type tasks needed during a ranger operation are the destruction of enemy facilities, equipment, and material by use of conventional explosives; and the interdiction of major lines of communication. To reduce the need for external support, the ranger regiment trains selected individuals in advanced demolition techniques. All members of the ranger regiment are well trained in the basic employment of conventional demolitions. Therefore, there is no need to attach combat engineers to conduct normal demolition missions. If specialized skill or knowledge is needed for a specific mission, qualified personnel can be attached.

c. During the occupation of the REMAB and the rehearsal for an operation, the ranger regiment needs construction engineer support to build simulated target areas. The simulated target areas should be camouflaged to prevent the enemy from recognizing the target. The construction engineers augment the ranger unit's self-help construction efforts. Limited-facilities construction and improvement may also be needed at the REMAB to provide security and isolation. The theater commander or JTF commander provides the construction engineer support.

d. Certain missions, such as those of interdiction, may need engineer special weapons teams. The teams are provided by the theater commander or JTF commander, or they may come from other special operations force assets. They are attached to the ranger force for a specific mission and released at the completion of the extraction and debriefing phase.

e. Some operations may need the ranger force to use captured enemy heavy construction equipment. Selected members of the ranger regiment are trained to operate such equipment.

However, construction engineer personnel may be attached to the ranger force for this mission.

CHAPTER 8

Combat Service Support

Combat service support consists of the logistical and administrative effort to maintain the battalion's ability to fight. The commander forecasts his needs for CSS. He employs CSS units to ensure accomplishment of the mission.

8-1. Ranger support element.

a. The ranger support element is a non-TOE organization formed from elements of CSS units collocated with CONUS-based ranger units augmented by selected personnel from the host installation. The mission of the RSE is to provide all the CSS needed to outload and deploy a ranger unit from CONUS. The personnel designated as part of the RSE have the same alert notification, reaction, and security requirements as members of the ranger unit they support. The RSE is normally built around a command and control group provided by a major CSS unit on the ranger unit's host installation. The remaining CSS elements and teams are then designated. An RSE consists of--

(1) An emergency operations center liaison group that provides timely coordination between the ranger unit and the functional elements of the RSE.

(2) A command and control group that coordinates and directs the actions of the functional elements.

(3) A supply section responsible for the timely issue of all Class II, III, IV, and VII items. This section maintains a stockage of contingency items marked for use by the resident ranger unit. It also issues items out of common stocks.

(4) A transportation section moves equipment and personnel from the unit area to the departure airfield and picks up equipment from local storage sites. It may also operate material-handling equipment and trailers to support loading of aircraft.

(5) An ammunition section responsible for the requisition, storage, issue, loading, and security of the ranger unit's basic load(s) of conventional munitions.

(6) A mess section provides all Class I support to the ranger unit. This includes operating a dining facility, providing food to remote sites, and drawing and issuing operational rations.

(7) A load team loads and unloads all equipment and supplies for the ranger unit. This team prepares pallets for loading, documents hazardous or dangerous cargo, and loads equipment pallets on USAF aircraft. It can also provide limited security for loads.

(8) A communications and electronics section provides secure communications among the RSE subelements, ranger unit, EOC, and DACG.

(9) An engineer platoon that is responsible for the construction of targets and rehearsal areas. This element of the RSE may or may not be activated, depending on the mission given the ranger unit. If the ranger unit deploys from CONUS to the objective area, this platoon has a key role to play in the premission preparation.

(10) A maintenance team whose primary mission is to ensure that the vehicles used by the RSE remain operational. It also provides predeployment maintenance and Class IX support for ranger unit vehicles and equipment.

(11) A medical team that provides unit-level medical care while the organic ranger medical section completes its outloading preparations. This team also provides any Class VIII supplies that the ranger unit may need.

(12) A security team responsible for security at the EOC, ranger unit location, and departure airfield. This team enforces stringent physical security precautions and maintains a constant counterintelligence (CI) surveillance of the outloading procedure. The security team provides armed guards for the protection of ammunition convoys, equipment movement, and vehicles.

(13) A rigger detachment that maintains, packs, rotates, and issues the personnel and equipment parachutes required by the ranger unit. It maintains enough stock of personnel parachutes to conduct a battalion-level parachute assault. The riggers lend technical skill to the rigging and aerial delivery of bulky, unusual, or delicate equipment.

b. The primary mission of the RSE is to provide the support necessary to quickly load and deploy the ranger units. The theater commander provides CSS to the ranger force once it has arrived at the OCONUS REMAB or ISB. The transition of support from the RSE to the CSS assets of the theater commander may be delayed for certain subelements in order to ensure a smooth deployment and enhance mission accomplishment. These subelements are released when the in-theater CSS assets can provide the support. Retained subelements of the RSE may be released from support requirements all at once, released in increments, directed to continue to provide support until mission completion, or augmented by in-theater CSS assets. The maintenance, medical, security, and rigger elements are those most likely to continue providing support OCONUS. This is likely if deployment is to an austere ISB pending immediate employment.

c. Once a ranger unit has been deployed OCONUS and has closed to a REMAB or ISB, the theater commander's staff coordinates CSS. If 1st SOCOM elements have deployed to the theater or operational area, the special operations support battalion (SOSB) of the 1st SOCOM coordinates CSS. Ranger logisticians at regiment and battalion level are mainly planners with limited logistics operations ability. Ranger logistical and personnel officers coordinate with CSS elements who provide such services.

d. The Theater Army Materiel Management Center (TAMMC) is the point of contact for information concerning in-theater logistical assets. In a developed theater of operations, a series of theater Army support groups will have been set up. One of these groups is assigned the mission of supporting the ranger force. The 1st SOCOM may also deploy certain assets available to provide CSS to the ranger force. The many CSS sources, the rapidly changing nature of SOF operations, and the unique CSS requirements of ranger units combine to make providing logistical and personnel support a challenge. The ranger unit's logistics officer must be included in the early phases of planning and deployment.

8-2. Logistics and personnel.

a. The echelons-above-corps CSS structure provides support to the ranger regiment as directed by the theater Army (TA) commander. In a low-intensity contingency involving a corps or less, there may be no echelons-above-corps logistics and personnel support elements. The corps support command (COSCOM) involved may be able to meet all needs from its own sources. However, in a combined operation in a high-intensity environment, the theater commander assembles a large CSS organization to provide support to the TA. This CSS base normally provides for all the ranger unit's requirements. During contingency operations in an area where there is no TA, the JTF commander provides the CSS elements. Until the buildup of these elements is complete, the ranger regiment must be prepared to operate from austere facilities with only the supplies it brings from CONUS.

b. The ranger regiment is normally assigned to the theater command for CSS. The TA commander, as directed by the theater commander, provides CSS to the ranger regiment and sets up support priorities. The EAC CSS organizations are under the command of the TA commander and consist of subordinate area and functional commands. The TA plans, coordinates, and executes its CSS functions through area-oriented support commands and specialized subordinate organizations. During the early stages of conflict, the TA may be organized as shown in Figure 8-1.

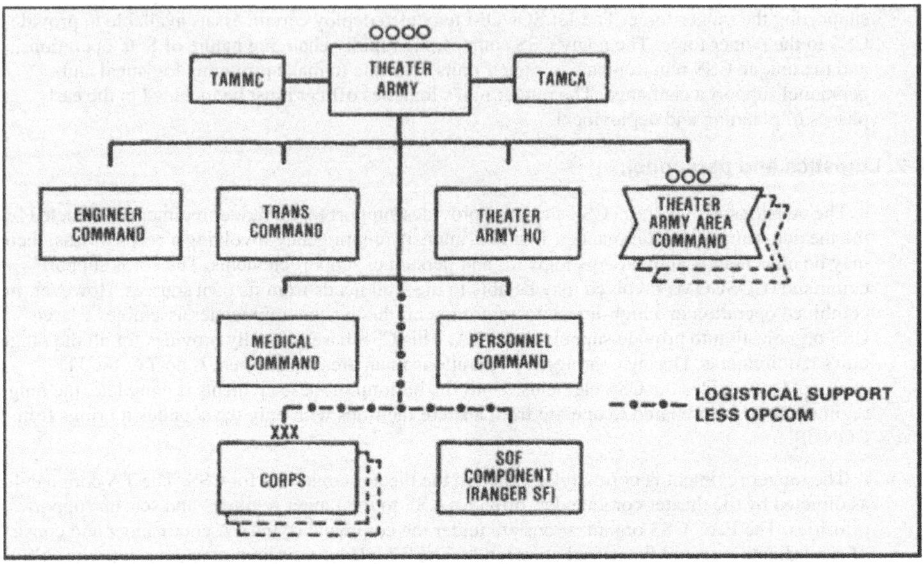

Figure 8-1. Ranger force relationship to theater Army.

c. The Theater Army Area Command (TAACOM) is the key CSS operator in the communications zone. It functions as a major subordinate command under the theater Army. The primary mission of the TAACOM is to provide direct support CSS to units such as the ranger regiment located in or passing through its assigned area. While located in an OCONUS REMAB, the ranger regiment receives most of its logistical support from the TAACOM.

d. The TAACOM area support group (ASG), which is subordinate to the TAACOM, is structured with operating units to provide direct support CSS to units located in or passing through its assigned geographical area of responsibility (see Figure 8-2). Area support groups are usually located along the lines of communication. The ASG is the prime source of CSS for the ranger regiment. It contains the units that actually provide services requested by the ranger unit logistics officer.

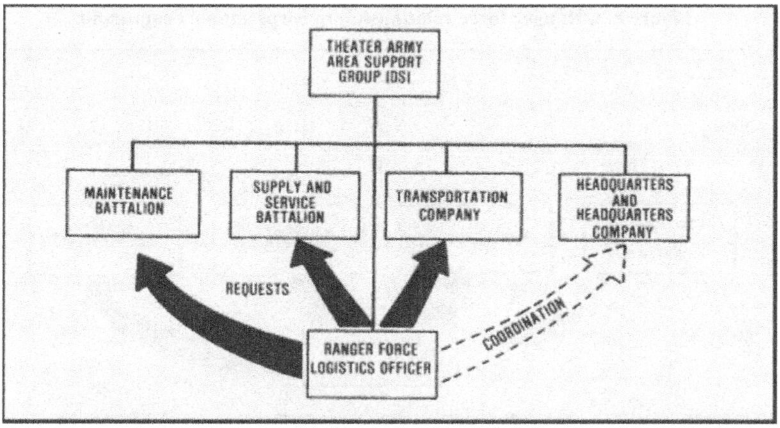

Figure 8-2. Ranger force relationship to theater Army area support group.

e. The ranger regiment is not normally assigned to a corps for employment. However, it may be placed under a corps' operational command for a specific mission. The ranger unit logistics and personnel officers must coordinate with the COSCOM that would then be providing CSS. (See Figure 8-3.) Corps support groups are major subordinate elements of the COSCOM. They provide command and control of COSCOM companies and battalions, which provide direct support (DS) and general support (GS) supply and field services to nondivisional units, such as the ranger battalion. (See Figure 8-4.)

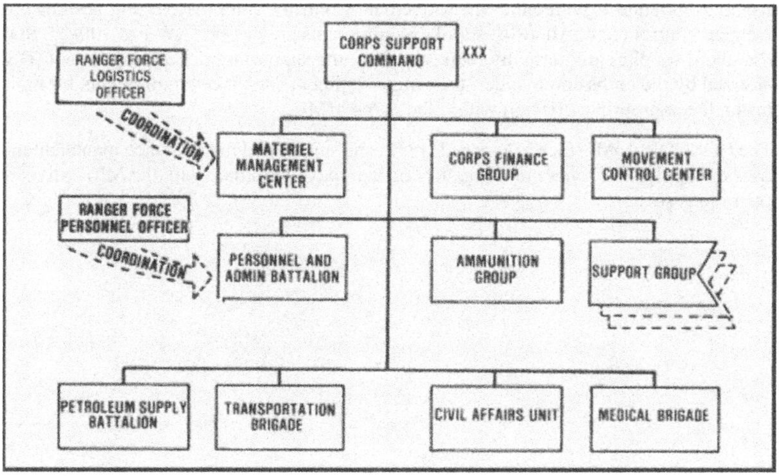

Figure 8-3. Ranger force relationship to corps support command.

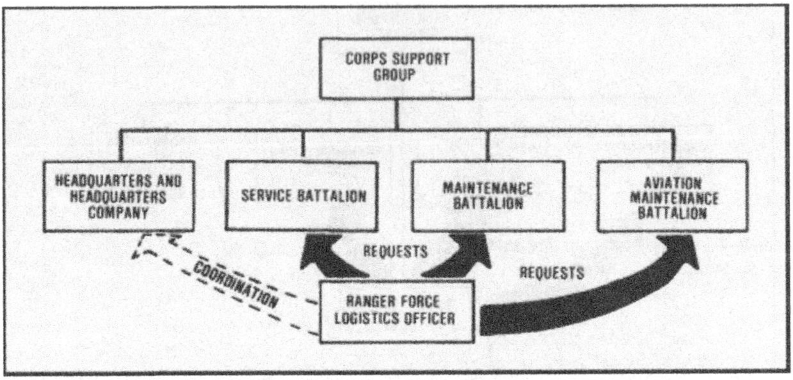

Figure 8-4. Ranger force relationship to corps support group.

f. Normally, the ranger regiment receives its logistical support from the corps support group responsible for the geographical area in which the REMAB or ISB is located. The corps support group provides all supplies except Classes V and VIII. The medical brigade provides Class VIII supply and medical maintenance support. The ammunition group provides Class V support. Financial and personnel services are provided by the corps finance group and the personnel and administration battalion, respectively.

g. Transportation needs are met by the transportation brigade. The ranger unit places ground transportation requests with either the Movement Control Center (MCC) or a designated movement control team. All airlift supply requirements are passed IAW FM 100-27. Standard and nonstandard supplies are normally requested from the supporting direct support unit (DSU). If authorized by the corps commander, the ranger regiment can place requirements for nonstandard items on the appropriate division within the corps MMC.

h. The COSCOM MMC (see Figure 8-5) performs supply and maintenance management for all classes of supply. The ranger unit logistics officer may coordinate with the MMC division chiefs or operations section.

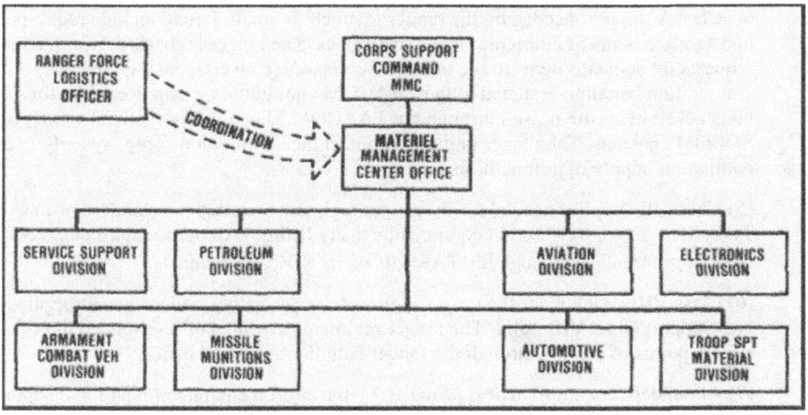

Figure 8-5. Ranger force relationship to COSCOM MMC.

i. Critical logistics operations, such as those supporting ranger and other SOF elements, need intensive management. They may be assigned to special staff sections that act as special operations force logistics managers. If designated, a logistics manager for special operations force elements would be a major point of contact for logistical support for the ranger regiment.

j. The ranger regiment may also receive logistical support from the SOSB. Normally, the SOSB provides administrative and logistical support to the Headquarters, Army Special Operations Forces (ARSOF). It complements the support provided by the Army component in the theater of operations by coordinating, augmenting, or assisting CSS forces. The SOSB provides unique quick-reaction support to the ARSOF or deploying SOF elements.

k. The ranger regiment does not make inordinate demands on the logistical system. However, a ranger unit may need certain nonstandard or low-density equipment. The concurrent need for nonstandard Class IX items requires management by logisticians to ensure support.

(1) Ranger battalion mess sections are responsible for Class I operations at the REMAB. The ranger regimental headquarters, which has no mess section, either collocates with one of the battalions or is supported by a mess facility operated by a nearby unit. It may share this facility on an area basis. The ranger unit personnel officer coordinates with the supporting DSU, which initiates the issue of Class I supplies.

(2) Class II, IV, and IX support to the ranger force is normally not of a great volume. It may consist of some nonstandard items. Depending on the need, the item, and the security, the SOSB may get some nonstandard items directly.

(3) Class III support needed by the ranger regiment is small and limited to standard items. The normal source of supply for Class III is the supply and service company providing area support.

(4) Class V support needed by the ranger regiment is small. It may include requests for nonstandard items of ammunition and explosives. The ranger logistics officer passes requests for standard items to the ordnance company (conventional ammunition) of the ammunition battalion assigned to the TAACOM ammunition group. Requests for nonstocked items are passed through the TAACOM Materiel Management Center or 1st SOCOM channels. The ranger unit must contact the ammunition company early to ensure a continuous supply of ammunition.

(5) Class VII supplies needed for the ranger regiment are often nonstandard or in excess of TOE. (See Appendix C for an equipment density listing.) All nonstandard or excess requests must be coordinated through the TAMMC or 1st SOCOM channels.

(6) Class VIII supplies for the ranger regiment are normally standard items supplied by the area support Class VIII point. The ranger regiment personnel officer sets up the Class VIII supply points of contact through the ranger battalion's medical officer.

(7) Aerial delivery equipment is provided by the airdrop equipment repair and supply company of the theater Army support group (TASG) (GS). There is normally one such company for each TAACOM. This unit supplies and repairs aerial delivery equipment. It does not pack and rig supplies for aerial delivery. The ranger regiment's requirements for aerial delivery supplies normally consist of standard items. Certain low-density items, such as HALO or HAHO parachute equipment, may also be provided through 1st SOCOM channels.

(8) The supporting motor transportation group of the transportation command is responsible for unit movement of the ranger regiment. The ranger regiment does not have any organic transportation assets. It depends on outside support whenever motor movement is needed.

8-3. Resupply.

a. Resupply operations for ranger units are normally covered during the planning and coordination phase of the mission. Ranger units try to complete their mission quickly and then be extracted or exfiltrate from the objective area. Ranger units carry into the objective area the equipment and munitions they need. They do not normally remain long enough to need extensive resupply. Certain operations may require the use of bulky or heavy items of equipment that cannot be carried by personnel. If the ranger force is airlanded on or near the objective, one or more of the aircraft may contain palletized supplies. These can be quickly off-loaded for future use.

b. If the ranger force is inserted into the objective area by parachute assault, aircraft can also drop resupply loads either just before or just after the personnel drop. These loads might include extra water, ammunition, demolitions, or barrier material. During peacetime, preloaded airdrop pallets are stored at CONUS depots for such contingency use. These mixed loads can be quickly moved to a departure airfield and inserted along with, or following, the ranger force. By dropping a resupply of water, food, equipment, and munitions with the ranger force insertion, the heavy-drop aircraft do not have to conduct a separate penetration of the enemy air defense zone. The ECM and SEAD programs used in the insertion of the personnel aircraft mask the penetration of the heavy-drop aircraft.

c. The aircraft dropping ranger personnel can also carry door bundles to deliver bulky or heavy items to the objective area. These door bundles, normally made up of A-7 and A-21 containers, may be color-coded to identify different loads or units. The most common door bundle loads are such items as 90-mm recoilless rifles and ammunition, 60-mm mortar ammunition, Dragon missiles, SHORAD missiles, or more demolition material. These bundles are pushed from the personnel aircraft just before the lead jumper's exit. They may be marked with lights or infrared sources for locating them at night.

d. If the tactical situation requires that a ranger force receive a resupply during the course of an operation, the most common method would be by airdrop. Airdrop could be needed if the operation takes longer than planned, initial supplies are lost, or changes occur in the operational plan. The ranger unit logistics officer and operations officers normally plan for resupply needs and schedule follow up loads of Classes I, II, IV, V, and VIII. There are two different airdrop requests channels for the ranger force to use:

(1) Nonemergency airdrop requests are sent through logistics channels. The ranger LOC sends requests for the airdrop of supplies and equipment to its supporting supply unit, and the supporting unit forwards the requests to the COSCOM or TAACOM MMC. The MMC coordinates with the movement control center (MCC). Jointly they direct the actions of the supply and transportation units. The MCC sends the request to the joint transportation board (JTB). The board validates it and assigns it a priority. The JTB then sends the request to the airlift control center for action. If the items requested are not on hand in the airdrop company, the MMC directs the proper supply activity to prepare the items for shipment to the airdrop unit. The MMC coordinates with the MCC, and the MCC ships the items. The airdrop unit prepares the supplies for the mode of delivery to be used. It also makes sure the supplies are delivered to the air terminal at the right time. (See Figure 8-6.)

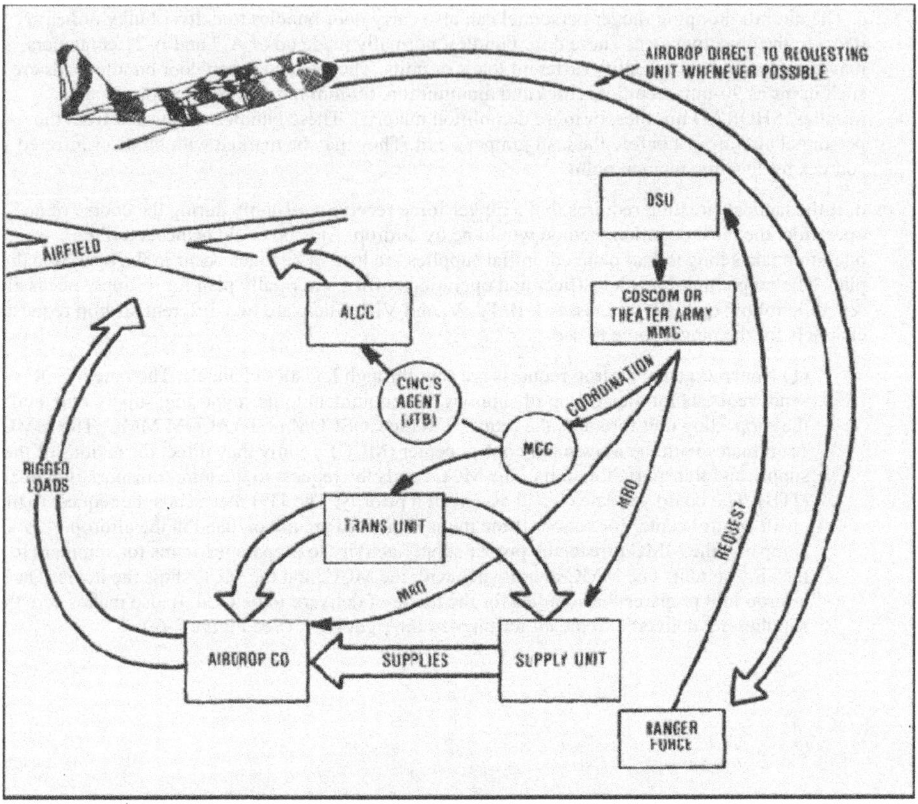

Figure 8-6. Ranger requests for nonemergency airdrop.

(2) Emergency requests for airdrop are passed through command channels in the quickest, most secure way.

(a) The request is usually sent to the command level that has authority to approve it, usually the tactical operations center at corps or above. The Air Force airlift officer at the direct air support center (DASC) coordinates with the Assistant Chief of Staff, Operations (G3) at the tactical operations center to determine whether airlift resources can be used for the mission.

(b) The TOC sends a request for an immediate airlift to the joint force commander's (JFC) designated agent for validation. At the same time, the airlift liaison officer advises the airlift control center (ALCC) that a request is coming through channels so plans can be made. When the ALCC receives the validated request from the JFC agent, it diverts or cancels less essential missions and provides airlift support. The TOC sends the request for supplies or equipment to the MMC.

(c) The MMC and MCC jointly direct the actions of the supply and transportation elements. The airdrop company stocks Classes I, III, and V supplies. It holds these supplies for emergency missions. If the supplies and equipment requested are not on hand in the airdrop company, the MMC directs the proper supply activity to prepare the items for shipment to the airdrop unit. The MMC coordinates with the MCC to ship the items. The airlift is arranged by the ALCC.

(d) The airdrop unit prepares the supplies according to the mode of delivery to be used. It coordinates with the MCC and delivers the supplies to the designated point at the air terminal (usually the cargo aircraft ramp). (See Figure 8-7.)

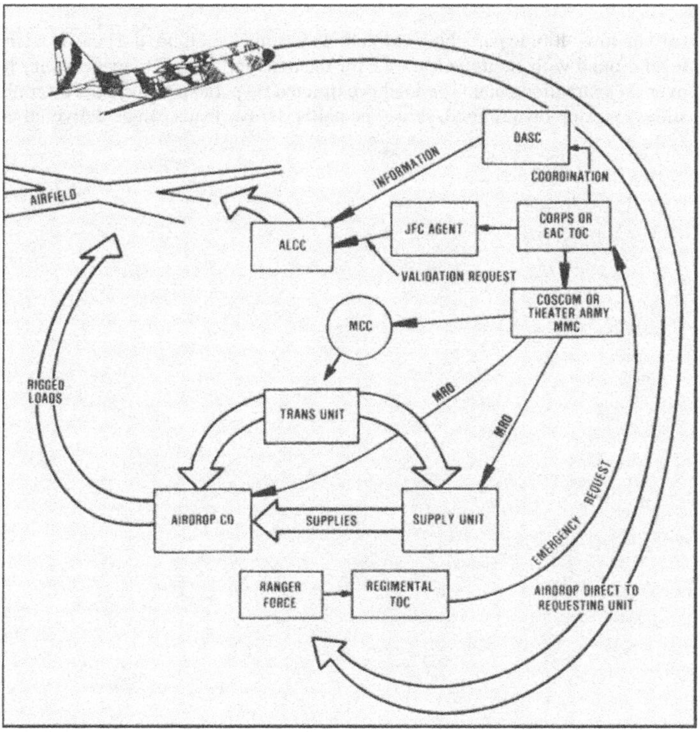

Figure 8-7. Request channels for emergency airdrop.

e. Aircraft delivering airdrop resupply deep behind enemy lines need to take precautions to avoid enemy detection and antiaircraft fire. The USAF is responsible for most of these precautions, such as ECM, suppression of enemy air defense (SEAD), and WILD WEASAL support. These do not concern the ranger commander. The choice of the type of parachute delivery system to use, however, may affect the ground commander's plan. The ranger force commander must specify any tactical restrictions that would require a specific type of delivery system. The ranger LOC and regimental logistics officer must advise the ranger force commander on the type of parachute system to be used. Conflicts must be quickly resolved so as not to interfere with the tactical operations.

(1) The safest way for the airdrop aircraft to penetrate enemy air defense and remain undetected is often by use of very low-level flight. There are four low-level parachute delivery systems that can be used without the aircraft being required to climb to a higher drop altitude and risk detection by enemy ADA.

(a) The low-altitude parachute extraction system (see Figure 8-8) needs a length of level ground with enough clearance for the delivery aircraft to make a very low pass over the extraction zone. The load is extracted by parachute from the aircraft and slides to a stop on a rugged, skid-type pallet. Heavy loads can be delivered using this system.

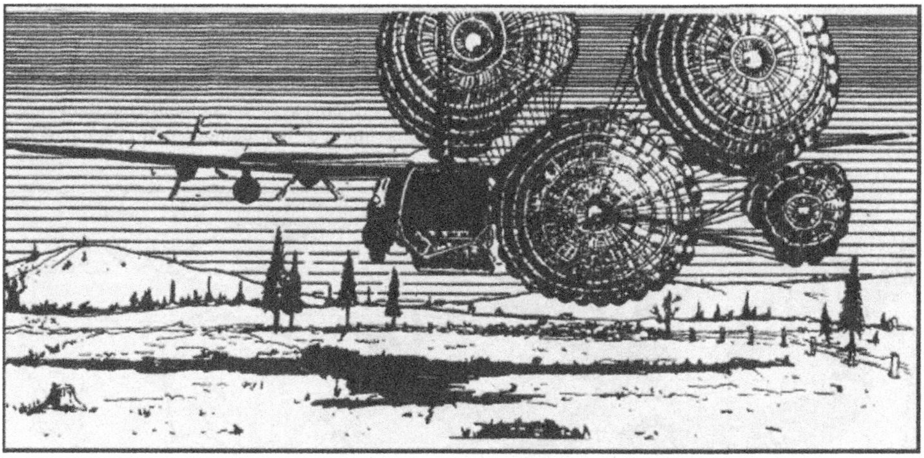

Figure 8-8. Low-altitude parachute extraction system.

(a) The containerized delivery system (CDS) (see Figure 8-9) consists of multiple, individually rigged A-22 containers, each with its own parachute, weighing up to 2,000 pounds each. The system provides single-pass delivery of up to 16 containers by C-130 aircraft and up to 28 containers by C-141 aircraft. The CDS loads may be delivered into drop zones using multiple points of impact to allow for tactical separation. The ranger force commander can direct the use of multiple points of impact, noting the advantages to be gained against the requirement for multiple passes.

Figure 8-9. Containerized delivery system.

(c) The high-speed, low-level, airdrop system (HSLLADS) (see Figure 8-10) is a single A-21 container specially rigged to withstand the parachute opening shock when airdropped from C-130 aircraft at high speed. This system can be used to deliver up to 600 pounds for each container, with a maximum of four containers for each pass, at speeds up to 250 knots.

Figure 8-10. High-speed, low-level, airdrop system.

(d) The CTU-2/A is a high-speed, aerial-delivery container (see Figure 8-11) that can be used to deliver supplies from high-performance aircraft flying at a minimum altitude of 300 feet above ground level and a maximum airspeed of 425 knots. The CTU-2/A is carried on the bomb racks of fighter or bomber-type aircraft. Upon release, a pilot parachute deploys the main chute and the container descends slowly.

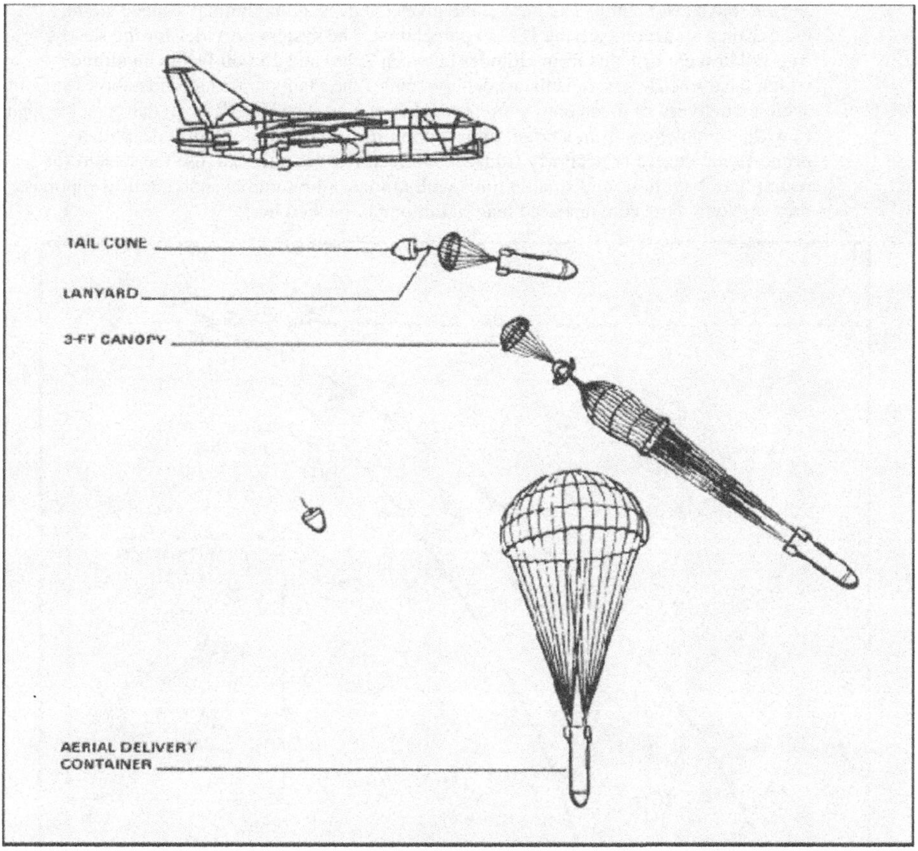

Figure 8-11. CTU-2/A high-speed, aerial-delivery container.

The CTU-2/A can be used to deliver up to 500 pounds of supplies such as weapons, water, food, or munitions. The container can be destroyed by burning. The main advantage of this system is that it can be delivered by high-performance aircraft deep behind enemy lines and in a dense air defense environment. The accuracy of this system is equal to that of a conventional bomb strike.

(2) Sometimes it may be better to drop resupply loads to the ranger force from a high altitude. The HAARS (see Figure 8-12) permits containerized unit loads weighing from 200 to 2,000 pounds to be delivered from aircraft at speeds up to 150 knots from up to 25,000 feet AGL. The HAARS consists of a cargo parachute, an airdrop container, an altitude

sensor, and a pilot chute. The pilot chute gives the descending bundle a speed slightly greater than an accompanying HALO parachutist. The system provides for the steady free-fall descent of loads from altitudes between 2,000 and 25,000 feet to an altitude at which a barometric sensor actuates deployment of the main parachute. This allows safe and accurate delivery of loads onto unprepared drop zones. The HAARS can deliver a payload to within 260 meters from a target impact point from a 10,000-foot altitude, with a proportional degree of accuracy from 25,000 feet. Ranger units can use the system for resupply of battalions and smaller units with rations, ammunitions, and medical supplies, breaking down the containerized material into man-packed loads.

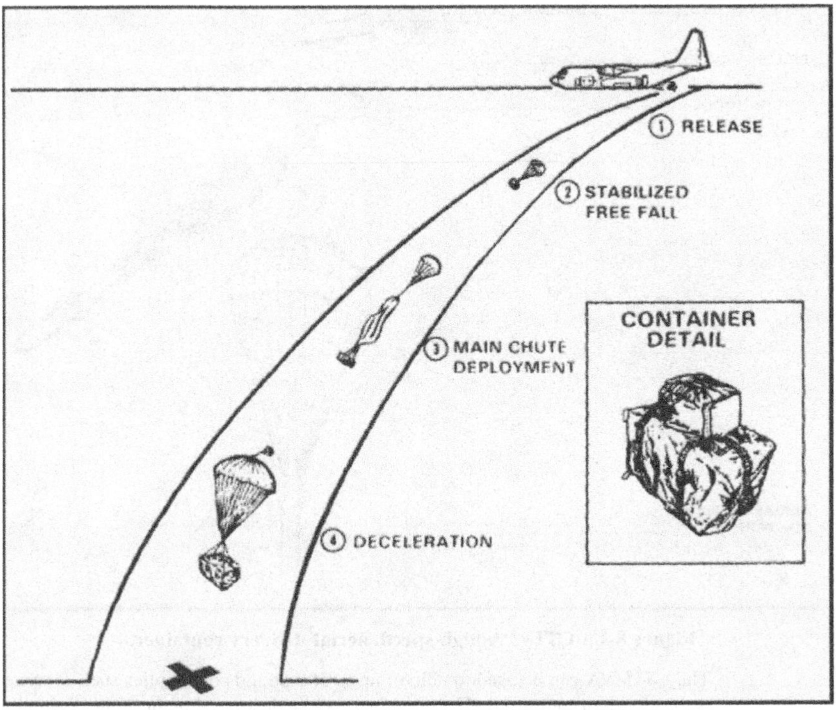

Figure 8-12. High-altitude, airdrop resupply system.

(3) The ranger force may need aircraft loads delivered during periods of poor visibility. The adverse weather aerial delivery system allows aircraft to drop resupply bundles accurately in adverse weather or darkness. These AWADS operations can be conducted safely and effectively with a minimum 300-foot ceiling AGL and a minimum visibility of less than half a mile. The minimum-drop altitude is 500 feet above the highest obstruction. These minimum ceiling or visibility restrictions do not apply to the Combat Talon aircraft.

f. Resupply to the ranger force can be provided by Army helicopters (see Figure 8-13). Normally this is done by SOA units. The danger to helicopters during long-range, low-level penetration of enemy rear areas requires that resupply be conducted when there is limited visibility.

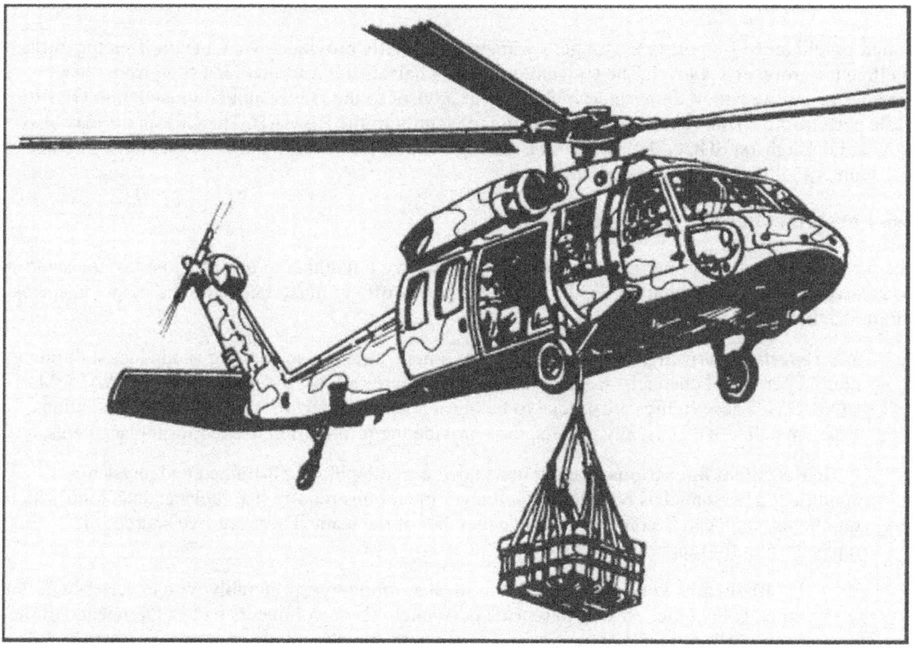

Figure 8-13. Resupply bt helicopter.

8-4. Military police.

a. The ranger regiment needs military police support to provide security at the REMAB. Although ranger units can provide their own security and local defense, using military police for this purpose allows the ranger unit to concentrate its manpower and training on mission preparation. A military police security company is normally designated to provide support to 1st SOCOM elements located in the communications zone. Military police provide security patrols of the REMAB, convoy security, and vehicular search. They provide entry and exit control of the sensitive compartmented information facility (SCIF).

b. A preemptive strike by enemy special operations forces is a threat to the ranger regiment while located in the REMAB; either by conventional attack or by terrorist-type bombing. The preparation and location of the headquarters, the type of buildings, the location of landing zones, the traffic flow, and the size of the local population all influence the size of the military police unit. A military police platoon, augmented with bomb and guard dog teams, should be enough security at the REMAB. These security forces would not normally deploy from the REMAB to an ISB with the ranger unit.

8-5. Counterintelligence.

Counterintelligence support to the ranger regiment is normally provided by a CI team from the military intelligence group at TA level. The CI team provides a liaison and informational flow from the CI agencies operating within the theater. It also acts as advisor to the ranger unit commander on OPSEC while performing CI surveillance operations in the vicinity of the REMAB. The CI support may also be provided through 1st SOCOM channels. It is through 1st SOCOM that liaison with national intelligence and counterintelligence agencies is achieved.

8-6. Personnel service support.

The personnel service support needs of a ranger unit are much like that of any other unit of the same size. To receive the support his unit needs, the ranger personnel officer must meet with the proper agencies within the theater of operations.

a. **Strength accounting.** Within the ranger regiment, strength accounting is conducted through normal personnel channels. Regimental strength figures are passed quickly to the TAACOM or COSCOM. These figures are crucial to the theater commander for planning future operations. When matched with casualty reports, they provide the information needed for replacements.

b. **Replacement operations.** Ranger units need exceptionally well-trained and physically conditioned personnel as replacements. Ranger operations require that replacements train with the unit before they can become an effective member of the team. There are five sources of replacements for ranger units:

(1) **In-theater volunteers.** Use of in-theater volunteers, preferably with experience in ranger units, is the quickest way to replace personnel. These volunteers have a knowledge of the basic skills required of rangers and have some operational experience. The ranger regimental personnel officer coordinates with the theater personnel officer to screen and select those volunteers with the required MOSs and grades. Volunteers from other special operations force components, or from other theaters of war, are identified by DA and screened for assignment to the ranger regiment. Even though these volunteers have individual training and experience, it takes several weeks to become fully integrated into the ranger unit's operations and to achieve the needed skill level.

(2) **Reorganization of units.** This method of reconstitution is used when two or more ranger units have high personnel losses. The remaining members of these units are reorganized into a single unit. This method provides the ranger regimental commander with enough ready-trained, physically fit personnel to form a full-strength unit. Extensive unit training must be conducted before such a reorganization regains the unity and esprit of its component parts. The high standards found among the subelements of the ranger regiment makes this a good method of reconstituting a ranger unit. The units that have lost all their personnel may be reconstituted later, using some other method, or left unreconstituted,

depending on the situation.

(3) **Hospital returnees.** These personnel, former members of the ranger regiment who were injured, evacuated, treated, and have now recovered, are valuable assets. They provide the unit with a major reconstitution ability. These personnel should be returned quickly to their original battalion. This enhances unity and increases morale and esprit. Returnees may not be physically able to return to full duty right away. While recuperating, they may be assigned to limited duty as instructors. The ranger unit uses the returnee's combat experience and individual skills to screen and train newly assigned volunteers. Lessons learned in combat can help prepare the unit for future operations.

(4) **Replacement volunteers from CONUS.** These personnel have training in the basic skills and may have completed the ranger course, but they do not have any ranger unit experience. They need several weeks of unit training to become prepared for combat operations.

(5) **Ranger indoctrination program.** During wartime, this program provides screening and training of newly assigned personnel. Volunteers who have neither served in a ranger unit before nor have graduated from the ranger course require extensive individual training before they are ready to begin unit training with a ranger battalion. This method of reconstitution of ranger units requires the longest training period. During this reconstitution phase, a ranger battalion needs extensive use of ranges, maneuver areas, and support by aviation and other special operations-type units. The ranger regimental commander may choose to move the program from CONUS to a location near the regiment's REMAB.

c. **Casualty reporting.** Casualty reports are handled through regimental personnel channels and must be timely and accurate. The reports are used to check strength accounting data, to assist graves registration personnel in the recovery and identification of remains, and to notify next of kin.

d. **Health service.** The health service support needed by the ranger units deployed OCONUS is usually small. Area medical support provided by host country or theater medical command assets usually meets their needs. Evacuation of wounded members of the ranger force begins upon extraction from the objective area. Personnel wounded in action are evacuated through USAF or Army channels to the nearest medical facility that can provide treatment. The ranger force medical officer and personnel officer coordinate with the area medical support facility providing preventive and routine medical care to the unit while at the REMAB. Vaccine needed for innoculation of the ranger force members and any special Class VIII supplies needed for the mission are drawn through normal medical channels.

e. **Financial service.** Financial support varies with the situation. In mid- and high-intensity conflicts, regular paydays are suspended in the theater of operations. As a nondivisional unit, the ranger regiment coordinates with the theater finance center or corps finance group for financial support. In the REMAB, ranger units are provided financial support from the closest finance support unit. The ranger unit personnel officer appoints Class A agents to help with the disbursement of funds. The ranger force may be given a specific amount of local currency or military scrip depending on the length of time in the REMAB and the operational situation. The ranger unit personnel officer also coordinates with the theater finance center to get any third-country currency that may be included as part of an escape-and-evasion kit provided to ranger elements.

f. **Postal service.** Postal service to the ranger unit is through normal channels. Mail for members of the deployed force is forwarded by the rear detachment commander to an established theater Army general support postal activity. This activity forwards the mail to the postal unit designated to support the REMAB site. The ranger unit commander appoints subordinate unit officers as censors. The exact location of the REMAB, and all aspects of the mission and its location, remains classified until released by an official Army spokesman. The ranger unit commander must ensure that the forwarding of mail, or the writing home of personal letters, does not endanger the ranger force mission. All outgoing mail is censored during wartime. For security reasons, outgoing mail may be held for a short time before it is forwarded. Mail from the ranger unit is never sent by host country or other nation's postal services.

g. **Chaplain activities.** Chaplain activities are provided by the chaplain assigned to the ranger battalion. He provides unit coverage through services, rites, and sacraments regardless of his or the unit member's religious affiliation.

h. **Legal service.** Ranger units do not need extensive legal service support. Legal service support is obtained through coordination with the theater Army staff judge advocate.

APPENDIX A

Organization and Equipment Introduction of the Ranger Regiment

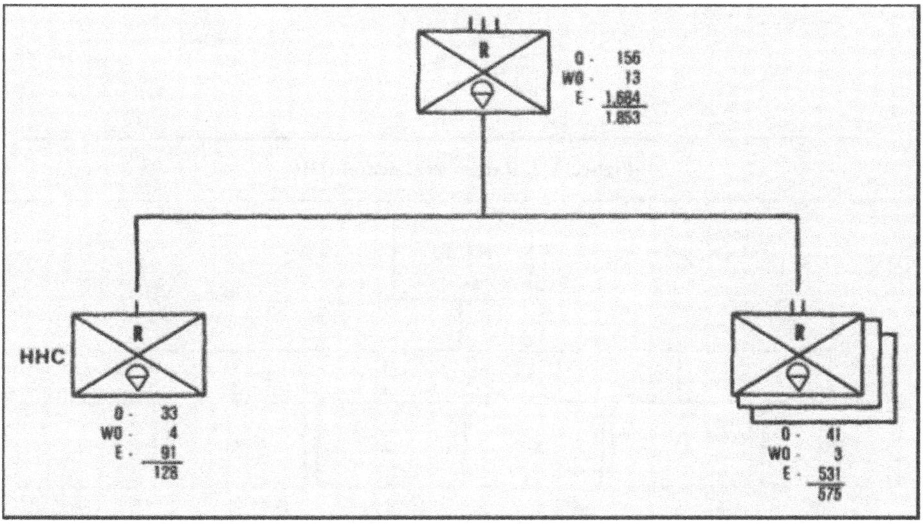

Figure A-1. Ranger regiment.

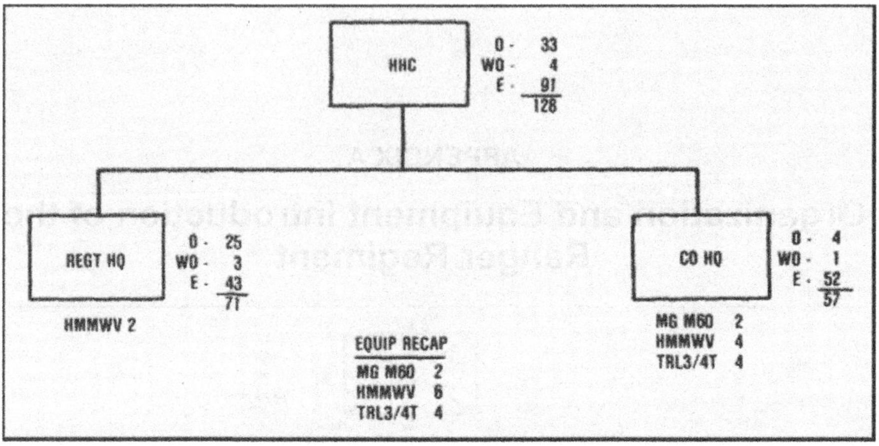

Figure A-2. Ranger regimental HHC.

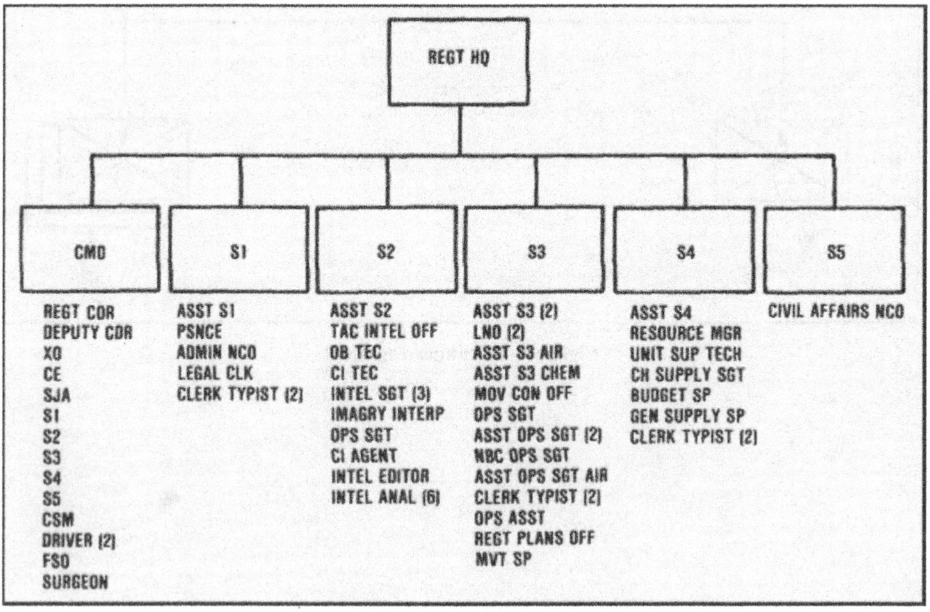

Figure A-3. Ranger regimental staff.

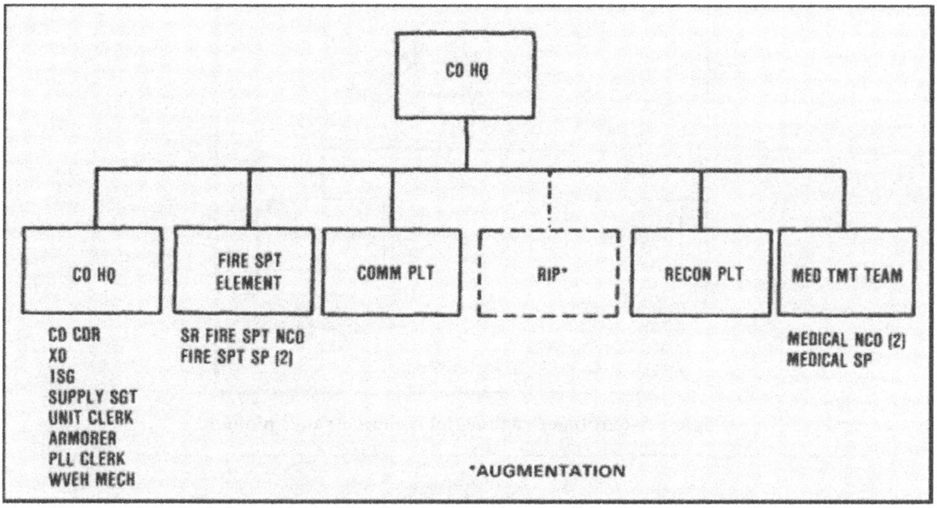

Figure A-4. Ranger regimental headquarters company.

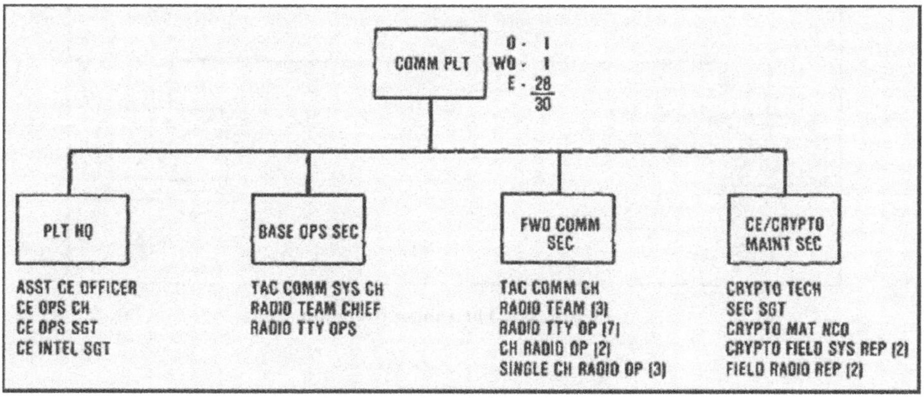

Figure A-5. Ranger regimental communications platoon.

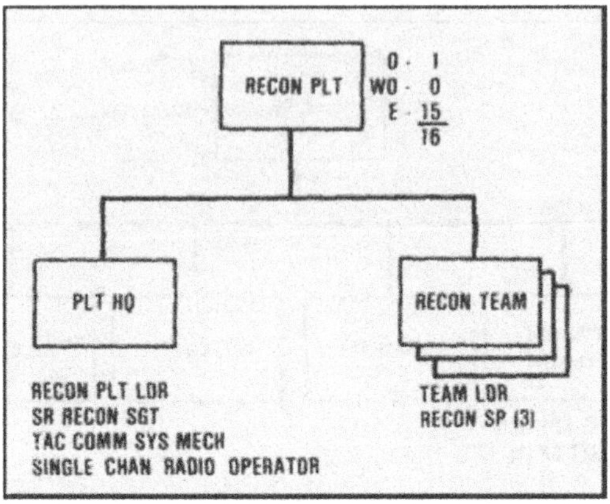

Figure A-6. Ranger regimental reconnaissance platoon.

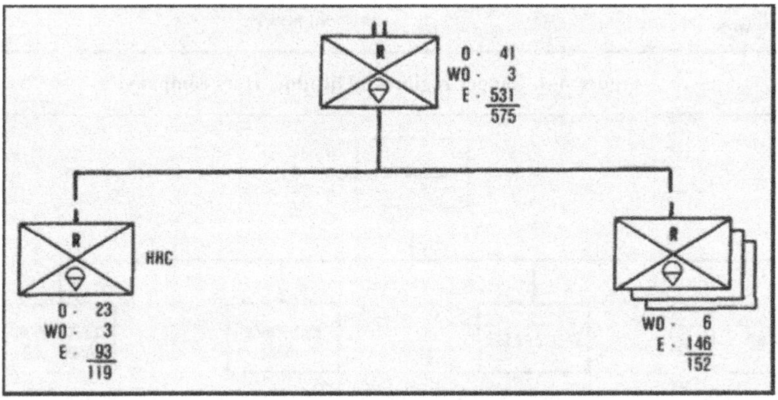

Figure A-7. The ranger battalion.

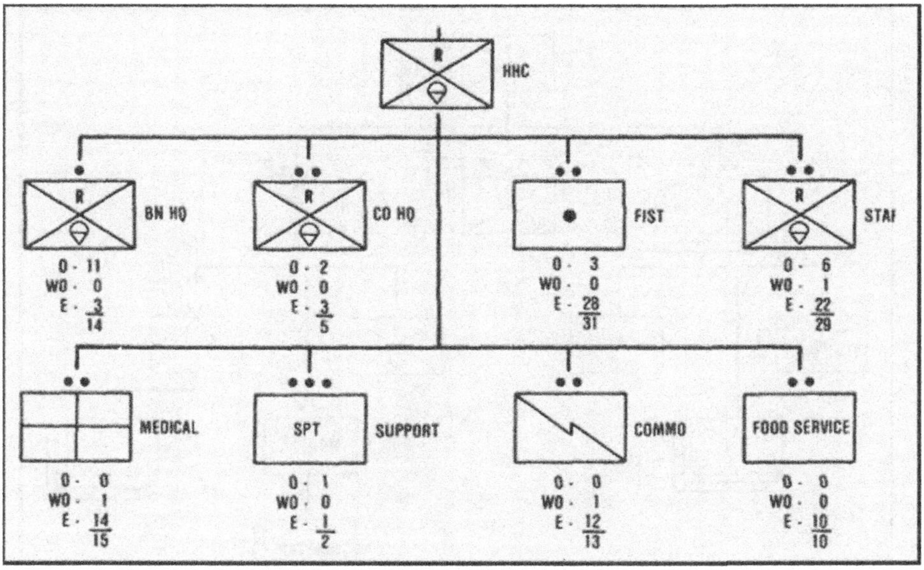

Figure A-8. Ranger battalion HHC.

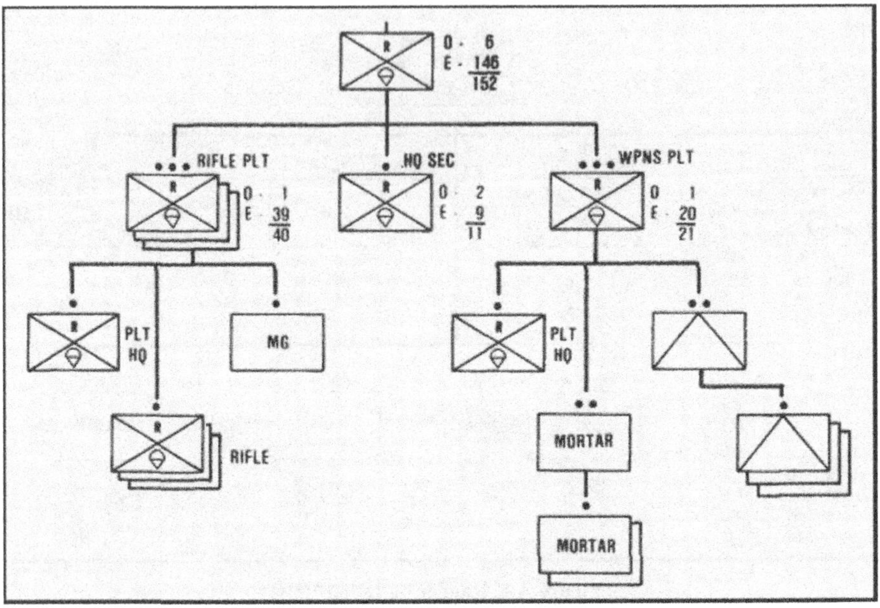

Figure A-9. Ranger rifle company.

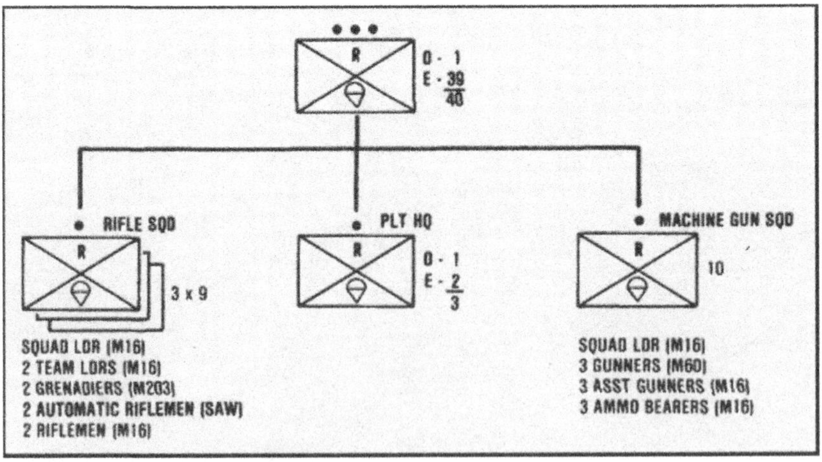

Figure A-10. Ranger rifle platoon.

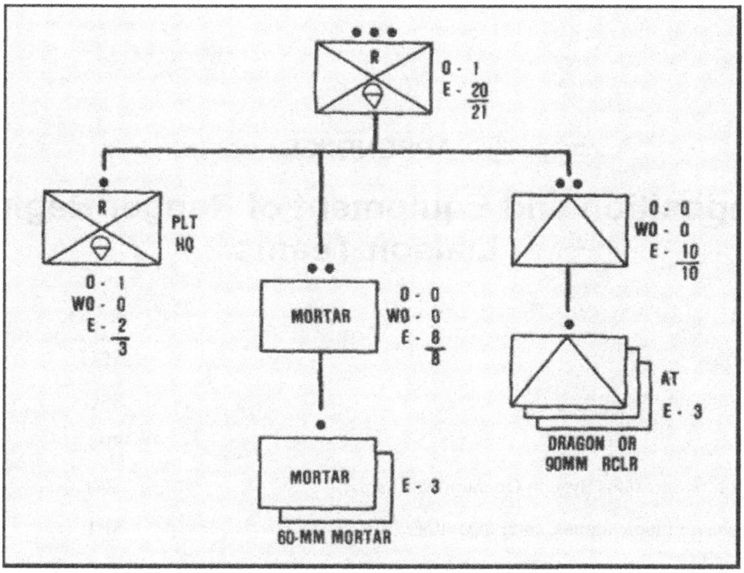

Figure A-11. Ranger weapons platoon.

APPENDIX B

Composition and Equipment of Ranger Regiment Liaison Teams

Team Red	**RANK**
Liaison Officer	MAJ
Senior Intelligence Analyst	SSG
Clerk-Typist	SP4
Tactical Communications System Operator/Mechanic	SGT
Combat Telecommunications Center Operator	SP4
Tactical Communications System Operator/Mechanic	SP4
Team White	
Liaison Officer	CPT
Senior Intelligence Analyst	SSG
Clerk-Typist	SP4
Tactical Communications System Operator/Mechanic	SGT
Combat Telecommunications Center Operator	SP4
Tactical Communications System Operator/Mechanic	SP4

Equipment (for Each Team)

HMMWV with 3/4-ton trailer	5-kw generator
Communications shelter	Tent, GP, small
AN/PCS-3 radio set (3)	AN/VSC-7 net control station (1)
MX 300 (4)	KY 57 with battery case (5)
HYX 57 with battery case (4)	HYP 57 (5)

KYK 13 (2)

KY 65 with battery case (1)

KG 84 (1)

H189 (2)

GRC 193 (1)

VGC-74 (1)

PP 1648 (1)

SB-22 switchboard (1)

KYI 18 (1)

KY 65 WLA (1)

Digital message device (1)

AN/PRC-77 radio (1)

Facsimile transmission device (1)

TA 312 with WD1 (5)

APPENDIX C

Equipment Density

C-1. Communication equipment.

	Regimental Headquarters	Battalions (x3)
AN/VSC-7 net control station	7	2
AN/PSC-3 radio set	23	13
AN/PRC-41 radio set	2	5
AN/PRC-104 radio set	9	26
AN/GRC-193A radio set	4	0
AN/PRC-70 radio set	6	0
AN/PRC-77 radio set	22	50
AN/VRC-46 radio set	2	0
AN/VRC-49 radio set	0	2
AN/PRC-68 radio set	0	66
TSEC/KY-57 speech security device	60	68
TSEC/KY-58 speech security device	6	0
TSEC/KY-65 speech security device	11	0
AO-8990/P digital message device group	15	23
AN/GXC-7A tactical FAX device	7	2
AN/UGC-74A(V)3 communications terminal	7	0
SB-22 switchboard	2	0
SB-3614(V)/TT switchboard	2	1

TSEC/KG-84	5	0
SB-993/GT switchboard	0	12

C-2. Weapons and equipment each battalion.

Infrared aiming light	96	Interrogator set, Stinger missile	3
Rifle, M16	451	Grenade launcher, M203	67
Rocket launcher, M202	13	Machine gun, M60	27
Mortar, 60-mm M224	6	Rifle, sniper, M21	24
Recoilless rifle, 90-mm	18	Viewer, infrared	9
Tracker, Dragon	9	Night vision devices (all types)	322
Truck, 1/4-ton	2	Laser target designators	11
Radio (all types)	164	Tactical satellite radios (all types)	15
Squad automatic weapon	54		
Laser range finder	14		

APPENDIX D

Standing Orders, Rogers' Rangers

1. Don't forget nothing.
2. Have your musket clean as a whistle, hatchet scoured, sixty rounds powder and ball, and be ready to march at a minute's warning.
3. When you're on the march, act the way you would if you was sneaking up on a deer. See the enemy first.
4. Tell the truth about what you see and what you do. There is an army depending on us for correct information. You can lie all you please when you tell other folks about the Rangers, but don't never lie to a Ranger or officer.
5. Don't never take a chance you don't have to.
6. When we're on the march we march single file, far enough apart so one shot can't go through two men.
7. If we strike swamps, or soft ground, we spread out abreast, so it's hard to track us.
8. When we march, we keep moving till dark, so as to give the enemy the least possible chance at us.
9. When we camp, half the party stays awake while the other half sleeps.
10. If we take prisoners, we keep 'em separate till we have had time to examine them, so they can't cook up a story between 'em.
11. Don't ever march home the same way. Take a different route so you won't be ambushed.
12. No matter whether we travel in big parties or little ones, each party has to keep a scout 20 yards ahead, 20 yards on each flank, and 20 yards in the rear so the main body can't be surprised and wiped out.
13. Every night you'll be told where to meet if surrounded by a superior force.
14. Don't sit down to eat without posting sentries.
15. Don't sleep beyond dawn. Dawn's when the French and Indians attack.
16. Don't cross a river by a regular ford.

17. If somebody's trailing you, make a circle, come back onto your own tracks, and ambush the folks that aim to ambush you.

18. Don't stand up when the enemy's coming against you. Kneel down, lie down, hide behind a tree.

19. Let the enemy come till he's almost close enough to touch. Then let him have it and jump out and finish him up with your hatchet.

MAJ Robert Rogers

APPENDIX E

The Ranger Creed

R Recognizing that I volunteered as a Ranger, fully knowing the hazards of my chosen profession, I will always endeavor to uphold the prestige, honor, and high "esprit de corps" of the Ranger Regiment.

A Acknowledging the fact that a Ranger is a more elite soldier who arrives at the cutting edge of a battle by land, sea, or air, I accept the fact that as a Ranger, my country expects me to move farther, faster, and fight harder than any other soldier.

N Never shall I fail my comrades. I will always keep myself mentally alert, physically strong, and morally straight and I will shoulder more than my share of the task, whatever it may be. One hundred percent and then some.

G Gallantly will I show the world that I am a specially selected and well-trained soldier. My courtesy to superiors, my neatness of dress, and my care of equipment shall set the example for others to follow.

E Energetically will I meet the enemies of my country. I shall defeat them on the field of battle, for I am better trained and will fight with all my might. Surrender is not a Ranger word. I will never leave a fallen comrade to fall into the hands of the enemy and under no circumstances will I ever embarrass my country.

R Readily will I display the intestinal fortitude required to fight on to the Ranger objective and complete the mission, though I be the lone survivor.

RANGERS LEAD THE WAY

APPENDIX G

Training

G-1. Application and guidance.

a. To prepare for combat, ranger units train on missions derived from their analysis of operational and contingency plans and the ranger mission training plans (MTPs). Training guidance is in doctrinal manuals, field circulars, and soldier's training publications. Training must ensure that the ranger unit is skilled in all basic light infantry skills and in ranger-unique skills. In the final analysis, good training depends upon the leadership and teamwork of ranger unit officers and NCOs.

b. Everything rangers do should prepare them to perform well in combat. Ranger units train IAW the battalion training management system (BTMS) concentrating on basic skills and on live-fire, hands-on training.

c. An effective training program breeds high morale and few discipline-related problems. Some qualities of ranger units that have good training programs are as follows:

(1) Leaders who can solve problems.

(2) A thorough and continuous analysis of the unit's mission.

(3) Validated training objectives.

(4) Attention to the basics (move, shoot, and communicate).

(5) Multiechelon sustainment training.

(6) Standardized procedures.

(7) Physical and mental strength.

(8) Live fire as an integral part of training.

(9) Safety as a basis to all operations.

(10) The use of night and adverse weather and difficult terrain as combat force multipliers,

d. When all training is planned with realism, rangers will be in training situations that require the same attention as in battle. Realism and stress in training result in faster learning and better retention of skills. Training objectives must be clear with measurable standards to determine proficiency. Rangers and ranger units must train to the standards that will be expected in combat.

e. Examples of unit training goals include:

(1) Developing each member of the chain of command to perform his duties with competence, confidence, and pride.

(2) Maintaining small unit integrity during all training and support activities. Accomplishing tasks as units.

(3) Ensuring that all tasks for training are keyed to mission needs and weaknesses. Assets--ranger's time, money, and materiel--are expended only after thorough planning.

(4) Training to standard.

(5) Developing and maintaining discipline and esprit to ensure the abilities of each ranger and all units are achieved.

(6) Emphasizing use of the ranger mission training plan (MTP) for each echelon of the organization.

(7) Emphasizing the basics.

(8) Maintaining a high level of physical fitness.

(9) Working together within the ranger regiment and other elements of the special operations forces community.

(10) Developing a unit atmosphere of quick reaction under stress. Encouraging and rewarding individual initiative. Accomplishing unit missions IAW the commander's intent.

G-2. Philosophy.

a. To be effective, training should be physically and mentally challenging. Each ranger will then have a sense of purpose, accomplishment, and satisfaction. Training is evaluated by the results rather than hours of training. Units that are well trained today need reinforcement training to sustain readiness. Training management must be continuous. Training managers must use the time available to improve training efficiency. Cyclic training will not produce the desired results. The "train-to-correct deficiencies" method is needed. The training managers must analyze weaknesses, conduct prescriptive and reinforcement training, and reevaluate their unit's training status. The process that is used to develop and maintain a continuous training and evaluation program is shown in Figure G-1.

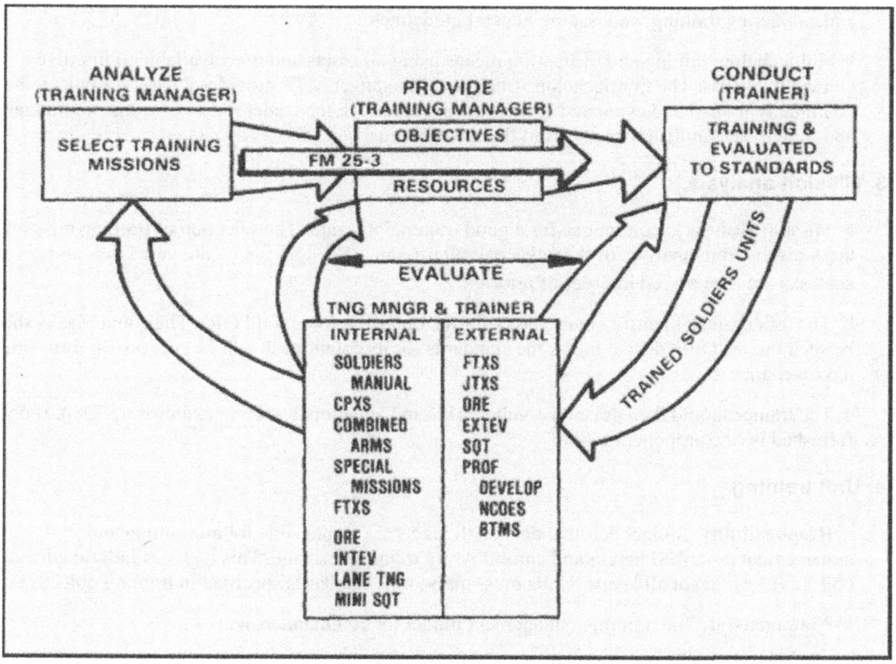

Figure G-1. The Army training and evaluation program.

b. Individual and collective training must be integrated within a unit training program. Training of the individual ranger, leaders, and units at each level must occur at the same time. The ranger commander or leader plans and conducts training for every member of his unit.

c. Ranger commanders analyze operational and contingency plans to determine unit missions and translate the associated tasks into training objectives. These objectives form the basis of the unit training program. The regimental headquarters provides training guidance in a formal regimental training plan. Commanders at lower levels program, supervise, and conduct training within their units.

d. All members of the chain of command are responsible for training. The NCO leader or first-line supervisor conducts individual and small unit collective task training. The officer plans individual training in his unit and directs and conducts collective training.

e. Centralized development of training objectives and decentralized training builds well-trained units and outstanding leaders. Authority and responsibility for the detailed planning, scheduling, conducting, evaluating, and supervising of training are delegated to the lowest element capable of managing this training. Higher headquarters is responsible for establishing objectives, monitoring and evaluating training, and issuing general guidelines.

f. Multiechelon training and evaluation means using all assets and time available to improve combat readiness. The multiechelon structure of the ranger MTP provides a basis for units to be continually trained and evaluated by internal evaluation. Each leader must know his mission tasks and train to accomplish those tasks to the set standards.

G-3. Mission analysis.

a. Mission analysis is continuous for a good training program. The selection of training missions or tasks results from analysis of the unit's overall mission. (See Figure G-1 above.) Tasks and subtasks are then placed in order of priority.

b. The trainer must identify countertasks for the opposing force (OPFOR). The countertasks should be such that an OPFOR that meets the standards for its countertask will be successful. Both sides have a chance to win.

c. The trainer should then develop combat drills and situational training exercises (STXs), if not furnished by the proponent school.

G-4. Unit training.

a. **Responsibility.** Ranger commanders at all levels are responsible for all training and management described herein and current Army training doctrine. This includes individual and collective training at all levels. Units must train to the standards specified in training objectives.

b. **Management.** The training management model for commanders is:

- Set objectives.
- Provide resources.
- Coach subordinates.
- Measure results.
- Provide feedback.

G-5. Individual training.

There are three elements basic to a well-trained individual ranger. He must be physically fit, be proficient in the use of his weapon, and have the psychological commitment to fight the enemy.

a. Individual training is the responsibility of the first-line supervisor under the supervision of the chain of command. It is conducted using performance-oriented techniques under battlefield conditions. New skills must be developed and practiced, and acquired skills must be reinforced throughout the training year. A high level of proficiency must be maintained through continuous training and practice.

b. Trainers should use peer instruction. Commanders ensure needed resources are available to conduct challenging individual training. Ranger units schedule and conduct this training and evaluate individual proficiency.

c. Flexibility in planning and scheduling is necessary. The goal of all individual training must provide for reinforcement training to offset the effects of the forgetting curve. (See Figure G-2).

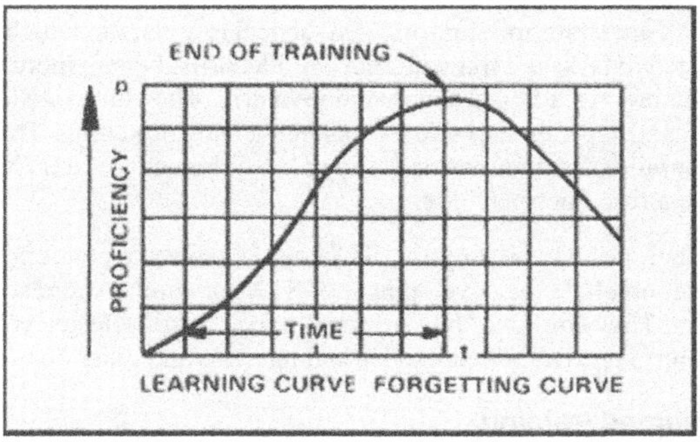

Figure G-2. The learning curve and forgetting curve.

d. The command sergeants major, unit first sergeants, and staff section sergeants are the keys to an effective training program. These senior NCOs must counsel, guide, and critique junior NCOs in training preparation and execution. Fire team leaders, section chiefs, and squad leaders coordinate and present individual training.

e. The NCO supervisors must ensure that the ranger is trained in the skills required for the specific military occupational specialty (MOS) and in those skills used daily. Not all training can be acquired in an office. For example, all rangers must maintain a high skill level with their weapons, be trained in NBC subjects, and be capable of performing contingency missions.

G-6. Collective training.

a. **Guidance.** Collective training trains crews, teams, squads, platoons, companies, and battalions to do their missions. The ranger MTP is used to plan training and evaluate the unit's weaknesses and strengths. If a ranger commander receives new training requirements or missions, they should become part of the collective training program. The ranger MTP describes the combat tasks to be performed by every echelon of the ranger battalion. It is a plan for organizing, conducting, and evaluating the year-round unit training program.

b. **Responsibility.** The ranger unit commander is responsible for the conduct of collective training and for all other training in his unit. Practical exercises emphasize the techniques of terrain use, movement, and effective employment of weapon systems. Training under stress will be conducted to move the unit to a higher plane of combat readiness.

c. **Drills.** Drills are written for small units--for example, fire team and squad, crew and section--and describe an action that is shorter than an entire STX. Drill training objectives cover both individual and collective skills (critical collective tasks) and list them together for effective group action. Drills must be mastered in order to react to rapidly changing situations in combat. The trainer can interrupt the execution of a drill to provide necessary training on individual or collective skills.

d. **Situational training exercise.**

(1) The STXs are short, mission-oriented tactical exercises in which a group of closely related collective tasks are trained. They are like drills but are more complex and flexible. They usually involve a larger unit than does a drill. They integrate leader training, leader tasks, drills, and separate individual tasks into a realistic setting. Trainers may execute STXs as written or tailor them to meet specific unit requirements. An STX is not standardized, but it must have a firm doctrinal base.

(2) Due to their design and purpose, STXs apply mainly to platoon- and company-size units. This does not preclude the development of STXs or similar exercises for unit staffs and headquarters. They combine "how-to-train" specifics for selected collective tasks with the "what-to-train" specifics found in doctrinal manuals and other documents.

G-7. Performance-oriented training.

a. Performance-oriented training emphasizes hands-on performance. Training is conducted to ensure a level of proficiency, rather than established number of hours of required training.

b. Performance-oriented training has precise training objectives and makes good use of resources. Training objectives serve as the basis for preparing, conducting, and evaluating all training.

G-8. Evaluation.

a. A training program must include an evaluation of training (planned and conducted). An evaluation plan must:

(1) Provide immediate feedback to the trainer.

(2) Recognize effective trainers.

(3) Evaluate all units in the conduct of training to provide for a comparison of training programs.

(4) Be flexible to unit needs; provide general and specific evaluations based on the level of training within the unit.

(5) Be in effect at all times.

b. Major considerations for training evaluations include:

(1) Ranger unit commanders establish and maintain an effective internal and external evaluation program.

(2) The director of each staff element is responsible for training evaluations within his area of staff responsibility with emphasis on low-density MOS groups.

(3) The operations officer of the ranger organization develops and validates training and evaluation plans.

c. Evaluation of training must be done vertically from the major task (mission) down through subtasks (collective tasks) to the leader and individual tasks. In selection of individual tasks for evaluation, priority should be given to critical individual tasks that impact on the success of the collective task or mission.

d. The critique and after-action review, after each major training event, are the most important aspects of training. The after-action review takes place immediately after completion of a training task (individual or collective). The individual(s) self-analyzes and critiques the training event. Then after a formal critique of the training, the individuals) trains to standard (not to time).

G-9. Field training.

a. **Considerations.** Ranger commanders schedule training under field conditions to permit maximum training opportunities. Field training emphasis is on mastery of a series of individual (soldier and leader) tasks and collective training tasks. These tasks are repeated until established standards are met. The goal of ranger field training is to ensure realistic, demanding, and meaningful training.

(1) When training is conducted under conditions of moderate stress, most combat critical skills are learned quicker and retained longer. (See Figure G-3 on the following page.) Stress can be incorporated in a training situation by physical deprivation (sleep, food), by live-fire exercises, by operation against an OPFOR under simulated battlefield conditions for an extended period, by the intensity of activity including demands for decision making, or by physically demanding activities. Ranger units train under stress so that they will react quickly in combat.

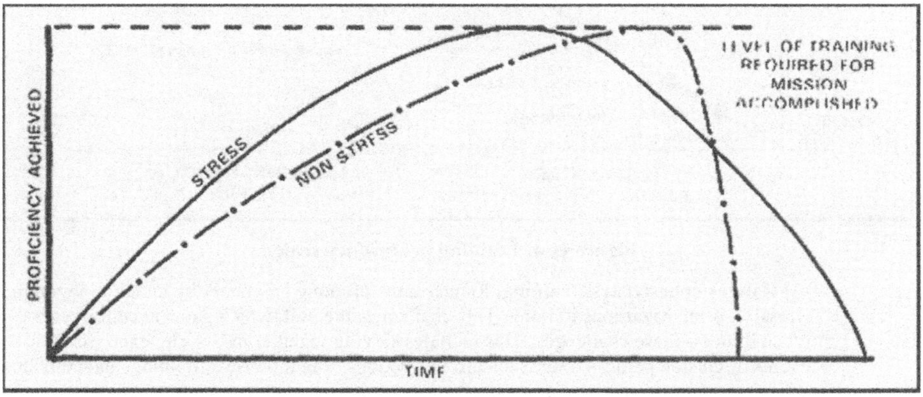

Figure G-3. The Army training and evaluation program.

(2) Antagonistic training situations are two-sided exercises conducted by opposing elements. Each element has a training objective (task and standards). The opposing force has a corresponding countertask. This training may be progressive--for example, starting with a

crawl-through, then a walk-through, and progressing to both elements operating at a fast pace. The use of multiple integrated laser equipment system (MILES) greatly enhances task and countertask training. The units should also be considered for a live-fire exercise against a realistic OPFOR target array when the basics have been mastered.

b. **Methods of training.**

(1) **Considerations.** The chance for success in combat can be increased by becoming proficient in training situations that depict realistic combat conditions. (See Figure G-4.) Meeting the standards of a realistic live-fire training exercise assures the commander that his unit could perform well in combat. Figure G-4 shows the relationship between realistic training and the challenge to the trainer.

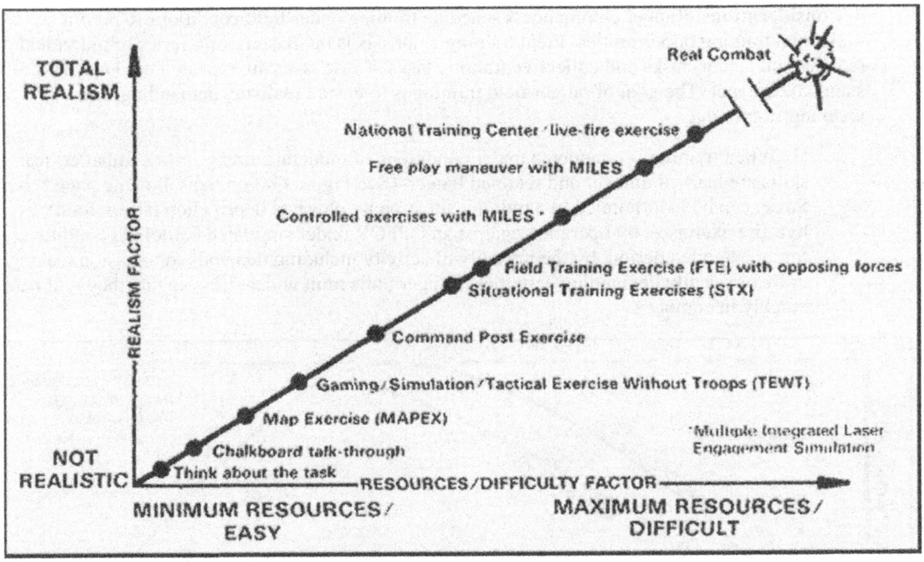

Figure G-4. Training proficiency scale.

(2) **Ranger cohesiveness training.** Ranger units promote teamwork by undergoing tough, stressful, often hazardous training. This challenges the unit, which gains in cohesiveness when it meets these challenges. This includes serving together in live-fire exercises, athletic events, precision parades, high-standard inspections, tough physical training, and combat.

(3) **Field training exercise.** An FTX is training normally conducted under simulated, realistic combat conditions. This training normally exercises command and control systems and the administrative and logistical needs of sustained combat. Stress training is part of this exercise, focusing on live-fire and MILES drills and STXS.

(4) **Combined arms training.** Combined arms training should include all means of support such as tactical air, armed helicopters, and indirect fire with emphasis on fire support coordination and airspace control.

(5) **Live-fire exercises.** Including live-fire exercises in drills and STXs offers a chance to evaluate units in their response to simulated combat missions. Ranger units train under pressure and always anticipate being selected to conduct a live-fire exercise. Live-fire exercises develop confidence and esprit, and reinforce unit discipline.

(6) **Opposing force.** Committing one unit against another adds a competitive spirit to a field training exercise (FTX). The use of MILES and squad combat operations exercises (simulation) (SCOPES) provides realism to two-sided field exercises. Trainers should use OPFOR equipment and training devices available, as well as making their own devices. (See FM 30-103, Appendix R, for more information.)

c. **Key functional training areas.**

(1) **Land navigation and patrolling.** Day and night navigation and patrolling exercises must be emphasized. Orienteering improves these skills, as well as physical fitness.

(2) **Movement techniques.** Infantry combat has two major components--to move and to hit the enemy first. Ranger units must be skilled in the use of traveling, traveling overmatch, and bounding overmatch. Units in contact must be skilled in moving under fire.

(3) **Fighting** positions. Training includes the building of concealed fighting positions.

(4) **Mine and countermine training.** Training includes how to defend against hostile mines, booby traps, and other demolitions. The identification, installation, neutralization, and removal of US mines and devices is also stressed.

(5) **Limited visibility operations.** Ranger units must be skilled in limited visibility operations and the use of night vision and thermal devices.

(6) **Operation security.** Operation security includes deception, physical security, signal security (SIGSEC), and information security. It is taught in all training and at all levels. Although all areas of OPSEC must be measured in training, techniques of camouflage and SIGSEC need special emphasis. (See AR 530-1 and TC 100-1 for more information.)

G-10. Professional development.

Ranger units at battalion and regimental levels conduct many officer and NCO professional development programs. This ensures that all officers and NCOs are skilled in their field, confident, and wellrounded. Professional readings and book reports, tactics seminars, and tactical exercises without troops are good development methods.

G-11. Physical training.

a. This program ensures that each ranger maintains the physical ability and stamina needed to do his mission. Other objectives are to promote unit cohesiveness and discipline. These are met by combining individual testing, formation, running, athletic competition, weight control, proper diet, teamwork drills, and freedom from drug and alcohol abuse. Mental fitness and toughness come from experience and success. Good physical conditioning increases a unit's chances of success.

b. Physical training (PT) programs are designed to include the elements of overload, progression, balance, variety, and regularity.

c. Ranger unit physical fitness standards are:

(1) Score 80 points for each Army physical readiness test (APRT) event and do six chinups.

(2) Pass the ranger swimming test.

(3) Complete an 8-kilometer run in 40 minutes.

(4) Complete a 12-kilometer road march in three hours (with rucksack, helmet, and weapon).

(5) Meet the Army height and weight standards.

G-12. Environmental and specialized training.

a. Rangers conduct training in environments as close to real combat missions as possible. This conditions rangers to survive and fight in a specific environment. Such training includes combat-critical tasks and technique training for that environment.

b. Each ranger battalion's environmental and specialized training includes at least the following:

(1) Jungle (once every 12 months).

(2) Desert (once every 12 months).

(3) Urban (once every 6 months).

(4) Extreme cold (once every 18 months).

(5) Mountain (once every 12 months).

(6) Amphibious (once every 18 months).

G-13. Ranger indoctrination program.

a. The RIP is a training and selection program. It is run by specially selected NCOs from the regimental headquarters. It has as its mission the training and indoctrination of newly assigned soldiers. It physically toughens and mentally sharpens the newly assigned soldier. It teaches basic skills and techniques. The RIP identifies and eliminates in three weeks any applicant who does not show dedication, motivation, physical fitness, and emotional stability.

b. An indoctrination program for newly assigned ranger-qualified officers and NCOs prepares them to assume duty as leaders within the ranger regiment. These personnel monitor and participate in RIP training, as well as undergoing individual ranger leadership training. The RIP provides them an overview of the regiment's history, its mission, and its goals. (See Appendixes C,

D, and E for further information about the history and goals of the ranger regiment.) The RIP also provides a review of the skills required in a ranger unit. Refresher training in airborne and pathfinder operations, use of indirect fires, and light infantry doctrine is conducted. During RIP all newly assigned officers and NCOs must meet the ranger physical fitness standard.

c. The RIP cadre also conducts a preranger program to prepare soldiers to attend the US Army Ranger School. It is a physically demanding course that teaches troop-leading procedures, operation orders, and small unit patrol actions.

Glossary

The following acronyms, abbreviations, and definitions are provided to aid the reader in understanding this publication.

Section I

ACRONYMS AND ABBREVIATIONS

AAT	Army assault team
ABCCC	airborne battlefield command and control center
ACL	allowable cargo load
ADA	air defense artillery
AFFOR	the USAF element of a joint task force
AGL	above ground level
AKZ	armor kill zone
ALCC	airlift control center
ALCE	airlift control element
AM	amplitude modulation
AMC	air mission commander
APERS	antipersonnel
APRT	Army physical readiness test
ARFOR	the Army element of a joint task force
ARSOC	Army special operations command
ARSOF	Army Special Operations Forces
ASG	area support group
ASPC	all-source production center

ATGM	antitank guided missile
AWADS	adverse weather aerial delivery system
BCE	battle coordination element
BTMS	battalion training management system
CA	civil affairs
CAS	close air support
CCT	combat control team
CDS	container delivery system
CE	communications-electronics
CI	counterintelligence
CINC	commander in chief
COMINT	communications intelligence
COMSEC	communications security
CONUS	continental United States
COSCOM	corps support command
CP	command post
CS	combat support; a chemical agent
CSS	combat service support
DACG	departure airfield control group
DACO	departure airfield control officer
DASC	direct air support center
DMDG	digital message device group
DOD	Department of Defense
DS	direct support
DSU	direct support unit
DZ	drop zone
EAC	echelons above corps

EACIC	Echelons Above Corps Intelligence Center
ECCM	electronic counter-countermeasure
ECM	electronic countermeasures
E&E	escape and evasion
EEFI	essential elements of friendly information
EOC	emergency operations center
EW	electronic warfare
FARP	forward area rearm and refuel point
FASCAM	family of scatterable mines
FAX	facsimile transmission
FDC	fire direction center
FEBA	forward edge of the battle area
FFL	friendly front lines
FIST	fire support team
FLOT	forward line of own troops
FM	field manual; frequency modulation
FO	forward observer
FORSCOM	US Army Forces Command
FPF	final protective fire
FPL	final protective line
FRAGO	fragmentary order
FSA	fire support area
FSCOORD	fire support coordination or coordinator
FSE	fire support element
FSO	fire support officer
FSS	fire support station
FTX	field training exercise

G3	Assistant Chief of Staff, Operations
GS	general support
GSR	ground surveillance radar
HAARS	high-altitude, airdrop resupply system
HAHO	high-altitude, high-opening
HALO	high-altitude, low-opening
HE	high explosive
HF	high frequency
HHC	headquarters and headquarters company
HSLLADS	high-speed, low-level airdrop system
HUMINT	human intelligence
IAW	in accordance with
IFF	identification-friend-or-foe
IP	initial point
IRP	initial rally point
ISB	intermediate staging base
JACC/CP	joint airborne communications center/command post
JCS	Joint Chiefs of Staff
JCSE	joint communications support element
JFC	joint force commander
JSOC	joint special operations command
JTB	joint transportation board
JTF	joint task force
kw	kilowatt
LAPES	low altitude parachute extraction system
LAW	light antitank weapon
LNO	liaison officer

LOC	logistical operations center
LRRP	long-range reconnaissance patrol
LRSU	long-range surveillance unit
LTD	laser target designator
LZ	landing zone
MAC	Military Airlift Command
MANPADS	man-portable air defense system
MCC	movement control center
METT-T	mission, enemy, terrain, troops and time available
MI	military intelligence
MILES	multiple integrated laser engagement system
mm	millimeter
MMC	Materiel Management Center
MMEE	minimum mission-essential equipment
MOS	military occupational specialty
MOUT	military operations on urbanized terrain
MRLS	multiple rocket launch system
MTP	mission training plan
NBC	nuclear, biological, chemical
NCA	National Command Authority
NCO	noncommissioned officer
NOD	night observation device
OCOKA	observation, concealment, obstacles, key terrain, and avenues of approach
OCONUS	outside continental United States
OPCOM	operational command
OPCON	operational control
OPFOR	opposing force

OPORD	operation order
OPSEC	operations security
ORP	objective rally point
PAC	personnel administration center
PDF	primary direction of fire
PIR	priority intelligence requirements
PLL	prescribed load list
POM	preparation for oversea movement
PSYOP	psychological operations
PT	physical training
PZ	pickup zone
QRE	quick-reaction element
RATELO	radiotelephone operator
REMAB	remote marshalling base
RFL	restrictive fire line
RIP	ranger indoctrination program
RP	rally point; release point
RPG	rocket-propelled grenade
RRF	ranger ready force
RRP	reentry rally point
RSE	ranger support element
RSTA	reconnaissance, surveillance, and target acquisition
S1	adjutant
S2	intelligence officer
S3	operations and training officer
S4	supply officer
S5	civil affairs officer

SAR	search and rescue
SATCOM	satellite communications)
SCIF	sensitive compartmented information facility
SCOPES	squad combat operations' exercises (simulation)
SCUBA	self-contained underwater breathing apparatus
SEAD	suppression of enemy air defense
SEMA	special electronic mission aircraft
SERE	survival, evasion, resistance, and escape
SF	special forces
SHORAD	short-range air defense
SIGSEC	signal security
SKE	station-keeping equipment
SLAR	side-looking airborne radar
SOA	special operations aviation
SOCB	special operations communications battalion
SOCOM	special operations command (usually, 1st SOCOM)
SOP	standing operating procedures
SOSB	special operations support battalion
SOTF	special operations task force
SPL	support platoon leader
STX	situational training exercise
TA	theater Army
TAACOM	Theater Army Area Command
TAC	Tactical Air Command
TACP	tactical air control party
TACSAT	tactical satellite terminal
TAMMC	Theater Army Materiel Management Center

TASG	theater Army support group
TDA	table of distribution and allowances
TOA	transfer of authority
TOC	tactical operations center
TOE	table(s) of organization and equipment
TRP	target reference point
TTY	teletype
UHF	ultrahigh frequency
US	United States
USAF	United States Air Force
USMC	United States Marine Corps
USN	United States Navy
VHF	very high frequency
WP	white phosphorus
XO	executive officer

Section II

DEFINITIONS

ADVANCE FORCE. An operation requiring the immediate deployment of US forces with the intent to negate a threat before the deployment of follow-on forces is required.

AIRLAND BATTLE DOCTRINE. An approach to military operations that realizes the full potential of US forces. Two notions (extending the battlefield and the ability to integrate conventional, nuclear, chemical, and electronics means) are combined to describe a battlefield where the enemy is attacked to the full depth of his formation.

AREA OF INFLUENCE. A geographical area wherein a commander is directly capable of influencing operations by maneuver or fire support systems normally under his command or control.

AREA OF INTEREST. That area of concern to the commander, including the area of influence, areas adjacent thereto, and extending into enemy territory to the objectives of current or planned operations. This area also includes areas occupied by enemy forces who would jeopardize the accomplishment of the mission.

AREA OF OPERATIONS. That portion of an area of conflict necessary for military operations. Areas of operations are geographical areas assigned to commanders for which they have responsibility and in which they have authority to conduct military operations.

CHALK. One specific aircraft load.

CLANDESTINE OPERATIONS. Activities to accomplish intelligence, counterintelligence, and other similar activities sponsored or conducted by governmental departments or agencies, in such a way as to ensure secrecy or concealment. (They differ from covert operations in that emphasis is placed on concealment of the operation rather than on concealment of the identity of the sponsor.)

COMMUNICATIONS INTELLIGENCE (COMINT). The interception and processing of foreign communications passed by radio, wire, or other electromagnetic means, and by the processing of foreign encrypted communications, however transmitted. Interception comprises search, intercept, operator identification, signal analysis, traffic analysis, crypto-analysis, study of plain text, the fusion of these processes, and reporting the results.

COMMUNICATIONS SECURITY. The protection resulting from all measures designed to deny to unauthorized persons information of value that might be derived from the possession and study of telecommunications, or to mislead unauthorized persons in their interpretations of the results of such study. COMSEC includes crypto-security, physical security, and transmission security.

COUNTERINTELLIGENCE. That aspect of intelligence activities, both offensive and defensive, designed to detect and neutralize or destroy the effectiveness of hostile foreign intelligence activities and to protect information against espionage, personnel against subversion, and installations and materiel against sabotage.

COVERT OPERATIONS. Operations that are so planned and executed as to conceal the identity of or permit plausible denial by the sponsor. They differ from clandestine operations in that emphasis is placed on concealment of identity of the sponsor rather than on concealment of the operation.

DECEPTION. Those measures designed to mislead the enemy by manipulation, distortion, or falsification of evidence to induce him to react in a manner prejudicial to his interests.

DEMONSTRATION. A show of force in an area where a decision is not sought. It is similar to a feint but does not make contact with the enemy.

DRILL. A drill focuses on a "chunk" or "slice" of battle and is targeted at small units. Drills are critical collective tasks that require a high degree of proficiency from a small unit. Generally, small units "drill" collective tasks that require rapid responses by unit members in the absence of detailed orders from unit leaders.

ELECTRONIC COUNTERMEASURES. That division of electronic warfare involving actions taken to prevent or reduce an enemy's effective use of the electromagnetic spectrum.

ELECTRONIC INTELLIGENCE (ELINT). The technical and intelligence information derived from foreign noncommunications electromagnetic radiations emanating from other than atomic detonations or radioactive sources.

ELECTRONIC WARFARE. Military action involving the use of electromagnetic energy to determine, exploit, reduce, or prevent hostile use of the electromagnetic spectrum and action that retains friendly use of the electromagnetic spectrum.

ELECTRONIC WARFARE SUPPORT MEASURES (ESM). That division of EW involving actions taken to search for, intercept, identify, and locate sources of radiated electromagnetic energy for the purpose of immediate threat recognition. ESM provides a source of information needed for immediate action involving electronic countermeasures, electronic counter-countermeasures, avoidance, targeting, and other tactical employment of forces.

ESPIONAGE. Actions directed toward the acquisition of information through clandestine operations.

EVACUATION/SAFEGUARD. The use of US forces to evacuate US citizens or allies from a hostile or potentially hostile area, or to safeguard US personnel or property.

EXFILTRATION. The removal of personnel or units from areas under enemy control by stealth, deception, surprise, or clandestine means. Exfiltration is a method of extracting ranger units.

EXTRACTION. The removal of personnel or units from areas under enemy control by any means, to include exfiltration, airland, airmobile, or amphibious.

FEINT. A diversionary supporting attack conducted to draw the enemy's attention from the main effort. It is normally executed by brigades and small units. Feints are usually shallow, limited objective attacks that go in before or during the main attack.

GUERRILLA WARFARE. Military and paramilitary operations conducted in enemy-held or hostile territory by irregular and predominantly indigenous forces.

HIGH-INTENSITY CONFLICT. A war between the US and any other nation or nations in which both belligerents use the most modern technology and all resources in intelligence, mobility, firepower (including NBC), command, communications and control, and service support.

HUMAN INTELLIGENCE. Intelligence derived from information provided by human resources.

INFILTRATION. The movement through or into an area of territory occupied by either friendly or enemy troops or organizations. The movement is made either by small groups or by individuals, at extended or irregular intervals. When used in conjunction with the enemy, it implies that contact is avoided. Infiltration is a subset of insertion.

INSERTION. Placement of troops and equipment into an operational area by any means.

INTERDICT. To isolate or seal off an area by military means; to prevent, hinder, or delay the use of an area or route by enemy forces.

INTERDICT LINE OF COMMUNICATIONS. An attack to seal off an area or to deny use of a route or approach.

JOINT TASK FORCE. A JTF may be constituted and designated by the Secretary of Defense or by the commander of a unified command, specified command, or an existing JTF. Normally, it performs missions having specific, limited objectives or missions of short duration. It dissolves when it has achieved its purpose. The joint task force commander is responsible to the commander, known as the establishing authority, who created the JTF. The JTF is composed of elements of two or more services operating under a single JTF commander. The JTF commander has operational control over the entire force and may have direct command of his own staff with representatives from the other services. He exercises logistics coordination or control only as necessary to meet his subordinate commanders' logistics needs.

LIAISON. That personal contact or communication maintained between elements of military forces to ensure mutual understanding and unity of purpose and effort.

LINE OF COMMUNICATIONS. All routes (land, water, and air) that connect an operating military force with one or more bases of operations and along which supplies and military forces move.

LOW-INTENSITY CONFLICT. Limited politico-military operations conducted to achieve political, social, economic, or psychological objectives. It is generally confined to a specific geographic area and is often characterized by operational constraints imposed on ranger units by political and economic considerations. It may involve intermittent combat against regular threat military forces, but it more often involves conflict with irregular, security, or paramilitary forces of a country or group involved in an operation that is counter to the best interests and national security of the United States. It is characterized by relatively low levels of combat and constraints placed on the weaponry, tactics, and movements of the ranger force. It may involve counterterrorist or antiterrorist operations and generally occurs before a pro forma declaration of war.

MID-INTENSITY CONFLICT. War, declared or undeclared, between the US and the regular forces of one or more nations and their respective allies, if any, in which the belligerents use the most modern technology and all resources in intelligence, mobility, firepower (excluding nuclear, chemical, and biological weapons), command, communications and control, and support. They are used for limited objectives under definitive policy limitations as to the extent of destructive power that can be used or the extent of geographical area that might be involved.

MILITARY STRATEGY. Art and science of using the armed forces of a nation to secure the objectives of national policy by application of force or the threat of force. Military strategy sets the fundamental conditions for operations.

NATIONAL COMMAND AUTHORITY. This includes the President and Secretary of Defense or the duly deputized alternates or successors. Commonly referred to as NCA.

OPERATIONAL COMMAND/CONTROL. Operational command (OPCOM) and operational control in joint force terminology both refer to the authority exercised by joint commanders over subordinate service components. Those terms are not interchangeable in joint operations. OPCOM applies to the authority exercised by commanders of unified commands. OPCON is the authority that subordinate joint task force commanders exercise in the conduct of specific operations. The authority that military departments exercise over their respective components is commonly referred to as command less operational command.

OPERATIONAL LEVEL OF WAR. The operational level of war uses available military resources to attain strategic goals within a theater of war. It is the theory of larger unit operations. It also involves planning and conducting campaigns. Campaigns are sustained operations designed to defeat an enemy force in a specified space and time with simultaneous and sequential battles. The disposition of forces, selection of objectives, and actions taken to weaken or to outmaneuver the enemy all set the terms of the next battle and exploit tactical gains. They are all part of the operational level of war. In AirLand Battle doctrine, this level includes the marshalling of forces and logistical support, providing direction to ground and air maneuver, applying conventional and nuclear fires in depth, and employing unconventional and psychological warfare.

OPERATIONS SECURITY. Those measures designed to protect information concerning planned, ongoing, and completed operations from unauthorized disclosure. It includes all actions a command takes to deny the enemy information about friendly units and their operations.

ORDER OF BATTLE (OB). The identification, strength, command structure, and disposition of the personnel, units, and equipment of any military force.

RAID. Usually a small-scale operation, involving a swift penetration of hostile territory to secure information, confuse the enemy, or destroy his installations. It ends with a planned withdrawal upon completion of the assigned mission.

RANGER MISSION. To conduct strike, raid, or special light infantry operations. Execution of ranger missions is normally directed by the National Command Authority. Rangers also support operational-level commanders during operations of limited duration when allocated or assigned by the theater commander.

RECONNAISSANCE. A mission undertaken to obtain, by visual observation or other detection methods, information about the activities and resources of an enemy or potential enemy; or to secure data concerning the meteorological, hydrographical, or geographical characteristics of a particular area.

SEARCH AND RESCUE. The use of aircraft, surface craft, submarines, specialized rescue teams, and equipment to search for and rescue personnel in distress on land or at sea.

SEIZURE. As an operational purpose, the capture of a voluntarily restricted portion of an enemy-controlled territory. Capture of an isolated land mass such as an island is usually categorized as a seizure.

SEIZE KEY FACILITY. An attack on an installation with the intent of holding it for a short period. Normally, rangers would make the assault to seize and secure an installation for follow-on forces.

SERVICE COMPONENT. Each service component commander is responsible for recommending the proper use of his forces and for accomplishing operational tasks assigned by the joint commander. He is also responsible for his service in matters of internal administration and discipline; training in own service doctrine, techniques, and tactics; designation of specific units to meet joint requirements; logistics functions normal to the component; tactical employment of service component forces; and service intelligence matters.

SHOW OF FORCE. A mission carried out to demonstrate US resolve, whereby US forces are deployed to defuse a situation that may be harmful to US interests or national objectives.

SIGNAL INTELLIGENCE (SIGINT). A generic term that includes COMINT and ELINT. Activities concerned with the collection and processing of foreign electromagnetic emissions for the purpose of producing intelligence information.

SITUATIONAL TRAINING EXERCISE. An STX is a short, mission-oriented exercise in which a group of closely-related collective tasks are trained, using a doctrinally preferred method of executing these tasks to established standards. An STX is drill-like in nature, but it is more complex and flexible than a drill. It usually involves a larger unit than a drill.

SPECIAL OPERATIONS. Military operations conducted by specially trained, equipped, and organized DOD forces against strategic or tactical targets in pursuit of national, military, political, economic, or psychological objectives. Special operations may support conventional military operations, or they may occur independently when the use of conventional forces is either inappropriate or infeasible. Sensitive peacetime operations, except for training, are normally authorized by the NCA and conducted under the direction of the NCA or designated commander. Special operations may include unconventional warfare, counter-terrorist operations, collective security, psychological operations, and civil affairs measures.

SPECIAL OPERATIONS FORCES. All civil affairs, psychological operations, ranger (active component), and special forces units within the US Army, US Army Reserve, and National Guard.

SPECIAL OPERATIONS COMMAND. The major subordinate command to FORSCOM exercising command and control of all active component Army special operations forces not placed under the operational command of unified or specified commands.

STRATEGIC LEVEL OF WAR. Military strategy uses the armed forces of a nation to secure the objectives of national policy by applying force or the threat of force. Military strategy sets the fundamental conditions for operations.

STRIKE. An attack that is intended to inflict damage on, seize, or destroy an objective. Sometimes referred to as direct action missions.

SURVEILLANCE. The systematic observation of aerospace, surface or subsurface areas, places, persons, or things by visual, aural, electronic, photographic, or other means.

TACTICAL. Pertaining to the employment of units in combat.

TACTICAL LEVEL OF WAR. Tactics are the specific techniques smaller units use to win battles and engagements that support operational objectives. Tactics employ all available combat, combat support, and combat service support. Tactics involve the movement and positioning of forces on the battlefield in relation to the enemy, the provisioning of fire support, and the logistical support of force before, during, and following engagements with the enemy. At corps and division, operational and tactical levels are not clearly separable, but they are guided by the same principles. An operation designed to defeat any enemy force in an extended area does so through operational maneuver and a series of tactical actions.

TACTICAL COMBAT INTELLIGENCE. That knowledge of the enemy, weather, and geographical features needed by a commander in the planning and conduct of combat operations. It is derived from the interpretation of information on the enemy and the environment.

TARGET ANALYSIS. A detailed and systematic examination of processed intelligence to identify and locate targets.

UNIFIED COMMAND. A unified command is established and designated by direction of the President. The command is organized under a single unified CINC and has a broad, continuing mission. The CINC is responsible to the Secretary of Defense for accomplishing his broad and continuing mission and operates under the strategic and operational direction of the JCS. A unified command is composed of assigned forces of two or more services. These forces, as service components, are under the operational command of the unified commander. The CINC exercises operational command authority through service component commanders. When tasked by higher authority, he may establish a subordinate JTF to conduct specific missions. Under emergency conditions, the CINC has the authority to use all facilities and supplies of assigned forces to accomplish his mission.

References

REQUIRED PUBLICATIONS

Required publications are sources that users must read in order to understand or to comply with this publication.

Field Manuals (FM)

FM 7-8	The Infantry Platoon and Squad (Infantry, Airborne, Air Assault, Ranger)
FM 7-10	The Infantry Rifle Company (Infantry, Airborne, Air Assault, Ranger)
FM 7-20	The Infantry Battalion (Infantry, Airborne, and Air Assault)

RELATED PUBLICATIONS

Related publications are sources of additional information. They are not required in order to understand this publication.

Army Regulations (AR)

AR 350-1	Army Training

Field Manuals (FM)

FM 10-52-1	Commander's Handbook for Water Usage in Desert Operations
FM 10-501	Airdrop of Supplies and Equipment: Rigging Containers
FM 10-512	Airdrop of Supplies and Equipment: Rigging Typical Supply Loads
FM 10-542	Airdrop of Supplies and Equipment: Rigging Loads for Special Operations
FM 10-547	Airdrop of Supplies and Equipment: Rigging the High Speed Aerial Delivery Container CTU-2/A
FM 10-553	Airdrop of Supplies and Equipment: Rigging Ammunition
FM 12-3-4	Echelons Above Corps (EAC) Personnel and Administrative Doctrine

FM 14-6	Comptroller/Finance Services in Theaters of Operations
FM 19-15	Civil Disturbances
FM 20-11-1	Military Diving (Volume 1)
FM 20-11-2	Military Diving (Volume 2)
FM 21-20	Physical Training Program
FM 21-26	Map Reading
FM 21-75	Combat Skills of the Soldier
FM 21-76	Survival, Evasion, and Escape
FM 21-77A(S)	Joint Worldwide Evasion and Escape Manual
FM 22-9	Soldier Performance in Continuous Operations
FM 23-67	Machinegun, 7.62-mm, M60
FM 25-1	Training
FM 25-2	Unit Training Management
FM 25-3	Training in Units
FM 25-4	How to Conduct Training Exercises
FM 25-5	Training for Mobilization and War
FM 25-7	Training Ranges
FM 26-2	Management of Stress in Army Operations
FM 29-47	Main Supply and Service Company, Forward Supply and Service Detachments, Supply and Service Battalion, Airborne Division
FM 29-51	Division Supply and Field Service Operations
FM 30-5	Combat Intelligence
FM 30-17	Counterintelligence Operations
FM 31-18	Long-Range Reconnaissance Ranger Company
FM 31-19	Special Forces Military Free-Fall Parachuting
FM 1-24	Special Forces Air Operations

FM 31-25	Special Forces Waterborne Operations
FM 31-70	Basic Cold Weather Manual
FM 34-1	Intelligence and Electronic Warfare Operations
FM 34-60A	Counterintelligence Operations (U)
FM 55-9	Unit Air Movement Planning
FM 57-38	Pathfinder Operations
FM 57-220	Basic Parachuting Techniques and Training
FM 63-3J	Combat Service Support Operations--Corps
FM 63-4	Combat Service Support Operations--Theater Army Area Command
FM 63-5	Combat Service Support Operations--Theater Army
FM 90-3	Desert Operations
FM 90-4	Airmobile Operations
FM 90-5	Jungle Operations
FM 90-6	Mountain Operations
FM 90-8	Counterguerilla Operations
FM 90-10	Military Operations on Urbanized Terrain (MOUT)
FM 90-10-1	An Infantryman's Guide to Urban Combat
FM 100-5	Operations
FM 100-20	Low Intensity Conflict
FM 100-26	The Air-Ground Operations System
FM 101-5	Staff Organization and Operations
FM 101-10-1	Staff Officers' Field Manual: Organization, Technical and Logistical Data (U)

Firing Table (FT)

FT 60-P-1	Firing Tables for Mortar, 60-mm: M224, Firing Cartridge, HE, M720, Cartridge, HE, M49A4, Cartridge, TP, M5OA3, Cartridge, WP, M302AI, Cartridge, Illuminating, M83A3

Soldier's Training Publication (STP)

STP 7-11B24-SM	Soldier's Manual MOS 11B, Infantryman, Skill Level 2/3/4

Technical Manuals (TM)

TM 10-1670-267-12&P	Operator's and Organizational Maintenance Manual Including Repair Parts and Special Tools List for High Altitude Airdrop Resupply System: 500 Pound Capacity

Training Circulars (TC)

TC 23-14	Sniper Training and Employment
TC 23-90	Mortar Training
TC 24-1	Communications-Electronics Operation Instructions, the CEOI
TC 31-20-5(C)	Special Forces Communications (U)
TC 90-6-1	Military Mountaineering

Other Related Publications

King, Michael J. Dr., *Leavenworth Papers, No. 11*, "Rangers Selected Combat Operations in World War II," Combat Studies Institute, USACGSC, Fort Leavenworth, KS

PROJECTED PUBLICATIONS

Projected publications are sources of additional information that are scheduled for printing but are not yet available. Upon print, they will be distributed automatically via pinpoint distribution. They may not be obtained from the USA AG Publications Center until indexed in DA Pamphlet 310-1.

Field Manuals (FM)

FM 7-70	Light Infantry Squad/Platoon
FM 7-71	Light Infantry Company
FM 7-72	Light Infantry Battalion
FM 7-73	Light Infantry Brigade

FM 23-36 60-mm Mortar, M224